Traditional Medicinal Plants and Malaria

Traditional Herbal Medicines for Modern Times

Each volume in this series provides academia, health sciences and the herbal medicines industry with in-depth coverage of the herbal remedies for infectious diseases, certain medical conditions or the plant medicines of a particular country.

Edited by Dr. Roland Hardman

Volume 1
Shengmai San, edited by Kam-Ming Ko

Volume 2
Rasayana, by H.S. Puri

Volume 3
Sho-Saiko-To and Related Formulations, by Yukio Ogihara and Masaki Aburada

Volume 4
Traditional Medicinal Plants and Malaria, edited by Merlin Willcox, Gerard Bodeker, and Philippe Rasoanaivo

Traditional Herbal Medicines for Modern Times

Traditional Medicinal Plants and Malaria

Edited by
Merlin Willcox, Gerard Bodeker,
and Philippe Rasoanaivo

CRC Press
Taylor & Francis Group
Boca Raton London New York

CRC Press is an imprint of the
Taylor & Francis Group, an **informa** business

CRC Press
Taylor & Francis Group
6000 Broken Sound Parkway NW, Suite 300
Boca Raton, FL 33487-2742

First issued in paperback 2019

ISBN-13: 978-0-415-30112-1 (hbk)
ISBN-13: 978-0-367-39411-0 (pbk)

Library of Congress Card Number 2004043569

Library of Congress Cataloging-in-Publication Data

Traditional medicinal plants and malaria / edited by Merlin Willcox, Gerard Bodeker, Philippe Rasoanaivo.
    p. cm. — (Traditional herbal medicines for modern times ; v. 4)
  Includes bibliographical references and index.
  ISBN 0-415-30112-2
  1. Antimalarials. 2. Herbs—Therapeutic use. 3. Traditional medicine—Tropics. I. Willcox, Merlin. II. Bodeker, Gerard. III. Rasoanaivo, Philippe. IV. Title. V. Series.

RC159.A5T735 2004
616.9'362061—dc22   2004043569

Visit the Taylor & Francis Web site at
http://www.taylorandfrancis.com

and the CRC Press Web site at
http://www.crcpress.comt

Dedication

*To those who use nature to prevent and treat,
to those who rely on this for their healthcare, and
to those who seek to understand and make
nature's wisdom available to all in the historic
and collective effort to combat malaria.*

Contents

PART 3   Ethnomedical Research

PART 4   Laboratory Research

PART 5   Clinical Research

PART 6   Repellence and Vector Control

Contributors

Dr. Jonathan Addae-Kyereme, M.Pharm., Ph.D.
Department of Pharmaceutical Chemistry
Faculty of Pharmacy
Kwame Nkrumah University of Science and
  Technology
Kumasi, Ghana
jaddaeky@netscape.net

Prof. Gerard Bodeker, M.Psych., Ed.M., Ed.D.
University of Oxford Medical School
Chair, Global Initiative for Traditional Systems
  (GIFTS) of Health, Oxford
Adjunct Professor of Epidemiology
Mailman School of Public Health
Columbia University, New York
gerry.bodeker@medschool.oxford.ac.uk

Dr. Christophe Boëte, Ph.D.
Laboratoire de Parasitologie Evolutive
CNRS UMR 7103
Université Pierre & Marie Curie
Paris, France
malariaplantproj@yahoo.fr

Dr. Geneviève Bourdy, Ph.D.
Chargée de recherche UMR-IRD-UPS 152
Pharmacochimie des Substances Naturelles et
  Pharmacophores Redox
Centre IRD (Institut de Recherche pour le
  Développement)
Guyane, France
yuruma@cayenne.ird.fr

Gemma Burford, M.Biochem., M.Sc.
Global Initiative for Traditional Systems
  (GIFTS) of Health
Green College, Oxford
and
Aang Serian Community School
Monduli, Arusha, Tanzania
gemmaburford@yahoo.co.uk

Maelle Carraz, M.Sc.
USM 0502-UMR 5154
CNRS Chimie et Biochimie des Substances
  Naturelles
M.N.H.N.
Paris, France
maelle.carraz@caramail.com

Joanne Chamberlain, D.Phil.
Principal Consultant
URS Sustainable Development
St. George's House, 2nd Floor
Wimbledon, London, United Kingdom
jo_chamberlain@urscorp.com,
  jochamber1@aol.com

Dr. Eric Deharo, Ph.D.
Institut de Recherche pour le Développement
  (IRD)
UMR-IRD-UPS 152
Pharmacochimie des Substances Naturelles et
  Pharmacophores Redox. Centre IRD, BP 165
Guyane, France
deharo@cayenne.ird.fr

Maria do Céu de Madureira
Institute of Health Sciences
Faculty of Pharmacy
Monte da Caparica, Portugal
mcmadureira@oninet.pt

Dr. Pedro Melillo de Magalhães, Dr.Sc.
Coordinator
Divisão de Agrotecnologia
CPQBA-UNICAMP
Campinas–SP, Brazil
pedro@cpqba.unicamp.br

Dr. Vikas Dhingra, Ph.D.
Assistant Research Scientist
Department of Veterinary Pathology
College of Veterinary Medicine
University of Georgia
Athens, Georgia
vdhingra@vet.uga.edu

Dr. Chiaka Diakité, M.D.
Médecin
Département de Médecine Traditionnelle
Centre Régional de Médecine Traditionnelle
  Bandiagara
Bandiagara, Mali
crmt@afribone.net.ml

Dr. Drissa Diallo, Ph.D.
Pharmacist
Department of Traditional Medicine
Bamako, Mali
dmtinrsp@technolab.com.ml

Sarika D'Souza, B.A.M.S.
Research Fellow
Foundation for Revitalisation of Local Health
  Traditions (FRLHT)
Bangalore, India
sarikads@yahoo.com

Prof. Nina L. Etkin, Ph.D.
Graduate Chair
Department of Anthropology and Department
  of Ecology and Health–Medical School
University of Hawai'i at Manoa
Honoluly, Hawai'i
etkin@hawaii.edu

Dr. Jacques Falquet, Ph.D.
Biochemist
Antenna Technologies
Meyrin, Switzerland
jfalquet@antenna.ch

Dr. Jorge F.S. Ferreira, Ph.D.
Phytochemist
U.S. Department of Agriculture
Agricultural Research Service
Beaver, West Virginia
jferreira@ars.usda.gov

J.F. Franetich
INSERM U511
Immunobiologie Cellulaire et Moléculaire des
  Infections Parasitaires
Centre Hospitalo–Universitaire
  Pitié-Salpêtrière
Université Pierre et Marie Curie
Paris, France
franetic@ext.jussieu.fr

**Prof. Karniyus S. Gamaniel, Ph.D.,
F.P.C.Pharm., O.O.N.**
Department of Pharmacology and
  Toxicology
National Institute for Pharmaceutical
  Research & Development
Idu, Abuja, Nigeria
ksgama@yahoo.com

Dr. Adebayo A. Gbolade, Ph.D.
Department of Pharmacognosy
Faculty of Pharmacy
Obafemi Awolowo University
Ile-Ife, Nigeria
adegbolade@yahoo.com,
  agbolade@oauife.edu.ng

Dr. Bertrand Graz, M.D. M.P.H., F.M.H.
Institut Universitaire de Médecine Sociale et
  Préventive, Faculty of Medicine, Lausanne
Federal Office of Public Health (Infectious
  Diseases)
Lausanne, Switzerland
Bertrand.Graz@hospvd.ch

Dr. Mark Harrison, D.Phil.
Wellcome Unit for the History of Medicine
University of Oxford
Oxford, United Kingdom
Mark.harrison@wuhmo.ox.ac.uk

Dr. Hans-Martin Hirt, Ph.D.
Coordinator of Action for Natural Medicine
  (ANAMED)
Winnenden, Germany
anamedhmh@yahoo.de

Mark Honigsbaum, M.A.
Freelance journalist and author
London, United Kingdom
honigsbaum@yahoo.com

Dr. Elisabeth Hsu, Ph.D.
Institute of Social and Cultural
  Anthropology
University of Oxford
Oxford, United Kingdom
elisabeth.hsu@anthropology.oxford.ac.uk

Prof. Dibungi T. Kalenda, Ph.D.
Centre d'Etudes des Substances Naturelles
  d'Origine Végétale (CESNOV)
Faculté de Pharmacie
Université de Kinshasa
Kinshasa, République Démocratique du Congo
td-kalenda@raga.net

Dr. Andrew Y. Kitua, M.Sc., M.D., Ph.D.
Director General
National Institute for Medical Research
Dar es Salaam, Tanzania
akitua@nimr.or.tz, akitua@hotmail.com

Dr. ir. Bart G.J. Knols, M.Sc., Ph.D.
Malaria Vector Group
Laboratory of Entomology
Wageningen University & Research Centre
Wageningen, The Netherlands
bknols@planet.nl

Dr. Sean Hsiang-lin Lei, M.Sc., Ph.D.
Associate Professor
Institute of History
National Tsing-Hua University
Tsinchu, Taiwan
hllei@mx.nthu.edu.tw

Annick D. Lenglet, B.Sc., M.Sc.
Disease Control and Vector Biology Unit
London School of Hygiene and Tropical Medicine
London, United Kingdom
Annick_ll@yahoo.com

Nzira Lukwa
Blair Research Institute
Causeway, Harare, Zimbabwe
nzira33@yahoo.co.uk

Dr. Ababacar Maïga, M.D.
Pharmacist
Department of Traditional Medicine
Bamako, Mali
dmtinrsp@technolab.com.ml

Hamisi M. Malebo, Dip.Ed. (Sc.), B.Sc., M.Sc.
Research Scientist (Medicinal and Bioorganic
  Chemistry)
Department of Traditional Medicine Research
National Institute for Medical Research
Dar es Salaam, Tanzania
Malebo@hotmail.com, hmalebo@nimr.or.tz

Dr. Motlalepula G. Matsabisa, Ph.D.
IKS (Health) Division
Division for Traditional Medicine and Drug
  Development
Medical Research Council — South Africa
Tygerberg, Cape Town, South Africa
Motlalepula.Matsabisa@mrc.ac.za

Prof. Dominique Mazier, M.D.
INSERM U511
Immunobiologie Cellulaire et Moléculaire des
  Infections Parasitaires
Centre Hospitalo–Universitaire
  Pitié–Salpêtrière
Université Pierre et Marie Curie
Paris, France
mazier@ext.jussieu.fr

Sarah J. Moore, B.Sc., M.Sc.
Disease Control and Vector Biology Unit
London School of Hygiene and Tropical
  Medicine
London, United Kingdom
Sarah.Moore@lshtm.ac.uk

**Dr. Idowu Olanrewaju, M.B., B.S.,
  F.M.C.G.P., D.T.M.&H.**
President
Olanrewaju Hospital
Ilorin, Nigeria
Idowu@hyperia.com

Dr. P. Pino, Ph.D.
INSERM U511
Immunobiologie Cellulaire et Moléculaire des
  Infections Parasitaires
Centre Hospitalo–Universitaire
  Pitié–Salpêtrière
Université Pierre et Marie Curie
Paris, France
pino@ext.jussieu.fr

Damien Provendier, M.Sc.
Biologist
Nomad RSI
Toulouse, France
damienpro@hotmail.com

Dr. Herintsoa Rafatro, Ph.D.
Research Pharmacologist
Laboratoire de Phytochimie et de
  Pharmacologie Cellulaire et Parasitaire
Institut Malgache de Recherches Appliquées
Antananarivo, Madagascar
rafatro@refer.mg

Dr. David Ramanitrahasimbola, Ph.D.
Pharmacologist
Laboratoire de Phytochimie et de
  Pharmacologie Cellulaire et Parasitaire
Institut Malgache de Recherches Appliquées
Antananarivo, Madagascar
hasimbola67@yahoo.fr

Dr. Jacques Ranaivoravo, M.D.
Département des Essais Cliniques
Laboratoire de Phytochimie et de
  Pharmacologie Cellulaire et Parasitaire
Institut Malgache de Recherches Appliquées
Antananarivo, Madagascar
miravo@syfed.refer.mg

Dr. Jean René Randriasamimanana, M.D.
Direction de l'Agence du Médicament
Ministère de la Santé
Antananarivo, Madagascar

Prof. Philippe Rasoanaivo, Dr.Sc.
Head
Laboratoire de Phytochimie et de
  Pharmacologie Cellulaire et Parasitaire
Institut Malgache de Recherches Appliquées
Antananarivo, Madagascar
rafita@wanadoo.mg

Prof. Suzanne Ratsimamanga-Urverg, Ph.D.
Director General
Laboratoire de Phytochimie et de
  Pharmacologie Cellulaire et Parasitaire
Institut Malgache de Recherches Appliquées
Antananarivo, Madagascar
Suzanne.Ratsimamanga@aventis.com
soamadi@wanadoo.mg

Darshan Shankar, B.Sc.
Director
Foundation for Revitalisation of Local Health
  Traditions (FRLHT)
Bangalore, India
darshan.shankar@frlht.org.in

Prof. V.P. Sharma, D.Sc.
Director
Malaria Research Centre
and
Additional Director General
Indian Council of Medical Research (Retired)
Satya Marg, New Delhi, India
v_p_Sharma@hotmail.com

Dr. O. Silvie, M.D.
INSERM U511
Immunobiologie Cellulaire et Moléculaire des
  Infections Parasitaires
Centre Hospitalo–Universitaire
  Pitié–Salpêtrière
Université Pierre et Marie Curie
Paris, France
silvie@ext.jussieu.fr

P.M. Unnikrishnan, B.A.M.S., M.S.
Senior Program Officer
Foundation for Revitalisation of Local Health
  Traditions (FRLHT)
Bangalore, India
unni.pm@frlht.org.in

S.N. Venugopal, B.A.M.S.
Research Officer
Foundation for Revitalisation of Local Health
  Traditions
Bangalore, India
venu.gopal@frlht.org.in

Prof. Charles O.N. Wambebe, Ph.D.
Division of Health Systems and Services
  Development
World Health Organization
Regional Office for Africa
Brazzaville, Republic of Congo
wambebec@afro.who.int

Dr. Merlin Willcox, B.A., B.M., B.Ch., M.R.C.G.P., D.C.H., D.R.C.O.G., D.F.F.P., D.G.M., Dip.GUM., D.T.M.&H.
Secretary
Research Initiative for Traditional Antimalarial Methods (RITAM)
Oxford, United Kingdom
merlinwillcox@doctors.org.uk

Dr. Colin W. Wright, B.Pharm., M.Sc., Ph.D., M.R.Pharm.S.
The School of Pharmacy
University of Bradford
West Yorkshire, United Kingdom
C.W.Wright@Bradford.ac.uk

Series Preface

Global warming and global travel are among the factors resulting in the spread of such infectious diseases as malaria, tuberculosis, heptatitis B, and HIV. All these are not well controlled by the present drug regimes. Antibiotics, too, are failing because of bacterial resistance. Formerly less well known tropical diseases are reaching new shores. A whole range of illnesses, for example cancer, occur worldwide. Advances in molecular biology, including methods of *in vitro* testing for a required medical activity, give new opportunities to draw judiciously upon the use and research of traditional herbal remedies from around the world. The re-examining of the herbal medicines must be done in a multidisciplinary manner.

Since 1997, 20 volumes have been published in the book series *Medicinal and Aromatic Plants — Industrial Profiles*. The series continues and is characterised by a single plant genus per volume. With the same series editor, this new series *Traditional Herbal Medicines for Modern Times* covers multiple genera per volume. It accommodates, for example, the traditional Chinese medicines (TCM), the Japanese Kampo versions of this, and the Ayurvedic formulations of India. Collections of plants are also brought together because they have been re-evaluated for the treatment of specific diseases such as malaria, tuberculosis, cancer, diabetes, etc. Yet other collections are of the most recent investigations of the endemic medicinal plants of a particular country, e.g. India, South Africa, Mexico, Brazil (with its vast flora), or Malaysia with its rainforests, said to be the oldest in the world.

Each volume reports on the latest developments and discusses key topics relevant to interdisciplinary health science research by ethnobiologists, taxonomists, conservationists, agronomists, chemists, pharmacologists, clinicians, and toxicologists. The series is relevant to all these scientists and will enable them to guide business, government agencies, and commerce in the complexities of these matters. The background to the subject is outlined below.

Over many centuries, the safety and limitations of herbal medicines have been established by their empirical use by the "healers" who also took a holistic approach. The "healers" are aware of the infrequent adverse effects and know how to correct these when they occur. Consequently and ideally, the pre-clinical and clinical studies of a herbal medicine need to be carried out with the full cooperation of the traditional healer. The plant composition of the medicine, the stage of the development of the plant material, when it is to be collected from the wild or when from cultivation, its post-harvest treatment, the preparation of the medicine, the dosage and frequency, and much other essential information is required. A consideration of the intellectual property rights and appropriate models of benefit sharing may also be necessary.

Wherever the medicine is being prepared, the first requirement is a well documented reference collection of dried plant material. Such collections are encouraged by organisations like the World Health Organisation and the United Nations Industrial Development Organisation. The Royal Botanic Gardens at Kew in the UK is building up its collection of traditional Chinese dried plant material relevant to its purchase and use by those who sell or prescribe TCM in the United Kingdom.

In any country, the control of the quality of plant raw material, of its efficacy, and of its safety in use are essential. The work requires sophisticated laboratory equipment and highly trained personnel. This kind of "control" cannot be applied to the locally produced herbal medicines in the rural areas of many countries, on which millions of people depend. Local traditional knowledge of the "healers" has to suffice.

Conservation and protection of plant habitats is required and breeding for biological diversity is important. Gene systems are being studied for medicinal exploitation. There can never be too many seed conservation "banks" to conserve genetic diversity. Unfortunately such banks are usually dominated by agricultural and horticultural crops with little space for medicinal plants. Developments such as random amplified polymorphic DNA enable the genetic variability of a species to be checked. This can be helpful in deciding whether specimens of close genetic similarity warrant storage.

From ancient times, a great deal of information concerning diagnosis and the use of traditional herbal medicines has been documented in the scripts of China, India, and elsewhere. Today, modern formulations of these medicines exist in the form of, e.g., powders, granules, capsules, and tablets. They are prepared in various institutions, e.g., government hospitals in China and Korea, and by companies such as Tsumura Co. of Japan, with good quality control. Similarly, products are produced by many other companies in India, the United States, and elsewhere with a varying degree of quality control. In the United States, the dietary supplement and Health Education Act of 1994 recognised the class of physiotherapeutic agents derived from medicinal and aromatic plants. Furthermore, under public pressure, the U.S. Congress set up an Office of Alternative Medicine, and this office in 1994 assisted the filing of several Investigational New Drug (IND) applications, required for clinical trials of some Chinese herbal preparations. The significance of these applications was that each Chinese preparation involved several plants and yet was handled as a *single* IND. A demonstration of the contribution to efficacy, of *each* ingredient of *each* plant, was not required. This was a major step forward towards more sensible regulations with regard to phytomedicines.

Something of the subject of Western herbal medicines is now being taught again to medical students in Germany and Canada. Throughout Europe, the United States, Australia, and other countries, pharmacy and health-related schools are increasingly offering training in phytotherapy. TCM clinics are now common outside of China, and an Ayurvedic hospital now exists in London with a degree course of Ayurveda available.

The term "integrated medicine" is now being used, which selectively combines traditional herbal medicine with "modern medicine." In Germany there is now a hospital in which TCM is integrated with Western medicine. Such co-medication has become common in China, Japan, India, and North America by those educated in both systems. Benefits claimed include improved efficacy, reduction in toxicity and the period of medication, as well as a reduction in the cost of the treatment. New terms such as adjunct therapy, supportive therapy, and supplementary medicine now appear as a consequence of such co-medication. Either medicine may be described as an adjunct to the other depending on the communicator's view.

Great caution is necessary when traditional herbal medicines are used by doctors not trained in their use, and likewise when modern medicines are used by traditional herbal doctors. Possible dangers from drug interactions need to be stressed.

Dr. Roland Hardman, B.Pharm., B.Sc., Ph.D.
Reader and Head of Pharmacognosy (Retired)
University of Bath, United Kingdom

Foreword

'Tis known I ever
Have studied physic, through which secret art,
By turning o'er authorities, I have,
Together with my practice, made familiar
To me and to my aid the blest infusions
That dwell in vegetives, in metals, stones;
And I can speak of the disturbances
That nature works, and of her cures.

— *Cerimon in Shakespeare's* Pericles, *Act III, Scene II*

After his invasion of the northwestern frontier of India in 326 B.C., Alexander The Great developed a high opinion of Indian Ayurvedic physicians and was particularly impressed by their skill in toxicology and the treatment of snakebite (Jaggi, 2000). Twenty-five centuries later, Western concepts of medical science are still confronted by alternative systems of therapy practiced in other parts of the world and based on unfamiliar and sometimes unacceptable concepts of anatomy, physiology, and pharmacology.

Although the use of plant products is common to both Western allopathic medicine and traditional herbal medicine, the way in which herbal ingredients are prepared and the evidence of their efficacy and safety are strikingly different. The West's scientific method, involving rigorous definitions and trial design with randomisation and blinding to minimise bias, is in marked contrast to the slow accumulation of experience over many generations, without clear comparison or controls, which has established local belief in many traditional remedies and encouraged their continued use over millennia.

In the West, the increasing popularity of herbal medicines was clearly illustrated by a survey, which found that in the U.S., the percentage of adults who had used a herbal medicine in the previous 12 months increased from 2.5 in 1990 to 12.1 in 1997. The cost of this treatment in 1997 was estimated at U.S. $5.1 billion (Eisenberg et al., 1998). However, among the medical community, scepticism and concern have increased, and it is clear that "just because a herb is natural does not mean that it is safe, and claims of remarkable healing powers are rarely supported by evidence" (Straus, 2002).

Serious consequences of adulteration and contamination of herbal remedies have been reported. Examples include dangerous or banned plants, microorganisms, animal and microbial toxins, pesticides, fumigants, metals, and drugs (De Smet, 2002). Recent causes célèbres have included the banned herb *Aristolochia* in weight-reducing traditional Chinese medicines, whose use in Belgium resulted in progressive renal interstitial fibrosis and urothelial cancer (IARC, 2003), and prescription-only medicines such as corticosteroids, glibenclamide, and fenfluramine in traditional remedies from China and Ghana (MCA, 2002).

In the absence of the surveillance for adverse effects insisted upon for drugs, side effects of herbal remedies may remain unrecognised. It has been calculated that a traditional healer would have to treat 4800 patients with a particular herb to have a 95% chance of spotting an adverse effect involving 1 in 1000 of them (De Smet, 1995). Embryotoxic, fetotoxic, and carcinogenic effects of these remedies have proved especially hard to detect. Problems arise through unsuspected toxicity, such as the severe hepatotoxicity of the anxiolytic kava-kava (from *Piper methysticum*); interaction with conventional drugs, typified by the antidepressant St. John's Wort (*Hypericum perforatum*), which induces cytochrome P-450 3A and P-glycoprotein; and inappropriate routes of administration, as in the case of ingestion of a topical aphrodisiac and the Chinese herbal tea Ch'an Su, both containing a toad skin bufadienolide cardiotoxin (Brubacher et al., 1996).

A recent review of herbal remedies recognised their popularity, but strongly recommended that they should not be prescribed without well-established efficacy (De Smet, 2002). It seems that in the U.S., the decision to class these substances as dietary supplements, under the Dietary Supplement and Health Education Act of 1994, has freed the herbal medicine industry from control by the Food and Drug Administration (FDA) (Marcus and Grollman, 2002).

Against this background of increasing concern, *Traditional Medicinal Plants and Malaria* is a particularly timely and important publication. Gerry Bodeker, his co-editors Merlin Willcox and Philippe Rasoanaivo, and the other distinguished and experienced contributors to this book search for methods and arguments by which proponents of what is now regarded as conventional science might be reconciled to the great wealth and wide popularity of traditional cures. They are aware that the greatest challenge is to find appropriate methods for establishing efficacy and to standardise these products. This has certainly been achieved, relatively recently, for the antimalarial Qing Hao Su (*Artemisia annua*). The use of this plant in the treatment of fever was recognised by the great Chinese herbalist, Li Shih Chen (1518–1593), whose classic "Ben Cao Kong Mu" was published in 1596 (Huang, 1998). Now, its derivatives, the artemisinins, are regarded as the most effective treatments for severe *Plasmodium falciparum* malaria (Warrell and Gilles, 2002). Many other candidate antimalarials, with impressive lineages of use and perceived efficacy, await an opportunity for definitive appraisal.

In a 1997 publication by this same group, I argued that:

> Whereas testing of individual compounds may lead to identification of the sole or major active component, possible synergism among the different ingredients or the special effects of the mode of preparation may be lost or obscured. I would advocate a more direct approach to the screening of antimalarial remedies in human patients. I believe that this can be entirely ethical if the subjects live in an area where a particular herbal remedy is the popular treatment of choice for symptoms attributed to malaria. (Warrell, 1997)

Clearly new approaches are needed in the search for antimalarials from plants, and it is the aim of this book to bring these to the attention of researchers.

Dialogue between practitioners of the venerable discipline of herbal medicine and the more recently evolved Western scientific medicine will surely be encouraged by this book, which, it is to be hoped, will help to resolve concerns about how the reputation, efficacy, and safety of herbal and other traditional medicines can be confirmed or refuted in the search for new approaches to combating malaria.

David A. Warrell
Professor of Tropical Medicine and Infectious Diseases
and Head, Nuffield Department of Clinical Medicine,
University of Oxford

REFERENCES

Brubacher, J.R., Ravikumar, P.R., Bania, T., et al. (1996). Treatment of toad venom poisoning with digoxin-specific Fab fragments. *Chest*, 110, 1282–1288.

De Smet, P.A.G.M. (1995). Health risks of herbal remedies. *Drug Saf.*, 13, 81–93.

De Smet, P.A.G.M. (2002). Herbal remedies. *N. Engl. J. Med.*, 347, 2046–2056.

Eisenberg, D.M., Davis, R.B., Ettner, S.L., et al. (1998). Trends in alternative medicine use in the United States, 1990–1997: results of a follow-up national survey. *JAMA*, 280, 1569–1575.

Huang, K.C. (1998). *The Pharmacology of Chinese Herbs*, 2nd ed. CRC Press, Boca Raton, FL, p. 452.

International Agency for Research on Cancer (IARC). (2003). Some traditional herbal medicines, some mycotoxins, naphthalene and styrene. IARC Monographs on the Evaluation of Carcinogenic Risks to Humans, 82.

Jaggi, O.P. (2000). Medicine in India: modern period. *Hist. Sci. Philos. Cult. Indian Civ.*, IX, 302–310.

Marcus, D.M. and Grollman, A.P. (2002). Sounding board. Botanical medicines: the need for new regulations. *N. Engl. J. Med.*, 347, 2073–2076.

Medicines Control Agency (MCA). (2002). Reminder: use of traditional Chinese medicines and herbal remedies. *Curr. Probl. Pharmacovigilence*, 28, 6.

Straus, S.E. (2002). Prospective herbal medicines: what's in the bottle? *N. Engl. J. Med.*, 347, 1997–1998.

Warrell, D.A. (1997). Herbal remedies for malaria. *Trop. Doct.*, 27 (Suppl. 1), 5–6.

Warrell, D.A. and Gilles, H.M. (2002). *Essential Malariology*, 4th ed. Arnold, London.

Introduction

At a time when potent drug cocktails, molecular research, and vaccine development are at the forefront in the global campaign to combat malaria, it might be asked, Why now a book on medicinal plants and malaria?

Malaria statistics have become familiar. It is estimated that 300 to 500 million malaria infections occur annually, 90% of these in sub-Saharan Africa. A third of those visiting rural dispensaries are seeking treatment for malaria. Malaria accounts for between 20 and 50% of all admissions in African health services, with the poorest countries bearing the greatest burden of morbidity and mortality. These statistics and the human suffering they represent drive the global effort to conquer malaria and focus the agendas of such major actors as the Global Fund, the Gates Foundation, and the World Health Organization (WHO).

Yet there are other statistics, less cited, that have profound implications for the viability of malaria control campaigns. First, there are the demographics of malaria. Fifty-eight percent of malaria deaths occur in the poorest 20% of the population (Barat, 2002). Studies have shown that 80% of febrile episodes are treated at home, frequently with herbal medicines. Then there is the target of global malaria eradication campaigns, namely, to ensure access to modern medicine for 60% of the population of at-risk countries by 2005. Even if this ambitious goal is achieved, what are the plans for the remaining 40%?

While current international expenditure on malaria stands at around $100 million a year, the drug bill for effective malaria prevention and treatment has been estimated to be $2.5 billion per year (Sachs and Malaney, 2002). There is no sign that the funds needed for full coverage will be available in the foreseeable future. It is clear, then, that the world's poorest countries and their people cannot rely on external aid to conquer their malaria burden. Indeed, they are continuing to do what has been done historically — for better or, in the absence of national and international efforts to accrue evidence, sometimes for worse.

Historically, communities in tropical regions have used local flora as a means of preventing and treating malaria. Indeed, over 1200 plant species are used throughout the world to treat malaria, and up to 75% of patients choose to use traditional medicines to treat malaria. Many of these medicines are gathered near the home and prepared by the family at minimal cost. They are available where modern drugs are not, and many have long been found to be useful in combating fevers we now know to be malarial.

This perspective is the starting point for this book. In addressing the evaluation of plants for preventing, treating, and controlling malaria, the authors' objective is to contribute rigorous and fresh perspectives to world malaria literature. The book brings together contributions by authors from many different disciplines who have been working on different aspects of medicinal plants for malaria. Although each discipline brings with it some technical terms and abbreviations, we have endeavoured to explain these in a glossary at the end of the book.

In December 1999, with support from WHO/Special Programme for Research and Training in Tropical Diseases (TDR), the Rockefeller Foundation, the Nuffield Foundation and others, the Global Initiative for Traditional Systems (GIFTS) of Health, an international research network on traditional medicine, established an international research partnership — the Research Initiative on Traditional Antimalarial Methods (RITAM). RITAM was established as a network of researchers and others active or interested in the study and use of traditional, plant-based antimalarials (Bodeker and Willcox, 2000).

FIGURE 1.1 Founder members of RITAM at the inaugural meeting in Moshi, Tanzania, in December 1999. (Copyright 1999 by Merlin Willcox.)

The inaugural RITAM meeting in Moshi, Tanzania, was designed to develop a strategy for more effective, evidence-based use of traditional medicines that could contribute to decisions on malaria control policy (Figure 1). The founding members of RITAM addressed the need for research and policy on the prophylactic and therapeutic effects of medicinal plants, as well as on vector control and repellence. There were five main outputs from these deliberations:

1. *Targets* for making a significant contribution to the control of malaria through the use of traditional antimalarial methods
2. *Methods* for achieving these targets, including ethical guidelines
3. An *implementation strategy* for moving this field ahead quickly and soundly, and for putting research findings into practice
4. *Linkages* established between researchers working on traditional antimalarial methods, based on agreed research priorities and designed to avoid unnecessary replication
5. Strengthening the RITAM *database* of current knowledge on traditional herbal antimalarial methods

To accomplish these objectives, four specialist groups were established in the following areas:

- Policy, advocacy, and funding
- Preclinical studies
- Clinical development
- Repellence and vector control

This program built on previous work by GIFTS, including an international symposium in 1995 at Oxford on traditional antimalarials and special journal issues with a collection of papers on this theme (Bodeker, 1996; Bodeker and Parry, 1997). In the intervening years, RITAM has hosted or participated in more than a dozen international meetings, including one with WHO on drug development and medicinal plants (Willcox et al., 2001). Nine years later, this book is a matured product. It is the fruit of more than 3 years of work by many of the 200 members of RITAM who have generated research and methodological standards for preclinical and clinical studies, as well as for research on vector control and repellents.

In designing new treatments, drugs, and public health programs in developing countries, it would be unscientific to cast aside traditional knowledge and wisdom after cursory review, on the assumption that modern methods of analysis and explanation are superior. Part 2 of this book records the history of prominent species — for example, *Cinchona*, *Artemisia annua* (Qing hao), *Dichroa febrifuga* (Changshan), and *Azadirachta indica* (neem). Each has a history of use, a known pharmacology, and substantial potential as an affordable herbal antimalarial. Each case study also offers lessons on methodology and risks. A starting point in a comprehensive approach to medicinal plants is to understand what has been used traditionally to combat fevers in communities in malarious regions. Accordingly, Part 3 of this book addresses ethnographic perspectives. This section underscores the importance of a social science contribution to understanding how plants have been used traditionally in combating malaria.

Then there are safety issues. Here also ethnography can be a starting point. Traditional medicine practitioners have an understanding of side effects and typically have knowledge of combinations of plants to offset these. Such knowledge can be documented and can serve as the source of hypotheses about risks, safety, and means of managing these. Yet, from a different angle, the issue of safety is also driving some of the international interest in plant-derived antimalarials. In RITAM, it is not uncommon for us to receive inquiries from members of the public who are alarmed about neurological and other side effects of some antimalarials and are in search of safer medicines. With the absence of evidence about herbal products, we are not in a position to advise. Yet the public is aware of research that shows that in the U.S., for example, 51% of Food and Drug Administration (FDA)-approved drugs have serious adverse effects not detected prior to their approval and that 1.5 million people are sufficiently injured by prescription drugs annually that they require hospitalisation (Moore et al., 1998). Once in the hospital, the incidence of serious and fatal adverse drug reactions (ADRs) is high. ADRs are ranked as between the fourth and sixth leading cause of death in the U.S., following heart disease, cancer, pulmonary disease, and accidents (Lazarou et al., 1998). Public concern is high. The public is demanding answers and new solutions. Among the general search for fresh approaches, they are seeking these for protection against malaria.

The chapters in Part 4 addressing preclinical methods and toxicological evaluation offer methodologies tailored specifically to accurately characterise the antimalarial effects of plants. They take into account the need to seek novel pathways and not to adhere simply to established assays and models for determining antiplasmodial effects. For example, it is recognised that some plants attack schizonts in the liver rather than the erythrocytic blood stage of the disease. Using only a conventional blood stage antiplasmodial assay could produce false negative results in such cases. These chapters seek to lay a foundation for generating the experimental data needed for proceeding to clinical investigation and understanding, which is discussed in Part 5.

The book incorporates an understanding of the pharmacological richness of individual plants and traditional complex mixtures. These traditional medicines, based on the use of whole plants with multiple ingredients or of complex mixtures of plant materials, could be argued to constitute combination therapies that may combat the development of resistance to antimalarial therapy. In the introduction to the special issue of *Tropical Doctor*, David Warrell (1997) noted that "Whereas testing of individual compounds may lead to identification of the sole or major active component, possible synergism among the different ingredients or the special effects of the mode of preparation may be lost or obscured." Williamson (2001) has reviewed the evidence for the occurrence of synergy in plant-based medicines. She notes that:

Synergistic interactions are of vital importance in phytomedicines, to explain difficulties in always isolating a single active ingredient, and explain the efficacy of apparently low doses of active constituents in a herbal product. This concept, that a whole or partially purified extract of a plant offers advantages over a single isolated ingredient, also underpins the philosophy of herbal medicine. Evidence to support the occurrence of synergy within phytomedicines is now accumulating.

With resistance of the malaria parasites to chloroquine and other antimalarial drugs, medical science has attached importance to synergistic drug combinations, such as co-artemether. White (2003) has argued that all combinations should contain an artemisinin derivative. Surprisingly, given that the lead drugs against malaria have their origins in traditional medicines (see Chapters 2 and 3), the call to combine antimalarial drugs overlooks the fact that combinations always existed in the traditional formulations, prior to the isolation and synthesis of active ingredients and their subsequent recombination. In addition, given the high cost of combination therapy, a very real question presents itself: Can resistance be controlled if people or countries cannot afford the combination therapy? Hence, there is an economic imperative to take into account the potential of natural synergies within plants and in traditional mixtures for creating a barrier to the development of resistance.

The final part of the book deals with vector control and repellence. Plants have been used historically for fumigation, rubbing on the skin to repel insects, and growing or scattering around living quarters for protection. In Vietnam, the eradication of *Aedes aegypti* was accomplished by community participation in treating breeding places with the freshwater shrimp *Mesocyclops* (Vu et al., 1998). Such successes have highlighted biological means of control as inexpensive and effective strategies for rural, malaria-endemic areas.

In summary, potential advantages of herbal treatments include the fact that they are inexpensive, are readily available, can be grown locally or in a domestic garden, and there is no evidence of resistance to whole plant extracts, possibly due to the synergistic action of many constituents. Medicinal plants may also help to minimise the risks of illness without eliminating the continued exposure to infection necessary for maintaining immunity, which is of central importance in the control of malaria in Africa.

This book, coming at the 10th anniversary of the establishment of GIFTS of Health, is the culmination of efforts by many from around the globe. Special recognition must go to RITAM secretary Merlin Willcox, who, starting not so long ago as a medical student doing a final-year elective on traditional antimalarials in Uganda, has gone on to help forge RITAM into a sizeable international partnership working with an important new body of knowledge. His focus, rigor, and insight have been hallmarks of the collaborative work on this book. RITAM Pre-Clinical Working Group chair and co-editor, Philippe Rasaonaivo of Madagascar has brought experience and inno-vativeness to bear in leading his group in the framing of cogent strategies for the preclinical evaluation of antimalarial plants. Their emphasis on novel pathways and their focus on the possible cytotoxic, immunostimulatory, and hepatic effects of antimalarial plants open up prospects for new classes of antimalarial drugs — both herbal and pharmaceutical. RITAM's Vector Control and Repellence Group, represented by Bart Knols and Dibungi Kalenda, has generated a coherent approach to evaluative models, research priorities, and end points in evaluating biological means of vector control and repellence. Their collected efforts and those of the members of their group — enhanced by very helpful review comments from Chris Curtis at the London School of Hygiene and Tropical Medicine — have brought fresh focus to research in this field. Above all, thanks go to the dedicated efforts of dozens of different contributors to this book — from professional writers to distinguished researchers, both northern and southern, some new to the field and many richly experienced. Their collected efforts have made this a unique volume, one that my co-editors and I have found offers a fascinating read while marking vital signposts for the way ahead.

While this volume is the first of its kind in bringing together a multidisciplinary approach to the study of plants used traditionally for malaria control, its publication is not an end in itself.

Rather, as with our 1997 special issue of *Tropical Doctor* on traditional medicine, it is a starting point. Now that standards have been articulated and promising research findings gathered and focussed, the challenge is for these to be applied — and then for products, projects, and programs to be developed. Inevitably, adequate funding will be needed — assigned by forward-thinking funders. At present, limited funding is constraining the development of this field. It is timely and important that the scientific, health policy, and funding communities both understand and give priority to this field, which has already contributed some of the most important antimalarials to date. Many more are already in use in tropical communities around the world, waiting for recognition and mobilisation. The work is just beginning.

Gerard Bodeker
*University of Oxford Medical School, UK
Chair, Global Initiative for Traditional
Systems (GIFTS) of Health, Oxford, UK
Adjunct Professor of Epidemiology,
Mailman School of Public Health,
Columbia University, New York, USA*

REFERENCES

Barat, L. (2002). Do Malaria Control Interventions Reach the Poor? A View through the Equity Lens. Paper presented at the 3rd Multilateral Initiative on Malaria Pan-African Conference, Workshop on "The Intolerable Burden of Malaria," Arusha, Tanzania, November 17.

Bodeker, G. (1996). Guest editor: special issue on traditional health systems and policy. *J. Altern. Complement Med.*, 2, 317–458.

Bodeker, G. and Parry, E.O., Eds. (1997). New approaches in traditional medicine. *Trop. Doct.*, 27 (Suppl. 1).

Bodeker, G. and Willcox, M. (2000). New research initiative on plant-based antimalarials. *Lancet*, 355, 761.

Lazarou, J., Pomeranz, B.H., and Corey, P.N. (1998). Incidence of adverse drug reactions in hospitalised patients: a meta-analysis of prospective studies. *JAMA*, 279, 1200–1205.

Moore, T.J., Psaty, B.M., and Furberg, C.D. (1998). Time to act on drug safety. *JAMA*, 279, 1571–1573.

Sachs, J. and Malaney, P. (2002). The economic and social burden of malaria. *Nature*, 415, 680–685.

Vu, S.N., Nguyan, T.Y., Kay, B.H., Martson, G.G., and Reid, J.W. (1998) Eradication of *Aedes aegypti* from a village in Vietnam, using copepods and community participation. *Am. J. Trop. Med. Hyg.*, 59, 657–660.

Warrell, D.A. (1997). Herbal remedies for malaria. *Trop. Doct.*, 27 (Suppl. 1), 5–6.

White, N.J. (2003). Antimalarial Combinations. Paper presented at the International Conference on Malaria: Current Status and Future Trends, Chulabhorn Research Institute, Bangkok, Thailand, February 16–19.

Willcox, M.L., Cosentino, M.J., Pink, R., Wayling, S., and Bodeker, G. (2001). Natural products for the treatment of tropical diseases. *Trends Parasitol.*, 17, 58–60.

Williamson, E.M. (2001). Synergy and other interactions in phytomedicines. *Phytomedicine*, 8, 401–409.

Part 1

Traditional Medicine and Malaria Control

1 Malaria Control in Africa and the Role of Traditional Medicine

Andrew Y. Kitua and Hamisi M. Malebo

CONTENTS

0-415-30112-2/04/$0.00+$1.50
© 2004 by CRC Press LLC

1.1 BACKGROUND

1.1.1 HISTORICAL DEVELOPMENT OF TRADITIONAL MEDICINE FOR MALARIA

Since prehistoric times man has gradually, through trial and error, recognised and used plants for the treatment of malaria and other ailments. Healers, elders, parents, or priests passed on orally the knowledge of efficacious traditional medicaments to some members of their family, community, and the next generation.

Early writings of over 6000 years ago in Egypt and China and those of the Vedic civilisation dated 1600 B.C. in India indicate that malaria has afflicted human beings since antiquity, and there is evidence that antimalarial traditional medicaments have been used in virtually all cultures as the mainstay for the treatment of the disease (WHO, 1986). In the fifth century B.C., Hippocrates discarded superstition as a cause for the fevers that afflicted ancient Greeks. He instead recognised the seasonality of fevers and described the early clinical manifestations and complications of malaria.

Over 5000 years ago, the Chinese emperor Shen Nung investigated medicinal plants and confirmed their pharmaceutical properties. As a result, over 11,000 herbal remedies were developed and used in China well before the advent of modern medicine. These include *Artemisia annua*, the source of artemisinin, which has recently been developed into a new antimalarial drug (see Chapter 3). In India, traditional medicine has also been in use for over 5000 years, and 8000 herbal remedies codified in the *Ayurveda* are still in clinical use today (see Chapters 5, 12, and 13). Other cultures that also have recorded the use of traditional medicine since antiquity include the Greeks, Romans, Arabs, Europeans, sub-Saharan Africans, and the indigenous peoples of North and South America (Huang et al., 1992).

The potential of traditional medicines in improving the health conditions of communities in developing countries by providing the needed medicaments at affordable costs has for a long time been acknowledged (Akerele, 1984). Currently, in malaria endemic tropical countries, modern medicine is not available, and when it is available, it is not affordable to most of the people living in rural areas. These people resort therefore to the use of traditional medicine as their centerpiece of primary health care. This has been so because traditional medicine is commonly available, culturally and socioeconomically acceptable, and affordable even in remote rural areas of such populations and communities. The World Health Organization (WHO) estimated that about 80% of the world's population rely on traditional medicine for their primary health care needs (Farnsworth et al., 1985). Over one third of the world's population lacks regular access to affordable drugs, such that for these people, modern medicine is unlikely to be a realistic mainstay of their primary health care needs.

In developing countries in Africa as well as in South America and Asia, traditional healers are still very often the only medical health care practitioners available to the majority of people living in remote areas (Akubue and Mittal, 1982). On this realisation the Organisation of African Unity's Scientific, Technical and Research Commission (OAU/STRC) organised a symposium in Dakar (Senegal) in 1968 to discuss, among other things, the contribution of traditional medicines to health care provision in Africa. The meeting resolved to intensify efforts for the creation of an inventory of medicinal plants, as well as to strengthen research on medicinal plants, with the primary aim of evaluating the therapeutic potential of traditional African plant remedies (Sofowora, 1993). It was also recommended at the Alma-Ata conference on primary health care in 1978 that governments give high priority to the promotion of traditional medicine and the integration of proven medicinal plant preparations into national drug policies and regulation (Gessler, 1994). The OAU Heads of Government meeting in July 2001 declared a decade for the development of African traditional medicine, from 2001 to 2011 (The Abuja Declaration, 2000).

WHO carried out a survey between 1977 and 1983 to assess the role of traditional medicine in member states. The findings indicated that the member states are interested in making use of efficacious and safe traditional medicaments and other resources in implementing their primary health care programs (Akerele, 1984).

1.1.2 The Contribution of Traditional Medicine to the Development of Modern Medicines for Malaria

The study of the use of plants by indigenous people (ethnobotany), followed by phytochemical, preclinical, and clinical studies, is an important approach toward the discovery and development of traditional medicines (see Parts 3 to 5 of this volume). The discovery of pure compounds as active principles in plants was first described at the beginning of the 19th century, and the art of exploiting natural products has become an important part of bio-organic chemistry. The discovery of quinine from *Cinchona succiruba* (Rubiaceae) and its subsequent development as a dependable antimalarial drug represented a milestone in the history of modern medicine for malaria (see Chapter 2). The discovery of quinine was followed by an era of synthetic organic chemistry that led to the development of synthetic antimalarial drugs, using the molecular framework of quinine as a template. A number of useful aminoquinoline-based antimalarials synthesised include pamaquine, chloroquine, amodiaquine, pentaquine, primaquine, and mefloquine (WHO, 1986).

Another important development in the chemotherapy of malaria occurred 30 years ago with the isolation of artemisinin (qinghaosu), the antimalarial principle of *Artemisia annua* (see Chapter 3). As for quinine, once the basic structure of the active compound is identified and isolated, it becomes relatively easy to develop or isolate related compounds that may be even more active or safer than the mother molecule. In this regard, a number of more active derivatives of artemisinin against malaria, such as artemether, arteether, and sodium artesunate, have been developed, based on the molecular framework of artemisinin.

1.2 ROLE OF TRADITIONAL MEDICINE IN PRIMARY HEALTH CARE FOR MALARIA

1.2.1 The Disease

Malaria in humans is caused by a protozoan of the genus *Plasmodium*, which is transmitted through bites by female mosquitoes of the genus *Anopheles*. Four subspecies, namely, *P. falciparum*, *P. vivax*, *P. malariae*, and *P. ovale*, are known to cause malaria in humans. The most severe malaria fevers and about 90% of malaria deaths are caused by *P. falciparum*, which is the predominant parasite species in Africa. It is also in Africa that the most efficient mosquito vector for malaria transmission, *Anopheles gambiae*, predominates.

1.2.2 Global Malaria Situation

The global malaria situation is deteriorating faster today than at any time in the past century. The number of new cases of the disease has quadrupled in the past 5 years, such that over 2 billion people, 40% of the world population, living in about 102 countries, are at risk of being infected, and half of these live in sub-Saharan Africa. The World Health Organization estimates that between 300 and 500 million new cases occur each year. In addition to causing untold suffering and disability, malaria ranks as one of the world's major killers, costing about 1 million people their lives annually. Children are especially vulnerable, as more children die from malaria than any other single disease (Figure 1.1). Pregnant women and especially primigravidae are the next highest risk group for malaria in malaria endemic areas. It is stated that malaria causes about 0.96 million deaths of children per year in Africa alone (WHO, 1999). Most malaria infections occur in Africa (see Figure 1.2). Countries in tropical Africa are estimated to account for 80% of all clinical cases and about 90% of all people who carry the parasite.

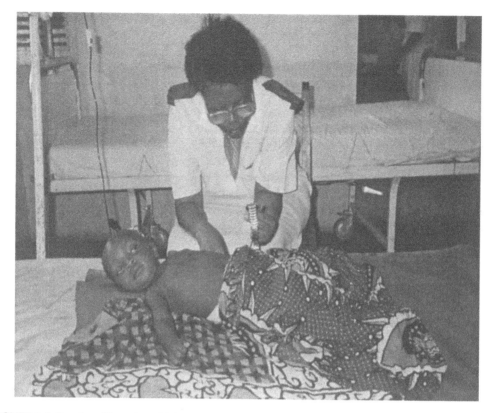

FIGURE 1.1 Inpatient Tanzanian child with severe complicated malaria and convulsion. Sister Nasemba taking the pulse. (Courtesy of Drs. A.Y. Kitua and M. Warsame.)

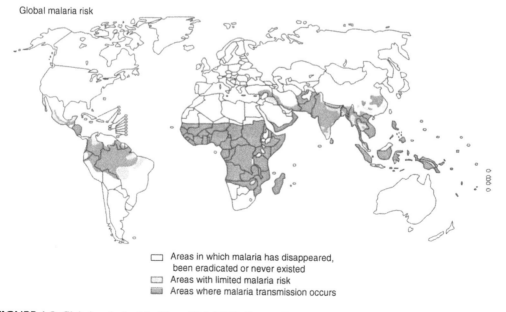

Global malaria risk

☐ Areas in which malaria has disappeared,
    been eradicated or never existed
▨ Areas with limited malaria risk
▧ Areas where malaria transmission occurs

FIGURE 1.2 Global malaria risk. (From 20th WHO Expert Committee Report on Malaria; WHO, 1998.)

The direct and indirect economic burden of malaria in Africa is estimated to be U.S. $2 billion annually. To the poor sub-Saharan countries of Africa, the WHO Roll Back Malaria has highlighted that:

- The cost of malaria control and treatment drains African economies. Endemic countries have to use scarce hard currency on drugs, nets, and insecticides.
- Africa's gross domestic product (GDP) today would be up to 32% greater if malaria had been eliminated 35 years ago, according to estimates from a Harvard study. Malaria endemic countries are among the world's most impoverished.
- In Thailand, malaria patients pay nine times their average daily wages for care.
- A malaria-stricken family spends an average of over one quarter of its income on malaria treatment, as well as paying prevention costs and suffering loss of income.
- Workers suffering an attack of malaria can be incapacitated for at least 5 days.
- Malaria-afflicted families on average can only harvest 40% of the crops harvested by healthy families.

1.2.3 FAILURE OF THE WORLDWIDE MALARIA ERADICATION AND CONTROL PROGRAM

The global malaria eradication principle was adopted in 1955 and WHO adopted a worldwide eradication program between 1957 and 1969. The eradication program was based on vector control by insecticides and chemotherapy of infected people aimed at termination of transmission and eradication of a human reservoir of infection in all malaria endemic areas in a short time. The program was initiated in all the malaria transmission areas of Africa, the Americas, Europe, and major areas of the countries in Asia. Within two decades malaria was eradicated from many parts of the countries in Europe and America and the intensity of transmission had been reduced significantly in Africa and Asia. The widespread use of DDT in India, Mauritius, and Madagascar almost eradicated malaria, whereas complete eradication was achieved in Rome and Sicily, Italy, and around the Mississippi areas in America.

This initial success in many tropical countries was interrupted mainly by the development of resistant mosquitoes and chloroquine-resistant *Plasmodium falciparum*, as well as, and probably most importantly, lack of strong political and financial support to the program once the problem was solved in Europe and America.

Such serious setbacks in the early 1970s caused a resurgence of the disease in areas of Asia and South America where the number of cases had been reduced to low levels.

In Africa, some pilot studies like the Pare-Taveta Scheme had produced a remarkable reduction in mortality and anemia (Bradley, 1991). However, since eradication was not achieved and this had been the goal, the scheme was abandoned and no attempts were made to replicate it elsewhere.

1.2.4 THE CURRENT MALARIA CONTROL STRATEGY

The malaria control strategy aimed at malaria eradication was reoriented in 1978 to focus on the reduction of malaria to a level at which it would no longer constitute a major public health problem (WHO, 1978). This strategy was based on the combined use of vector control measures and effective treatment of malaria patients.

In the interregional meeting on malaria control in Africa held in 1991 in Congo Brazzaville, analysis and ultimately revision of the malaria control strategy were made after considering that it was difficult to implement it in the majority of African countries. The new strategy consisted of three crucial elements: early diagnosis and effective treatment of malaria cases; selective vector control in areas where it is suitable and cost-effective; and early detection and rapid control of epidemics.

The present malaria control strategy, which was adopted by the Ministerial Conference on Malaria in Amsterdam in 1992, is based on the prevention of death, reduction of illness, and reduction of social and economic loss due to malaria (WHO, 1994). The practical implementation of the strategy requires two main approaches:

- Malaria chemotherapy for early and effective treatment of malaria cases, management of severe and complicated cases, and prophylaxis for the most susceptible population (particularly pregnant women and nonimmune travelers)
- Use of insecticide-treated nets for protection against mosquito bites

This strategy has, however, been increasingly confronted by serious setbacks due to the continued spread of drug-resistant *P. falciparum* strains, insecticide-resistant species of the mosquito vectors, and political and socioeconomic problems. The malaria situation has continued to worsen as evidenced by increased frequency of malaria epidemics and levels of parasite resistance to the most affordable drug, chloroquine (WHO, 1996).

1.2.5 LIMITATIONS OF THE CURRENT MALARIA CONTROL STRATEGY

Currently, chemotherapy of malaria to those who need it most worldwide is based on nine key drugs: chloroquine, amodiaquine, primaquine, mefloquine, quinine, sulfadoxine/pyrimethamine, pyrimethamine/dapsone, halofantrine, and artemisinin derivatives. The availability of chloroquine, amodiaquine, sulfadoxine/pyrimethamine, and quinine in Africa has considerably improved the treatment of malaria, but unfortunately, three decades ago *P. falciparum* was shown to have developed some resistance to most of the cheaper antimalarials. Since then, the incidence of drug-resistant *P. falciparum* has been increasing at a faster rate than that of the efforts for development of new drugs (see Figure 1.3).

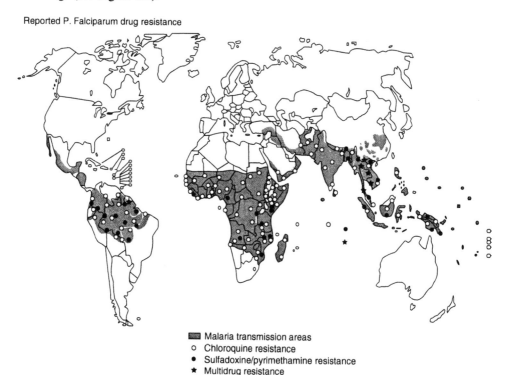

Reported P. Falciparum drug resistance

- ▨ Malaria transmission areas
- ○ Chloroquine resistance
- ● Sulfadoxine/pyrimethamine resistance
- ★ Multidrug resistance

FIGURE 1.3 Reported *P. falciparum* drug resistance. (From 20th WHO Expert Committee Report on Malaria; WHO, 1998.)

Against this background of increasing drug resistance is the unfortunate situation whereby new effective antimalarial drugs coming into the market are completely unaffordable to the majority of the affected populations. Africa south of the Sahara has populations who are dying for lack of malaria treatment, not because there are no effective antimalarial drugs, but because these drugs are beyond their affordability.

The increasing spread of chloroquine-resistant *P. falciparum* malaria is a major public health problem in Africa south of the Sahara. Several countries have already abandoned chloroquine as the first-line therapy. Tanzania, Kenya, Malawi, Botswana, and South Africa have switched to sulfadoxine/pyrimethamine, while Cameroon has switched to amodiaquine. Sulfadoxine/pyrimethamine seems to have a short therapeutic life span, as resistant strains of *P. falciparum* are widespread in Southeast Asia and South America. Recent research findings from Kenya and the United Republic of Tanzania indicate a decline in parasite sensitivity to sulfadoxine/pyrimethamine in East Africa. *P. falciparum* is also reported to have developed resistance to mefloquine in the border areas of Thailand with Cambodia and Myanmar. Parasite resistance to quinine has already been reported in Southeast Asia and in the Amazon region. Furthermore, there is growing evidence that *P. vivax* has also evolved resistance to chloroquine in Indonesia (Irian Jaya), Myanmar, Papua New Guinea, and Vanuatu (WHO, 1998).

Recently, the reemergence of malaria in areas where it had been eradicated, such as Tadjikistan, the Democratic People's Republic of Korea, and the Republic of Korea, and the spread of malaria in countries where it was almost eradicated, such as northern Iraq, Azerbaijan, and Turkey, are indications of the increase in malaria risk areas in the world. The current malaria epidemics in the majority of these countries and in Africa are the result of a rapid deterioration of malaria prevention and control operations, due mainly to the lack of sufficient funds for malaria control and the paucity of effective control tools.

The search for a vaccine has been plagued by a number of shortcomings. Many of the shortcomings are related to antigenic variation, antigenic diversity, and immune evasion mechanisms exhibited in various stages of the complex life cycle of malaria parasites. In addition, malaria research and the search for vaccines require large sums of money, and it can be said that malaria research has been greatly underfunded.

While efforts are being made to overcome the hurdles for vaccine development, people are dying, and the only available effective means of reducing the number of deaths is the provision of affordable and effective medicines. Many young people are already dying of HIV/AIDS in Africa for lack of cure and affordable life-prolonging drugs. If, in addition to this, malaria is not controlled using effective drugs, Africa may see the loss of generations of youths and a huge economical setback to the extent that poverty eradication will remain but a dream for ages.

The absence of new effective and affordable antimalarials is a formidable limitation to the current malaria control strategy, and there is an urgent need to search for traditional medicines to boost the dwindling number of treatments for malaria.

1.2.6 TRADITIONAL MEDICINE: TO USE OR NOT TO USE

Studies of plants used in traditional medicine for the treatment of malaria in various cultures have yielded important drugs that are critical to modern medicine. Two of the most effective drugs for malaria originate from traditional medicine: quinine from bark of the Peruvian *Cinchona* tree, and artemisinin from the Chinese antipyretic *Artemisia annua* (see Chapters 2 and 3). While new efficacious antimalarial compounds and phytomedicines are being rediscovered in medicinal plants, one can only wonder about potential drugs that have not yet been discovered. Plants used in traditional medicine may hold keys to the secrets of many potent antimalarial drugs. Pharmacological investigations already carried out on crude extracts and pure compounds for antimalarial activity have shown that most plants used traditionally for the treatment of malaria are efficacious, and some of them are even more effective than some currently used antimalarials in clinical use (Gessler, 1994).

In Africa south of the Sahara and probably in many parts of the tropical world, populations use and rely on traditional medicines more than on modern medicine. This is because traditional medicines and traditional health care are easily accessible to the majority of the populations whether urban or rural. In addition, because traditional healers live within and are part of the community, they have a higher distribution and a lower patient–healer ratio in rural areas than modern medical practitioners. The fact that traditional healers live with and grow within the community makes them more understanding and understandable to the communities surrounding them. They spend more time with their clients, and knowing the background information of the clients puts them in a better position to satisfy the patient psychologically.

Modern medicine is a comparatively foreign culture and is only well known to the scholars and doctors who practice it. Because such doctors are few and are often practicing in communities within which they did not grow up and live, they have little understanding of the customs and beliefs and tend to reject these customs and practices as foolish and ignorant. They therefore tend to be rejected by the communities and are usually visited as the last resort. This fact makes modern practice difficult in rural areas because many patients are brought in too late.

The other advantage of traditional medicine is that it usually approaches diseases in a holistic manner, combining spiritual and physical care of the patient. Modern medicine is therefore viewed as a strange way of healing in many communities and appears like a mechanical garage rather than a healing process.

The spiritual needs have made mission hospitals more appealing to communities than government hospitals wherever they are, because they conduct prayers in between treatment sessions, reassuring and consoling the soul of the sick body.

In terms of cost, traditional medicine is often more expensive than modern medicine. However, this is not seen as a burden by the patient and caretaker because payment can be made in many forms, often in kind. It can also be given in installments, and through this process, the patient and the traditional healer are linked together and see each other more often and end up developing even stronger trust.

Modern medicine will only accept money for treatment, and in many rural areas in poor countries, money is hardly available to many poor households. The parents may have some property like a goat or chicken and may be willing to give it for treatment, but often at the time of sickness of the child there may be no buyer and they may have no money.

In many cases, deaths that occur in the hands of traditional healers are considered to be caused by failure of the sick or relatives to abide by the rules. This puts traditional healers in the advantageous position of always being blameless, while deaths occurring in hospitals, even though the patients are brought very late, are more likely to be blamed upon the incumbent doctors. It is therefore impossible to stop traditional medical practice, and yet the practice often lacks clear evidence for safety, efficacy, and quality control in the process of making the product (Table 1.1).

1.2.7 INTRODUCING SCIENTIFIC APPROACHES TO TRADITIONAL MEDICINE

Most contemporary research concerning modernising traditional medicine is aimed at confirming the safety and effectiveness of certain traditional methods and treatments through scientific experiments designed on the basis of modern scientific theories and approaches. In the development of antimalarial drugs or phytomedicines from traditional medicine, the emphasis is on the identification of new therapeutic leads over a variety of traditional medicinal plant sources. Four basic methods are generally useful in the selection of traditional medicinal plants to be investigated for drug discovery and development: (1) random selection of traditional medicinal plants followed by mass screening; (2) selection based on ethnomedical uses; (3) leads from literature searches and review of databases; and (4) chemotaxonomic approaches. One of the most difficult tasks in the

TABLE 1.1
Comparison between Traditional Medicine and Modern Medicine

Traditional Medicine	Modern Medicine
1. Based on belief and empirical evidence for cure; safety, efficacy, and quality of drugs often not well known	1. Based on scientific proof and clearly demonstrated safety, efficacy, and quality of drugs
2. Highly accessible to individuals and communities	2. Has limited access to individuals and communities, especially those in rural areas
3. Highly acceptable and understood by the communities and individuals in both rural and urban places	3. Only highly acceptable and understood by urban populations and those with basic education
4. Highly affordable even if it is often more costly than modern medicine	4. Depends on money and highly unaffordable by many rural populations
5. Reputation and credibility spread by way of mouth and individual/community experience	5. Reputation based on scientific fact and spread by advertisement and education
6. Holistic and psychologically more effective	6. Often focuses on physical illness only
7. Uses understandable language and explanations and takes time with patient	7. Often uses a foreign language and doctors have little time with patient
8. Prescription not specified or restricted and can be provided by anybody at any level, including household	8. Prescription specified and has restrictions depending on the type of drug

development of new antimalarial drugs from traditional medicine is the selection of the lead having the highest probability to yield safe and efficacious antimalarial drugs or phytomedicines.

The currently available traditional remedies (phytomedicines) for malaria, according to the guidelines of WHO, the rules of the European Community, the Food and Drug Administration (FDA) in the U.S., and any other licensing body, must have evidence of the product's safety, efficacy, and quality to be registered for human use. However, long-term traditional use is indicative of lack of immediate and acute toxicity, but this does not rule out unforeseeable long-term or latent toxicity (WHO, 2000).

1.2.8 ISSUES OF ETHICS, PATENTS, AND INTELLECTUAL PROPERTY RIGHTS

Issues concerning ethics, patent, intellectual property rights, compensation for loss of finance-rich traditional resources, and the acquisition and safeguarding of traditional health care knowledge need to be addressed in any program that aims to use traditional medicine as a basis for new drug development. The issues of concern are the relationship between the community/traditional healer possessing the healing knowledge and researchers, and the relationship between researchers and pharmaceutical firms. Bioprospecting of new antimalarial drugs from traditional medicinal plants and the exploitation of unprotected traditional knowledge should now be the focus of monitoring measures. Similar concerns that call for closer observation of cultural and intellectual property rights have been addressed in the *Chiang-Mai* and *Kari-Oca* Declarations. The first countries to seriously address these issues were China and India. Programs dealing with traditional medicinal plant conservation, cultivation, community involvement, and sustainable development being initiated elsewhere could benefit greatly from the Chinese and Indian experiences (World Bank, 1997).

Recently the case of the *Hoodia* cactus in South Africa is important evidence of how indigenous knowledge may be exploited and put to better care; however, approaches to share profits emanating from the use of indigenous knowledge still pose ethical problems. From time immemorial, South African Kung bushmen who live around the Kalahari desert have eaten the *Hoodia* cactus to stave off hunger and thirst on long hunting trips. They used to cut off a stem of the cactus about the size of a cucumber and munch on it over a couple of days, and according to their tradition, they must eat together what they caught, so they brought it back and did not eat while hunting. Now the *Hoodia* cactus is at the center of a biopiracy row, as it is the source of the patented P57, the appetite-suppressing drug. P57 is an appetite suppressant with novel pharmacological properties that has no effects on behaviour. Pfizer, the U.S. pharmaceutical giant, bought the right to license the drug for $21 million from Phytopharm. However, it appears that the drug companies had forgotten to inform the bushmen, whose traditional knowledge they had used and patented. In addition to that, Phytopharm has six other patents in progress covering the use of the plant and its active molecules, derivatives, and mode of action. *Hoodia* cactus has been successfully planted in greenhouses, and today plantations have been established worldwide in collaboration with South Africa's Council for Scientific and Industrial Research (CSIR) and Pfizer, and a clinical supplies unit dedicated to the manufacture of the material has been opened in South Africa.

Currently, Phytopharm and CSIR are facing demands for compensation. The Kung tribe claims that their traditional knowledge has been stolen. The Kung people also assert that they were never consulted on the matter, and have accused Phytopharm and Pfizer of biopiracy of their ancient medicinal knowledge of the plant. However, surprisingly, Phytopharm claims that it never consulted the Kung, as it believed the tribe was extinct (Barnett, 2001).

The genetic biodiversity of traditional medicinal plants is continuously under the looming threat of extinction due to ever-growing exploitation, environmental degradation, unsustainable plant harvesting techniques, loss of plant growth habitats, and uncontrolled trade in medicinal plants. Currently, the industrial uses of traditional medicinal plants are many. These range from traditional remedies, herbal teas, nutraceuticals, galenicals, phytopharmaceuticals, and industrially produced pharmaceuticals. In addition, traditional medicinal plants constitute a valuable source of valuable foreign exchange for most developing countries, as they are a dependable source of drugs such as quinine and artemisinin. The traditional medicinal plants market in the U.S. is estimated at U.S. $1.6 billion per year. China is leading with exports of over 120,000 tonnes per year, followed by India with some 32,000 tonnes per year and Madagascar with 8198 tonnes per year. It is estimated that Europe imports about 400,000 tonnes annually of traditional medicinal plants from Africa and Asia with an average market value of U.S. $1 billion (Hoareau and DaSilva, 1999).

The production and commercialisation of traditional medicinal plant-based products in developing countries are dependent upon the availability of resources and information concerning the downstream bioprocessing, phytochemical extraction, and marketing of the phytopharmaceutical products.

1.2.9 THE POSITION OF TRADITIONAL MEDICINE IN THE MULTILATERAL INITIATIVE ON MALARIA

1.2.9.1 The Concept and Goal of MIM

The Multilateral Initiative on Malaria (MIM) is a global alliance of organisations and individuals concerned with malaria in Africa, created in January 1997, aimed at maximising the impact of scientific research on malaria in Africa through the promotion of capacity building and facilitation of global collaboration and coordination. The initiative creates a forum for exchange of resources, information, and ideas between African malaria researchers and those working in developed countries, and thus fosters global collaboration among experts researching malaria to focus on common problems and reinforces the need to collaborate more closely.

1.2.9.2 The Objectives of MIM

- To raise awareness of the problem of malaria and identify key research priorities with a view to mobilising necessary resources and actions
- To develop research capacity in Africa through facilitating a global scientific partnership

- To promote communication, coordination, and collaboration between individuals and organisations involved in malaria research activities in order to maximise the impact of these activities
- To strengthen bridges between the research and implementation communities to ensure that research leads to practical benefit

The role of MIM in traditional medicines falls under three categories in accordance with the above objectives. MIM must recognise that the majority of the high-risk groups for malaria in Africa reside in rural areas that are difficult to access and where the distribution of modern health services is poor or nonexistent. In such communities, the availability of modern health services is a dream, and where they may exist, they are too expensive for poor populations.

These communities, which may account for between 60 and 80% of African populations, rely heavily on traditional medicine for malaria because of the reasons given above.

Within such communities one can find well-established traditional practices and practitioners whose network and expertise range from household and village herbalists to highly reputable healers whose services extend beyond one single community to as vast an area as a district or region.

In Tanzania, a recent study has established that there are five or six healers per village (district), and this is far above the number of modern doctors and health workers available per district (Mhame, 2001).

MIM must also recognise that this situation is necessary because without the presence of providers of alternative medicine and medical practices, the majority of these communities would be left to succumb to the undesirable effects and high death rates due to tropical infectious diseases. For this reason, there is credibility that their medicines and medical practices do work; otherwise, individuals and communities would not use their services at the rate they currently do.

Many African governments, knowing too well that they are unable to provide sufficient efficacious and affordable medicines and medical services to their populations, have either given blanket permission for use of traditional medicines or turned a blind eye to what is happening, hoping that it is all well and good.

It is difficult and will be unethical to stop any of the existing traditional medical practices in the absence of a better substitute, and yet it is equally unethical to let practitioners administer to individuals and populations medicines and medical practices for which there is no scientific evidence of efficacy and safety. It is even more difficult to deal with the innumerable claims of practitioners of alternative medicine in Africa and those of the outer world wishing to sell their products in Africa because of nonspecificity of the claimed efficacy.

Recently, there is an increasing tendency for importers of traditional medicines from China, South Africa, and elsewhere to claim that they are food additives and therefore may not require proof of efficacy. Many of these are making lucrative businesses by advertising that they heal more than 100 diseases, and others claim that they have a cure for HIV/AIDS and many other diseases. Ngoka and Ngetwa are herbal medicines currently circulating in Tanzania that are claimed to cure 120 diseases, including malaria, hypertension, and diabetes. However, the definition of disease is often confused with symptoms like fever, headaches, and feeling weak

Countries must therefore have competence to confirm or deny medical claims and safeguard the health of needy populations, and it is unethical not to have such capacities and to do nothing to help the poor people.

1.3 CHALLENGES FOR THE MULTILATERAL INITIATIVE ON MALARIA

The Multilateral Initiative on Malaria therefore has the following challenges to face if it is to attain a high level of achievement of its goals and objectives.

The most important challenges of each objective are as follows.

1.3.1 Objective 1: Raise Awareness of the Problem of Malaria and Identify Key Research Priorities with a View to Mobilising Necessary Resources and Actions

- What will be effective strategies to raise awareness of the problem to a level that will facilitate fund mobilisation? This poses several additional questions: What strategies will be used? Will the strategies depend on the target group?
- How will the research priorities be identified and who will be the partners identifying such priorities when it comes to traditional (alternative) medicine?
- What actions should be taken, and with what resources, to ensure that poor and isolated populations can access better medicaments and malaria treatment services at affordable costs?

The Multilateral Initiative on Malaria will find it hard to achieve the first objective without seriously considering the identification, promotion, and provision of evidence-based traditional medicines and practices at affordable costs.

Currently, most of the new antimalarials are beyond the affordability of much of the sub-Saharan populations living in malaria endemic zones.

If these same populations have been using traditional antimalarials for a long time and knowledge of these uses have been passed from generation to generation, this indicates that there must be some empirical evidence for their safety and efficacy.

The challenge to MIM will be how to raise sufficient funds for research aimed at confirming the safety and efficacy of these medicaments and to support production of standard treatment packages using these medicaments.

1.3.2 Objective 2: Develop Research Capacity in Africa through Facilitating a Global Partnership

The major challenge to MIM is how to develop research capacity to screen and identify efficacious medicaments and carry them through toxicological and safety studies (preclinical and clinical) and develop usable products.

This requires strong partnership between the North and the South since much of this knowledge and capacity for this activity are northern based. It will require recognition of the value of improving the health of the poor for global peace and prosperity and the willingness of the North to share and transfer this knowledge to the South.

It is a formidable challenge that requires particular strategies to change the current attitudes that the South can be fed and does not need to know how to grow the food it requires. Support must be given to create southern-based capacities and southern-produced medicaments that will be affordable to the majority poor.

Creation of southern-based industries will greatly help to reduce poverty and enhance development by providing employment and income to farmers who will be encouraged to grow the necessary medicinal crops. This will at the same time provide sure means of ensuring preservation of rare crops from extinction and protection of biodiversity in African forestry.

1.3.3 Objective 3: Promote Communication, Coordination, and Collaboration between Individuals and Organisations Involved in Malaria Research Activities in Order to Maximise the Impact of These Activities

MIM must foster communication and exchange of information between the established capacities in Africa and ensure that they are supported to have better South–South and North–South links and do not feel isolated. MIM must promote the exchange of knowledge so that newly developed capacities do not suffer isolation and neglect. The existence of an international network on traditional antimalarial methods (Research Initiative on Traditional Antimalarial Methods, or RITAM) whose objective, among others, is to foster collaboration among those working in traditional antimalarials is a positive aspect, which is complementing the efforts of MIM. It is therefore important that the network is recognised and supported accordingly.

The challenge is how to use science to evaluate the current knowledge and promote what is scientifically valid as safe and efficacious medication. The guidelines in other chapters of this book aim to provide a starting point for such scientific evaluation.

An additional challenge is how to educate traditional medical practitioners about modern treatments that are superior to their practices, and encourage them to accept and introduce modern medicine in their package for better disease management. If most of the traditional medical practitioners would refer cases with convulsions (cerebral malaria) for fast administration of quinine, mortality due to such a complication would be considerably reduced. There is also the additional challenge of educating modern medical practitioners to accept that traditional medicine may play an important role in disease management, rather than outright discrediting it.

1.3.4 OBJECTIVE 4: STRENGTHEN BRIDGES BETWEEN THE RESEARCH AND IMPLEMENTATION COMMUNITIES TO ENSURE THE RESEARCH LEADS TO PRACTICAL BENEFITS

The challenge is how to create effective links between modern and traditional medical practitioners in order to foster better education, and to agree on effective practices to be implemented by both parties. In this process confidence and partnership must be built between the parties, rather than exploitation being the driving attitude.

MIM must collaborate with other organisations to ensure that the research conducted on indigenous knowledge is not pirated but is used to support and develop the communities from which it was derived.

The issue of property rights is critical. Strategies should be made to ensure that the providers of traditional medicines are involved and own the benefits of the research and are not merely used to provide information. Indeed, Section 8j of the Convention on Biological Diversity states that the traditional use of a product by a community contributes to ownership of such a product by the community. This must be strongly supported and advocated as a universal binding principle to be respected by all nations (Convention on Biological Diversity, 2001).

MIM must also ensure that there are effective strategies to involve industry and the private sector without making the end product unaffordable to the populations from which knowledge was derived.

1.4 RESEARCH ISSUES

1.4.1 RESEARCH ISSUES OF IMMEDIATE CONCERN

- Ascertain *in vitro* and *in vivo* activity of claimed traditional antimalarials
- Ascertain safety and efficacy of the claimed traditional antimalarials
- To determine the level of toxicity of identified traditional medicines and establish a dosage, including the minimal lethal dose (LD_{10} and LD_{50})
- To determine the clinical efficacy of the single and multiple combinations in human volunteers
- To determine the most efficient ways of drug production with the aim to maintain affordable costs

- To determine ways of patenting that would provide maximal benefit to the owners of the traditional knowledge
- To determine and develop national or regional production units involving the private sector
- To establish a market and promote agricultural production of medicinal plants
- To develop strong national and regional capacities for undertaking the above activities in poor malaria endemic countries with special emphasis on preclinical and clinical capacities

1.4.2 LONG-TERM ACADEMIC/PHARMACEUTICAL RESEARCH ISSUES

- Identify and isolate the active ingredients/chemicals and their mode of action
- To synthesise the active compounds and develop synthetic routes for large-scale production
- To establish functional combinations and optimal combination ratios
- To determine the preclinical efficacy of the single or multiple combinations of compounds in animal malaria
- To determine genetic markers for *P. falciparum* resistance of such compounds

1.5 CONCLUSION

The search for new drugs through the evaluation and validation of traditional medicines offers a good opportunity and a highly credible channel for the discovery and development of better medicines. The advantage of such drugs is that their precursors are plants that are often widely available in rural areas of Africa. Traditional medicine research can therefore provide information and new clues regarding the effectiveness of combination therapy in curing the disease and preventing development of drug resistance.

The development of traditional medicines for malaria, and of African-based pharmaceuticals, would provide the following major benefits to the poorest and worst deprived populations in the world as far as health and economic development is concerned:

- Provision of affordable and effective drugs
- Prevention of a large number of deaths of children and pregnant women
- Alleviation of poverty by reducing the burden of malaria and offering the populations alternative commercial crops
- Creation and strengthening of capacities for drug production
- A replicable approach to the provision of effective and affordable medicines for other diseases

The only fear is that if no serious actions are taken to handle the harvesting of such plants, they may soon become extinct because of overharvesting, which is already happening because of the increased cost of antimalarials.

In this respect and following the above argument, there is no better role for MIM than that of identifying means and ways of developing capacity in Africa for preclinical and clinical research in the use of traditional medicines.

MIM should support and catalyse the promotion and development of pharmaceutical industries in Africa that will use natural products to produce locally cheap but effective medicaments. Resources for such developments will be hard to get because the North will always fear losing its monopoly over the pharmaceutical industry. However, its creation will provide a good source of traditional medicines at a cheap price even for northern populations, and allow for safe tourist havens in the tropics.

MIM can and should foster the creation of effective networks in Africa in the field and exchange of information and expertise between the leading researchers in Africa on traditional medicine. The development of African-based medicines and especially medicines against malaria is the best recipe for poverty eradication, and MIM cannot afford not to support and be the driving force for the promotion of these activities.

The World Bank health sector, United Nations Development Programme (UNDP), Rockefeller Foundation, Gates Foundation, and other donors should support MIM in this important but difficult endeavour because accelerating malaria control efforts requires money and poor African countries do not have the level of money required for effective malaria control. Maintaining a comprehensive alliance in developing traditional medicines as a component of malaria control is needed in order not only to ensure future availability of affordable and effective antimalarials in Africa, but also to enable the development of a better economic base for Africa.

The recently established Medicines for Malaria Venture (MMV), which finds its origins in the failure of the market system to provide the required incentives for wide-scale R&D in new medicines to treat malaria, is another area for stronger collaboration between MIM and RITAM. MMV's approach is in the combination of the pharmaceutical industry, with its knowledge and expertise in drug discovery and development, and the public sector, with its depth of expertise in basic biology, clinical medicine, field experience, and above all its public remit. The development of African traditional antimalarial medicines has great potential for poverty alleviation. This must be considered as a priority issue for Africa south of the Sahara. RITAM in collaboration with MIM and MMV could be a vehicle for the development of traditional antimalarial medicines and insecticides affordable to the communities of low-income countries.

REFERENCES

The Abuja Declaration (2000). The African Head of States Summit on Roll Back Malaria, Abuja, 25 April 2000 (WHO/CDS/RBM/2000.17). WHO/CDS/RBM, Abuja, Nigeria.

Akerele, O. (1984). WHO's traditional medicine programme: progress and perspectives. *WHO Chronicle*, 38, 76.

Akubue, P.I. and Mittal, G.C. (1982). Clinical evaluation of a traditional herbal practice in Nigeria: a preliminary report. *J. Ethnopharmacol.*, 6, 355.

Barnett, A. *The Observer* (London), June 17, 2001.

Bradley, D.J. (1991). Morbidity and mortality at Pare-Taveta, Kenya and Tanzania, 1954–1966: the effects of a period of malaria control. In *Disease and Mortality in Sub-Saharan Africa*, Feachem, R.G. and Jamison, D.T., Eds. Oxford University Press, Oxford, p. 248.

Convention on Biological Diversity, Article 8j, 2001.

Denke, A. (1989). Antimalarial Effects of Eight African Medicinal Plants. *J. Ethnopharmacol.*, 25, 115.

Farnsworth, N.R., Akerele, O., Bingel, A.S., Soejarto, D.D., and Guo, Z. (1985). Medicinal plants in therapy. *Bull. WHO*, 63, 965.

Gessler, M. (1994). The Antimalarial Potential of Medicinal Plants Traditionally Used in Tanzania, and Their Use in the Treatment of Malaria by Traditional Healers. Ph.D. dissertation, University of Basel, Basel, Switzerland.

Hoareau, L. and DaSilva, E.J. (1999). Medicinal plants: a re-emerging health aid. *El. J. Biotech.*, 2, 1.

Huang, P.L., Huang, P.L., Huang, P., Huang, H.I., and Lee-Huang, S. (1992). Developing drugs from traditional medicinal plants. *Chem. Ind.*, 290.

Krentzman, B.Z. (2001). P57: A Naturally Occurring Appetite Suppressant. *The Krentzman Obesity Newsletter*, November, Vol. 8, p. 23.

Mhame, P.P. (2001). The Role of Traditional Healers in Oral Health Promotion to the Population. A report submitted to NIMR on research done in Morogoro rural and Kilosa districts, Tanzania.

Multilateral Initiative on Malaria. (1999). *Strengthening Health Research in the Developing World: Malaria Research Capacity in Africa*. The Welcome Trust, London.

Sofowora, A. (1993). Recent trends in research into African medicinal plants. *J. Ethnopharmacol.*, 38, 209.

UNESCO. (1996). Culture and Health, Orientation Texts: World Decade for Cultural Development 1988–1997, Document CLT/DEC/PRO — 1996, 129. UNESCO, Paris.

WHO. (1978). Primary health care. Report of the International Conference on Primary Health Care, Alma Ata, USSR, 16–20 September 1978. Geneva, World Health Organization.

WHO. (1986). Bruce-Chwatt, L.J. 1986. *Chemotherapy of Malaria*, 2nd ed., WHO, Geneva, p. 211–233.

WHO. (1994). Control of Tropical Diseases: 1, Progress Report, Report CTD/MIP/94.4. The Organisation, Division of Control of Tropical Diseases, Geneva.

WHO. (1996). Assessment of therapeutic efficacy of antimalarial drugs for uncomplicated *falciparum* malaria in areas with intense transmission. WHO/MAL/96. 1077, Geneva.

WHO. (1998a). Global Malaria Situation, 20th Report. WHO Expert Committee on Malaria.

WHO. (1999). The World Health Report 1999. WHO, Geneva.

WHO. (2000). General Guidelines for Methodologies on Research and Evaluation of Traditional Medicine, WHO/EDM/TRM/2000.1. WHO, Geneva.

World Bank. (1997). Medicinal Plants: Rescuing a Global Heritage, Technical Paper 355, Lambert, J., Srivastava, J., and Vietmeyer, N., Eds. World Bank, p. 61.

Part 2

Case Studies of Plant-Based Medicines for Malaria

2 Cinchona

Mark Honigsbaum and Merlin Willcox

CONTENTS

2.1 INTRODUCTION

Cinchona bark is one of the most important naturally occurring drugs in the medical pharmacopoeia. Although the circumstances surrounding cinchona's discovery in Peru in the early part of the 17th century are clouded by unreliable sources, apocryphal ornament, and botanical confusion, there is little doubt that cinchona bark was the first specific treatment for malaria, or indeed for any other disease, in Western medicine. Given the wide geographic distribution of malaria, and the high rates of morbidity and mortality associated with the disease, it is probably no exaggeration to say that cinchona is the remedy that has spared, or at least ameliorated, the greatest number of lives in human history. Indeed, cinchona bark is one of the most enduring antimalarials ever discovered, and will probably continue to have benefits for the treatment of severe and drug-resistant malaria well into the 21st century.

A member of the Rubiaceae (madder) family, *Cinchona*, according to the latest classification, comprises 23 species of plants (Andersson, 1998). Broadly speaking, the genus consists of evergreen shrubs or trees distinguished by their ovate or lanceolate finely veined leaves, arranged in pairs at right angles to one another, and by their capsular fruit containing numerous winged albuminous seeds. The fruit dehisces from the base into two valves that are held together at the apex by a thick permanent calyx, while the flowers are tubular and can be rose coloured or yellowish white (see Figure 2.1). Unfortunately, species of cinchona are so much alike that they can only be distinguished by resorting to a number of morphological characteristics that, taken singly, are of no great significance. Historically, this has led to taxonomical confusion, particularly since until the early 18th century no European botanist had described the tree and naturalists had to attempt a classification solely on the basis of bark imported to Europe from South America. Even after the studies of Ruiz and Pavon (1792) and Weddell (1849, 1871), confusion remained with some botanical authorities listing as many as 36 species, while others listed 33.

FIGURE 2.1 *C. ledgeriana* in flower. (Copyright 2000, Merlin Willcox.)

Although cinchonas are now cultivated throughout the tropics (Figure 2.2), the genus is native only to South America, occurring between 10° North latitude and 22° South latitude, a range corresponding roughly to Merida in Venezuela and Santa Cruz in Bolivia. The average altitudinal range is 1200 to 3000 meters, but species of cinchona have also been found below and above these limits. The most important species from a pharmaceutical and commercial point of view are *Cinchona ledgeriana* of Bolivia, *Cinchona succirubra* of Ecuador, and *Cinchona officinalis* of Ecuador and Peru.

2.2 HISTORICAL BACKGROUND: THE DISCOVERY OF CINCHONA

Variously known in earlier historical periods as quinquina, Peruvian bark, or Jesuits' bark, cinchona bark is indelibly associated with the name of the earliest and most illustrious patient to have supposedly benefited from its antimalarial action — Francisca Henriquez de Ribera, the fourth Condesa (Countess) de Chinchón. The tree was assigned the botanical nomenclature *Cinchona* by the Swedish botanist Carl von Linné, or Linnaeus, in his *Genera Plantarum* of 1742, presumably to memorialise the story of the Countess of Chinchón's cure in Lima with a maceration of powdered bark. In the process, Linnaeus inexplicably left out the first *h* in *Chinchón*, resulting in the spelling

FIGURE 2.2 Cinchona plantation in Mandraka, Madagascar. (Copyright 2000, Merlin Willcox.)

that persists to this day. However, as the medical scholars Haggis (1942) and Jaramillo-Arango (1949) have shown, the story of the countess's cure is almost certainly apocryphal, and the origins of the discovery and entry of the bark into the *materia medica* remain a matter for conjecture.

The earliest reference to the febrifugal properties of the cinchona tree comes in a book by an Augustin monk, Father Antonio de la Calancha, published in Barcelona in 1638, but bearing a Lima imprimatur dated 1633. De la Calancha (1638) writes:

> There is a tree of "fevers" in the land of Loja, with cinnamon-coloured bark of which the Lojans cast powders which are drunk in the weight of two small coins, and [thereby] cure fevers and tertians; [these powders] have had miraculous effects in Lima.

The next reference comes in Father Bernabé Cobo's *History of the New World*, first published in 1653 (Cobo, 1943). Cobo devotes a whole chapter of his book to the *Arbol de Calenturas* (Fever Tree), agreeing with de la Calancha that the tree is found in Loja, has a cinnamon-coloured bark, and that the febrifuge is taken as a powder "to the weight of two *reales*, in wine or some other liquid." Cobo also writes that the bark is coarse and very bitter and that its powders "are now very well known and esteemed, not only in all the Indies but in Europe, and are urgently sent for and demanded from Rome."

However, from the point of view of the history of taxonomical and etymological confusion surrounding the identification of cinchona, the most significant entry in Cobo's book is contained in another chapter, entitled "Quina-Quina." Cobo's description of *quina-quina* makes it clear that this was the original Quechua term for the Peruvian balsam tree, *Myroxylon peruiferum*, and not, as later European naturalists erroneously argued, for cinchona. The resin of the balsam tree achieved popularity as a remedy for various ailments at a much earlier date than cinchona (Haggis, 1942). Monardes of Seville (1574) wrote about its medicinal properties, and the French savant Charles-Marie de la Condamine (1738) reports that the Jesuits of La Paz exported the bark to Rome in the early 17th century, where it was distributed as a febrifuge. However, de la Condamine mistakenly claimed that *quina-quina* was the original Peruvian term for the cinchona tree, and that *quina* was the Quechua term for "cloak" or "bark." In French, the compound Indian word was written as *quinquina*, hence the term commonly applied to cinchona bark from the late 17th century onward.

In fact, *quina* does not appear in any Quechua dictionary. However, there are entries for *quinua-quinua*, meaning "a certain leguminous plant so-called," which in all probability refers to the seed pods of the balsam tree that at first sight have the appearance of a legume (Jaramillo-Arango, 1949). In all probability the reason for the confusion is that for some time the medicinal properties of cinchona were little appreciated and merchants illicitly exported it to Europe in substitution for, or mixed with, the bark of the balsam tree, where it was distributed under the same name, *quina-quina*, or quinquina. Indeed, the merchants of Loja soon began referring to cinchona bark as *Cascarilla*, meaning "little bark," a term that implies comparison with the bark of another, more superlative tree, presumably *Myroxylon* (Haggis, 1942). The later substitution of the popular terms *Jesuits' bark* and *Peruvian bark* for cinchona did little to clear up the confusion, as both could also be applied to the balsam barks that, like cinchona, the Jesuits had introduced to Europe from Peru.

The origins of the discovery of cinchona's febrifugal powers are similarly fraught with confusion and error. The early Spanish religious chroniclers locate the discovery of cinchona to Loja, a province in the far north of what was once the Spanish Viceroyalty of Peru, now southern Ecuador. But they are mute as to who first applied the bark to the treatment of malarial fevers, leaving open the possibility that the knowledge may originally have been transmitted to the Jesuits by native Indian healers. Unfortunately, there are two major problems with this theory. The first is the complete absence of cinchona in the hieroglyphics and archaeological remains of the Inca, or in the lists of Indian remedies recorded by Spanish chroniclers of the conquest, such as Nicolas Monardes (1574). The second is that it is widely accepted that malaria almost certainly did not exist in South America prior to Columbus, but was imported from the middle 16th century onward, first by Spanish colonists and later by black Africans brought to the Americas as slaves by the Spanish, Portuguese, and British. Although cinchona bark has other medical applications — to this day it is used by native healers in Ecuador to treat diarrhoea, to induce labor, and as a dentrifice* — it is thought that the probable absence of malaria prior to the Spanish conquest means that native American Indian peoples could not have known of its specific application to the disease. Moreover, it is also considered unlikely that they would have used cinchona bark to treat other types of fevers, as quinine initially raises the body's temperature, thus seeming to exacerbate the symptoms of an infection (Hobhouse, 1999). It is only in the specific case of malaria that quinine is also an antipyretic. But even then it acts slowly, and its effect is not apparent for at least a couple of days. De la Condamine, Humboldt, and Spruce — scientists who traveled in Loja and other parts close to the time of cinchona's discovery — all commented on the native Ecuadorians' aversion to *Cinchona* precisely because its chemotherapeutic action was counterintuitive to their medical prejudices, namely, that fevers should be treated with calafrios (chilled drinks) and colds with hot infusions.

On the other hand, it is known that the Yanomami in the northern Amazon region of Brazil responded quickly to malaria introduced in the 1980s by gold miners. They had never seen it before and quickly tried their various febrifuges, finding that the most bitter worked best. This information spread throughout the Yanomami region (Milliken and Albert, 1997). It is possible that this empirical approach may have led Ecuadorian traditional healers to try known remedies for the new disease and to find that quina-quina worked. If this were the case, they would probably have kept the remedy secret from their Spanish enemies until the Jesuits had converted them to Catholicism. However, there is no evidence for this, and it is equally likely that the Jesuits first discovered the application through their own experimentation. They are known to have been excellent empirical scientists, and it was their custom to examine native remedies to see if they could be applied to other diseases.

In 1801, while passing through Loja, Ecuador, the German geographer and naturalist Alexander von Humboldt (1821) observed that while

---

* Information supplied to M.H. by Jorge Mendietta, an Ecuadorian herbalist, during a visit to the cinchona forests of Hinuma, near Vilcambamba, Ecuador.

agues are extremely common ... the natives there ... would die rather than have recourse to cinchona bark, which, together with opiates, they place in the class of poisons exciting mortification. The Indians cure themselves by lemonades, by oleaginous aromatic peel of the small green wild lemon, by infusions of *Scoparia dulcis*, and by strong coffee.

Von Humboldt (1821) insists categorically that the antimalarial properties of cinchona bark were discovered by the Jesuits. He claims this occurred by chance when they were testing various trees that had been felled in the forest in Loja:

> There being always medical practitioners among the missionaries, it is said they had tried an infusion of the cinchona in the tertian ague, a complaint which is very common in that part of the country.

Unfortunately, Humboldt does not say how he came by his information, and his conclusion that "this tradition is *less improbable* than the assertion of European authors ... who ascribe the discovery to the Indians" [my italics] suggests that he was unsure of his source.

Ironically, given the subsequent doubts that have been cast on the story, the best evidence that South American medical historians can produce for the pre-Colombian knowledge of cinchona is the legend of the Countess of Chinchón's cure. The story was first recorded by Sebastiano Bado (1663), physician to the influential Spanish cardinal Juan de Lugo and later the head of Genoa's two city hospitals. According to Bado, he derived his information from a letter written by an Italian merchant, Antonius Bollus, who had lived for many years in Peru and who had told him that the countess had fallen ill in Lima "thirty or forty years before" — thus placing the time of her cure in 1623–1633. Unfortunately, Bollus's original letter is missing or lost, so the only account that we are left with is the one contained in Bado's book. According to Bado, the countess was suffering a "tertian fever, which in that part is by no means mild but severe and dangerous."* When news of her illness reached Loja, the corregidor (governor) of the province wrote to her husband, the Viceroy of Peru, to inform him of a remedy for fever prepared from the bark of the tree that grew there. The viceroy summoned the governor to Lima, where he verbally confirmed what he had written in his letter in the presence of the countess. "Having heard this, she decided to take the remedy, and after taking it, to the amazement of all, she recovered sooner than you can say it," writes Bado.

Bado continues, quoting Bollus, that "the bark was known to the Indians and that they used it upon themselves in disease; but that they always tried with all means in their power to prevent the remedy becoming known to the Spaniards, who of all Europeans particularly aroused their ire" — presumably because of their leading role in the conquest of Peru and subjugation of Indian culture. The other significant details he gives are that after the countess's recovery, she ordered large quantities of the bark to be sent to her in Lima, whereupon she distributed it free to the sick and, after her return to Spain, to tenant farmers on her husband's estates at Chinchón, near Madrid. As in Lima, the remedy was distributed in powdered form, gaining the familiar name *pulvis comitisae* (countess's powder).

Subsequent to Bado a number of important details were added to Bollus's account. In 1737, de la Condamine and the French botanist Joseph de Jussieu arrived in Ecuador on a scientific mission for the French Academy of Sciences. Although their main object was to measure one degree of latitude at the equator, de la Condamine and de Jussieu also made a study of the cinchona forests and questioned the Lojans about the origins of the remedy. According to de la Condamine (1738), local legend had it that the Indians had discovered the bark's febrifugal powers after observing

---

* Before the identification of malaria with specific plasmodia, the disease was defined by the interval between attacks: thus *Plasmodium vivax* was known as benign tertian because of the relative mildness of the paroxysms and the fact that the fever occurred every second day, with the remission coming on the third day; *Plasmodium malariae* was known as quartan because the attacks occurred every fourth day; while *Plasmodium falciparum* was known as malignant tertian or pernicious fever because of the severity of the attacks, lasting anywhere from 24 to 36 hours, with only about 12 hours of intermission.

pumas, who suffered their own form of intermittent fever, drinking from a lake that had been suffused with the trunks of fallen quinine trees. De Jussieu (1936) was more specific, writing that "it is certain" that the discovery was owed to the Malacatos Indians, a tribe who had settled in a malarious valley near Vilcabamba, some 60 miles south of the town of Loja:

> Since they had endured much suffering from the hot and humid climate and from intermittent fevers, they were forced to find a remedy against the *maladie aussi importune*. After experimenting on various plants, they discovered that the bark of quinquina was the last and almost unique remedy against intermittent fevers.

According to de Jussieu, the Malacatos Indians called the cure *yarachucchu carachucchu*, meaning "bark of the tree for the cold of fevers," or simply *ayac cara* — "bitter bark." The Spanish had discovered the secret around 1600 when a Jesuit was passing through Malacatos and fell ill with fever. Taking pity on him, the local *cacique* (chief) offered to cure him and went to fetch some bark from the nearby mountains. On his return he prepared an infusion and the missionary recovered. According to de Jussieu, the Jesuits later noticed similar trees growing throughout Peru and exported the bark to Spain.

De Jussieu makes no mention of the legend of the countess's cure. However, de la Condamine incorporated the story as told by Bado into his account, adopting at the same time the usage *quina-quina*, which had first been used by Bado. In addition, de la Condamine added new details, including that it was not the corregidor but the viceroy's physician, Juan de Vega, who suggested the countess take the powder and who, on her recovery and return to Spain in 1640, brought a quantity of the bark with him, selling it at Seville, a then very malarious area of Spain, for an English sovereign an ounce (about £75 an ounce in today's money).

In fact, as Haggis (1942) has shown, neither Bado's nor de la Condamine's versions can be true for three reasons. First, Doña Francisca Henriquez de Ribera, the fourth countess of Chinchón, died on January 14, 1641, in Cartagena. In other words, she never returned to Spain and thus could not have brought cinchona with her to Madrid. The records do not give the cause of her death but suggest that it was from some epidemic disease then sweeping the Colombian port — most likely yellow fever. Second, from 1630 to 1638 the viceroy kept a meticulous diary in which he never once refers to his wife suffering a tertian or any other recognisably malarial fever, although he does mention her other ailments. Finally, records show that de Vega never left Peru but continued to practice medicine in Lima until at least 1650.

The confusions and contradictions surrounding the discovery of cinchona's therapeutic properties and the correct identification of its barks inevitably fueled medical scepticism about the remedy. This scepticism was compounded by the fact that for the first 100 years of knowledge of the cure no European botanist had seen the tree in nature. All that was known was that cinchona barks came in four distinct colours — red, yellow, orange, and gray — and that each produced powders of varying efficacy.

The first reference to what may have been the use of cinchona in medical practice came in Belgium in 1643, when a public health official in Ghent recommended a powder, *pulvis indicus*, for the treatment of tertian fevers (Jarcho, 1993). The first indisputable reference, however, is the *Schedula Romana*, a handbill issued by the Pharmacy of the Collegio Romano in 1649 and again in 1651, containing precise instructions on dosage and administration. Entitled "Instructions for the Use of the Bark Called Fever Bark," the *Schedula* says the bark has been imported from Peru under the name *China Febris* and should be administered, in the case of tertian fevers, as a finely ground or sieved powder to the amount of 2 drachms (a quarter of an ounce or 8 g) in a glass of strong white wine "three hours before the fever is due and as soon as the shivers begin, or the first symptoms are noted" (Jaramillo-Arango, 1949). The *Schedula* also says that constant use of powdered bark has "cured practically all the patients who have taken it." But while it cautioned that the remedy should be administered only by competent physicians, other doctors were far from ready to accept the Roman claims that cinchona was a specific treatment for tertian and other intermittent fevers.

For instance, John Jacob Chifflet (1653), physician to Leopold William, Archduke of Austria, Belgium, and Burgundy, penned a critique, *Exposure of the Febrifuge Powder from the American World*, describing the archduke's failed treatment the previous year with cinchona for a double quartan fever. Remarkably, Chifflet managed to both defend his treatment of the archduke and attack the use of cinchona bark at the same time. Chifflet had treated the archduke with cinchona, but when his malaria had relapsed a month later, instead of taking more bark the archduke became angry and ordered Chifflet to warn the public against the powdered febrifuge. Accordingly, Chifflet explained how the bark merely lengthened the intervals between fever attacks and did not cure the disease. Challenging Rome's advocacy of cinchona, he pointed out that though the bark had been introduced into Belgium by the Jesuits, it was not needed, as many other febrifuges were available.

Chifflet's book was interpreted as an affront to the authority of the influential Spanish cardinal Juan de Lugo, then the leading advocate of cinchona therapy. In 1644, de Lugo had instructed the Pope's physician, Gabriel Fonseca, to subject cinchona to empirical study. Fonseca's report was favorable, and as a result, de Lugo had begun distributing the bark *gratis* to the poor of Rome from his palace, and at the Hospital of Santo Spirito, where it was known as *pulvis cardinalis* or *pulvis Jesuiticus* — "Jesuits' bark." In 1649, at a gathering in Rome of the Jesuit Order, de Lugo recommended the powder to the assembled delegates, ensuring the remedy's dissemination to missions throughout Europe (Duran-Reynals, 1946). The meeting also coincided with the arrival in Seville of large, regular shipments of cinchona from Peru, and under the supervision of the Jesuits it was soon being traded on mercantile exchanges where it acquired the popular name Peruvian bark.

In 1655, Honoré Fare, a Jesuit priest and nonphysician who used the pseudonym Antimus Conygius, was instructed to reply to Chifflet, possibly at the instigation of de Lugo and Fonseca, both of whom approved his work. Fare described how the powder had cured more than 100 people in Rome, including cardinals, princes, and high government officials, as well as many poor people.

More important, from the point of view of convincing sceptical physicians elsewhere in Europe, may have been the defense of the bark mounted by Roland Sturm, a native of Louvain who practiced medicine in Delft, the center of a large outbreak of quartan fever in 1657–1658. Sturm (1659) listed 13 cases in Belgium of quartan and tertian fevers, describing how in most instances the Jesuit prescription of two drachms of powdered bark had sufficed to quell the attacks, although relapses were frequent. Sturm mentioned the difficulty of determining from which plant the drug was obtained, but reproduced the *Schedula Romana* in both its original Italian and a Latin translation, commenting that the authorities in Rome would not have allowed it to be printed unless a proper investigation had been made (Jarcho, 1993).

Despite these eloquent defenses, however, the correct administration and optimum dosage remained largely a matter of guesswork. Although some physicians recommended administering cinchona in the intervals between fever attacks, others prescribed it only at the height of the paroxysm. Moreover, the efficacy of a dosage could vary widely depending on the way the febrifuge was prepared and the amount of cinchona alkaloids in a particular species of bark.

In England, the adoption of cinchona therapy was further hampered by Puritan prejudices against so-called Popish remedies and the deliberate obfuscation by quack practitioners such as Robert Talbor. Talbor enjoyed considerable success promoting a cure for the ague — as malaria was known in England — the formula for which he deliberately kept secret. Born in Cambridge, he had settled on the coast of Essex — then one of the most malarious regions of England — to experiment with different combinations of bark, gradually perfecting his own mixture of powders. In 1671 he moved to London and the following year published his *Pyretologia, or a Rational Account of the Cause and Cure of Agues* in which he boasted that he had developed a secret remedy for tertian and quartan agues consisting of four ingredients, two of which were native to England, the other two from abroad (Dobson, 1988). After curing Charles II of fever in 1679, Talbor was made physician to the royal household and granted a royal patent.

Pandering to the prejudices of his native countrymen, Talbor cautioned against Jesuits' powder, warning that while it was an excellent cure when properly administered, it could also lead to convulsions. This was disingenuous to say the least, as it gave the impression that cinchona bark was absent from Talbor's own mixture. In fact, as was shown after his death in 1681, it was the major ingredient — along with a little wine and opium. Indeed, in his book *The English Remedy*, published posthumously by Louis XIV — a lifelong malaria sufferer who had paid Talbor 2000 *louis d'or* for his proprietary secret — Talbor wrote that "quinquina ... is without contradiction the surest of all the simple febrifuges, so is it the only basis of the English Remedy" (Jarcho, 1993).

In fact, cinchona was almost certainly used in England as early as 1655 (Baker, 1785). The *Mercurius Politicus*, one of the earliest English newspapers, contains in several of its editions for 1658 — a year remarkable for the prevalence in England of an epidemic remittent fever — advertisements offering for sale "the excellent Powder known by the name of Jesuit's Powder, which cureth all manner of Agues, Quotidian, Tertian and Quartan," brought over by James Thompson, merchant of Antwerp. And in 1677, cinchona bark was officially listed in the London pharmacopoeia under the name *Cortex peruanus* (Jaramillo-Arango, 1949).

Nonetheless, Talbor was probably the first to carry out clinical trials. One reason for his success may have been his access to the best species of Peruvian bark via his smuggling contacts in Essex. Indeed, some physicians appear to have suspected Talbor of cornering the market in red bark — one of the species now known to be richest in quinine (Dobson, 1988). Unfortunately, Talbor left no record of his purchases or details of his studies, so we do not know if the conjecture is true and whether as part of his clinical observations he included control groups.

No doubt in part due to Talbor's obfuscation and deceit, other English physicians were reluctant to embrace the implications of his results. Thomas Sydenham, often described as the Hippocrates of English medicine, who began practicing in Westminster around 1656, rejected the use of Peruvian bark in the treatment of tertian fevers in 1666 and again in 1676. Sydenham considered that the bark was only valid for the treatment of quartan fevers, although even in these cases he argued its effects were temporary. It was not until 1679 that Sydenham declared himself unequivocally in favor of Peruvian bark, describing it as "his sheet anchor," and in 1683 as "that great specific for intermittent fevers." Sydenham's method consisted of administering an ounce of finely powdered bark made into a pill and divided into 12 parts, 1 part being taken every 4 hours, beginning immediately after the paroxysm (Jarcho, 1993).

However, it fell to an Italian, Francesco Torti, to conduct the first systematic studies of the action of cinchona on different types of fever. In his *Therapeutice specialis* of 1712, Torti clearly identified cinchona as a specific therapy for intermittent, as opposed to continued, fevers, which he held were rarely treatable with cinchona. In particular, Torti emphasised the importance of cinchona in the treatment of pernicious intermittents or severe quartan fevers and carried out a series of trials in which he experimented with different dosages and methods of administration. As a result of these observations, Torti concluded that while the procedure described in the *Schedula Romana* was appropriate for simple cases of tertian fever, in the case of more complicated tertians and pernicious fevers the dosage should be increased. He held that this was particularly important at the start of the illness, with up to 6 drachms being given in divided doses when the paroxysm was most intense, up to a total of 20 drachms (2½ ounces) spread over 3 weeks in the most severe cases (Jarcho, 1993).

At the beginning of the 19th century more rigorous study of the therapeutic action of cinchona became possible thanks to the isolation of alkaloids via new chemical techniques. The first to succeed was Bernardino Antonio Gomez, a Portuguese naval surgeon. In 1812 he soaked powdered gray bark in alcohol and added caustic potash (potassium hydroxide) to the solution to crystallise out an alkaloid to which he gave the name *cinchonino*. But the real breakthrough came 8 years later in France when the chemists Pierre Pelletier and Joseph Caventou subjected the yellow and red cinchona barks to similar analysis. By further precipitation and crystallisation, they discovered that the base febrifuge was not *cinchonino* but two alkaloids that occurred separately or together in different kinds of bark. The first they called cinchonine, the second they called quinine.

The isolation of the cinchona alkaloids opened the way for comparative studies of their antimalarial action, although initially medical interest focused almost exclusively on quinine. In 1820, a French physician, F.J. Double, treated several patients diagnosed with different types of intermittent fever with quinine sulfate. In each case, he reported, they recovered. In 1821, another doctor, Auguste F. Chomel, experimented with quinine sulfate, cinchonine sulfate, and cinchona bark. He reported that of 13 individuals struck with intermittent fever and treated with quinine sulfate, 10 were cured. The three who failed to respond to quinine were then treated with cinchona bark only for it also to prove ineffective (Smith, 1976).

Although these trials were hardly scientific — not only did the dosages vary, but there were no controls — such reports quickly convinced the medical profession that quinine was superior to cinchona bark, and by 1822, quinine had superseded nearly all other remedies for intermittent fevers in both Europe and the U.S. Even when faced with results that were seemingly contradictory, the medical profession preferred quinine. In 1823, for instance, Francis Baker of Dublin reported on 30 cases of quinine therapy he had collected with the assistance of four other Dublin physicians. In four cases an effort had been made "to ascertain whether the sulphate of [quinina] would prove efficacious in cases which resisted the use of cinchona," and on each occasion the quinine succeeded where the bark failed. Yet in another case, where his supply of quinine had been exhausted, Barker reported that "the cure was completed by cinchona" (Smith, 1976).

Interestingly, it was commercial and political considerations, rather than medical ones, that would prompt further examination of the cinchona alkaloids. In 1860, the India Office, concerned that the South American forests were being overexploited and that supplies might soon be exhausted, sent a series of collectors to the Andes to gather the five most valuable species of cinchona for transplantation to the government plantations in India. These missions, which were coordinated by Clements Markham, a civil servant and historian attached to the India Office, met with mixed success. Although Markham's intention had been to proceed to Bolivia to gather the highest quinine-yielding species, known at the time as *Cinchona calisaya*, at the last moment he was forced to change his plans. Instead, he proceeded to Peru to collect an inferior type of *C. calisaya*, and a species known as *C. officinalis* from southern Ecuador. But neither did well in the Indian plantations, and instead, government horticulturists concentrated on propagating *C. succirubra* — the species of red bark that had been forwarded to the plantations from central Ecuador by another British collector, Richard Spruce (Honigsbaum, 2001). This species contained near equal amounts of the four antimalarial cinchona alkaloids, and as the Indian government's priority was the production of a cheap febrifuge that would be affordable to the mass of Indians affected by malaria, a decision was made to conduct clinical trials to ascertain whether the various alkaloids were as efficacious as quinine at treating infections.

In 1866 the secretary of state for India ordered the establishment of cinchona commissions in Calcutta, Madras, and Bombay, staffed by medical officers tasked with conducting the experiments. Scientists had yet to identify the malaria parasite, so instead of microscopic examination of blood smears, the medical officers had to rely on their readings of patients' symptoms to determine whether the alkaloids had cleared the disease. There were also no controls. Instead, each of the alkaloids was administered to a different set of patients and the results tabulated (see Table 2.1).

The commissions' conclusion was that quinidine and quinine possessed "equal febrifugal power," while cinchonidine was "only slightly less efficacious." Even cinchonine, "though considerably inferior to the other alkaloids, [was] notwithstanding a valuable remedial agent in fever" (Markham, 1980).

Markham used the commissions' findings to call for the manufacture of a "mixture of all the alkaloids which would combine cheapness and efficacy in the highest degree." Other quinologists, such as John Eliot Howard, a leading quinine manufacturer, were even more enthusiastic, arguing

TABLE 2.1
Results of Treatment of Fevers with Cinchona Alkaloids by the Madras Commission

Alkaloid	Number Treated	Failures (N)	Failures (%)
Quinine	846	6	0.7
Quinidine	664	4	0.6
Cinchonine	559	13	2.3
Cinchonidine	403	4	1.0

Source: Markham, C.R., (1980), *Peruvian Bark: A Popular Account of the Introduction of Chinchona Cultivation into British India 1860–1880*, John Murray, London.

that all ordinary Indians needed to do was to prepare a decoction from the bark — a process so simple "it may be carried on by every family that has a few trees planted at its door" (Gramiccia, 1988).

In fact, in some parts of India, this was fast becoming common practice. In 1872, a member of the Royal Pharmaceutical Society of Great Britain observed, while passing through Ceylon, that Indian coolies had such confidence in *C. succirubra* as a remedy for malaria that they did not "hesitate [to help themselves] to a strip of bark from the nearest tree when occasion necessitates it" (Duran-Reynals, 1946).

2.3 PRECLINICAL STUDIES

2.3.1 ETHNOBOTANY

Cinchona bark, in the form of a decoction, infusion, or maceration, is still used for the treatment of malaria today in several countries (Table 2.2). The bark is used not only in Ecuador, but also in Peru (Ortiz, personal communication). The most popular species are *C. officinalis* (Madagascar and Sudan) and *C. succirubra* (Madagascar and São Tome). *C. ledgeriana* is also used in Madagascar: trees, including hybrids with a good yield of quinine, had been brought there from the Dutch plantations in Java in 1900 (Prudhomme, 1902) and grown in several large plantations (Figure 2.2). A processed powder of *C. ledgeriana* bark is sold as a phytomedicine under the name of totaquina by the Institut Malgache de Recherches Appliquées (Figure 2.3).

2.3.2 ETHNOPHARMACOLOGY

If the indigenous peoples of South America ever used cinchona, it is not known how they prepared it. Several preparations were used in Britain before quinine became available, as have been described above. Talbor used an infusion of cinchona bark, an alcoholic tincture, and port wine in which powdered bark had been macerated for a week (Grier, 1937). In about 1755, Dr. John Huxham devised *Tinctura Cinchonae Composita*, prepared by macerating the following ingredients in a closed vessel for some days, shaking it three or four times, and then straining: red cinchona bark (2 ounces), bitter orange peel (1 ounce), serpentary ($\frac{1}{2}$ ounce), saffron (3.5 g), cochineal (1.8 g), and 1 pint of proof spirit (Hale White, 1897). Warburg's antiperiodic tincture (c. 1840) contained not only cinchona bark but also aloes, rhubarb, camphor, a little opium, and aromatic drugs; it was reputed to be very successful in both Europe and India (Grier, 1937).

Garrod (1875) and Hale White (1897) describe several other preparations of cinchona bark: a decoction, an infusion, an acid–alcohol extract, and a tincture. These were still mentioned in the British Pharmaceutical Codex as late as 1949. There is no indication as to which of these is preferable and what the duration of treatment should be. Garrod (1875) states that quinine is better

Table 2.2
Contemporary Use of Cinchona Bark for the Treatment of Malaria

Species	Preparation	Country	Reference
Cinchona sp.	Infusion	Guyana	Johnston and Colquhoun, 1996
	Mixed with brandy	Ecuador	Milliken, 1997
		Colombia	Montes Giraldo, 1981
Cinchona sp. (dirita)	Not specified	Philippines	Espino et al., 1997
C. ledgeriana Moens	Decoction	Madagascar	Rasoanaivo et al., 1992
C. officinalis L	Infusion	Sudan	El-Kamali and El-Khalifa, 1997
	Decoction	Bolivia	Rea, 1995
		Madagascar	Rasoanaivo et al., 1992
C. pubescens Vahl	Decoction	Nicaragua	Coe and Anderson, 1996
(= *C. grandiflora*)	Various	Colombia	Blair et al., 1991
C. succirubra Pavon et	Decoction	Madagascar	Rasoanaivo et al., 1992
Kiutzsch	Macerate	São Tome and Principe	Madureira et al., 1998

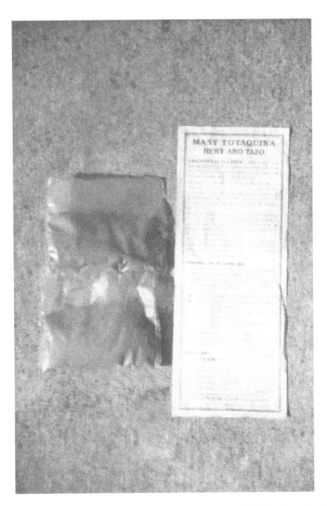

FIGURE 2.3 Totaquina, a phytomedicine produced from the bark of *C. ledgeriana* by the Institut Malgache de Recherches Appliquées. (Copyright 2003, Merlin Willcox.)

TABLE 2.3
Cinchona Preparations and Their Alkaloid Content (%)

Preparation	Species	Crystallisable Alkaloids	Qn	Qd	Cn	Cd	Uncrystallisable Alkaloids[a]
Quinetum (= cinchona febrifuge = totaquina type I)	C. succirubra, C. robusta	54.6–87	7.4–32	1–22.8	11–18.6	5.8–34	10–45.3
Residual alkaloids (= totaquina type II)	C. ledgeriana	62.8–83	11.5–28.5	0.3–5.4	20–55	6.6–26	12.7–37.2
Quinimax™			71.4	18.6	5	0	

Note: Qn = quinine; Qd = quinidine; Cn = cinchonine; Cd = cinchonidine.

[a] Quinoidine.

than the bark for the treatment of severe acute cases requiring large doses because the volume to be administered is smaller and less likely to "upset the stomach"; however, he goes on to recommend the use of bark for more chronic cases "with great debility and weak heart."

After quinine was first isolated in 1820, it became increasingly more popular than the crude bark. However, it was observed on several occasions that the bark could be even more effective than the isolated quinine. *Northnagel's Encyclopaedia of Practical Medicine* (Mannaberg and Leichtenstern, 1905) states:

Experience teaches us that sometimes an obstinate infection which resists quinin [*sic*] will yield to bark. In this case 10–15g of the bark or its decoction should be given during the apyrexia.

Several crude extracts of cinchona bark were in common use in the first half of the 20th century (Table 2.3). Totaquina type I (or quinetum) was prepared as a total alkaloid extract of *C. succirubra* or *Cinchona robusta* by dissolving the soluble constituents of powdered bark with hydrochloric acid, then precipitating the alkaloids with sodium hydroxide and drying the crude deposit. Totaquina type II (or residual alkaloids) was left over from the bark of *C. ledgeriana* after the extraction of quinine. Sometimes cinchona febrifuge was made from the total alkaloids of *C. ledgeriana*, without the prior extraction of quinine. The composition varied widely in different batches and factories (Fletcher, 1923; Hicks and Diwan Chand, 1935; Hofman, 1923; MacGilchrist, 1915; Pampana, 1934). This variation unfortunately makes it difficult to compare studies on totaquina unless the composition of the batch is known. A minimum standard was recommended by the Malaria Commission of the League of Nations in 1927 that the preparation must contain no less than 70% of crystallisable cinchona alkaloids, of which not less than 20% should be quinine. This was included in the British pharmacopoeia in 1932 (Clark, 1933). For comparison, the composition of the modern preparation Quinimax (Sanofi-Synthelabo) is given in the table.

The sum of uncrystallisable substances left after soluble alkaloids were crystallised out, a dark treacly substance, was called quinoidine. It contains both uncrystallisable alkaloids and other substances described below. It can be made into tablets or dispensed as a solution in alcohol, diluted with glycerine, and was also successfully used as an antimalarial in India (Waters, 1913, 1916).

Although cinchona did not exist in India prior to its importation by the British, it has now been incorporated into several ayurvedic remedies for malaria, for example that sold under the name of Vishamajvarunthaka gudika by Kottakal Arya Vaidya Sala, an ayurvedic pharmaceutical company from Kerala state. The recipe for a batch of pills contains 840 g of cinchona extract; this is mixed with *Piper longum* (70 g), *Piper nigrum* (70 g), *Zingiber officinalis* (70 g), *Allium sativum* (70 g),

TABLE 2.4
Average Alkaloid Contents (% Dry Weight) of Commercially Important Native Andean Cinchona Barks

Species and Type	Source	Cinchonine	Cinchonidine	Quinine	Quinidine	Total Alkaloids
C. calisaya	Bolivia	0.55	0.83	3.35	0.07	4.80
C. nicrantha	Northern Peru	4.08	0.05	0	0	4.13
C. officinalis	Southern Ecuador	1.16	1.12	0.41	0	2.69
C. pubescens	Northern Ecuador	1.96	1.57	1.48	0	5.01

Source: Hodge, W.H., (1948), *Econ. Bot.*, 2, 229–257.

Cuminum cyminum (70 g), Loha bhasma, a purified ferrous compound (35 g), and sankha bhasma, a purified arsenic compound (35 g). All the ingredients are ground together in a decoction of *Swertia chirata* and made into pills the size of a seed of *Abrus precatorius*. The recommended dose is one pill two or three times per day, along with ginger juice (D'Souza, personal communication).

2.3.3 CHEMICAL CONSTITUENTS AND THEIR ACTIVITY

Almost 30 alkaloids have now been described in cinchona bark (OUA, 1985), several of which are active against plasmodia *in vitro*, and some of which are not (Karle and Bhattacharjee, 1999). The four most well-known alkaloids are quinine with its D-isomer quinidine, and cinchonine with its L-isomer cinchonidine, all of which have antiplasmodial activity. They are found in varying proportions in different barks (see Table 2.4). *Cinchona pubescens* (red bark) contains near equal amounts of cinchonine, cinchonidine, and quinine, while *C. succirubra* contains equal amounts of all four alkaloids. *C. calisaya* (yellow bark) contains all four alkaloids, but *C. ledgeriana*, the most commercially important variety, contains much higher levels of quinine and none of the other alkaloids except for cinchonine. *C. officinalis* (red bark) contains mainly cinchonine and cinchonidine and a little quinine. *Cinchona nicrantha* (gray bark) contains cinchonine and cinchonidine, but never quinine (Gramiccia, 1988; Hodge, 1948).

Interestingly, quinine is not the most potent of the alkaloids: quinidine, dihydroquinidine, and cinchonine all have a consistently lower inhibitory concentration 50% (IC_{50}) *in vitro* (Table 2.5). Druihle et al. (1988) showed that the combination of quinine with quinidine and cinchonine was 2 to 10 times more effective against quinine-resistant strains, and that the mixture of alkaloids had a more consistent effect than any of the alkaloids used singly.

Cinchonine also inhibits P-glycoprotein, a transmembrane pump that pumps toxins (including drugs) out of cells and is overexpressed in cancer cells, contributing to multidrug resistance. Cinchonine has been used to good effect in combination with chemotherapy for patients with lymphoproliferative syndromes (Solary et al., 2000). It is possible that similar mechanisms are involved in malaria drug resistance (Frappier et al., 1996), and cinchonine may therefore be able to counteract this.

Aside from the alkaloids, the bark also contains tannins. There has been no published scientific investigation of these tannins and their possible synergy with cinchona alkaloids in the treatment of malaria. However, Garrod (1875) states that "Although the efficacy of the bark is chiefly due to the alkaloids it contains, yet it possesses certain properties of its own. The cincho-tannic and red cinchonic acids are powerfully astringent — like tannic and gallic acids; and they contribute in some measure to the total effect." Mitchell Bruce (1884) states that the astringent effect may be useful for cases with "passive diarrhoea, sweating, chronic mucous discharges, and purulent formations," but may be deleterious in those with "dyspepsia, constipation or irritability of the bowels."

TABLE 2.5
IC$_{50}$ (nmol/l) of Cinchona Alkaloids against Two Clones
of *P. falciparum*

Alkaloid	D-6 clone[a]	W-2 clone[b]
Quinidine	13.4	43.7
Dihydroquinidine	10.4	74.1
Cinchonine	18.3	70.8
Quinine	29.3	103.2
Dihydroquinine	21.3	151.7
Cinchonidine	69.5	206.8

[a] Sierra Leone D-6 clone, chloroquine sensitive.
[b] Indochina W-2 clone, chloroquine resistant.

Source: Karle, J.M. and Bhattacharjee, A.K., (1999), *Bioorg. Med. Chem.*, 7, 1769–1774.

TABLE 2.6
Minimum Lethal Dose of Cinchona Alkaloids
in Guinea Pigs

Substance	MLD (mg per kg)
Cinchonidine	600
Quinine hydrochloride	550
Quinine sulfate	525
Cinchonine sulfate	425
Quinidine sulfate	400
Quinoidine	350

There are a few other components of the bark, namely, chinovin (a glucoside), cinchona red (the insoluble colouring matter of the bark), and a volatile oil giving the bark its smell (Hale White, 1897).

Table 2.6 shows the minimum lethal dose (MLD) of the different constituents in guinea pigs, determined by MacGilchrist (1915). Although the MLD for quinoidine is less than that for quinine, it is still far in excess of any doses given to patients (the recommended dose of quinine is 30 mg/kg daily, in three divided doses). In humans, the average fatal dose of quinine for an adult is 8 g, although deaths have been reported from as little as 1.5 g in an adult and 900 mg in a child. The fatal dose of quinidine may be as little as 250 mg in cases of hypersensitivity, but is usually much greater (Toxbase, 2002).

2.4 CLINICAL STUDIES

2.4.1 CLINICAL PHARMACOLOGY

Quinine and quinidine are both well absorbed orally, peak levels being reached within 6 hours for quinine and 1 hour for quinidine. Twenty percent of the drug is excreted unchanged by the kidneys, and 80% is cleared by the liver. The main metabolite, 3-hydroxyquinine, is also active against malaria parasites (White, 1996).

The terminal elimination half-life of quinine is 9.9 hours, compared to 5.7 hours for quinidine (Karbwang et al., 1993). However, the terminal elimination half-life is affected by several factors. It is much shorter in children with global malnutrition (6.3 hours), so a more frequent dosing is recommended (Tréluyer et al., 1996). It is longer in patients with malaria, because systemic clearance is reduced in proportion to disease severity: so it is 16 hours in uncomplicated malaria and 18 hours in severe malaria (White, 1996).

The recommended dose of both quinine and quinidine is 30 mg/kg/day in three divided doses for 7 days. However, Kofoed et al. (1999, 2002) have shown that the minimum effective dose of quinine for the treatment of falciparum malaria is 15 mg/kg/day in a single dose, with only a 9% recrudescence rate at day 28 (some or most of which may have been attributable to reinfection).

2.4.2 SAFETY AND TOLERABILITY

Some early observers felt that quinine caused less gastric irritation than cinchona bark (Smith, 1976). The smaller dose and more easily concealed taste are also advantages. However, cinchona febrifuge can be made more palatable in the form of gelatine capsules, pills, or tablets, rather than as a solution whose "bitter taste is enough to make many healthy people vomit, let alone those ill with malaria" (Fletcher, 1934). Acton (1920) states that "Cinchona febrifuge taken in tablet form is more pleasant to take and less toxic than quinine." Waters (1916) also reports that in general quinoidine produces less gastric irritation and less cinchonism than quinine.

Hicks and Diwan Chand (1935) found a greater incidence of vomiting on days 1 and 2 in patients taking tablets of totaquina type I or II compared with those taking quinine (13.2 to 15.7% in the totaquina groups, compared to 7.4% in the quinine group), but this was probably due to the disease rather than the drug, as the vomiting decreased after day 2, and no patients were complaining of vomiting by the fourth day of treatment. Vertigo was equally common in those taking quinine and totaquina type I (up to a maximum of 9% on day 3), but less so in those taking totaquina type II (maximum of 4% on day 2).

In its multicentre trial of totaquina, the League of Nations concluded, "As regards toxicity, the case records contain no cogent evidence that Totaquina is more toxic than quinine in the doses given" (Fletcher, 1934).

Large doses of quinine may cause reversible tinnitus and visual disturbances. The symptom complex of tinnitus, high-tone hearing impairment and nausea, is called cinchonism. Quinine causes a greater degree of hearing loss than quinidine (Karbwang et al., 1993), while auditory side effects are much less common with cinchonine and cinchonidine (MacGilchrist, 1915). Cinchonism does, however, occur with higher doses of cinchonine (Solary et al., 2000). Visual disturbances have also been observed with cinchonine, but not with quinidine or cinchonidine (MacGilchrist, 1915; White et al., 1981). Blindness and deafness occur following quinine overdose, but are rare in malaria treatment.

Quinine and quinidine stimulate pancreatic β-cells, causing hypoglycemia. This may occur in severe malaria, especially in pregnant women (White, 1996). Large doses of cinchonine may cause diarrhoea, but cinchonidine in isolation, even in large doses, causes very few adverse effects (MacGilchrist, 1915).

Both quinine and quinidine cause prolongation of the corrected QT interval* (QTc), the maximum effect being 1.5 to 2.0 hours after intravenous administration of the drug (Karbwang et al., 1993). The maximum effect of quinine is about half that of quinidine (see Table 2.7). Cinchonine has been used in high doses (30 mg/kg/day) intravenously as a resistance reverser with chemotherapy for leukemia (Solary et al., 2000). It was well tolerated, but prolonged cardiac repolarisation was the dose-limiting toxic effect. The QTc prolongation appeared within the first few hours of the infusion and returned to baseline after 8 to 10 hours of stopping the drug. QTc prolongation

---

* "QT interval" is a cardiological term referring to the time delay between the Q wave and the T wave on an electrocardiogram.

TABLE 2.7
Maximum Prolongation of QTc

Drug	% Increase in QTc over Baseline	Reference
Quinine	20	Karbwang et al., 1993
Quinidine	37	Karbwang et al., 1993
Cinchonine	>20	Solary et al., 2000
Quinimax	20–25	Sowumni et al., 1990

also occurs in patients taking Quinimax (Sowumni et al., 1990), but less than that observed with quinidine. Although arrhythmias are very uncommon, these drugs are contraindicated in patients with cardiac conduction disturbances.

In summary, it seems that a combination of the alkaloids at the same total dose (but a lesser dose of each individual component) reduces the frequency of at least some of the adverse effects. Quinimax produces less cinchonism than quinine alone (Bunnag et al., 1989) and less QTc prolongation than quinidine (Sowumni et al., 1990).

2.4.3 EFFICACY

As described above, early observers reported that quinine cured some cases that had resisted the cinchona bark (Smith, 1976). However, there was no verification of the quality of the bark used or of the dosage administered, and of course, there was no confirmation that the diagnosis was malaria, because the malaria parasite was not discovered until 1880. Even after that, it was many years before researchers routinely examined blood films of suspected malaria patients. This section will examine those studies of cinchona bark in which parasitaemia was an inclusion criterion and an outcome measure.

Waters (1916) reports that treatment with quinoidine gave results as good as quinine. Quinoidine was used in preference to quinine in several hospitals in India, for several thousands of patients, with apparently good results, in the treatment of both vivax and falciparum malaria.

Hicks and Diwan Chand (1935) investigated the initial response of patients with *P. vivax* and *P. falciparum* infections to treatment with either quinine, totaquina type I, or totaquina type II. They found no significant difference in parasite or fever clearance times among the groups, with 100% clearance of both fever and parasites by day 4 of treatment (Table 2.8). They even had a control group infected with *P. vivax* but receiving no treatment to prove that the cinchona alkaloids were effective in treating malaria. Unfortunately, there was no longer term follow-up to look at relapse rates.

Acton (1920) investigated the use of cinchona febrifuge for the treatment of vivax malaria (Table 2.8). His patients were observed for at least 2 months after treatment in a hilltop hospital with little risk of reinfection, and all cases were confirmed with microscopy. A cure was defined as the absence of a relapse during the period of follow-up. The low cure rates should not come as a surprise to modern readers, knowing that *P. vivax* can reactivate from liver hypnozoites a long time after the successful treatment of an acute attack; the only drug known to prevent this is primaquine (which is still not available in most developing countries due to its high cost). Quinine alone, even when administered for up to 2 months for the treatment of a single attack, produced cure rates of only 18 to 30%. Acton concludes:

> The administration of cinchona febrifuge in benign tertian infections is better than quinine. A three weeks' course gives about the same curative results as a four months' course of quinine.... The immediate results are also slightly better.

TABLE 2.8
Summary of Clinical Trials of Totaquina

Reference	Totaquina Type	N	Dose	Species	Efficacy
Acton, 1920	I	110	1.3 g daily for 10 or 21 days	*P. vivax*	52–53% parasite-free at 2 months
Fletcher, 1923	I	21	640 mg bd for 4 weeks	*P. malariae*	2 relapses within 3 weeks
Sinton and Bird, 1929	I	110	1.9 g daily for 7 days, then 1.3 g for 21–24 days	*P. vivax*	27% parasite-free at 2 months
Pampana, 1934	I	175	600 mg daily for 5 days	*P. vivax*	98–100% clearance by day 5
	II	290			
Pampana, 1934	I	155	1.2 g daily for 5 days	*P. falciparum*	92–97% clearance by day 5
	II	261			
Hicks and Diwan Chand, 1935	I	48	600 mg daily for 5 days	*P. vivax*	100% clearance by day 4
	II	36			
Hicks and Diwan Chand, 1935	I	42	1.2 g daily for 5 days	*P. falciparum*	100% clearance by day 4
	II	31			

Note: N = number of patients treated; bd = twice daily; tds = three times daily; qds = four times daily.

He goes on to state that the cinchona febrifuge is considerably cheaper than quinine, and that this would save the government a considerable sum of money if used.

Sinton and Bird (1929) report a similar series of 110 patients with chronic *P. vivax* treated with cinchona febrifuge at a dose of 1.9 g per day for 7 days, followed by 1.3 g per day for a further 21 to 24 days. They were also followed up for 8 weeks after completion of the treatment, but the cure rate was low (26.9%) and similar to that observed with quinine. The same authors report a smaller series of 34 patients with acute malaria (nonspecific), treated with 1.9 g of cinchona febrifuge per day for 4 days. The cure rate in this group was higher (53%).

Fletcher (1923) observed the efficacy of cinchona febrifuge in 42 patients with *Plasmodium malariae* (21), *P. vivax* (5), *P. falciparum* (5) and mixed infections (11). The dose was 640 mg twice daily for 4 weeks. Only five relapses were observed, all of *P. vivax* or *P. malariae*. Dominguez et al. (1933) observed 36 patients with *P. vivax* or *P. malariae* (or a mixed infection), taking a preparation of cinchona alkaloids and chinovin with other herbs, namely, *Berberis flexuosa* and *Aspidospermum quebracho blanco*. The dose of cinchona alkaloids was of 254 mg qds for 5 days, followed by 254 mg tds for 5 days, then a break of 5 days, and 254 mg bd for a few more days. No relapses were observed in the 2-month follow-up period.

A large multicentre trial of totaquina was conducted under the auspices of the League of Nations in 1933 (Pampana, 1934). Four hundred twenty-one patients were given totaquina type I, and 634 were given totaquina type II. The outcome measurement was parasite and fever clearance on the fifth and last day of treatment; there was no further follow-up to look for relapses. There was an attempt at standardising the composition of the totaquina used and the doses administered (600 mg per day for *P. vivax*, 1200 mg per day for *P. falciparum* and *P. malariae*), but there was still quite a wide variation in the actual doses given. In spite of the variation in dosage, the parasite clearance achieved was surprisingly uniform, and as good as quinine (see Table 2.8).

Much more data exist on the efficacy of the purified alkaloids than for their mixtures. There have been many trials demonstrating the efficacy of quinine, although its efficacy has been declining in some areas over several decades due to the evolution of resistant parasites (WHO, 1973). In 1988, Giboda and Denis reported that 47% of cases in Cambodia were resistant to quinine at the RI level, and a further 16% at the RII and RIII levels. In a study of pregnant women in Thailand, the cure rate with quinine in *P. falciparum* infections was about 67% at day 63 of follow-up

(McGready et al., 2000); this cure rate can be improved by combination with tetracycline or clindamycin, although these are contraindicated in pregnancy.

Quinidine can also be used for the treatment of malaria (White et al., 1981). It is at least as effective as quinine, and more so in the case of resistant infections. Cinchonine and cinchonidine, when used in isolation, are also as effective as quinine (MacGilchrist, 1915; Sinton and Bird, 1929).

A standardised mixture of cinchona alkaloids has been developed and marketed as Quinimax. It contains 71.4% quinine, with 18.6% quinidine and 5% cinchonine (Table 2.3). It is as effective as quinine in the treatment of malaria, but with a lower incidence of side effects (Bunnag et al., 1989). In some areas, but not others, a 3-day treatment course has been as effective as a 7-day course, which is advantageous in terms of cost and compliance (Deloron et al., 1989; Rogier et al., 1996; Kofoed et al., 1997). There have been no published studies comparing its efficacy with that of quinine in areas with significant quinine resistance.

In summary, all the available evidence suggests that crude extracts of cinchona bark, in spite of variability in composition, are as effective as quinine in the treatment of both vivax and falciparum malaria. Furthermore, crude bark extracts, with the synergism of many active ingredients, may be more effective in areas where *P. falciparum* has evolved resistance to quinine.

2.5 PUBLIC HEALTH POTENTIAL

Cinchona trees are still cultivated in many parts of the tropics, for example, the Congo, Madagascar, Guatemala, the Philippines, Vietnam, Indonesia, and India. Cinchona bark is often exported as a source of raw materials for the commercial production of quinine. In Kivu (Democratic Republic of Congo) alone, *C. ledgeriana* plantations produce 4000 to 5000 tonnes of bark a year, and this in turn produces 400 to 450 tonnes of consumable alkaloids (van Leeuwen, personal communication). In Indonesia, the state-owned Sinkona Indonesia Lestari company cultivates 10,000 hectares of cinchona, yielding 115 tonness of quinine and 40 tonnes of cinchonine per year (see http://sinkona.sundanet.com/company-profile.htm).

In India, the cultivation of cinchona has declined considerably. At present it is only grown in the states of West Bengal and Tamil Nadu. The total area of plantations in 1988 amounted to just over 5000 hectares. Although several species of cinchona are grown in India, the bulk of the commercial supplies of bark is derived from a few cultivated species: *C. calisaya* Wedd., *C. ledgeriana* Moens, *C. officinalis* Linn., *C. succirubra* Pav. ex Klotzsch, and their hybrids. Due to natural hybridisation over a long period, true-to-type species are no longer found in the commercial plantations.*

It would be relatively simple for such plantations to provide some bark at low cost to local communities for their own use, and if there is sufficient demand, to prepare a totaquina-type mixture for larger-scale use. This is likely to be cheaper than the use of second-line drugs for resistant falciparum malaria. Totaquina is still manufactured on a small scale in Madagascar; while a course is more expensive than chloroquine, it is cheaper than sulfadoxine-pyrimethamine or other second-line drugs. In some areas of the country, the bark of the tree is more readily available than modern drugs.

Unfortunately, cinchona is becoming rarer due to deforestation and overexploitation in some countries. This was already a problem in Bolivia in the 19th century (Honigsbaum, 2001) and continues today in Colombia (Blair, personal communication). In these cases, communities could be encouraged to plant cinchona trees around their villages and taught how to harvest and prepare the bark.

Although modern antimalarials may have a more clearly defined safety profile and a more precise dose, cinchona bark is safe and has been effective for several hundred years. Such a treatment

---

* Information from Directorate of Cinchona and other Medicinal Plants, West Bengal; and Cinchona Department, Udhagamandalam, Tamil Nadu.

is indisputably better than nothing in areas where modern drugs are unavailable, unaffordable, or ineffective due to multidrug resistance. Produced and harvested sustainably, it could still have a valuable impact on malaria morbidity and mortality.

REFERENCES

Acton, H.W. (1920). Researches on the treatment of benign tertian fever. *Lancet*, 1920, 1257–1261.

Andersson, L. (1998). A revision of the genus Cinchona (Rubiaceae-Cinchoneae). *Mem. N.Y. Bot. Garden*, 80, 25–67.

Bado, S. (1663). *Anastasis Corticis Peruviae seu Chinae Chinae defensio*. P. Calenzani, Genoa.

Baker, G. (1785). Observations on the late intermittent fevers; to which is added a short history of the Peruvian bark. *Med. Trans. Coll. Physicians London*, 3, 141–216.

Blair, S., Correa, A., Madrigal, B., Zuluaga, C.B., and Franco, H.D. (1991). *Plantas antimaláricas: una revisión bibliográfica*. Universidad de Antioquia, Medellín, Colombia.

Bunnag, D., Harinasuta, T., Looareesuwan, S., Chittamas, S., Pannavut, W., Berthe, J., and Mondesir, J.M. (1989). A combination of quinine, quinidine and cinchonine (LA 40221) in the treatment of chloroquine resistant falciparum malaria in Thailand: two double-blind trials. *Trans. R. Soc. Trop. Med. Hyg.*, 83, 66.

Chifflet, J.J. (1653). *Pulvis febrifugus orbis Americani ... ventilatus ratione, experientia, auctoritate*. Louvanii, Brussels.

Clark, A.J. (1933). *Applied Pharmacology*, 5th ed. J&A Churchill, London.

Cobo, B. (1943). *Historia del Nuevo Mundo*. Ediciones Atlas, Madrid.

Coe, F.G. and Anderson, G.J. (1996). Ethnobotany of the Garífuna of eastern Nicaragua. *Econ. Bot.*, 50, 71–107.

De Jussieu, J. (1936). *Description de l'arbre a quinquina: memoire inédit de Joseph de Jussieu*. Société du traitement des Quinquinas, Paris.

De La Calancha, A. (1638). *Coronica moralizada del Orden de San Agustín en el Perú*. Pedro Lacavalleria, Barcelona.

De La Condamine, C.M. (1738). Sur l'arbre du quinquina. *Mem. Acad. R. Sci. Paris*, p. 226–243.

Deloron, P., Lepers, J.P., Verdier, F., et al. (1989). Efficacy of a 3-day oral regimen of a quinine-quinidine-cinchonine association (Quinimax) for treatment of falciparum malaria in Madagascar. *Trans. R. Soc. Trop. Med. Hyg.*, 83, 751–754.

Dobson, M. (1988). Bitter-sweet solutions for malaria: exploring natural remedies from the past. *Parassitologia*, 40, 69–81.

Dominguez, J.A., Mazza, S., and Soto, N.A. (1933). *El Yara Chucchu (Cinchona sp) y sus alcaloides en el tratiamiento del paludismo*, Trabajos del Instituto de Botanica y Farmacologia No. 51. Instituto de Botanica y Farmacologia, Buenos Aires.

Druihle, P., Brandicourt, O., Chongsuphajaisiddhi, T., and Berthe, J. (1988). Activity of a combination of three Cinchona bark alkaloids against *Plasmodium falciparum* in vitro. *Antimicrob. Agents Chemother.*, 32, 250–254.

Duran-Reynals, M.L. de A. (1946). *The Fever Bark Tree: The Pageant of Quinine*. Doubleday, New York.

El-Kamali, H.H. and El-Khalifa, K.F. (1997). Treatment of malaria through herbal drugs in the Central Sudan. *Fitoterapia*, 68, 527–528.

Espino, F., Manderson, L., Acuin, C., Domingo, F., and Ventura, E. (1997). Perceptions of malaria in a low endemic area in the Philippines: transmission and prevention of disease. *Acta Trop.*, 63, 221–239.

Fletcher, W. (1923). Notes on the Treatment of Malaria with the Alkaloids of Cinchona. Studies from the Institute for Medical Research, Kuala Lumpur, Federated Malay States.

Fletcher, W. (1934). Critical analysis of the results achieved. *League of Nations Q. Bull. Health Organ.*, III, 344–358.

Frappier, F., Jossang, A., Soudon, J., et al. (1996). Bisbenzylisoquinolines as modulators of chloroquine resistance in Plasmodium falciparum and multidrug resistance in tumor cells. *Antimicrob. Agents Chemother.*, 40, 1476–1481.

Garrod, A.B. (1875). *The Essentials of Materia Medica and Therapeutics*, 5th ed. Longmans, Green & Co., London.

Giboda, M. and Denis, M.B. (1988). Response of Kampuchean strains of *Plasmodium falciparum* to antimalarials: in-vivo assessment of quinine and quinine plus tetracycline; multiple drug resistance in vitro. *J. Trop. Med. Hyg.*, 91, 205–211.

Gramiccia, G. (1988). *The Life of Charles Ledger 1818–1905: Alpacas and Quinine.* Macmillan, Basingstoke.

Grier, J. (1937). *A History of Pharmacy.* Pharmaceutical Press, London.

Haggis, A.W. (1942). Fundamental errors in the early history of cinchona. *Bull. Hist. Med.*, 10, 586–592.

Hale White, W. (1897). *Materia Medica: Pharmacy, Pharmacology and Therapeutics.* J&A Churchill, London.

Hicks, E.P. and Diwan Chand, S. (1935). The relative clinical efficacy of totaquina and quinine. *Rec. Malaria Surv. India*, 5, 39–50.

Hobhouse, H. (1999) *Seeds of Change: Six Plants That Transformed Mankind.* Macmillan, London.

Hodge, W.H. (1948). Wartime cinchona procurement in Latin America. *Econ. Bot.*, 2, 229–257.

Hofman, J.J. (1923). Cinchona Febrifuge, Quinetum et Quinine. In *Chininum: Scriptiones Collectae.* Bureau Tot Bevordering van Het Kinine-Gebruik, Amsterdam.

Honigsbaum, M. (2001). *The Fever Trail: The Hunt for the Cure for Malaria.* Macmillan, London.

Jaramillo-Arango, J. (1949). *A Critical Review of the Basic Facts in the History of Cinchona.* Academic Press for the Linnean Society of London, London, pp. 272–311.

Jarcho, S. (1993). *Quinine's Predecessor, Francesco Torti and the Early History of Cinchona.* Johns Hopkins University Press, Baltimore.

Johnston, M. and Colquhoun, A. (1996). Preliminary ethnobotanical survey of Kurupukari: an Amerindian settlement of Central Guyana. *Econ. Bot.*, 50, 182–194.

Karbwang, J., Davis, T.M.E., Looareesuwan, S., Molunto, P., Bunnag, D., and White, N.J. (1993). A comparison of the pharmacokinetic and pharmacodynamic properties of quinine and quinidine in healthy Thai males. *Br. J. Clin. Pharmacol.*, 35, 265–271.

Karle, J.M. and Bhattacharjee, A.K. (1999). Stereoelectronic features of the cinchona alkaloids determine their differential antimalarial activity. *Bioorg. Med. Chem.*, 7, 1769–1774.

Kofoed, P.E., Có, F., Poulsen, A., et al. (2002). Treatment of *Plasmodium falciparum* malarial with quinine in children in Guinea-Bissau: one daily dose is sufficient. *Trans. R. Soc. Trop. Med. Hyg.*, 96, 185–188.

Kofoed, P.E., Lopes, F., Johansson, P., Dias, F., Sandström, A., Aaby, P., and Rombo, L. (1999). Low-dose quinine for treatment of *Plasmodium falciparum* in Guinea-Bissau. *Trans. R. Soc. Trop. Med. Hyg.*, 93, 547–549.

Kofoed, P.E., Mapaba, E., Lopes, F., Pussick, F., Aaby, P., and Rombo, L. (1997). Comparison of 3, 5, and 7 days' treatment with Quinimax for falciparum malaria in Guinea-Bissau. *Trans. R. Soc. Trop. Med. Hyg.*, 91, 462–464.

MacGilchrist, A.C. (1915). The relative therapeutic value in malaria of the cinchona alkaloids quinine, cinchonine, quinidine, cinchonidine and quinoidine, and the two derivatives: hydroquinine and ethylhydrocupreine. *Indian J. Med. Res.*, III, 1–53.

Madureira, M.C., Martins, A.P., Gomes, M., et al. (1998). Antimalarial Activity of Medicinal Plants from São Tomé and Príncipe Islands. Clone, Cure and Control: Tropical Health for the 21st Century. Liverpool School of Tropical Hygiene and Medicine, European Congress Office, Liverpool. Poster 64, abstract, p. 137.

Mannaberg, J. and Leichtenstern, O. (1905). *Northnagel's Encyclopaedia of Practical Medicine: Malaria, Influenza and Dengue.* W.B. Saunders & Co., Philadelphia.

Markham, C.R. (1980). *Peruvian Bark: A Popular Account of the Introduction of Chinchona Cultivation into British India 1860–1880.* John Murray, London.

McGready, R., Brockamn, A., Cho, T., Cho, D., van Vugt, M., Luxemburger, C., Chongsuphajaisiddhi, T., White, N., and Nosten, F. (2000). Randomised comparison of mefloquine-artesunate combination versus quinine in treatment of multi-drug resistant falciparum malaria in pregnancy. *Trans. R. Soc. Trop. Med. Hyg.*, 94, 689–693.

Milliken, W. (1997). *Plants for Malaria, Plants for Fever. Medicinal Species in Latin America: A Bibliographic Survey.* Kew: Royal Botanical Gardens.

Milliken, W. and Albert, B. (1997). The use of medicinal plants by the Yanomami Indians of Brazil, part II. *Econ. Bot.*, 51, 264–278.

Mitchell, B.J. (1884). *Materia Medica and Therapeutics: An Introduction to the Rational Treatment of Disease.* Cassell & Co., London.

Monardes, N. (1574). *Historia medicinal de las cosas que se traen de nuestras Indias Occidentales*. A. Escribano, Seville.

Montes Giraldo, J.J. (1981). *Medicina popular en Colombia: vegetales y otras sustancias usadas como remedios*. Instituto Caro y Cuervo, Bogotá (quoted in Blair et al., 1991).

OUA. (1985). *Pharmacopée Africaine*, Vol. 1. OUA, Lagos.

Pampana, E.J. (1934). Clinical tests carried out under the auspices of the Malaria Commission. *League of Nations Q. Bull. Health Org.*, III, 328–343.

Prudhomme, E. (1902). Le Quinquina à Madagascar. *Revue de Madagascar*, 10, 129–143.

Rasoanaivo, P., Petitjean, A., Ratsimamanga-Urverg, S., and Rakoto-Ratsimamanga, A. (1992). Medicinal plants used to treat malaria in Madagascar. *J. Ethnopharmacol.*, 37, 117–127.

Rea, L. (1995). Cinchona y la tribu Conchonae (Rubiaceae) en Bolivia, actualización sistematica, fitoquimica y actividad antimalarica. Thesis, UMSA, La Paz.

Rogier, C., Brau, R., Tall, A., Cisse, B., and Trape, J.F. (1996). Reducing the oral quinine-quinidine-cinchonin (Quinimax) treatment of uncomplicated malaria to three days does not increase the recurrence of attacks among children living in a highly endemic area of Senegal. *Trans. R. Soc. Trop. Med. Hyg.*, 90, 175–178.

Ruiz, D.H. and Pavon, D.J. (1792). *Quinolgia, ó tratado del árbol de la Quina o Cascarilla, con su descripcion y la de otras especies de quinos nuevamente descubiertas en el Perú*. Widow & Son of Marin, Madrid.

Sinton, J.A. and Bird, W. (1929). Studies in malaria, with special reference to treatment. Part XI: The cinchona alkaloids in the treatment of benign tertian malaria. *Indian J. Med. Res.*, 16, 725–746.

Smith, D.C. (1976). Quinine and fever: the development of the effective dosage. *J. Hist. Med. Allied Sci.*, 31, 343–367.

Solary, E., Monnone, L., Moreau, D., et al. (2000). Phase I study of cinchonine, a multidrug resistance reversing agent, combined with the CHVP regimen in relapsed and refractory lymphoproliferative syndromes. *Leukaemia*, 14, 2085–2094.

Sowumni, A., Salako, L.A., Loye, O.J., and Aderounmu, A.F. (1990). Combination of quinine, quinidine and cinchonine for the treatment of acute falciparum malaria: correlation with the susceptibility of *Plasmodium falciparum* to the cinchona alkaloids in vitro. *Trans. R. Soc. Trop. Med. Hyg.*, 84, 626–629.

Sturm, R. (1659). *Febrifugi Peruviani Vindiciarum*. Oosterhout, Delft.

Toxbase. (2002). See http://www.spib.axl.co.uk/Toxbase.

Tréluyer, J.M., Roux, A., Mugnier, C., Flouvat, B., and Lagardère, B. (1996). Metabolism of quinine in children with global malnutrition. *Pediatr. Res.*, 40, 558–563.

Von Humboldt, A. (1821). *An Account of the Cinchona Forests of South America Drawn up during Five Years Residence and Travels on the South American Continent*, Lambert, A.B., Trans. Longman, London.

Waters, E.E. (1913). The value of amorphous cinchona alkaloid in malaria. *Indian Med. Gaz.*, 48, 89–91.

Waters, E.E. (1916). The value of quinoidine in malaria. *Indian Med. Gaz.*, 51, 335–338.

Weddell, H.A. (1849). *Histoire Naturelle des Quinquinas*. V. Masson, Paris.

Weddell, H.A. (1871). *Notes on the Quinquinas*, 8 vols., Markham, C.R., Trans. India Office, London, p. 65.

White, N.J. (1996). Malaria. In *Manson's Tropical Diseases*, 12th ed., Cook, G.C., Ed. W.B. Saunders, London.

White, N.J., Warrell, D.A., Bunnag, D., Looareesuwan, S., Chongsuphajaisiddhi, T., and Harinasuta, T. (1981). Quinidine in Falciparum malaria. *Lancet*, 14th Nov. 1981, 1069–1071.

WHO. (1973). Chemotherapy of Malaria and Resistance to Antimalarials, WHO Technical Report Series 529. WHO, Geneva.

3 Artemisia annua as a Traditional Herbal Antimalarial

Merlin Willcox, Gerard Bodeker, Geneviève Bourdy,
Vikas Dhingra, Jacques Falquet, Jorge F.S. Ferreira,
Bertrand Graz, Hans-Martin Hirt, Elisabeth Hsu,
Pedro Melillo de Magalhães, Damien Provendier,
and Colin W. Wright

CONTENTS

3.1 INTRODUCTION

The species *Artemisia annua* L (Asteraceae) is native to China. Its ancient Chinese name, Qing Hao, literally means "green herb." The genus *Artemisia* comprises over 400 species, many of which have an aromatic, bitter taste. There are two theories as to the origin of its name. Ferreira et al. (1997) say that it is named after the Greek goddess Artemis, meaning literally "she who heals sickness," who was in fact goddess of the hunt, of forests, and was thought to be responsible for sudden death in women (Guirand, 1959). Apparently plants of this genus, probably *Artemisia absinthium*, were used to control pain in childbirth and to induce abortions. Bruce-Chwatt (1982) says that *Artemisia* was named after Queen Artemisia of Caria (Turkey), who lived in the fourth century B.C. She was so aggrieved on the death of her husband and brother, King Mausolus of Halicarnassus, that she mixed his ashes with whatever she drank to make it taste bitter.

A. annua is so named because it is almost the only member of the genus with an annual cycle. It is a shrub, often growing over 2 m high (Ferreira et al., 1997; see Figure 3.1 and Figure 3.2). Its leaves and flowers contain artemisinin, first isolated in China in 1971; this is the constituent with the greatest antimalarial activity (see Table 3.3). Artemisinin has been found in only two other species, *Artemisia apiacea* and *Artemisia lancea* (Tan et al., 1998).

FIGURE 3.1 *A. annua.* (Copyright 2001, Merlin Willcox.)

FIGURE 3.2 *A. annua* hybrid growing in Brazil. (Copyright 1996, Pedro Melillo de Magalhães.)

Artemisinin is poorly soluble in oil or water, so is usually administered orally, although it can be given rectally (Ashton et al., 1998) and, when suspended in oil, intramuscularly (Titulaer et al., 1990). Synthetic derivatives that are water soluble (artesunate) and oil soluble (artemether) have been developed to enable intravenous and intramuscular administration, respectively (Van Agtmael et al., 1999a). It is now universally accepted that this family of compounds is among the most powerful antimalarial drugs ever discovered. The pharmacological and clinical evidence is well documented (Wright and Warhurst, 2002; Wilairatana and Looareesuwan, 2002).

Artemisinin cannot be synthesised cost effectively, so it is still extracted from *A. annua* aerial parts. Therefore, the science of commercial cultivation of *A. annua*, to maximise artemisinin yields, is already well developed (Laughlin et al., 2002). However, the end product is often unaffordable for the poor, especially as it is recommended to only be used in combination with other drugs (Nosten, 2002). The lowest cost of a course of Co-artem (artemether-lumefantrine), provided by the manufacturers to some countries in Africa, is $2.40, and a course of artesunate costs $1.20 (White, 2003). This is above the commonly quoted affordable limit of $1 (but even that is more than many people can afford). Furthermore, its value makes it an attractive target for forgery. A survey in 2000–2001 in Southeast Asia found that 38% of shop-bought oral artesunate samples were fake and contained no artesunate (Newton et al., 2001). A more recent survey has found that some fake artesunate packs are now indistinguishable from the genuine ones (Newton et al., 2003).

Therefore, some organisations have already set up programs for small-scale cultivation and local utilisation of the herbal formulation of *A. annua*. For example, the non-governmental organisations (NGOs) Anamed (see below) and Medicos Descalzos (barefoot doctors) in Guatemala promote its use.

However, the science of small-scale cultivation and local production is far from perfected, and has not been studied to the same extent as that of large-scale cultivation and use of its pharmaceutical end product. This chapter sets out to address some of these practical considerations as far as is possible within the current state of knowledge, and to highlight key research priorities for the immediate future. Its aims are to review the literature on the classical ethnomedical use of *A. annua*, literature on the pharmacokinetics and pharmacodynamics of Herba *A. annua* and its constituents, and experience in its horticulture and cultivation. Common concerns about the use of herbal preparations of *A. annua* will be discussed.

3.2 ETHNOPHARMACOLOGY

3.2.1 *A. annua* as a Traditional Antimalarial

The earliest record of the medicinal use of the shrub *A. annua* (Qing Hao) dates back to 168 B.C. in the "Fifty-two prescriptions" discovered in one of the Han Dynasty tombs in Mawangdui; this advocated the use of *A. annua* for the treatment of hemorrhoids. It is likely that the plant had been used for some time before this, as it appears in the *Shen Nong Ben Cao Jing*, the foundation text of Chinese herbal medicine. It is considered that this was first written in about 200 A.D., but represents the cumulative knowledge of herbal medicine transmitted orally over many centuries (Shou-zhong, 1997). This text claims that among other properties, Qing Hao "relieves lodged heat in the joints," which could be interpreted as treating rheumatoid arthritis, or possibly a number of febrile conditions. Li Shizhen (1596) described the use of Qing Hao for malarial fevers in his *Bencao Gangmu*.

Zhang Ji (150–219 A.D.) in his classic text *On Cold Damage* (*Shang Han Lun*) recommends a decoction of *A. annua* to treat fevers with sweating and jaundice (Mitchell et al., 1999). He recommends that fruits of *Gardenia jasminoides* and roots of rhubarb (*Rheum palmatum*) should be added to the decoction. The *Handbook of Prescriptions for Emergency Treatment* (*Zhouhou Beiji Fang*), written in 340 A.D., also recommends the use of Qing Hao for intermittent fevers. The

TABLE 3.1
Ingredients with Which *A. annua* Is Combined for Different Types of Fever

Type of Fever	Added Ingredient	Part
No sweating, but dizziness and stifling sensation in chest	*Dolichos lablab* L (Leguminosae)	Seed
	Talcum	Powder
Night sweats and anaemia	*Lycium barbarum* L (Solanaceae)	Stem
	Cynanchum atratum Bge (Asclepiadaceae)	Root
Heat smoldering in *yin* regions of body	*Rehmannia glutinosa* (Scrophulariaceae)	Root
	Amyda sinensis	Tortoise shell
Damp summer heat with nausea, stifling sensation in chest, and intense fever	*Scutellaria baicalensis* Georgi (Labiatae)	Root
	Pinellia ternata (Thunb) Breit (Araceae)	Rhizome

Source: Bensky, D. and Gamble, A., (1993), *Chinese Herbal Medicine Materia Medica*, Eastland Press, Inc., Seattle, WA.

mode of preparation was a cold aqueous extraction: a handful of the aerial parts of the plant should be soaked in two Sheng (approximately 1 to 2 l) of water, and the juice should all be drunk (Tu, 1999). Different preparations have subsequently been recommended in different texts: a decoction of 30 g of *A. annua* daily, pills, powdered dried leaves in a dose of 3 g per day for 5 days, and fresh plant juice at a dose of 3 g daily (QACRG, 1979; Foster and Chongxi, 1992).

Mixtures with other medicines have also been used; for example, in 1798 Wenbing Tiaobian (quoted in Laughlin et al., 2002) used *A. annua* combined with *Rehmannia glutinosa* (Scrophulariaceae), *Anemarrhena asphedeloides* (Liliaceae), *Paeonia suffruticosa* (Ranunculaceae), and *Carapax Amydae sinensis* (tortoise shell). There exist several other combinations, according to the type of fever (see Table 3.1). The leaves of *A. annua* are also burned in China as a fumigant insecticide to kill mosquitoes (Foster and Chongxi, 1992).

3.2.2 ANAMED RECOMMENDATIONS

Anamed (Action for Nature and Medicine) is an NGO promoting the use of traditional medicines (www.anamed.org). It distributes seeds of a recently developed artemisinin-rich genotype of *A. annua* (Delabays, 1997; De Magalhães, 1996) for cultivation and preparation as an herbal antimalarial. Two hundred and forty partner organisations in developing countries are participating in this program, and their feedback has helped the organisation to refine its recommendations, which are summarised below.

Hirt (2001) recommends an infusion of 5 g of dried leaves on which 1 l of boiling water is poured and left to cool for 15 minutes (for a 60-kg adult; 2.5 g of leaves for a 30-kg child and 1.25 g for a 15-kg child). This method extracts 55% of the artemisinin into the water, and 35 to 40% remains in the leaves. Only 5% is lost, in contrast to a decoction (when the plant is boiled in water for several minutes), where 50% of the artemisinin is lost, because it is not heat stable. An infusion in full fat milk can increase the proportion of artemisinin extracted to 80% (De Magalhães et al., 2003). Anamed advises that as artemisinin reacts with iron, the tea should be prepared in pots made of other materials.

Anamed recommends the dose of 250 ml of the infusion, taken every 6 hours for 7 days; this dose is based on that in the Chinese pharmacopoeia, which recommends a dose of 4.5 to 9 g daily. An alternative is to swallow 1 g of dried leaves three times a day, but here Anamed has made only a few positive observations. Anamed is also observing patients treated with an enema using double the dose of leaves and half the amount of water; there have been some positive results, but further research is awaited (Hirt, personal communication).

3.2.3 *A. annua* and Artemisinin

Artemisinin is a sesquiterpene lactone, containing an unusual endoperoxide group, which is believed to be responsible for the antimalarial activity of *A. annua*. Unlike quinine-related drugs and antifolate drugs, artemisinin and its derivatives are gametocytocidal and reduce the transmission potential of falciparum malaria (Price et al., 1996). The mechanism of action is explained in detail in Chapter 15.

Artemisinin was first isolated from cold-ether (rather than hot-water) extracts of the plant (Yu and Zhong, 2002). Aqueous decoctions of *A. annua* were found to be ineffective against *Plasmodium berghei* in mice, and the serum of rabbits treated with this decoction was ineffective against *Plasmodium falciparum in vitro*, but ether extracts were active (Lwin et al., 1991). This is not entirely surprising because decoctions contain less artemisinin than infusions (see above). Furthermore, preparations containing whole *Artemisia* leaves are likely to have a higher artemisinin content than filtered decoctions or infusions, if equivalent doses are used, because much of the artemisinin will remain in the leaves rather than be dissolved in the water. Räth et al. (2004) compared the artemisinin content of *Artemisia annua* tea prepared by three different methods. They found that the greatest concentration (94.5 mg/L) was obtained by adding one litre of boiling water to 9 g of *A. annua* leaves, briefly stirring, leaving the mixture to stand for 10 minutes, then filtering out the plant material and gently squeezing it to release residual water. This method extracted 76% of the artemisinin. Secondly, the mouse model is not always an accurate predictor of effectiveness in humans. Absorption and metabolism of artemisinin and other active ingredients may differ, and *P. berghei* may have different properties and sensitivities from *P. falciparum*.

The dose of artemisinin in 5 g of dried leaves (with a relatively high content of about 1.4% artemisinin) would be no more than 70 mg. One litre of infusion made from 5 g of leaves contains 60 mg of artemisinin, and a litre of infusion prepared as described above contains 94.5 mg of artemisinin (Räth et al., 2004). The dose recommended by the World Health Organization (WHO) is 20 mg/kg as a divided loading dose on the first day, followed by 10 mg/kg once daily for 6 days (Phillips-Howard, 2002). For a 60-kg adult, this works out as 1200 mg on the first day, followed by 600 mg daily on subsequent days.

The large discrepancy between the artemisinin content of the infusion and the recommended doses of artemisinin has led to two important concerns:

1. Herbal preparations of *A. annua* could not possibly be effective.
2. Subtherapeutic concentrations of artemisinin will favor the evolution of resistant parasites.

A third concern is whether herbal preparations are safe and well tolerated. Each of these points will be addressed in detail below.

3.3 CLINICAL EFFICACY

Artemisinin and its derivatives are the most potent and rapidly acting antimalarial drugs: the parasite biomass is reduced by 10,000-fold per asexual life cycle, compared to 100- to 1000-fold for other antimalarials. They also decrease gametocyte carriage by 90%, thus reducing transmission of malaria (White et al., 1999). A meta-analysis of trials, with data on 1919 patients, has shown that artemether is at least as effective as quinine and is associated with fewer serious adverse effects (Artemether-Quinine Meta-analysis study group, 2001).

There are fewer data on the efficacy of the whole herb. The larger clinical trials of different *A. annua* preparations are summarised in Table 3.2. Chang and But (1986) report on a trial of the dilute alcohol extract of the herb, given as a total dose of 72 g of crude extract over 3 days in divided doses. The cure rate was said to be 100% in 485 cases of *Plasmodium vivax* and 105 cases of *P. falciparum*. However, the definition of cure was not clear; it would be impossible to achieve a cure rate of 100% with a 3-day course of pure artemisinin.

TABLE 3.2
Clinical Trials of *A. annua* Preparations for Patients with Malaria

Efficacy	Recrudescence Rate at 28	N^a	Species	Preparation	Total Dose (g)	Days	Reference
100%	?	485	Vivax	Crude alcohol	72	3	Chang and But, 1986
100%	?	105	Falciparum	Crude alcohol	72	3	Chang and But, 1986
100%	8%	50	Vivax	Oil-based capsules	129	6	Yao-De et al., 1992
100%	?	5	Falciparum	Aqueous infusion	20	5	Mueller et al., 2000
93%	13%	254	Falciparum + others	Aqueous infusion	35	7	Hirt and Lindsey, 2000
92%	?	48	Falciparum	Aqueous decoction	20	4	Mueller et al., 2000
77%	37%	39	Falciparum	Aqueous infusion	35 g	7	Mueller et al., 2004
70%	39%	33	Falciparum	Aqueous infusion	63 g	7	Mueller et al., 2004

[a] N = number of subjects.

TABLE 3.3
Clinical Trial of *A. annua* in Gelatine Capsules

Drug	N^a	Days of Treatment	Total Dose (g)	FCT[b] (hours)	PCT[c] (hours)	Recrudescence (%)
COEA	53	3	73.6	18.3	33.7	12/36 (33)
COEA	50	6	128.8	18.4	32.9	4/50 (8)
QHET	41	3	80.0	21.8	33.0	13/41 (32)
Chloroquine	20	3	1.2	23.7	49.9	0/20 (0)

[a] N = number of subjects.
[b] FCT = fever clearance time.
[c] PCT = parasite clearance time.

Source: Yao-De, W. et al., (1992), *Chung Kuo Chi Sheng Chung Hsueh Yu Chi Sheng Chung Ping Tsa Chih*, 10, 290–294.

Hirt and Lindsey (2000) reported a 93% parasite clearance rate in 254 patients who had taken a 7-day course of *A. annua* infusion in the Democratic Republic of Congo, mainly for falciparum malaria, although some had infections with other *Plasmodium* species. In a subset of 31 patients followed long-term, the recrudescence rate was 13% after 1 month (Hirt, 2001).

Yao-De et al. (1992) report on a clinical trial of two types of gelatine capsule (called COEA and QHET) containing the herb *A. annua* (Table 3.3). One of these, called COEA, contained oil to enhance absorption of the artemisinin. All the patients were infected with *P. vivax*; 53 were treated with COEA for 3 days, and 50 were treated for 6 days. The dosage was 36.8 g on the first day, followed by 18.4 g on subsequent days, to give the total doses recorded in Table 3.3. Parasite clearance times and fever clearance times were significantly faster for all the herbal preparations than for chloroquine. Patients were followed up for 30 days, with a blood film examined for parasitaemia every 10 days. About one in three patients who had taken the capsules for only 3 days experienced a recrudescence, but this was reduced to only 8% in those who had taken the capsules for 6 days.

Mueller et al. (2000) evaluated two preparations of *A. annua* for patients with falciparum malaria. Although numbers were small, they found that both the infusion (1 l of boiling water added to 5 g of dried leaves, and mixture left to cool for 15 minutes before filtering) and the decoction (5 g of dried leaves placed in 1 l of water, boiled for 5 minutes, and then filtered) were effective. In each case, the dose was 250 ml 4 times a day, although the infusion group was treated for 5 days and the decoction group for 4 days. The results were good in both groups (see Table 3.2) — this is surprising, given that the decoction contains less artemisinin and was ineffective in mice (see above). Although in this study patients were not followed up beyond day 7, a further trial by the same group showed that recrudescence rates were high (Mueller et al., 2004; see Table 3.2). However, it also showed that the *Artemisia annua* infusion was significantly more effective than chloroquine in an area of chloroquine resistance. The definition of cure was parasite clearance at day 7, rather than the clinical definition adopted by WHO (see Chapter 21).

None of the above studies adhere to an ideal design (see Chapter 21). Nevertheless, they all suggest that herbal preparations of *A. annua* may be effective against malaria in humans. Yet even in the Chinese studies, using relatively large amounts of *A. annua*, the dose of artemisinin would be less than half the WHO-recommended dose (assuming artemisinin content of 1%, the 129-g total dose would contain 1290 mg of artemisinin). There is an evident discrepancy between the reported clinical efficacy of traditional formulations of *A. annua* and their low artemisinin content. Several possible mechanisms will be discussed below.

1. *Efficacy of lower doses of artemisinin*
   Ashton et al. (1998) have suggested that oral doses of 500 mg of artemisinin daily are unnecessarily high. Rectal doses of 500 mg are equally effective, although their relative bioavailability was only 30% (equivalent to an oral dose of 160 mg). There was a high degree of variability in the artemisinin plasma concentrations after both oral and rectal administration. Interestingly, clinical end points such as parasite and fever clearance times do not correlate with drug exposure. Räth et al. (2004) found that after ingestion of one litre of *A. annua* tea, containing 94.5 mg of artemisinin, plasma concentrations of artemisinin remained at an effective antiplasmodial level for over 4 hours.

2. *Improved bioavailability of artemisinin*
   Pure artemisinin has poor oral bioavailability, about 32% (Titulaer et al., 1990). This is not affected by food intake (Dien et al., 1997). Artemisinin is poorly soluble in water and oil, but dissolves in organic solvents. Infusions of *A. annua* contain greater amounts of artemisinin than an infusion made by pouring hot water onto pure artemisinin, implying that other plant constituents, for example, flavonoids, are helping to dissolve artemisinin by acting as detergents (Mueller et al., 2000). Saponins from other plants used in traditional mixtures may further amplify this effect. Artemisinin is absorbed much faster from an infusion than from capsules, with the plasma concentration peaking after 30 minutes (compared to 2.3 hours after ingestion of capsules), but the area under the concentration–time curve is lower, reflecting the lower total artemisinin content of the tea. Overall, the bioavailability of artemisinin from a simple infusion of *A. annua* is now known to be very similar to that from pure artemisinin capsules (Räth et al., 2004).

3. *Inhibition of artemisinin catabolism*
   The bioavailability of artemisinin is reduced by a factor of 6.9 by the fifth day of repeated treatment (Sidhu et al., 1998). This may be due to autoinduction of liver enzymes (Svensson et al., 1998), although the half-life of artemisinin remains unchanged. The plasma concentration of the active metabolite dihydroartemisinin increases with repeated treatment, so this may compensate for the loss in artemisinin concentration (Van Agtmael et al., 1999b). *In vitro*, the enzymes mediating the catabolism of artemisinin are cyto-chrome P450 2B6 and 3A4 (Svensson and Ashton, 1999). There is a large variation of the level of these enzymes in human subjects, which may explain the interindividual

variation in artemisinin kinetics. It is conceivable that other components of the plant (or of other plants used in traditional mixtures) may inhibit these enzymes. No studies have yet been performed on the pharmacokinetics of artemisinin from traditional preparations of the whole herb.

4. *Immunostimulation*

Artemisinin stimulates the phagocytic activity of macrophages in the mouse abdominal cavity *in vivo* (Ye et al., 1982). Furthermore, after treatment with artemisinin, mouse macrophages were better at phagocytosing *P. berghei*-infected mouse erythrocytes *in vitro*. The activity of macrophage acid phosphatase was also enhanced. This property of artemisinin may boost the human immune response to malaria.

5. *Other constituents with antimalarial activity*

Other *Artemisia* species have antimalarial activity without containing artemisinin (Valecha et al., 1994). *A. absinthium* and *A. abrotanum* were used to treat malaria in Europe, but their activity is attributable to other constitutents (Cubukcu et al., 1990; Deans and Kennedy, 2002). *Artemisia afra* extracts are effective against *P. falciparum in vitro*, and this activity is attributable to a complex mixture of flavonoids and sesquiterpene lactones, rather than to a single compound (Kraft et al., 2003). Some of these phytochemicals may also contribute to the activity of *A. annua*.

A. annua contains many different classes of compounds: at least 28 monoterpenes, 30 sesquiterpenes, 12 triterpenoids and steroids, 36 flavonoids, 7 coumarins, and 4 aromatic and 9 aliphatic compounds (Bhakuni et al., 2002; Tang and Eisenbrand, 1992). Artemisinin is not the only antimalarial compound in *A. annua*. The callus of the plant has some antimalarial activity even though it contains no artemisinin (François et al., 1993). Furthermore, the water-soluble fraction of *A. annua*, after extraction of artemisinin, has an antipyretic effect (Chang and But, 1986).

6. *Synergistic activity of other constituents of* A. annua

Artemisinin is only 1 of 29 sesquiterpenes in *A. annua*. Some of these are in much greater concentrations than artemisinin in wild strains of the plant: arteannuin B (two to four times) and artemisinic acid (seven to eight times). Both of these have antibacterial and antifungal properties (Dhingra et al., 2000). Arteannuin B used alone was found to be ineffective and toxic in rat malaria, but it potentiated the effect of artemisinin (Chang and But, 1986). However, in hybrid plants with a high artemisinin content, the concentration of artemisinic acid is much lower (Magalhães, personal communication).

In addition, *A. annua* produces at least 36 flavonoids. Many of these have antimalarial activity *in vitro*, although the inhibitory concentration 50% (IC_{50}) is much higher than that of artemisinin (Table 3.4). Five of these, artemetin, casticin, chrysoplenetin, chrysosplenol-D, and cirsilineol, have been shown selectively to potentiate the *in vitro* activity of artemisinin against *P. falciparum* (Liu et al., 1992). Casticin, at a concentration of 5 µmol/l, induced a three- to fivefold reduction in the IC_{50} for artemisinin (Elford et al., 1987). Chrysosplenol-D has the strongest potentiating effect, and this is also the most abundant flavone in plant material (Liu et al., 1992). Interestingly, the flavones do not potentiate the antimalarial activity of chloroquine (Elford et al., 1987). Although they have no effect on hemin themselves, they do catalyse the reaction of artemisinin with hemin (Bilia et al., 2002) and may also help to solubilise artemisinin (see above).

The effect of all the flavones in combination with artemisinin has not been investigated. Other flavones, and indeed other components of *A. annua*, may have a similar effect; they have not all been tested because it is difficult to purify them. The antimalarial properties of the traditional preparation of *A. annua* most probably reside in the combination of many constituents, not just artemisinin.

TABLE 3.4
In Vitro **Antimalarial Activity of Constituents of *A. annua***

Constituent	IC$_{50}$ (μM)
Artemisinin	0.03
Artemetin	26
Casticin	24
Chrysoplenetin	23
Chrysosplenol-D	32
Cirsilineol	36
Eupatorin	65

Source: Liu, K.C.-S. et al., (1992), *Plant Cell Rep.*, 11, 637–640.

3.4 EVOLUTION OF RESISTANCE

One of the principle concerns regarding the use of *A. annua* for malaria is that the ingestion of low doses of artemisinin may accelerate the development of resistance to this drug. There is little evidence to support or refute this claim, and further research is needed.

It could be argued that some degree of resistance to artemisinin has already evolved in China. The IC$_{50}$ of artemisinin varies according to the strain of *P. falciparum* and can be as low as 6 nM (Wongsrichanalai et al., 1997). Chinese strains (IC$_{50}$ = 630 nM) are much less sensitive to artemisinin than African strains (IC$_{50}$ = 25 nM). This could be explained by the long-standing local use of *A. annua*, or of artemisinin itself over the last 30 years (Wernsdorfer, 1999). The parasites would find it easier to develop resistance to a single agent than to a battery of antimalarial compounds such as those contained in whole plant extracts.

There is undoubtedly a high recrudescence rate when artemisinin is used alone, and if it is given in a short course; there are at least three factors that may contribute to this. First is the time-dependent reduction in bioavailability, although an escalating dose of artemisinin did not reduce the frequency of recrudescences (Gordi et al., 2002).* Second, low concentrations of dihydroarte-misinin have a static rather than cidal effect on early trophozoites *in vitro*; these then remain in a metabolic resting state for up to 6 days followed by renewed growth. Third, the short half-life of artemisinin (approximately 2 hours) means that although parasitaemia is reduced below detectable levels, the drug is not present in the plasma long enough to eliminate all the parasites. Parasite clearance would require effective drug levels during three to four life cycles (each one lasting 48 hours), which can be achieved by combination with a long-acting antimalarial drug (Van Agtmael et al., 1999a) or by administration of the drug every 4 to 6 hours for 7 days.

On the other hand, these pharmacokinetic properties will help to protect artemisinin against the evolution of resistance. The shorter the half-life, the less time parasites are exposed to subther-apeutic concentrations, and so the less the selective pressure for evolution of resistance. As arte-misinin has a short half-life, parasites would be exposed to subtherapeutic doses for only a few hours. Drugs such as sulfadoxine-pyrimethamine have a much longer half-life, and parasites are exposed to subtherapeutic concentrations for 37 days after a single treatment (Watkins and Mosobo, 1993).

The traditional preparation of the herb may be effective in spite of its low artemisinin content (as discussed above) and thus may kill parasites before they can develop resistance. Studies in China have shown that the recrudescence rate can be reduced by combining *A. annua* with the

---

* This may have been because the initial dose was low (100 mg) and was metabolised faster than the higher dose of 500 mg, which may have saturated first-pass metabolism.

TABLE 3.5

Therapeutic Index of *Artemisia* Extracts and Artemisinin in *P. berghei*-Infected Mice

Substance	LD_{50}[a] (mg/kg)	ED_{50}[b] (mg/kg)	Therapeutic Index (LD_{50}/ED_{50})
Crude plant	162,500	11,900	13.7
Ether extract (neutral fraction)	7425	2646	2.80
Dilute alcohol extract	4162	2526	1.64
Artemisinin	5105	139	36.80

[a] LD_{50} = dose lethal in 50% of mice.

[b] ED_{50} = dose effective in curing 50% of mice.

Source: Yao-De, W. et al., (1992), *Chung Kuo Chi Sheng Chung Hsueh Yu Chi Sheng Chung Ping Tsa Chih*, 10, 290–294; Chang, H.M. and But, P.P.H., (1986), *Pharmacology and Applications of Chinese Materia Media*, Vol. 1, World Scientific Publishing, Singapore.

roots of two other plants, *Astragalus membranaceus* (Leguminosae) and *Codonopsis pilosa* (Campanulaceae) (Chang and But, 1986). The mechanism for this has not yet been elucidated, but *A. membranaceus* is known to have immunostimulant properties (Foster and Chongxi, 1992). Modern drug combinations may unwittingly be mimicking the combinations of phytochemicals (and sometimes plant species) with synergistic activities contained in the traditional preparations of *A. annua*. Indeed, the pharmaceutical industry may find it beneficial to further investigate combinations of artemisinin with other compounds produced by *A. annua*, which may improve its antimalarial efficacy and reduce the risk of resistance.

3.5 SAFETY AND TOLERABILITY

The presence of multiple chemical constituents in herbal preparations of *A. annua* raises the question as to their safety. The lethal dose 50% (LD_{50}) and therapeutic index have been measured for the crude plant and are reassuringly high (see Table 3.5).

Artemisinin has a slightly negative chronotropic effect on the heart, causing mild hypotension. Toxicological studies on artemisinin are reviewed in detail later in this book (see Chapter 15), but clinically it seems to be very safe. The chief concerns over the safety of artemisinin originated from a study that showed that high-dose intramuscular arteether and artemether in dogs and rats caused a clinical neurological syndrome with gait disturbances and loss of some reflexes, and prominent brainstem lesions (Brewer et al., 1994). However, artemisinin has now been used in several million patients, with only one report of neurological side effects following artesunate treatment (WHO, 1998b; White et al., 1999; Wilairatana and Looareesuwan, 2002). In patients who died of severe malaria, neuropathological findings were similar in recipients of quinine and artemether, and there was no evidence for artemether-induced toxic effects (Hien et al., 2003).

Artemisinin is generally considered to be safe in the second and third trimesters of pregnancy (WHO, 1998b). However, early studies showed that relatively low doses of artemisinin (13 to 25 mg/kg, or 1/200 to 1/400 of the LD_{50}) cause fetal resorption in rodents; therefore, use in the first trimester is not recommended (WHO, 1994). However, in a series of 16 patients treated with artesunate in the first trimester of pregnancy, the rate of abortion (20%) was similar to that of the general population (McGready et al., 1998).

A literature search has not revealed any animal toxicity studies, but the herb extract has been evaluated in China and was deemed to be of low toxicity. Five hundred and ninety patients were treated with the herb extract, and of these, 3.4% developed gastrointestinal symptoms such as nausea, vomiting, abdominal pain, and diarrhoea. No adverse effects were observed in patients with cardiac, renal, or hepatic dysfunction, or in pregnant women (Chang and But, 1986). Interestingly, the pharmacokinetics of artemisinin are not affected in patients with cirrhosis of the liver (De Vries et al., 1997), but artemisinin does induce certain liver enzymes, and thus interacts with some other drugs such as omeprazole (Svensson et al., 1998). Observational studies in Africa found that 25% of malaria patients being treated with *A. annua* infusion had nausea, although none vomited. Other mild adverse events during treatment included dizziness, tinnitus, pruritus, and abdominal pain (Hirt, 2001). Nausea and tinnitus were much less common in patients taking oral *Artemisia annua* infusion than in those taking quinine (Mueller et al., 2004). In a group of 14 healthy volunteers given a high-dose oral infusion of *A. annua*, only one patient experienced an adverse event, which was diagnosed as an allergic reaction. The patient's cough and skin rash resolved quickly and did not require treatment (Räth et al., 2004).

A. *annua* can be regarded as an established traditional medicine, as it has been widely used and is included in the pharmacopoeia of the People's Republic of China (Mueller et al., 2000). Nevertheless, any future clinical trials of *A. annua* preparations should carefully monitor subjective side effects as well as end-organ function (using a protocol such as that described in Chapter 21 of this volume).

3.6　CULTIVATION

A. *annua* is native not only to China but also to Japan, Korea, Vietnam, Myanmar, northern India, and southern Siberia through to eastern Europe (WHO, 1998a). It has been introduced to many other countries, in Europe, North America, and the Tropics (Laughlin et al., 2002). Seed varieties have been adapted by breeding for lower latitudes, and cultivation has been successfully achieved in many tropical countries, for example, in the Congo (Mueller et al., 2000), India (Mukherjee, 1991), and Brazil (Milliken, 1997; Carvalho et al., 1997; De Magalhães et al., 1997).

The tiny seeds succeed best when sown on top of well-aerated soil, as they germinate in the light (Hirt and Lindsey, 2000); in areas with a heavy soil, the plants can first be developed in a greenhouse. In order to maximise the yield of artemisinin, the critical factor is day length, because the plant usually grows in the long summer days at high latitudes and flowers when the day length shortens. The concentration of artemisinin peaks around the time of flowering, although in some cases this may be just before flowering, and in other cases during full flowering (Ferreira et al., 1995a; Laughlin et al., 2002). The artemisinin concentration peaks at a slightly different time in different areas, and identifying this will help to maximise the yield of artemisinin in harvested plants (Chang and But, 1986). In the tropics, where days are shorter than in northern summers, flowering occurs earlier, reducing the biomass achieved. However, yields can be maximised at higher altitudes and with late-flowering varieties (Laughlin et al., 2002), or by artificially lengthening hours of daylight to over 13.5 hours (Ferreira et al., 1995a).

In wild-type plants, the greatest concentration of artemisinin is found in the inflorescence, although it occurs in all other aerial parts of the plant, except the seed (Ferreira et al., 1997). In artemisinin-rich plants, the greatest concentration of artemisinin occurs at the beginning of the flowering season (De Magalhães, personal communication). It used to be thought that sun and oven drying reduced the artemisinin content and that it was best to air-dry leaves in the shade (Laughlin et al., 2002). However, Simonnet et al. (2001) found that sun-drying plants in the field increased the artemisinin content (perhaps by promoting conversion of some precursors to artemisinin), but that if drying continued for more than a week, leaves were lost, decreasing the overall yield. The optimum would therefore seem to be drying in the field for 1 week, followed by air-drying in the shade.

Although artemisinin content is affected by climate and time of harvesting, the main influence is genetic variation. Ferreira et al. (1995b) evaluated the same 23 clones of *A. annua*, which varied from 0.001 to 0.35% artemisinin, under tissue culture, greenhouse, and field conditions. Broad-sense heritability analyses indicated that artemisinin was mainly under genetic, not environmental, control. Delabays et al. (2002) confirmed that genes outplay the environment by studying different varieties, which yielded from 0.02 to about 1.4% artemisinin. Efforts have been made to increase the artemisinin content as far as possible, by exploring the natural variability. This has been achieved in a hybrid (*A. annua* var. *Artemis*, seeds available from www.anamed.org) and in a nonhybrid strain collected from Vietnam (Sutakavatin, 2002, personal communication). However, artemisinin yield depends not only on its concentration, but also on the total number of leaves and branches. The Institute of Materia Medica in Vietnam has been breeding plants for all three of these characteristics, to optimise artemisinin yield (Dong and Thuan, 2003).

Once hybrids have germinated from the original seeds, the genotypes can only reliably be propagated by taking cuttings or by repeating the cross between the parents of the hybrid. A stock of progenitor plants needs to be maintained under artificial lights to mimic long days and prevent flowering, until a new progenitor plant is established. If the breeding between the parent plants is not repeated, and second generation seeds are taken from the hybrid plants, only some will germinate, yielding weaker plants that contain 30% less artemisinin (Hirt, 2001).

Seeds are the easiest method of propagation and maintain their vigor for 3 years if stored in dry, cool conditions (Ferreira et al., 1997). High artemisinin yield was passed on to subsequent generations in seed-propagated plants in the breeding program of the Institute of Materia Medica in Vietnam (Dong and Thuan, 2003).

Artemisinin content can be measured semiquantitatively using thin-layer chromatography (TLC) (Box 3.1). This could potentially be used in resource-poor settings as a form of quality control.

Box 3.1: Low-Cost Method for Quality Control of *A. annua*: Semiquantitative Dosage of Artemisinin by TLC and Densitometry

Developed by Jacques Falquet (personal communication), based on Delabays (1997)

1. Mix 100 mg of powdered dried leaves with 2 ml of ethyl acetate; leave for 1 hour at 40°C.
2. Filter the solution.
3. Evaporate 1 ml of the solution in a stream of nitrogen or carbon dioxide gas if possible or, if not, in air.
4. Redissolve the solute in 0.2 ml of ethyl acetate.
5. Place 5 µl on a silica gel plate.
6. Use a mixture of ethyl acetate and cyclohexane (3:7) as the eluant; allow to migrate for 8 cm.
7. Spray with 5% anisaldehyde and 0.5% sulfuric acid in glacial acetic acid, then heat for 5 minutes at 110°C.
8. For a semiquantitative reading, there should only be one spot, corresponding to artemisinin. This should be near the center of the chromatogram (rf = 0.5). This can be compared to results from a reference standard (leaves of a known concentration, in several dilutions) applied to the same chromatogram. Accuracy of ±20% may be achievable.
9. For a more quantitative reading, the plate can be scanned into a computer and interpreted by densitometry software such as Image-J. This is available free on the Internet at http://rsb.info.nih.gov/ij/index.html.

3.7 PUBLIC HEALTH POTENTIAL

The development of artemisinin-rich genotypes of *A. annua* has encouraged some to promote cultivation and local use of the plant to treat malaria. However, others argue that this is premature, and that further research is needed before *A. annua* infusions can be promoted as part of a malaria control program.

Although the results of the clinical studies are promising, none of them was conducted with the most rigorous methodology. Often, the definition of cure was not clear, and the high cure rates may not be replicated if patients are followed up for longer periods of time. It will be important to conduct a controlled trial according to strict guidelines, such as those suggested in Chapter 21, before promoting the wider use of *A. annua*.

There are some contradictory results, such as the clinical efficacy of a decoction of *A. annua*, whereas this preparation was ineffective in mice. One reason for this discrepancy may be that subjects had some degree of immunity to malaria. Another is that the mouse model does not always accurately predict outcomes in humans. There are only anecdotal reports of efficacy in nonimmune patients, for example, a 2 1/2-year-old Caucasian boy with laboratory-confirmed malaria who was treated with *A. annua* tea only and made a rapid recovery (Hans-Martin Hirt, personal communication). A priority for future research would be to test whether use of *A. annua* teas can reduce the mortality rate in African children and other nonimmune patients with malaria. It would be ethical to do this if efficacy and safety has been demonstrated in a rigorous clinical trial in a semi-immune population.

Perhaps the most important finding by Mueller et al. (2000) was that local cultivation and preparation of *A. annua* are feasible in Africa. If effectiveness in nonimmune patients is demonstrated, local cultivation and preparation of *A. annua* could be considered part of a malaria control strategy, especially in remote areas with poor access to health facilities and poor availability of effective antimalarial drugs. Such remote areas (such as the Brazilian Amazon and remote rural areas of Africa) are particularly problematic for malaria control programs and are often neglected. Herbal medicines may not be as perfect as the exact dosages administered in industrially produced formulations, but may be better than no treatment, or treatment with fake artesunate tablets, which are widespread in Southeast Asia. In an area with significant chloroquine resistance, a herbal infusion of *A. annua* proved more effective than chloroquine in semi-immune patients (Mueller et al., 2004). If there are no effective local treatments for malaria, early treatment with an herbal preparation of *A. annua* may in some circumstances prove to be life saving. The risk of resistance evolving is likely to be small, for reasons discussed above.

An important concern is the potential ecological impact of introducing an alien plant species into a new fertile environment where it may spread quickly and endanger local species. In these circumstances it may be preferable to first explore the possibility of using other local plants as antimalarials, or cultivating *A. annua* under controlled conditions, not allowing it to go to seed. In any case, the plant is harvested before it goes to seed, and the seeds from the second-generation hybrid plants are of no use for propagation (vegetative propagation is used instead). In areas like the Brazilian Amazon, there is very active competition among species, and it is unlikely that *A. annua* would survive outside of cultivated areas, where it would be protected by farmers (P. M. de Magalhães, personal communication). In arid areas, for example, in parts of Africa, it is unlikely that *A. annua* would be able to spread rapidly.

On balance, the potential benefits of local *A. annua* cultivation and preparation in poor areas with no other antimalarial treatments may outweigh the risks of such a program; therefore, some NGOs are already promoting this. However, others argue that it is premature to promote the use of Herba *A. annua* until there are better data on its safety and efficacy. Furthermore, research is needed on the optimal preparation, dose, and length of treatment, and on the best variety to cultivate.

For this reason, the Research Initiative on Traditional Antimalarial Methods (RITAM) has established an *A. annua* Task Force to investigate the feasibility of using a traditional formulation of Herba *A. annua* as a more affordable and accessible option for the early treatment of malaria in poor and remote areas of developing countries. This chapter represents the first step in this process and is mainly an overview of the published literature. There are probably many more unpublished sources to be explored. It is hoped to build on this review, and to use it to plan further clinical evaluation of *A. annua* for the treatment of malaria. Priorities for future research are summarised in Box 3.2.

BOX 3.2: PRIORITIES FOR FUTURE RESEARCH ON *A. ANNUA*

1. To determine the best genotype of *A. annua* to cultivate in each region of interest (in function of its environmental characteristics)
2. To determine the most effective method of preparation, dose, and length of treatment
3. To test the clinical effectiveness of this *A. annua* preparation for treating falciparum malaria in nonimmune patients
4. To determine whether use of this preparation increases the risk of *P. falciparum* developing resistance to artemisinin
5. To test whether the use of this preparation reduces mortality from malaria

---

ACKNOWLEDGMENTS

Members of the RITAM *A. annua* Task Force contributed to and commented on this chapter. We are grateful to Harald Noedl for some helpful comments. Dr. Carsten Flohr kindly translated the paper by Yao-De et al. (1992), and Dr. Phantip Vattanaviboon of Mahidol University, Thailand, kindly translated the paper by Ye et al. (1982).

REFERENCES

Artemether-Quinine Meta-Analysis Study Group. (2001). A meta-analysis using individual patient data of trials comparing artemether with quinine in the treatment of severe falciparum malaria. *Trans. R. Soc. Trop. Med. Hyg.*, 95, 637–650.

Ashton, M., Duy Sy, N., Van Huong, N., et al. (1998). Artemisinin kinetics and dynamics during oral and rectal treatment of uncomplicated malaria. *Clin. Pharmacol. Ther.*, 63, 482–493.

Bensky, D. and Gamble, A. (1993). *Chinese Herbal Medicine Materia Medica*. Eastland Press, Inc., Seattle, WA.

Bhakuni, R.S., Jain, D.C., and Sharma, R.P. (2002). Phytochemistry of *Artemisia annua* and the development of artemisinin-derived antimalarial agents. In *Artemisia*, Wright, C.W., Ed. Taylor & Francis, London.

Bilia, A.R., Lazari, D., Messori, L., Taglioli, V., Temperini, C., and Vinvieri, F.F. (2002). Simple and rapid physico-chemical methods to examine action of antimalarial drugs with hemin: its application to *Artemisia annua* constituents. *Life Sci.*, 70, 769–778.

Brewer, T.G., Grate, S.J., Peggins, J.O., et al. (1994). Fatal neurotoxicity of arteether and artemether. *Am. J. Trop. Med. Hyg.*, 51, 251–259.

Bruce-Chwatt, L.J. (1982). Qinghaosu: a new antimalarial. *Br. Med. J.*, 284, 767–768.

Carvalho, J.E., Dias, P.C., and Foglio, M.A. (1997). Artemísia. *Rev. Racine*, 36, 56–57.

Chang, H.M. and But, P.P.H. (1986). *Pharmacology and Applications of Chinese Materia Media*, Vol. 1. World Scientific Publishing, Singapore.

Cubukcu, B., Bray, D.H., Warhurst, D.C., Mericli, A.H., Ozhatay, N., and Sariyar, G. (1990). *In vitro* antimalarial activity of crude extracts and compounds from *Artemisia abrotanum* L. *Phytother. Res.*, 4, 203–204.

Deans, S.G. and Kennedy, A.I. (2002). *Artemisia absinthium*. In *Artemisia*, Wright, C.W., Ed. Taylor & Francis, London.

Delabays, N. (1997). Biologie de la reproduction chez l'*Artemisia annua* L. et génétique de la production en artémisinine. Thèse de Doctorat, Faculté des Sciences, Université de Lausanne.

Delabays, N., Darbellay, C., and Galland, N. (2002). Variation and heritability of artemisinin content in *Artemisia annua* L. In *Artemisia*, Wright, C.W., Ed. Taylor & Francis, London.

De Magalhães, P.M. (1996). Seleção, Melhoramento e Nutrição de *Artemisia annua* L. para cultivo em região inter tropical. Ph.D. thesis, UNICAMP-IB, Campinas-SP, Brazil.

De Malaghães, P.M., Debrunner, N., and Delabays, N. (2003). Aqueous Extracts of *Artemisia annua* L. Paper presented at the International Conference on Malaria: Current Status and Future Trends, Chulabhorn Research Institute, Bangkok, Thailand, February 16–19.

De Magalhães, P.M., Delabays, N., and Sartoratto, A. (1997). New hybrid lines of antimalarial species *Artemisia annua* L. guarantee its growth in Brazil. *Ciên. Cult.*, 49, 413–415.

De Vries, P.J., Nguyen, X.K., Tran, K.D., et al. (1997). The pharmacokinetics of a single dose of artemisinin in subjects with liver cirrhosis. *Trop. Med. Int. Health*, 2, 957–962.

Dhingra, V., Pakki, S.R., and Narasu, M.L. (2000). Antimicrobial activity of artemisinin and its precursors. *Curr. Sci.*, 78, 709–713.

Dien, T.K., de Vries, P.J., Khanh, N.X., et al. (1997). Effect of food intake on pharmacokinetics of oral artemisinin in healthy Vietnamese subjects. *Antimicrob. Agents Chemother.*, 41, 1069–1072.

Dong, N.H. and Thuan, N.V. (2003). Breeding of a high leaf and artemisinin yielding *Artemisia annua* variety. Paper presented at the International Conference on Malaria: Current Status and Future Trends, Chulabhorn Research Institute, Bangkok, Thailand, February 16–19.

Elford, B.C., Roberts, M.F., Phillipson, J.D., and Wilson, R.J.M. (1987). Potentiation of the antimalarial activity of qinghaosu by methoxylated flavones. *Trans. R. Soc. Trop. Med. Hyg.*, 81, 434–436.

Ferreira, J.F.S., Simon, J.E., and Janick, J. (1995a). Developmental studies of *Artemisia annua*: flowering and artemisinin production under greenhouse and field conditions. *Planta Med.*, 61, 167–170.

Ferreira, J.F.S., Simon, J.E., and Janick, J. (1995b). Relationship of artemisinin content of tissue-cultured, greenhouse-grown, and field-grown plants of *Artemisia annua*. *Planta Med.*, 61, 351–355.

Ferreira, J.F.S., Simon, J.E., and Janick, J. (1997). *Artemisia annua*: botany, horticulture, pharmacology. *Horticult. Rev.*, 19, 319–371.

Foster, S. and Chongxi, Y. (1992). *Herbal Emissaries: Bringing Chinese Herbs to the West*. Healing Arts Press, Rochester, VT.

François, G., Dochez, C., Jaziri, M., and Laurent, A. (1993). Antiplasmodial activities of sesquiterpene lactones and other compounds in organic extracts of *Artemisia annua*. *Planta Med.*, 59 (Suppl.), A677–A678.

Gordi, T., Huong, D.X., Hai, T.N., Nieu, N.T., and Ashton, M. (2002). Artemisinin pharmacokinetics and efficacy in uncomplicated malaria patients treated with two different dosage regimens. *Antimicrob. Agents Chemother.*, 46, 1026–1031.

Guirand, F. (1959). *Larousse Encyclopedia of Mythology*. Paul Hamlyn, London.

Hien, T.T., Turner, G., Mai, N.T.H., et al. (2003). Neuropathological assessment of artemether-treated severe malaria. *Lancet*, 362, 295–296.

Hirt, H.M. (2001). Document: *Artemisia annua* Anamed; A Plant with Anti-malarial Properties. Anamed, Winnenden.

Hirt, H.M. and Lindsey, K. (2000). *Natural Medicine in the Tropics: Experiences*. Anamed, Winnenden, Germany. See also http://www.anamed.org.

Kraft, C., Jenett-Siems, K., Siems, K., et al. (2003). *In vitro* antiplasmodial evaluation of medicinal plants from Zimbabwe. *Phytother. Res.*, 17, 123–128.

Laughlin, J.C., Heazlewood, G.N., and Beattie, B.M. (2002). Cultivation of *Artemisia annua* L. In *Artemisia*, Wright, C.W., Ed. Taylor & Francis, London.

Liu, K.C.-S., Yang, S.-L., Roberts, M.F., Elford, B.C., and Phillipson, J.D. (1992). Antimalarial activity of *Artemisia annua* flavonoids from whole plants and cell cultures. *Plant Cell Rep.*, 11, 637–640.

Lwin, M., Maun, C., and Aye, K.H. (1991). Trial of antimalarial potential of extracts of *Artemisia annua* grown in Myanmar. *Trans. R. Soc. Trop. Med. Hyg.*, 85, 449.

McGready, R., Cho, T., Cho, J.J., et al. (1998). Artemisinin derivatives in the treatment of falciparum malaria in pregnancy. *Trans. R. Soc. Trop. Med. Hyg.*, 92, 430–433.

Milliken, W. (1997). Traditional anti-malarial medicine in Roraima, Brazil. *Econ. Bot.*, 51, 212–237.

Mitchell, C., Ye, F., and Wiseman, N. (1999). *Shang Han Lun (On Cold Damage): Translation and Commentaries*. Paradigm Publications, Brookline, MA.

Mueller, M.S., Karhagomba, I.B., Hirt, H.M., Wernakor, E., Li, S.M., and Heide, L. (2000). The potential of *Artemisia annua* L. as a locally produced remedy for malaria in the Tropics: agricultural, chemical and clinical aspects. *J. Ethnopharmacol.*, 73, 487–493.

Mueller, M.S., Runyambo, N., Wagner, I., Borrmann, S., Dietz, K., Heide, L. (2004). Randomised controlled trial of a traditional preparation of *Artemisia annua* L. (annual wormwood) in the treatment of malaria. *Trans. Roy. Soc. Trop. Med. Hyg.*, 98(5), 318–321.

Mukherjee, T. (1991). Antimalarial herbal drugs. A review. *Fitoterapia*, 62, 197–204.

Newton, P., Dondorp, A., Green, M., Mayxay, M., and White, N.J. (2003). Counterfeit artesunate antimalarials in Southeast Asia. *Lancet*, 362, 169.

Newton, P., Proux, S., Green, M., et al. (2001). Fake artesunate in Southeast Asia. *Lancet*, 357, 1948–1950.

Nosten, F. (2002). Artemisinin Based Combination Therapy: The Way Forward. Paper presented at the Third European Congress on Tropical Medicine and International Health, Lisbon, Portugal, September 8–11.

Phillips-Howard, P. (2002). Regulation of the quality and use of artemisinin and its derivatives. In *Artemisia*, Wright, C.W., Ed. Taylor & Francis, London.

Price, R.N., Nosten, F., Luxemburger, C., et al. (1996). Effects of artemisinin derivatives on malaria transmissibility. *Lancet*, 347, 1654–1655.

QACRG — Quinine Antimalaria Coordinating Research Group. (1979). Antimalaria studies on Qinghaosu. *Chin. Med. J.*, 92, 811–816.

Räth, K., Taxis, K., Walz, G., Gleiter, C.H., Li, S.M., Heide, L. (2004). Pharmacokinetic study of artemisinin after oral intake of a traditional preparation of *Artemisia annua* L. (annual wormwood). *Am. J. Trop. Med. Hyg.*, 70(2), 128–132.

Shou-zhong, Y. (1997). *The Divine Farmer's Materia Medica: A Translation of the Shen Nong Ben Cao Jing*. Blue Poppy Press, Boulder, CO.

Sidhu, J.S., Ashton, M., Huong, N.V., et al. (1998). Artemisinin population pharmacokinetics in children and adults with uncomplicated falciparum malaria. *Br. J. Clin. Pharmacol.*, 45, 347–354.

Simonnet, X., Gaudin, M., Hausamann, H., and Vergères, Ch. (2001). Le fanage au champ d'*Artemisia annua* L: élever la teneur en artémisinine et abaisser les coûts de production. *Rev. Suisse Vitic. Arboric. Hortic.*, 33, 263–268.

Svensson, U.S. and Ashton, M. (1999). Identification of the human cytochrome P450 enzymes involved in the *in vitro* metabolism of artemisinin. *Br. J. Clin. Pharmacol.*, 48, 528–535.

Svensson, U.S.H., Ashton, M., Hai, T.N., et al. (1998). Artemisinin induces omeprazole metabolism in human beings. *Clin. Pharmacol. Ther.*, 64, 160–167.

Tan, R.X., Zheng, W.F., and Tang, H.Q. (1998). Biologically active substances from the genus *Artemisia*. *Planta Med.*, 64, 295–302.

Tang, W. and Eisenbrand, G. (1992). *Chinese Drugs of Plant Origin*. Springer-Verlag, Berlin.

Titulaer, H.A.C., Zuidema, J., Kager, P.A., Wetsteyn, J.C.F.M., Lugt, C.H.B., and Merkus, F.W.H.M. (1990). The pharmacokinetics of artemisinin after oral, intramuscular and rectal administration to volunteers. *J. Pharm. Pharmacol.*, 42, 810–813.

Tu, Y.Y. (1999). The development of new antimalarial drugs: qinghaosu and dihydro-qinghaosu. *Chin. Med. J.*, 112, 976–977.

Valecha, N., Biswas, S., Badoni, V., Bhandari, K.S., and Sati, O.P. (1994). Antimalarial activity of *Artemisia japonica*, *Artemisia maritima* and *Artemisia nilegarica*. *Indian J. Pharmacol.*, 26, 144–146.

Van Agtmael, M.A., Cheng-Qi, S., Qing, J.X., Mull, R., and van Boxtel, C.J. (1999a). Multiple dose pharmacokinetics of artemether in Chinese patients with uncomplicated falciparum malaria. *Int. J. Antimicrob. Agents*, 12, 151–158.

Van Agtmael, M.A., Eggelte, T.A., and van Boxtel, C.J. (1999b). Artemisinin drugs in the treatment of malaria: from medicinal herb to registered medication. *Trends Pharmacol. Sci.*, 20, 199–205.

Watkins, W.M. and Mosobo, M. (1993). Treatment of *Plasmodium falciparum* malaria with pyrimethamine-sulfadoxine: selective pressure for resistance is a function of long elimination half-life. *Trans. R. Soc. Trop. Med. Hyg.*, 87, 75–78.

Wernsdorfer, W.H. (1999). The Place of Riamet® in Dealing with Drug-Resistant Falciparum Malaria. Paper presented at Novartis Satellite Symposium: Controlling Malaria in Non-immune Travellers: Riamet (Artemether and Lumefantrine) as Standby Emergency Treatment, Novartis Pharma AG, Basel, June 9.

White, N.J., Nosten, F., Looareesuwan, S., et al. (1999). Averting a malaria disaster. *Lancet*, 353, 1965–1967.

White, N.J. (2003). Malaria. In *Manson's Tropical Diseases*, Cook, G.C. and Zumla, A. (Eds.). London, Elsevier Science.

WHO. (1989). *Medicinal Plants in China*, WHO Regional Publications, Western Pacific Series No 2. WHO Regional Office for the Western Pacific, Manila.

WHO. (1994). *The Role of Artemisinin and Its Derivatives in the Current Treatment of Malaria (1994–1995)*, WHO/MAL/94.1067. WHO, Geneva.

WHO. (1998a). *Medicinal Plants in the Republic of Korea*, WHO Regional Publications, Western Pacific Series No. 21. WHO Regional Office for the Western Pacific, Manila.

WHO. (1998b). *The Use of Artemisinin and Its Derivatives as Anti-malarial Drugs*, WHO/MAL/98.1086. Malaria Unit, Division of Control of Tropical Diseases, WHO, Geneva.

Wilairatana, P. and Looaresuwan, S. (2002). The clinical use of artemisinin and its derivatives in the treatment of malaria. In *Artemisia*, Wright, C.W., Ed. Taylor & Francis, London.

Wongsrichanalai, C., Dung, N.T., Trung, T.N., Wimonwattrawatee, T., Sookto, P., Heppner, D.G., and Kawamoto, F. (1997). In vitro susceptibility of *Plasmodium falciparum* isolates in Vietnam to artemisinin derivatives and other antimalarials. *Acta Trop.*, 63, 151–158.

Wright, C.W. and Warhurst, D.C. (2002). The mode of action of artemisinin and its derivatives. In *Artemisia*, Wright, C.W., Ed. Taylor & Francis, London.

Yao-De, W., Qi-Zhong, Z., and Jie-Sheng, W. (1992). Studies on the antimalarial action of gelatin capsule of *Artemisia annua*. *Chung Kuo Chi Sheng Chung Hsueh Yu Chi Sheng Chung Ping Tsa Chih*, 10, 290–294.

Ye, X.S., Cheng, D.X., and Wang, Y.Q. (1982). Effect of Qinghaosu on macrophage phagocytosis in the mouse abdominal cavity. *J. Beijing Med. Coll.*, 14, 141–142.

Yu, H. and Zhong, S. (2002). *Artemisia* in Chinese medicine. In *Artemisia*, Wright, C.W., Ed. Taylor & Francis, London.

4 Changshan (*Dichroa febrifuga*) — Ancient Febrifuge and Modern Antimalarial: Lessons for Research from a Forgotten Tale

Sean Hsiang-lin Lei and Gerard Bodeker

CONTENTS

4.1 INTRODUCTION

Changshan, an ancient traditional Chinese febrifuge, was developed by the Chinese government early in the 20th century as a national antimalarial drug. Despite its efficacy as an antimalarial, Changshan was abandoned less than three decades later. The process by which Changshan was modernised, the associated problems with compliance resulting from side effects not present in the traditional formulation, and the subsequent abandonment of a viable antimalarial constitute an important historical case study in the development of antimalarial drugs from traditional medicine.

At the heart of this exercise are two scientific debates, each of which is still of central significance in the scientific evaluation of traditional medicines and which stand as important themes in the medical and scientific history of drug development.

The first of these relates to the appropriate sequence and process for testing traditional medicines for efficacy. These issues are addressed in this book in the chapters dealing with preclinical evaluation (Part 4), clinical evaluation (Part 5), and safety (Chapter 18). The core issue here is whether the long-standing customary use of a traditional medicine qualifies it for fast-tracking to clinical trials with basic toxicology data but without extensive prior analysis to establish its chemical structure and pharmacological properties.

The second scientific issue is that of the commitment to the identification of single active agents from plants as opposed to an approach that recognises that there may be intended therapeutic synergies between the multiple constituents of a plant and among the many plants contained in traditional complex mixtures.

Underlying the way in which the science has proceeded has been the consistent tension between tradition and modernity in medicine and the disparity in power relationships between modern medical exponents and the custodians of traditional medical knowledge. These issues recur in other chapters in this book. The story of Changshan provides a striking case study of how they have played out in recent times and what the gains and losses of particular strategies and professional relationships have been.*

4.2 WHAT IS CHANGSHAN?

There are several versions of Changshan recorded in classical Chinese texts — in each case it is a complex mixture containing as a central ingredient the plant now known to be *Dichroa febrifuga* Lour. While there have indeed existed several versions of Changshan, as a rule this name refers to the root of the plant *D. febrifuga*, rather than to a complex mixture.

D. febrifuga is a small genus of 12 species of evergreen shrubs in the saxifrage family, *Saxifragaceae*. *D. febrifuga* ranges from India and Nepal eastward to southern China and into Southeast Asia, growing at forest edges at altitudes of between 900 and 2400 m. The plant resembles a small-leaved hydrangea, and its flowers, blue in colour, are formed at the tips of the shoots from summer onward (see Figure 4.1).

Agreement on the active species of Changshan was only reached in the early 20th century after a period of confusion over whether *D. febrifuga*, a plant of Chinese origin, or *Orixa japonica* Thunb., a species mostly found in Japan, was the key plant ingredient. Subsequently, the alkaloid febrifugine was isolated from *D. febrifuga* (see Figure 4.2).

Febrifugine has been described as being "a hundred times more active than quinine as an antimalarial drug, but its use as such is limited by its toxicity" (Harbourne and Baxter, 1993, p. 274). The toxicity of febrifugine contrasts with the historical use of Changshan with no reported toxicity. These represent two important counterpoints in the history of use and development of this potent antimalarial and will be considered in detail in this chapter.

---

* This chapter draws on historical research on Changshan by Sean Lei (1999) and incorporates insights from Gerard Bodeker into the potential that classical formulations may offer for reversing the failure of research into Changshan to produce a viable antimalarial. In telling the Changshan story, the authors have attempted to combine two forms of scholarship and perspective — historical research and biomedical research — and seek the indulgence of readers in bridging any resultant dissonances in style and language.

FIGURE 4.1 *D. febrifuga.* (Copyright 2002, Bruce Peters, California Horticultural Society, http://www.cal-hortsociety.org.)

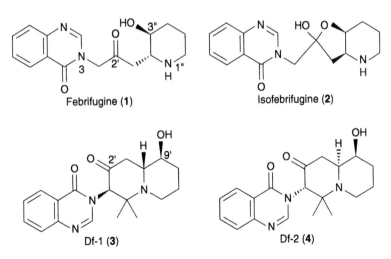

FIGURE 4.2 Molecular structures of febrifugine and its derivatives. (From Harbourne, J.B. and Baxter, H., 1993, *Phytochemical Dictionary: A Handbook of Bioactive Compounds from Plants*, Taylor & Francis, London.)

4.3 ORIGINS OF CHANGSHAN

The earliest record of Changshan comes from the *Shen Nong Ben Cao*, dated around 200 A.D. (Unschuld, 1992). The *Shen Nong Ben Cao* (Liao, 1625) lists the known pharmacopoeia of the day and reports Changshan to be helpful in treating fevers. Toward the end of the 6th century A.D., 1 of the more than 100 prescriptions carved into the walls of the Buddhist caves of Longmen (also known as Dragon's Gate), south of Luoyang in Henan Province, contains Ge Hong's use of *D. febrifuga*.

Table 4.1 summarises the frequency of appearance of Changshan in Chinese medical texts dating from the 4th to the 18th centuries A.D.

TABLE 4.1

Changshan in Chinese Medical Literature from the 4th to 18th Centuries A.D.

Date	Name of Text	Author	No. of Citations of Changshan for Treating Malaria/Fever
4th century	*Zhou-hou bei-chi fang* (*Prescriptions for Emergencies*)	Ge Hong	14
650–659	*Pei-chi ch'ien-chin yao-fang* (*Emergency Prescriptions Worthy of 1,000 Pieces of Gold*)	Sun Ssu-mo	15
752	*Wai-t'ai pi-yao fang* (*Arcane Essentials from the Outer Tribunal*)	Wang T'ao	54
992	*T'ai-ping sheng-hui fang*	Government	66
1118	*Sheng-chi tsung-lu*	Government	126
1132	*P'u-chi pen-shih fang*	Hsu Shu-wei	1
1133	*Chi-feng pu-chi fang*	Chang Kuang	14
1174	*San-yin chi-i ping-cheng fang-lun* (*Pathological Discussions of the Three Causes*)	Ch'en Yen	2
1178	*Yang-shi cha-tsang fang* (*Secret Prescriptions Kept by the Yang Family*)	Yang Tan	4
1184	*Wei-shung chia-pao*	Hsu An-kuo	3
1196	*Shih-ch'ai pai-i hsuan-fang* (*One Hundred and One Prescriptions Selected by Shih-ch'ai*)	Wang Ch'iu	3
1253	*Yen-shih chi-sheng fang*	Yen Yung-ho	1
1264	*Jen-chai chih-chih fang* (*Prescriptions Directly Prescribed by Jen-chai*)	Yang Shih-ying	7
1337	*Shih-i te-hsiao fang* (*Effective Prescriptions for the Use of Hereditary Doctors*)	Wei I-lin	3
1396	*Yu-chi wei-i* (*The Secrets of Medicine*)	Hsu Yung-ch'eng	0
1482	*I-lin lei-cheng chi-yao* (*Classified Book of Medicine*)	Wang Hsi	6
1515	*I-hsueh cheng-ch'uan* (*The Orthodox Tradition of Medicine*)	Yu T'uan	6
1587	*Wan-ping hui-ch'un* (*For the Recovery from Various Diseases*)	Kung T'ing-hsien	5
~1640	*Ching-yueh ch'uan-shu* (*Complete Works of Ching-yueh*)	Chang Chieh-pin	3
1658	*I-men fa-lu* (*Rules for Medicine*)	Yu Ch'ang	1
1766	*Lin-cheng chih-nan i-an* (*Classified Guide to Clinical Chart Collection*)	She Kuei	3

Source: Adapted from Miyasita, S., (1979), *Acta Asiat.*, 36, 90–112.

In 1885, the Customs lists of China refer to the diversity of ingredients known as Changshan: "Several plants supply drugs of this name, which are used as febrifuges, as *Dichroa febrifuga* Lour., *Hydrangea* sp., and an unknown herbaceous plant" (Stuart, 1976, p. 293). Early records also note the emetic effects of *D. febrifuga* and provide a method of correcting this: "the drug is steeped in a decoction of liquorice root to correct its nauseant properties" (Stuart, 1976, p. 293).

At different times in China's history different versions of Changshan were in use and the rationale for their use has varied. The modern story of Changshan centers on three interrelated research efforts:

1. The confirmation of the antimalarial efficacy of Changshan
2. The identification of the effective Changshan as *D. febrifuga*
3. The subsequent search for the active principles of *D. febrifuga*

The formula that was the basis for research and development in early-20th-century China came from a prescription circulated in a newspaper, reportedly from classical sources (Guoyao Yanjiushi, 1944). This formulation is a complex mixture of different species consisting of seven ingredients: *D. febrifuga* Lour., areca nut (*Areca catechu* L.), tortoise shell (*Carapax Amydae Sinensis*), liquorice (*Glycyrrhiza glabra* L.), black plum (*Prunus mume* (Sieb.) Sieb. et Zucc.), red jujube (*Ziziphus jujuba* Mill.), and raw ginger (*Zingiber officinale* Rosc.).

While *D. febrifuga* was used in combination with several other plants in the classical Chinese formulations, in the Himalayas a decoction of the leaves, bark, and roots of *D. febrifuga* has been taken traditionally for fever. Although the plant does not appear to have a Sanskrit name and is not described in classical Ayurvedic texts, it is known in folk medical traditions by the Hindi name *basak* (Foundation for the Revitalisation of Local Health Traditions, India, personal communication, 2002). In the preparation of the medicine, caution is exercised, as the fresh juices are known to be emetic. These effects are reported elsewhere, where *D. febrifuga* is listed as a known emetic (http://www.floridaplants.com/Med/emetic.htm).

4.4 CONTRASTING PARADIGMS

The choice of plant ingredients in the formula including Changshan was guided by classical Chinese medical theory (see the *Yellow Emperor's Classic*, c. 300 B.C., Ni, 1995), which would favor the selection of multiple — rather than single — ingredients that were known to be suited to the type of fever and the type of individual to be treated.

By contrast, the Western medical search for new antimalarials from plants has followed a different track (Kirby, 1997). Typically, the search has begun with a clue from the indigenous use of plants in the management of fevers, such as, for example, the use of Changshan found in or reported from ancient records. Often, no further action is taken and the traditional medicine is simply not developed or applied within a modern medical context (see Chapter 12). However, when the traditional medicine has been investigated, a subsequent step in development is for reports to be gathered from patients about fevers being relieved by the medicine. Early nonindigenous accounts of use can be a source of hypothesis generation as well, such as travelers' tales of miraculous cures (see Chapter 2). If data from these sources indicate that a plant or plant mixture seems to be producing results that may show potential for malaria, a path of scientific investigation begins. Botanists work to identify the correct species that are described in classical literature or in folk oral traditions, as some plants may be used interchangeably due to regional variations in plant availability and in the use of a single common name for different species. Experimental studies are conducted to screen plant extracts for antiplasmodial activity, and then to isolate possible active ingredients (see Chapters 16 and 18).

Clearly, the two paradigms are so different from one another that it is easy to understand why each might seem incompatible with the other.

4.5 CHANGSHAN IN CONTEMPORARY CHINA

By way of background to understanding how Changshan came to be developed as an antimalarial in 20th-century China, it is instructive to consider the rise of modern medical research into traditional Chinese medicine (TCM) and the relationship between the modern and traditional sectors in terms of influence in the direction of this research.

From the early 20th century, the destiny of TCM has been strongly associated with the history of the Chinese state. In 1928, when the Kuomintang (Nationalist Party, KMT) ended the political chaos of the warlord period (1911–1928) and unified China, it quickly established a Ministry of Health at Nanjing. For the first time in its history, China had a national administrative center to take charge of all health care-related issues.

The next year, the first National Public Health Conference, dominated by Western-trained physicians, unanimously passed a proposal to abolish the practice of Chinese medicine. However, this resolution had the unanticipated effect of mobilising the largely unorganised traditional Chinese doctors into a massive National Medicine Movement. In the following 20 years, Chinese doctors demanded that the KMT state should grant them equal status to that of their Western-trained colleagues — to set up an official state organ run by themselves, to establish their own state-sanctioned system of licensing, and to incorporate TCM into the national school system.

While Western-style doctors demanded the wholesale abolition of Chinese medicine, they exempted Chinese drugs as distinctly different from other elements of traditional medicine and developed a research program, which they called Scientific Research on Nationally Produced Drugs (Guochan Yaowu Kexue Yanjin). The term *Nationally Produced Drugs* actually meant those drugs produced in China but the use of which was unrelated to traditional Chinese medical theory. Western-trained scientists asserted that Scientific Research on Nationally Produced Drugs could be carried out only by scientists, and essentially claimed a monopoly of scientific competence (Yu, 1936, p. 190). It was made clear that the success of their project did not depend in any way on Chinese medical theories.

Western-style doctors also repudiated the need to cooperate with Chinese doctors and took the view that Chinese drugs were just natural raw material and should be studied as such. They refused to treat such drugs as a part of the traditional Chinese medicine conceptual and professional framework.

4.6 SCIENTIFIC RESEARCH ON NATIONALLY PRODUCED DRUGS

In 1945, the leading Chinese scientific magazine *Kexue* (*Science*) began publication of a series of special issues to review the past three decades of development of each scientific discipline in China. In the issue on "The Recent 30 Years of Scientific Research in Chinese Drugs," the author of the review praised the research on Changshan as being second only to the world-famous work on mahuang (ephedrine) in the 1920s, as the major achievement of Chinese drugs studies in the 1940s (Zhang, 1949).

Unlike most other Scientific Research on Nationally Produced Drugs, which was directed and controlled by Western-style doctors, the research on Changshan was inaugurated, recorded, and officially supported by an enthusiastic advocate of Chinese medicine — Chen Guofu. Chen Guofu and his younger brother, Chen Lifu, were long-term political allies of the KMT leader Chiang Kai-shek. The Chen brothers also provided strong political support for the National Medicine Movement organised by the TCM doctors. Chen Lifu served as the first president of the Guoyi Guan (National Medicine Institute) established in 1931, and Chen Guofu was on that institute's board of directors. At the same time, both were deeply involved in constructing a German-style national health administration. From Chen Guofu's side, an important personal dimension was that, having been a victim of TB for 40 years, he had developed a strong interest in medicine. He had consulted more than 100 doctors, including both Chinese and Western trained, and had written a multitude of booklets on medical matters.

While the Chen brothers' interests in medical matters were simultaneously medical, political, and personal, they had no formal medical education at all. Chen Lifu held a master's degree in mining engineering from the University of Pittsburgh, and Chen Guofu's formal education had ended with military high school. Therefore, even though they both had been lifelong supporters of Chinese medicine, their support was generally perceived as being ideological and nationalistic rather than scientific (Croizier, 1968, pp. 92–99). As a result, historians have never taken seriously the Chen brothers' influence on the practice of Chinese medicine. According to Chen Guofu, however, the discovery of the antimalarial efficacy of Changshan was built completely on his own audacious personal experiment. From Chen Guofu's point of view, he played a groundbreaking role in this discovery.

For him, the history of discovering Changshan began in 1940 when he sent the Chinese prescription for treating malaria to the clinic of the KMT's Central Politics School (Zhongyang Zhengzhi Xuexiao). Once Changshan, one of the seven drugs in this prescription, had entered the school clinic, it was handled almost exclusively by Western-trained scientists and physicians.

4.7 "DISCOVERING" CHANGSHAN AND OVERCOMING BARRIERS TO ENTRY

The initial stage of discovering Changshan looked clear and simple. As Chen Guofu recalled, the starting point was when one of the school guards found in the local newspaper a prescription for treating malaria and distributed copies of this prescription to the staff members of the Central Political School. In a screenplay that he subsequently wrote to ensure public recognition of his own contribution, Chen Guofu described what followed:

> At that time, the director of the school clinic, Doctor Cheng Peizhen, was in Chen Guofu's office. While Doctor Cheng was working, he mentioned to Chen Guofu that because the price of quinine had risen steeply, he was very worried about a potential shortage. Then, Chen Guofu asked Doctor Cheng why he did not use traditional Chinese drugs instead. Doctor Cheng responded that since he had no idea which Chinese drug was effective for treating malaria, he could not take advantage of Chinese drugs. (Chen Guofu, 1952, p. 264)

Two things were clear from this dialogue. First of all, Chen Guofu became interested in the problem of malaria because he saw it as an opportunity for Chinese herbal medicine to make a contribution to national medical problems. If he succeeded in finding an effective traditional Chinese drug, his success would demonstrate that, contrary to common wisdom, Chinese medicine could also contribute to solving state medical problems. As Chen Guofu's mission was to look for a substitute for quinine, the success of his project inevitably resulted in translating Chinese drugs into just another therapeutic technique in the Western-style doctors' armamentarium, rather than drawing on the theoretical framework of TCM in understanding and managing febrile conditions, of which malaria is one specific type.

Second, although the director of the school clinic claimed his ignorance as the reason why he did not take advantage of Chinese drugs, lack of information on TCM for malaria was by no means the real problem. The problem was that Western-style doctors could not accept the classical explanations for use of traditional antimalarials and considered it necessary to translate the applications and outcomes of Chinese drugs into their own conceptual framework.

The following episode illustrates the point. Some years after the antimalarial efficacy of Changshan had been established scientifically, Xu Zhifang, a chemist in the Chinese Academy of Sciences (Academia Sinica), recalled how he had cured his own malaria with jienue pills, which included Changshan as their key component. After taking the pills, which he made according to the prescription in the *Bencao Gangmu* (*Systematic Materia Medica*, 1552–93 A.D.), Xu Zhifang recovered completely. However, as he wrote:

> [Because] there were no physiological and pharmacological experiments [on the efficacy], I did not dare to announce my result to the public. Moreover, since I could not figure out the chemical composition of the pill, I could not trust it myself. (Xu, 1948, pp. 31–34)

Xu Zhifang was clearly not lacking information. He not only knew of the reputed curative efficacy of the jienue pill, but had also conducted a self-experiment, which confirmed its efficacy to him. Unfortunately, this single-case experiment did not mean anything to Western-trained scientists. As long as he could not tell his colleagues and would not write a paper on his success, his success remained only a personal experience. Most importantly, Xu Zhifang's account made it clear that he could not make his personal experiment public unless the Chinese drug first went through a series of experimental stages — from preclinical through to clinical.

Chinese prescriptions for treating malaria were easily accessible for anyone who bothered to look for them. In the opening summary of his *Elementary Introduction to Chinese Drugs* (*Zhongyao Qianshuo*) (1930), Fubao Ding (1873–1950) mentioned a Chinese antimalarial prescription, including the herbal drug Changshan, and suggested that it was marvelously effective and inexpensive. In one of the most often consulted English studies of Chinese *Materia Medica*, F. Porter Smith (1969, p. 293) also recognised that "there is no form of the malarial fever for which Changshan is not recommended [by Chinese *Materia Medica*]." Furthermore, when the *National Medical Journal of China*, the Western-style doctors' leading journal, published a special issue on malaria, it included an article on "Historic Study of Malaria in Chinese Medicine" (Li and Woguo, 1932). In that paper, Li Tao, a historian of Chinese medicine, pointed out that a prescription of Chinese drugs (again, including Changshan) was one of four major traditional methods for treating malaria. Nevertheless, Li Tao was quick to add that "no one is sure about the curative efficacy of this prescription." It was not difficult at all to discover bits of information about Chinese drugs. However, Western-style doctors were unable to trust Chinese drugs until they had translated them into their own framework.

While Xu Zhifang did not dare to report his experience at the time, 4 years later, when he found that many poor villagers were sick from malaria and could afford neither Western nor Chinese drugs, Xu decided to produce the pills for sale. Within 8 months, Xu had sold more than 100,000 pills. Xu's inconsistency in not reporting his personal experience due to concern about the need for scientific evidence and his contrasting commercial action in selling the product to the public reflects his personal dilemma. As a Western-trained chemist, he was not permitted to proclaim the efficacy of Changshan without first translating it into a scientific framework. However, considerable financial and technical resources would have been required to do this. Xu ended up opting to use Changshan within its traditional context — i.e., making the jienue pills according to the *Systematic Materia Medica* and then selling them to Chinese patients accustomed to taking traditional medicines.

The director of the school clinic was under similar constraints. He was not ignorant about Chinese antimalarials but was required by his professional standards to assert ignorance of efficacy in situations where the average Chinese would think otherwise.

4.8 CURING MRS. CHU

To return to Chen Guofu's history of Changshan, after receiving the prescription, Chen Guofu described how he tested this prescription on a visitor to his family, Mrs. Chu, who was sick with malaria. When she recovered after taking this prescription, Chen Guofu became keenly interested in it and asked the director of the school clinic to undertake a program of systematic experimentation on the Chinese drugs contained in the Changshan mixture.

Chen Guofu arranged for a clinical trial to be conducted on 50 students suffering from malaria. When the Changshan mixture produced full recovery and blood tests showed that the malaria parasites were being killed by the drug, the director of the school clinic conducted further research that found that the single plant known as Changshan was equally effective in clearing parasites as the complex mixture. No detailed data exist from this study as the trial was done in a poorly equipped school clinic and no paper was published on the findings. As the study clearly lacked scientific credibility, Chen Guofu asked the government to establish a team of the best-trained Chinese scientists to study Changshan.

At this point, Chen Guofu reported his discovery directly to the leader of the KMT, Chiang Kai-shek. According to Chen Guofu, Chiang Kai-shek immediately appropriated special aid to fund the Research Laboratory for National Drugs in the Central Political School and ordered the National Health Administration to support Chen Guofu's project. With Chiang's full support, Chen Guofu not only set up a well-equipped research laboratory but also recruited a team of Western-style doctors and scientists to join his project. The work of the team was divided into four areas: (1) pharmacognosy, by Guan Guangdi; (2) chemistry, by Jiang Daqu; (3) pharmacology, by Liu Chenru; and (4) clinical research, by Chen Fangzhi (1944) and his colleagues.

4.9 IDENTIFYING CHANGSHAN

Before pharmacognostic research could be conducted on Changshan, Guan Guangdi had to solve a tricky problem: What was the thing called Changshan? For Guan Guangdi, the problem of identifying Changshan consisted of two subproblems. First of all, what was the historical identity of the herbal drug Changshan in the tradition of Chinese *Materia Medica*? Second, in terms of modern pharmacognosy, which plant was the source of this herbal drug Changshan described in the Chinese *Materia Medica*?

By comparing descriptions of Changshan documented in the Chinese *Materia Medica*, Guan Guangdi classified Changshan into three types of herbal drugs: jigu-Changshan, haizhou-Changshan, and tu-Changshan. Then, citing remarks on Changshan by the famous 19th-century scholar of Chinese *Materia Medica* Wu Qijun (1789–1847), Guan Guangdi concluded that "at that time there were so many types of Changshan that Wu Qijun could not differentiate the genuine from the fake ones" (Guoyao Yanjiushi, 1944). In terms of a rigorous scientific evaluation, the research by Chen Guofu and Cheng Peizheng meant very little since they did not know the scientific identity of the herbs that they were using. Yet the research was given political weight by their self-proclaimed success as Chen Guofu was able to convince President Chiang to fund a team of scientists to further study Changshan. The rigorous research came later. Similarly, Guan Guangdi in subsequent research was unable to conclude positively which was the genuine article among the three possibilities. Throughout his paper in the report, he called his tentative conclusion — jigu-Changshan — a hypothesis.

Historical documents provided only partial information for identifying Changshan. However, the clinically confirmed antimalarial activity of Changshan turned out to be the most useful guide (or constraint) in determining its identity. For example, Guan Guangdi eliminated the possibility that tu-Changshan was the genuine Changshan, because it was reported to have no efficacy in curing malaria.

Assuming jigu-Changshan to be the Changshan described in the *Materia Medica*, Guan Guangdi found two possible plant sources: *O. japonica* Thunb. and *D. febrifuga* Lour. The former was produced mostly in Japan and had been thoroughly studied by Japanese chemists, pharmacologists, and pharmacognosists. It was known definitively that while *O. japonica* demonstrated antipyretic activity, it had no efficacy for treating malaria. Since Western-trained scientists generally accepted the Japanese identification of the Changshan source as *O. japonica*, they were suspicious about the antimalarial efficacy of Changshan as recorded in Chinese *Materia Medica*.

The truth was that the Japanese Changshan was completely different from the Chinese Changshan. Although Japanese scholars had traced Changshan to *O. japonica* since 1827, the root of this plant was morphologically different from the one described in Chinese *Materia Medica*. In fact, because Changshan was not produced in Japan, Japanese doctors had used various native substitutes, including kokusaki *(O. japonica)* from ancient times. Due to the leading position held by Japanese scholars in scientific research into Chinese *Materia Medica*, Chinese doctors had just copied the Japanese scholars' conclusion that *O. japonica* was the botanical identity of Chinese Changshan. For example, in the popular *Dictionary of Chinese Pharmacopoeia, O. japonica* is simply entered as the "foreign name" for Changshan.

Now that the local Chinese Changshan was demonstrated to be effective in treating malaria, the pharmacognosist discarded the ineffective *O. japonica* and took *D. febrifuga* to be the plant source of Changshan. The ultimate evidence came from juxtaposing the sliced specimen of the root of *D. febrifuga* (grown in the laboratory) and jigu-Changshan (bought from the local drug market in Chongqing) under a microscope.

4.10   THE RESEARCH PROCESS

In terms of research procedures, there is a sharp contrast between the Changshan research and the research that led to the world-famous discovery of ephedrine from the Chinese herbal drug Mahuang (Chen and Schmidt, 1930). To a very large degree, a research process known as 12345 captured the procedures that K.K. Chen's group used to develop ephedrine. The so-called 12345 process of Mahuang research went in the following sequence: chemical analysis, retest and artificial synthesis, structural modification, animal experiment, and clinical experiment (Reardon-Anderson, 1991, pp. 149–151).

In contrast, the Reverse-Ordered Program, which Chen Guofu called 54321, went in the following sequence: clinical experiment, animal experiment, chemical analysis, retest and artificial synthesis, and structural modification (Yu, 1952). The key difference between the two competing programs depended on whether to put clinical experimentation with human subjects as the first step. The second program would be, more accurately, 32145 (see Table 4.2).

In fact, in the 1930s, clinical experimentation with Chinese drugs on the human body was a highly charged issue. Western-style doctors repeatedly and strongly attacked Chinese doctors for treating their patients as guinea pigs by feeding them "life-threatening drugs" (Wang, 1935). Zhao Juhuang, the first and probably the most important Chinese pharmacognosist at that time, specifically argued against any research program that conducted clinical experiments on the human body before chemical analysis and pharmacological experiments. In his proposal for establishing the

---

TABLE 4.2

Comparison of the Received Program of Research (12345) Used to Develop Ephedrine from *Mahuang* and Chen Guofu's Reverse-Ordered Programme Used in the Development of Changshan

Received Program (12345)	Reverse-Ordered Program
Chemical analysis	Clinical experiment
Animal experiment	Animal experiment
Retest and artificial synthesis	Chemical analysis
Structural modification	Retest and artificial synthesis
Clinical experiment	Structural modification

Research Institute of Chinese Drugs in the Academia Sinica, Zhao did not include clinical experiments among the tasks of the experiment section (Zhao, 1929).

By accepting customary use as a starting point to justify clinical experimentation, Chen Guofu and his colleagues first conducted a clinical experiment with human subjects and thus established the traditionally believed curative effect of Changshan. In so doing, they were of the view that they had crucially shortened the time and labor needed to escape the conventional research cycle, and had thus been able to fast-track into national production an antimalarial drug used historically for related purposes. As it turns out, this is consistent with the current policy of the World Health Organization on the evaluation of traditional medicines that have been in long-standing use in a culture (see Chapter 18 on safety evaluation).

4.11 THE ETHICS AND POLITICS OF CLINICAL EXPERIMENTS WITH CHANGSHAN

Chinese doctors had clear reasons to adopt the clinical-experiment-first strategy. First, if clinical experimentation were generally accepted as the first step for studying Chinese drugs, to a certain degree the autonomy of Chinese medicine would be maintained. Since clinical experimentation came first, it would have to be conducted more or less in the traditional fashion. On the other hand, if the research procedure on Chinese drugs followed the received program, then immediately after the first step of chemical analysis, the extracted active component would be tested and this would be totally foreign to Chinese doctors.

Second, as revealed in the study of Changshan, precisely because Chen Guofu's research did not follow the received program but started with a clinical experiment, it had the subversive effect of demonstrating that Japanese and Western scholars had been wrong for decades in assuming that the botanical identity of Chinese Changshan was the root of the *O. japonica*.

Third, inasmuch as the received program presupposed that the efficacy of Chinese drugs could be traced to one active chemical, a reductionist framework was built into it. As a result, as long as researchers worked within this received program, there was no way that they could empirically prove this reductionist presupposition to be wrong or counterproductive. If the extracted principal component of a Chinese drug turned out not to have traditional curative efficacy, this failure would be taken as a refutation of the traditional belief, just as in the case of Mahuang before 1924. On the contrary, if the curative efficacy of a Chinese drug were clinically verified from the outset, chemists' failure to find the active component would call into question the reductionistic approach.

After clinical experiments on the human body showed *D. febrifuga* to be active (Chen Fangzhi, 1944), the distinguished Cambridge scholar of Chinese medicine Joseph Needham sent the same local herbal drug to K.K. Chen's research group at Eli Lilly & Company (Henderson et al., 1948) and to another American group for animal experiments (Tonkin and Work, 1945). When these U.S. animal experiments confirmed the antimalarial efficacy of *D. febrifuga*, some samples of the same herbal drug were sent to the University of London for pharmacognostic research (Fairbairn and Lou, 1950; Lou, 1954). The other Chinese governmental laboratory that pioneered antimalarial drugs also requested an aqueous extract of Changshan from the Research Laboratory of National Drugs. The clinical success of Chen Guofu and his Western-trained colleagues made the local variety of Changshan and its extract the obligatory passage point for further research. In the end, Chen Guofu did not wait for the Western-style doctors to isolate the active agents or to synthesise the chemical component through artificial means. While Western-style doctors were busy conducting experiments and writing papers, which never led to a useful substitute for the extract of Changshan, Chen Guofu worked on the mass cultivation of *D. febrifuga* on the Jingfo Mountain (Chen Guofu, 1952, p. 267). Lei (1999) gives a more detailed analysis of this history.

4.12 ISOLATED ACTIVE INGREDIENT OR SYNERGISTIC ACTIVITY WITHIN A COMPLEX MIXTURE?

Despite the promising experimental work with *D. febrifuga* and its derivatives and the promising clinical findings from the research initiated by Chen Guofu, the well-known emetic effect of *D. febrifuga*, in the absence of the accompanying plants used classically to offset nausea, was to prove a central factor in Changshan not being adopted for widespread contemporary use as an antimalarial. Patients simply would not take it.

Febrifugines, alkaloids derived from *D. febrifuga*, were analysed and synthesised during World War II in a program to protect American forces from malaria in the Pacific and other tropical campaigns (Koepfli, 1947). Despite advances in understanding the chemistry of Changshan, the chemists engaged in the research were unable to separate out the nausea that the drugs produced, and the result was that the Changshan-derived antimalarials did not prove viable (http://www.chamisamesa.net/drughist.html).

Evidently, the vomiting and nausea produced by *D. febrifuga* were intensified when the active alkaloids were used in isolation from the other chemical constituents of the plant. These effects were further compounded by the isolation of *D. febrifuga* itself from the companion ingredients used in the classical formulation of Changshan.

It is this outcome that raises the second scientific issue of great importance in the history of Changshan and its modern development as an antimalarial drug, that is, the issue of the search for activity of a single ingredient or the identification of synergistic activity among the multiple components of traditionally used plant-based medicines.

In today's scientific community, the debate over what the Chinese called 54321 or 12345 is still current in the evaluation of traditional medicines. Likewise, the paradigm of isolating and synthesising so-called active ingredients from plants is also pitched against an alternative view. The latter holds that the multiple ingredients in traditional remedies, in themselves, constitute a form of polypharmacy or combination therapy with intrinsic therapeutic validity.

As noted in the introduction to this book, evidence to support the occurrence of synergy within phytomedicines is now accumulating (Williamson, 2001). The current emphasis on combining antimalarial drugs to combat the development of resistance, including those drugs derived from traditional medicine, does not take into account that combination always existed in the traditional formulations, prior to the isolation and synthesis of active ingredients and their subsequent recombination.

Thus, an important principle of traditional or indigenous pharmacology — namely, that multiple ingredients serve a therapeutic role that is greater than the sum of their so-called active ingredients — is consistent with the most advanced thinking in the treatment of multi-drug-resistant diseases. Through discarding the intended polypharmacy that is intrinsic to the use of multiple ingredients of whole plants and the complexity of traditional herbal mixtures, the active-ingredient model of pharmacological research is implicated in the development of resistant strains of diseases such as malaria. Through natural selection, the parasite may have a higher probability of developing resistance to a single active ingredient than to complex mixtures, containing multiple ingredients with diverse activities, where these combine to produce an effective dose.

The issue of the appropriate sequence for research into traditional antimalarials and the issue of synergy have been addressed earlier by David Warrell, a distinguished clinical malariologist, who as noted in the Foreword to this book, has taken the view that:

> Whereas testing of individual compounds may lead to identification of the sole or major active component, possible synergism among the different ingredients or the special effects of the mode of preparation may be lost or obscured. I would advocate a more direct approach to the screening of antimalarial remedies in human patients. I believe that this can be entirely ethical if the subjects live in an area where a particular herbal remedy is the popular treatment of choice for symptoms attributed to malaria. (Warrell, 1997)

4.13 POLYPHARMACY AS A MEANS TO OFFSET NAUSEA INDUCED BY *D. FEBRIFUGA*

In the case of Changshan, it is not resistance that has been the problem leading to disuse of the drug, as we have seen, but nausea and vomiting associated with the known emetic effects of febrifugine. Not surprisingly, however, at least three of the other ingredients in the classical formulation studied in China in the mid-20th century would seem to be candidates for offsetting the emetic properties of Changshan. They are *Zingiber officinale* (ginger), *Glycyrrhiza glabra* (liquorice), and *Areca catechu* (betel nut).

Zingiber officinale, ginger, is an ingredient in the classical formula and has been found through a large body of experimental and clinical research to be effective against nausea. A systematic review of the evidence from randomised controlled trials (RCTs) of ginger for nausea and vomiting found that six studies met all inclusion criteria. Of three RCTs on postoperative nausea and vomiting, two suggested that ginger was superior to placebo and as effective as metoclopramide. RCTs examining the effects of ginger on seasickness, morning sickness, and chemotherapy-induced nausea also supported the other data to collectively favor ginger over placebo in the reduction of nausea (Ernst and Pittler, 1998). A cautionary note comes from rodent studies investigating reproductive toxicity from ginger. In rats fed ginger in their drinking water, it was found that while there was no maternal toxicity, there was increased early embryo loss, with increased growth in surviving fetuses (Wilkinson, 2000).

Z. officinale is used as an antimalarial in Zambia (Vongo, 1999); in India, where the rhizome is mixed with goat's milk into a paste, especially to treat malaria during pregnancy (Singh and Anwar Ali, 1994; Sharma, 1999); and in Sierra Leone, where a stem infusion is mixed with pepper to treat malaria (Lebbie and Guries, 1995). The root decoction is used as an antipyretic in Nicaragua (Coe and Anderson, 1996).

Chang and But (1987, p. 1076) note that the emetic effect of Changshan may be reduced "by stir-frying the herb in ginger juice or yellow wine or using it with the tuber of *Pinellia ternate*, the whole plant of *Pogostemon cablin*, or *Citrus* peel."

Glycyrrhiza glabra (Leguminosae), liquorice, has been the basis for an aqueous solution in which *D. febrifuga* has been soaked, in a classical Chinese method to remove the nausea-inducing effects. A simple reason for using liquorice may have been to overcome the unpalatable taste of the bitter *D. febrifuga*. It has certainly been used for similar purposes in Western herbal and medical traditions, including offsetting the taste of bitter antimalarials. A classic Victorian textbook of *Materia Medica* (Hale White, 1897, p. 552) states that:

> Liquorice or its preparations are contained in many preparations, generally to cover their nauseous taste. They hide very well that of *aloes*, *cascara sagrada*, chloride of ammonium, *hyoscyamus*, *senega*, *senna*, turpentine, and bitter sulphates, as sulphates of quinine.... It is used to hide the taste of nasty medicines, and as a basis for pills.

Of specific interest is the fact that *G. glabra* is one of the traditional means of treating malaria in Ayurveda, the classical health care system of India (Sharma, 1999). Liquorice is also known to have gastrointestinal effects and is used in Ayurvedic medicine to treat gastric and duodenal ulcers (www.herbmed.org). It has been found in purified form to stimulate and accelerate gastric mucus formation and secretion (van Marle et al., 1981). These gastrointestinal effects of liquorice would suggest some effect on nausea. In clinical research in Japan, glycyrrhizin, the aqueous extract of liquorice root, has been shown to prevent hepatotoxic responses to chemicals and is used in treating chronic hepatitis (Shibayama, 1989). It also reduces the side effects, including nausea, resulting from chemotherapy in postoperative breast cancer patients (Akimoto et al., 1986). Liquorice is also known to have synergistic effects with other ingredients of plant-based medicines (Williamson, 2001) and may serve such a role in facilitating the mechanism of action of other ingredients in the traditional formula.

Areca catechu, betel nut, another ingredient of the traditional formula, is commonly used in Asian medical traditions. It also has widespread social usage, as it is thought to soothe the digestion and to be a stimulant. It is reportedly chewed regularly by at least 10% of the world's population, is the fourth most widely used addictive substance, and, with regular use, is associated with increased tumor risks (Boucher and Mannan, 2002).

Throughout Asia, betel chewing is believed to produce a sense of well-being, euphoria, warm sensation of the body, sweating, salivation, palpitation, heightened alertness, and increased capacity to work. This collective experience across time would suggest that *A. catechu* consumption affects the central and autonomic nervous systems.

Specific arecal alkaloids act as competitive inhibitors of gamma aminobutyric acid (GABA) receptors and have widespread effects in the body, including actions on the brain, cardiovascular system, lungs, gut, and pancreas (Boucher and Mannan, 2002). The pro-convulsant properties of GABA antagonists may account for some of the brain stimulation. A dichloromethane fraction from *A. catechu* has been found to inhibit monoamine oxidase type A isolated from the rat brain (Dar and Khatoon, 2000). There is evidence indicating that *A. catechu* interacts with central nervous agents and can produce exacerbation of extrapyramidal effects with neuroleptic drugs (Fugh-Berman, 2000). There is also evidence to suggest that *A. catechu* consumption enhances extrapyramidal functioning in schizophrenic patients (Sullivan et al., 2000).

The combined central and autonomic effects of *A. catechu* make it an additional candidate, with ginger and liquorice, in offsetting any emetic effects of *D. febrifuga* in the classical formula.

The remaining three ingredients are used classically in a range of traditional medicines:

Ziziphus jujuba, the red date, has, like *D. febrifuga*, been found to stimulate nitric oxide generation (Tamaoki et al., 1996) and hence to have possible antimalarial relevance. An extract of its seeds enhances the permeability of cell membranes to assist with cross-membrane transport of drugs and other substances (Eley and Dovlatabadi, 2002), which may be important in the context of synergism in the case of Changshan and the traditional pharmacological use of potentiating agents.

Prunus mume (Sieb.) Sieb. et Zucc., the black plum, is widely used in the Chinese pharmacopoeia as an additive to herbal formulae to prolong, enhance, and harmonise the effect of the other ingredients. One of its specific applications is to moderate the toxicity of potent drugs (http://www.foodsnherbs.com/new_page_47.htm). It has been found to have hepatoprotective effects (Sinclair, 1998), increase blood flow (Chuda et al., 1999), and activate complement via the alternative pathway (Dogasaki et al., 1994). On its own, complement activation could also be associated with such adverse effects as toxic shock, anaphylaxis, and tissue damage.

Amydae carapax, the soft-shelled Chinese tortoise (*Trionyx sinensis* Wiegmann, family Trionychidae) has been used historically in TCM to correct what is known as yin deficiency. The shell (*A. carapax*), which contains colloid, keratin, iodine, and several vitamins, may serve to inhibit the hyperplasia of connective tissue and enhance the function of plasma proteins (Ou Ming, 1989; http://botanicum.com/singles/biejia.htm). Its direct relevance to malaria cannot be determined from the very limited amount of pharmacological research available, but is best understood within the classical framework of being applied by TCM practitioners to correct what they assess to be yin deficiencies. Naturally, the use of animal species in contemporary applications of traditional medicine raises both ethical issues on animal welfare and issues of conservation of endangered species and their protection as addressed under the Convention on the International Trade in Endangered Species (CITES).

4.14 CARDAMOM AND CHANGSHAN

As noted in Table 4.1, there are many classical antifebrile formulations containing Changshan other than that used to cure Mrs. Chu. One simple formula that is currently under study (Jiang, 2003) adds cardamom seeds to *D. febrifuga* to offset nausea. The seeds of two species of cardamom have been used traditionally:

1. Cao Guo or *Amomum tsaoko* Crevost et Lemaire
2. Sha Ren (grains-of-paradise fruit) or *Amomum vilosum* Lour.

Both *A. tsaoko* and *A. vilosum* are used in the treatment of abdominal pain and congestion, lack of appetite, nausea, and vomiting. *A. tsaoko* has also been used to treat malarial patients (Li, 1593; Yao et al., 1995). *A. vilosum* was used in classical Chinese medicine to treat nausea and vomiting, abdominal pain, diarrhoea, indigestion, gas, loss of appetite, morning sickness, pain and discomfort during pregnancy, and involuntary urination (Time-Life, 1997). Clearly there are a range of options open to researchers seeking a traditional source for the management of nausea associated with *D. febrifuga* and its derivatives.

4.15 CONTEMPORARY RESEARCH ON *D. FEBRIFUGA* AND ITS CHEMICAL CONSTITUENTS

The modern medical analysis of Changshan began early in the 20th century. From the roots of *D. febrifuga*, Chinese chemists isolated two compounds (the dichroines), one of which later proved to control avian malaria. The leaves of *D. febrifuga* were also found to contain antimalarial chemicals — the febrifugines — one of which is identical to one of the dichroines. The alkaloids identified as the active principles of *D. febrifuga* were isolated in China by Jang et al. (1946, 1948) — reported on in the journals *Science* and *Nature*, respectively — and Fu and Jang (1948), and in the U.S. by Koepfli (1947). Jang et al. initially named the alkaloids as dichroines A, B, and C. These were later named α-dichroine, β-dichroine, and γ-dichroine. The U.S. group led by Koepfli called the alkaloids febrifugine and isofebrifugine. Febrifugine is identical to β-dichroine and isofebrifugine is identical to α-dichroine. These two alkaloids are isomeric quinazolin derivatives (Koepfli et al., 1950). The stereochemistry of febrifugine has been determined (Barringer et al., 1973a, 1973b), as has its absolute configuration (Hill and Edwards, 1962).

Early findings on the antiplasmodial effects were reported by Spencer et al. (1947): "Extracts of leaves, roots or stems of *D. febrifuga* showed significant suppressive activity against *Plasmodium gallinaceum* in the chick." These were both aqueous and chloroform extracts, administered both orally and subcutaneously, and were some of the most active of the 600 plants screened by Spencer et al.

Chang and But (1987), in their work on the *Pharmacology and Applications of Chinese Materia Medica*, report on a series of studies on Changshan, including studies carried out over a 25-year period in the People's Republic of China — from 1953 to 1978. Tang and Eisenbrand (1992) have also summarised research on the chemistry and pharmacology of Changshan. The following draw on these two works.

4.15.1 CHEMICAL CONSTITUENTS

Changshan contains three tautomers, α-dichroine (i.e., isofebrifugine), β-dichroine (i.e., febrifugine), and γ-dichroines — dichroidine, 4-quinazolone, and umbelliferone. The root is reported to contain about 0.1% of alkaloids, whereas the leaf contains about 0.5%.

4.15.2 ANTIMALARIAL EFFECT

Zhang (1943) found that *Plasmodium gallinaceum* (chicken malaria) is very susceptible to an aqueous extract of Changshan, and also that the antimalarial activity of the leaf was five times that of the root.

American researchers found that febrifugine was approximately 100 times as active as quinine against *Plasmodium lophurae* in ducks (Koepfli, 1947; Hewitt et al., 1952) and 50 times as active as quinine against *P. gallinaceum* in chicks and against *P. cynomolgi* infection in rhesus monkeys (Koepfli et al., 1950).

4.16 TOXICITY

A number of toxicity studies have been conducted on the alkaloids of Changshan. Acute lethal dose 50% (LD_{50}) values of orally administered alkaloids of Changshan (Chang and But, 1987) follow:

- α-Dichroine (isofebrifugine), 570 mg/kg
- β-Dichroine (febrifugine), 6.57 mg/kg
- γ-Dichroine, 6.45 mg/kg
- Total alkaloids, 7.79 mg/kg

Acute LD_{50} values by intravenous administration in mice (Chang and But, 1987) follow:

- α-Dichroine, 18.5 mg/kg
- β-Dichroine, 6.5 mg/kg
- γ-Dichroine, 5 mg/kg

Nausea, vomiting, diarrhoea, and gastrointestinal mucosal congestion and hemorrhage occurred in dogs following oral administration of the aqueous extract of Changshan. This was also found with intramuscular injection of the ethanolic extract or with subcutaneous injection of α-dichroine (Fujian Medical College, 1959).

Febrifugine appears to have primarily stimulated the vagal and sympathetic nerve endings of the gastrointestinal tract, causing reflex vomiting.

Febrifugine has subsequently been found to promote nitric oxide production (Kim et al., 2000; Murata et al., 1998), providing a potentially important role in host defense against malaria.

The historic interest in Changshan by Japanese researchers — noted earlier in the chapter — has continued to the present day. There is ongoing research by Japanese scientists to investigate the potential of *D. febrifuga* as a source of compounds for new antimalarial drugs. While noting the potential of febrifugine to lead to important new antimalarials, they recognise that "the strong side effects such as the emetic effect have precluded their clinical use against malaria" (Kikuchi et al., 2002, p. 2563).

Japanese and Korean studies have compared febrifugine derivatives with chloroquine, have reported on efforts to synthesise febrifugine analogues, and have compared synthetic forms with the natural febrifugines. Some of these are discussed below.

In ongoing research at Tohoku University into the antimalarial effects of *D. febrifuga*, it has been found that adducts of febrifugine and isofebrifugine showed significant activity (Takaya et al., 1999). Most recently, Kikuchi and co-workers (2002) at Tohoku University in Japan have evaluated the *in vitro* antimalarial activity of the analogues of febrifugine and isofebrifugine. The research team found that the 3' '-keto derivative of febrifugine exhibited potential antimalarial activity with high selectivity against *P. falciparum in vitro* (inhibitory concentration 50% (IC_{50}) = 20 nM). They found that the *in vitro* activities of the reduction product were strongly active and selective. Kikuchi and co-workers also found that the Dess-Martin oxidation product was strongly active with high selectivity against *P. falciparum*. A structure–activity relationship study (SAR) demonstrated that the 4-quinazolinone ring plays an essential role in the appearance of activity (Kikuchi et al., 2002).

Febrifugine has been successfully synthesised from 2,3-diacetoxy-N-benzyloxycarbonylpiperidine on the basis of diastereoselective nucleophilic substitution reactions (Okitsu et al., 2001). However, whether synthetic or natural forms of febrifugine are more effective is under question. In a study at the University of Tokyo, antimalarial activities of the synthesised febrifugine and isofebrifugine and their enantiomers were examined. The Tokyo team found that the activities and selectivities of natural febrifugine and isofebrifugine were much higher than those of the enantiomers (Kobayashi et al., 1999), a finding that raises again the question of whether a natural form of Changshan might hold promise as an effective and economical antimalarial. Despite this ongoing research in Japan, there appears to be no research effort to evaluate the antinausea effects of the companion ingredients in the classical formulation of Changshan for offsetting the problematic emetic effects of *D. febrifuga*.

At the same time, a World Health Organization regional publication (2002) reports that a synthetic chemical derived from febrifugine (Df-1) is one of the two most powerful antimalarial compounds of 4700 screened. The other is an endoperoxide synthesised from artemisinin (http://www.wpro.who.int/malaria/docs/shanghai.pdf).

4.17 DISCUSSION

The history of the development of modern antimalarials is one that has been intertwined with the exploration of traditional medical knowledge from indigenous and ancient societies. As the chapter on *Cinchona* highlights (Chapter 2), secrecy, Western extractive priorities, and lack of acknowledgment of traditional origins and ownership rights have characterised this exercise.

The Changshan story highlights the fundamental scientific differences between traditional medical paradigms and modern or Western scientific assumptions about mechanisms of action and therapeutic potential of plant-based medicines. It also highlights an ongoing issue — that of the appropriate sequence for evaluating traditional herbal medicines that have been in long-standing customary use. Of fundamental importance is the difference between the active-ingredient approach of Western biomedical science and the polypharmacy approach used traditionally in Chinese medicine. As we have seen, the case of Changshan and the varied activities of its multiple botanical ingredients serve to illustrate that ingredients that reduce the overall toxicity of a medicine or alter the host immune response to pathogens are potentially significant in determining the safety and efficacy of traditional formulations. Both of these aspects of phytotherapy, however, are overlooked in the active-ingredient approach. The current emphasis on combination drug therapy in malaria treatment calls for reassessment of the whole complex mixture classically prescribed in ancient Chinese antimalarials. This in turn carries implications for the development of appropriate research methodologies for detecting the significance of the multiple ingredients and their interactions. As noted in Chapter 18, WHO guidelines on the evaluation of traditional herbal medicines allow that proceeding directly from common use to clinical evaluation is ethical where drugs have long been in traditional use, provided that there is no evidence of toxicity (WHO, 1993).

Of central importance in any program of scientific evaluation of traditional herbal medicines is an open-minded and respectful dialogue between modern medical scientists and the custodians of traditional medical knowledge. Such dialogue offers the potential to recognise within traditional explanatory frameworks new pathways and modes of action, potential interactive effects of the multiple ingredients in traditional complex herbal mixtures, and methods for capturing these effects in appropriately and sensitively designed research.

In the case of Changshan, a vital dialogue that should have taken place to refine and focus the research questions that guide research and development appears not to have occurred. As the modern history of Changshan illustrates, intrinsic power imbalances, compounded by mutual incomprehension, led to traditional Chinese explanatory models being overlooked as sources of valid research hypotheses, in favor of the more conventional assumptions made by modern medical investigators.

The result was not just one of a dominant scientific interest group prevailing over a weaker group. Rather, the resultant outcome was a significant scientific oversight, resulting in the loss of an important new, affordable, and easily produced antimalarial treatment. The simple addition of traditionally used ingredients may have offset the emetic effects of *D. febrifuga* or its purified alkaloids

Clearly, Changshan deserves to be reappraised in light of its traditional formulation, including the antinausea agents and processes classically employed. The case of Changshan also highlights the possible plethora of false negatives that may exist in the many plants rejected in contemporary antimalarial screening programs. It raises the prospect that many other plants may merit renewed evaluation according to culturally and paradigmatically sensitive research methods.

ACKNOWLEDGMENTS

The authors acknowledge the kind permission of Sage Publications Ltd. of Social Studies of Science, who have allowed sections of Sean Hsiang-lin Lei's 1999 paper "From Changshan to a New Antimalarial Drug" (copyright © 1999, Sage Publications Ltd.) to be drawn on here. Thanks are offered also to Bruce Peters, California Horticultural Society (http://www.calhortsociety.org), for the image of Changshan, and to Taylor & Francis for the authorisation to reproduce the representation of the molecular structure of febrifugine and dichroine. Special thanks also to Gemma Burford for substantial contributions to figures and tables; to Major Suping Jiang, Ph.D., for his helpful insights into the use of cardamom to offset Changshan-induced nausea; to Prof. Roland Hardman, Dr. Merlin Willcox, Prof. Philippe Rasoanaivo, and Dr. Carsten Flohr for their helpful editorial contributions; and to Dr. Elizabeth Hsu for introducing the two authors into this collaboration.

REFERENCES

Akimoto, M., Kimura, M., Sawano, A., Iwasaki, H., Nakajima, Y., Matano, S., and Kasai, M. (1986). Prevention of cancer chemotherapeutic agent-induced toxicity in postoperative breast cancer patients with gly-cyrrhizin (SNMC). *Gan No Rinsho*, 32, 869–872.

Barringer, D.F., Jr., Berkelhammer, G., Carter, S.D., Goldman, L., and Lanzilotti, A.E. (1973a). Stereochemistry of febrifugine. I. Equilibrium between *cis*- and *trans*-(3-substituted 2-piperidyl)-2-propanones. *J. Org. Chem.*, 38, 1933–1937.

Barringer, D.F., Jr., Berkelhammer, G., and Wayne, R.S. (1973b). Stereochemistry of febrifugine. II. Evidence for the transconfiguration in the piperidine ring. *J. Org. Chem.*, 38, 1937–1940.

Boucher, B.J. and Mannan, N. (2002). Metabolic effects of the consumption of *Areca catechu*. *Addict. Biol.*, 7, 103–110.

Chang, H.M. and But, P.P.H. (1987). *Pharmacology and Applications of Chinese Materia Medica*. World Scientific Publishing, Singapore.

Chen Fangzhi. (1944). The preliminary report on clinical research on Changshan for treating malaria (II). In *Changshan Zhinue Chubu Yanjiu Baogao* [*The Preliminary Research Report on Changshan for Treating Malaria*]. Guoyao Yanjiusi [Research Laboratory on National Drugs], Central Politics School, Nanjing.

Chen Guofu. (1952). *Changshan Zhinue* [Changshan for treating malaria] (screen play for an educational film). In *Chen Guofu Xiansheng Quanji* [*Complete Works of Mr. Chen Guofu*], Vol. 8. Zhengzhong Press, Hong Kong.

Chen, K.K. and Schmidt, C.F. (1930). Ephedrine and related substances. *Medicine*, 9, 1–7.

Chuda, Y., Ono, H., Ohnishi-Kameyama, M., Matsumoto, K., Nagata, T. and Kikuchi, Y. (1999). Mumefural, citric acid derivative improving blood fluidity from fruit-juice concentrate of Japanese apricot (*Prunus mume* Sieb. et Zucc). *J. Agric. Food Chem.*, 47, 828–831.

Coe, F.G. and Anderson, G.J. (1996). Ethnobotany of the Garífuna of eastern Nicaragua. *Econ. Bot.*, 50, 71–107.

Croizier, R. (1968). *Traditional Medicine in Modern China: Science, Nationalism, and the Tensions of Cultural Change*. Harvard University Press, Cambridge, MA.

Dar, A. and Khatoon, S. (2000). Behavioural and biochemical studies of dichloromethane fraction from the *Areca catechu* nut. *Pharmacol. Biochem. Behav.*, 65, 1–6.

Ding, F. (1930). *Zhongyao Qianshuo* [*Elementary Introduction to Chinese Drugs*]. The Commercial Press, Shanghai.

Dogasaki, C., Murakami, H., Nishijima, M., Ohno, N., Yadomae, T., and Miyazaki, T. (1994). Biological activity and structural characterisation of alkaline-soluble polysaccharides from the kernels of *Prunus mume* Sieb. et Zacc. *Biol. Pharm. Bull.*, 17, 386–390.

Eley, J.G. and Dovlabatadi, H. (2002). Permeability enhancement activity from *Ziziphus jujuba*. *Pharm. Biol.*, 40, 149–153.

Ernst, E. and Pittler, M.H. (2000). Efficacy of ginger for nausea and vomiting: a systematic review of randomised clinical trials. *Br. J. Anaesth.*, 84, 367–371.

Fairbairn, J.W. and Lou, T.C. (1950). A pharmacognostical study of *Dichroa febrifuga* Lour.: a Chinese antimalarial plant. *J. Pharm. Pharmacol.*, 2, 162–177.

Fu, F.Y. and Jang, C.S. (1948). Chemotherapeutic studies on chang shan *Dichroa febrifuga*. III. Potent antimalarial alkaloids from changshan. *Sci. Technol. China*, 1, 56–61.

Fugh-Berman, A. (2000). Herb-drug interactions. *Lancet*, 355, 134–138.

Fujian Medical College (1959). Report on Changshan.

Guoyao Yanjiushi. (1944). *Changshan Zhinue Chubu Yanjiu Baogao* [*The Preliminary Research Report on the Anti-malarial Drug Changshan*]. Guoyao Yanjiushi [Research Laboratory on National Drugs], Central Politics School, Nanjing.

Hale White, W. (1897). *Materia Medica: Pharmacy, Pharmacology and Therapeutics*. J.&A. Churchill, London.

Harbourne, J.B. and Baxter, H. (1993). *Phytochemical Dictionary: A Handbook of Bioactive Compounds from Plants*. Taylor & Francis, London.

Henderson, F.G., Rose, C.L., Harris, P.N., and Chen, K.K. (1948). g-Dichroine, the antimalarial alkaloid of chang shan. *J. Pharmacol. Exp. Ther.*, 95, 191–200.

Hewitt, R.I., Wallace, W.S., Gill, E.R., and Williams, J.H. (1952). An antimalarial alkaloid from hydrangea. XII. The effects of various synthetic quinazolones against *Plasmodium lophurae* in ducks. *Am. J. Trop. Med. Hyg.*, 1, 768–772.

Hill, R.K. and Edwards, A.G. (1962). Absolute configuration of febrifugine. *Chem. Ind. (Lond.)*, 858.

Jang, C.S., Fu, F.Y., Huang, K.C., and Wang, C.Y. (1948). Pharmacology of chang shan (*Dichroa febrifuga*), a Chinese antimalarial herb. *Nature*, 161, 400–401.

Jang, C.S., Fu, F.Y., Wang, C.Y., Huang, K.C., Lu, G., and Chou, T.C. (1946). Chang shan, a Chinese antimalarial herb. *Science*, 103, 59.

Jiang, S. (2003). Personal communication from Major Suping Jiang, Ph.D. U.S. Armed Forces Research Institute on Malaria (AFRIM), Bangkok, Thailand.

Kikuchi, H., Tasaka, H., Hirai, S., Takaya, Y., Iwabuchi, Y., Ooi, H., et al. (2002). Potent antimalarial febrifuge analogues against the *Plasmodium* malaria parasite. *J. Med. Chem.*, 45, 2563–2570.

Kim, Y.H., Ko, W.S., Ha, M.S., Lee, C.H., Choi, B.T., Kang, H.S., and Kim, H.D. (2000). The production of nitric oxide and TNF-alpha in peritoneal macrophages is inhibited by *Dichroa febrifuga* Lour. *J. Ethnopharmacol.*, 69, 35–43.

Kirby, G.C. (1997). Malaria and vector control. *Trop. Doct.*, 27 (Suppl.), 5–25.

Kobayashi, S., Ueno, M., Suzuki, R., Ishitani, H., Kim, H.S., and Wataya, Y. (1999). Catalytic asymmetric synthesis of antimalarial alkaloids febrifugine and isofebrifugine and their biological activity. *J. Org. Chem.*, 64, 6833–6841.

Koepfli, J.B. (1947). An alkaloid with high antimalarial activity from *Dichroa febrifuga*. *J. Am. Chem. Soc.*, 69, 1837.

Koepfli, J.B., Brockman, J.A., Jr., and Moffat, J. (1950). Structure of febrifugine and isofebrifugine. *J. Am. Chem. Soc.*, 72, 3323.

Lebbie, A.R. and Guries, R.P. (1995). Ethnobotanical value and conservation of sacred groves of the Kpaa Mende in Sierra Leone. *Econ. Bot.*, 49, 297–308.

Lei, S.H. (1999). From Changshan to a new antimalarial drug. *Soc. Stud. Sci.*, 29, 3, 323–358.

Li, S. (1593). *Ben Cao Gang Mu*. China Archive Press, Beijing (reprinted in 1999).

Li, T. and Woguo, N.K. (1932). [Historic study of malaria in Chinese medicine]. *Natl. Med. J. China*, 18, 415–419.

Liao, X. (1625). *Shen Nong Ben Cao Jing Shu*. China Press of Traditional Medicine. Beijing (reprinted in 1997).

Lou Zhicheng. (1954). Changshan de shenyao jianding [The pharmacognostic identification of Changshan]. *Zhonghua Yixue Zazhi* [*Chin. Med. J.*], 869–870 .

Miyasita, S. (1979). Malaria in Chinese medicine during the Chin and Yuan periods. *Acta Asiat.*, 36, 90–112.

Murata, K., Takano, F., Fushiya, S., and Oshima, Y. (1998). Enhancement of NO production in activated macrophages *in vivo* by an antimalarial crude drug, *Dichroa febrifuga*. *J. Nat. Prod.*, 616, 729–733.

Ni, M., Trans. (1995). *The Yellow Emperor's Classic of Medicine. A New Translation of the Neijing Suwen with Commentary*. Shambhala Press, Boston, MA.

Okitsu, O., Suzuki, R., and Kobayashi, S. (2001). Efficient synthesis of piperidine derivatives. Development of metal triflate-catalysed diastereoselective nucleophilic substitution reactions of 2-methoxy- and 2-acyloxypiperidines. *J. Org. Chem.*, 66, 809–823.

Ou Ming. (1989). *Chinese-English Manual of Common-Used Prescriptions in Traditional Chinese Medicine*. Joint Publishing Co., Hong Kong.

Porter Smith, F. (1911). *Chinese Materia Medica: Vegetable Kingdom*, 2nd rev. ed. Gu Ting Book House, Taipei.

Porter Smith, F. and Stuart, G.A., Trans. (1973). *Chinese Medicinal Herbs*, Shih-Chen, L., Comp. Georgetown Press, San Franscisco.

Reardon-Anderson, J. (1991). *The Study of Change: Chemistry in China, 1840–1949*. Cambridge University Press, Cambridge, U.K.

Sharma, K.D. (1999). An Antimalarial Herbal Preparation. Paper presented at the First International Meeting of the Research Initiative on Traditional Antimalarials, Moshi, Tanzania, December. Available at http://mim.nih.gov/english/partnerships/ritam_program.pdf.

Shibayama, Y. (1989). Prevention of hepatotoxic responses to chemicals by glycyrrhizin in rats. *Exp. Mol. Pathol.*, 51, 48–55.

Sinclair, S. (1998). Chinese herbs: a clinical review of Astragalus, Ligusticum, and Schizandrae. *Altern. Med. Rev.*, 3, 338–344.

Singh, V.K. and Anwar Ali, Z. (1994). Folk medicines in primary health care: common plants used for the treatment of fevers in India. *Fitoterapia*, 65, 68–74.

Spencer, C.F., Koniuszy, F.R., Rogers, E.F., et al. (1947). Survey of plants for antimalarial activity. *Lloydia*, 10, 145–174.

Stuart, G.A. (1976). *Chinese Materia Medica: Vegetable Kingdom*. Southern Materials Center. Inc., Taipei.

Sullivan, R.J., Allen, J.S., Otto, C., Tiobech, J., and Nero, K. (2000). Effects of chewing betel nut (*Areca catechu*) on the symptoms of people with schizophrenia in Palau, Micronesia. *Br. J. Psychiatry*, 177, 174–178.

Takaya, Y., Tasaka, H., Chiba, T., Uwai, K., Tanitsu, M., Kim, H.S., et al. (1999). New type of febrifugine analogues, bearing a quinolizidine moiety, show potent antimalarial activity against *Plasmodium* malaria parasite. *J. Med. Chem.*, 42, 3163–3166.

Tamaoki, J., Kondo, M., Tagaya, E., Takemura, K., and Konno, K. (1996). *Zizyphi fructus*, a constituent of antiasthmatic herbal medicine, stimulates airway epithelial ciliary motility through nitric oxide generation. *Lung Res.*, 22, 255–266.

Tang, W. and Eisenbrand, G. (1992). *Chinese Drugs of Plant Origin*. Springer-Verlag, Berlin.

Time-Life Books. (1997). *The Drug and Natural Medicine Advisor: The Complete Guide to Alternative and Conventional Medications*. Time-Life Custom Publishing, Alexandria, VA.

Tonkin, I.M. and Work, T.S. (1945). A new antimalarial drug. *Nature*, 156, 630.

Unschuld, P. (1992). *Medicine in China: A History of Ideas*. University of California Press, Berkeley, CA.

van Marle, J., Aarsen, P.N., Lind, A., and van Weeren-Kramer, J. (1981). Deglycyrrhizinised liquorice (DGL) and the renewal of rat stomach epithelium. *Eur. J. Pharmacol.*, 72, 219–225.

Vongo, R. (1999). The Role of Traditional Medicine on Antimalarials in Zambia. Paper presented at the First International Meeting of the Research Initiative on Traditional Antimalarials, Moshi, Tanzania, December. Available at http://mim.nih.gov/english/partnerships/ritam_program.pdf.

Wang Qizhang. (1935). *Zheng Yizhe Wu Yiren Shishu dong Yishu Huoren [Advice to the Doctors: Don't Test Your Skill on Patients but Cure Them with Your Skill]: The Preliminary Comments on the Recent Twenty Years' Medical Matters in China.* Zgengliao Yibao Press, Shanghai, pp. 61–62.

Warrell, D.A. (1997). Herbal remedies for malaria. *Trop. Doct.*, 27 (Suppl.), 5–6.

Wilkinson, J.M. (2000). Effect of ginger tea on the fetal development of Sprague-Dawley rats. *Reprod. Toxicol.*, 14, 507–512.

Williamson, E.M. (2001). Synergy and other interactions in phytomedicines. *Phytomedicine*, 8, 401–409.

World Health Organization. (1993). *Research Guidelines on Evaluating the Safety and Efficacy of Herbal Medicines.* World Health Organization Regional Office for the Western Pacific, Manila, Philippines.

World Health Organization. (2002). *Report: Meeting on Antimalarial Drug Development, Shanghai, China, 16–17 November 2001*, (WP)MVP/ICV/MVP/1.4/001.E. World Health Organization Regional Office for the Western Pacific, Manila, Philippines.

Xu, Z. (1948). Guochan jienueyao zhi yanjiu [Research on the nationally produced antimalarial drug]. *Yiyaoxue [Med. Pharm.]*, 1, 31–34.

Yao, D., Zhang, J., Chu, L., Bao, X., Shun, Q., and Qi, P. (1995). *A Colored Atlas of the Chinese Materia Medica Specified in Pharmacopoeia of the People's Republic of China.* Guangdong Science and Technology Press, Guangzhou.

Yu Yunxiu. (1952). Xianzai gai yanjiu zhongyao le! [Now is the time to study Chinese drugs!]. In *Zhongguo Yaowu de Kexue Yanjiu [The Scientific Research on Chinese Drugs]*, Lansun, H., Ed. Qianqingtang Press, Shanghai, pp. 6–11.

Yu Yunxiu. (1976). Preliminary discusssion of studying nationally produced drugs. In *Yixue Geming Lunwunxuan [Collected Essays on Medical Revolution]*. Yiwen Press, Taipei.

Zhang Changshao. (1949). Sanshi nianlai zhongyao zhi kexue yanjiu [The recent thirty years of scientific research on Chinese drugs]. *Kexue [Science]*, 31, 99–116.

Zhang Changshao and Zhou Tingchong. (1943). Guochan kangnue yaocai zhi yanjiu [Study on nationally produced drugs]. *Zhonghua Yixue Zazhi [Natl. Med. J. China]*, 29, 137–142.

Zhao Juhuang. (1929). Zhongyang yanjiuyuan nishe zhongyao yanjiusuo jihuashu [A plan for establishing the Research Institute of Chinese Drugs in the Academia Sinica (Chinese Academy of Science)]. *Yiyao Pinglun [Med. Rev.]*, 1, 44–47.

5 Ayush-64

V.P. Sharma

CONTENTS

5.1 BACKGROUND AND HISTORY

Widespread resurgence of malaria in India during the 1970s prompted the Central Council for Research in Ayurveda and Siddha (CCRAS) to develop an Ayurvedic remedy for malaria. CCRAS is an autonomous institution of the Ministry of Health and Family Welfare, Government of India. In Ayurvedic literature, malaria fever is known as *Vishamajwara* (intermittent fever with rigor). The Ayurvedic pharmacopoeia and experience in the treatment of various fevers, including malaria, was the basis for the antimalarial herbal mixture Ayush-64.

In 1980 scientists of CCRAS selected four plants used in the treatment of fever, including malaria, and prepared a formulation named Ayush-64. CCRAS further developed the drug in collaboration with 20 laboratories in the country and patented Ayush-64 as a new antimalarial herbal compound. So, it is not really an Ayurvedic treatment for malaria. Rather, it is a phytomedicine produced by scientists based on ethnobotanical leads. No research has been done on classical Ayurvedic treatments for Vishamajwara. It seems a little unfair to label this product of modern drug R&D methods as Ayurvedic and then note its failure. It is more correctly characterised as a failure of modern phytomedicine development, not of classical Ayurveda. The drug was subsequently licensed for production and clinical trials. CCRAS has been producing Ayush-64 in its pilot plant for clinical trials and limited distribution. Twenty-two pharmaceutical companies had purchased the know-how from CCRAS, and at least seven of these are currently manufacturing Ayush-64. The drug is marketed throughout India by the private sector and dispensed by the central and state governments in the treatment of malaria. Ayush-64 is prepared by mixing the following four herbs (for properties of herbs, refer to CCRAS, 1987) in the proportions mentioned below.

1. Katuka (*Picrorhiza kurroa* Royale ex. Benth): The dried rhizomes and roots are considered a valuable bitter tonic and produce beneficial action in fever, infective hepatitis with jaundice, urinary tract disorders, epilepsy, rheumatism, dyspepsia, and purgative preparations. One part of aqueous extract of the rhizome is used in Ayush-64.

2. Saptachada (*Alstonia scholaris* R.Br.): Bark is a bitter tonic, alternative, and febrifuge. It is useful in the treatment of diarrhoea and dysentery. Milky juice is applied to ulcers and ripe fruits are used in the treatment of syphilis, insanity, and epilepsy, but lacks antimalarial activity. One part of aqueous extract of bark is used in Ayush-64.

3. Chirata (*Swertia chirata* Ham): An infusion of whole plant is used in the treatment of many diseases such as the skin diseases, astringent, fever, cough, scanty urine, dropsy, and sciatica. It possesses the stomachic, febrifuge, and antidiarrhoeic action. It is a bitter tonic and antihelmintic. One part of aqueous extract of the whole plant is used in Ayush-64.

4. Kuberakshi (*Caesalpinia bonducella* Fleming): One part of the aqueous extract of the rhizome is used in Ayush-64. Seeds are used in the preparation of drugs. β-caesalpin, found in the seeds, has antimalarial activity against *Plasmodium berghei* in albino rats. Two parts of fine powder of seed pulp are used in Ayush-64.

Aqueous extracts of the above plants are prepared separately in the ratios mentioned above. The plant extract decoction is boiled to reduce the volume, cooled, and filtered. The filtrate is then evaporated to dryness by steam. The powders obtained from the aqueous extract are meshed and mixed with adhesives like starch and gum acacia, etc. Tablets of 250 and 500 mg are strip-packed. CCRAS recommends Ayush-64 tablets in the treatment of *Plasmodium vivax*, *Plasmodium falciparum*, and mixed infections. The recommended dosages are as follows. Adults should take 1 g (two tablets) three times a day for 5 to 7 days. For children aged 5 to 12 years, the dose is 500 mg twice daily for 5 to 7 days, and for infants below 5 years, a half tablet of 500 mg powdered and mixed with honey twice a day for 5 to 7 days. Ayush-64 is also recommended in the prevention of malaria at the dosage of 500 mg twice daily for 7 days; for children 5 to 12 years, one tablet twice daily for 7 days; and for infants below 5 years, a half tablet of 500 mg powdered and mixed with honey twice daily for 7 days.

5.2 REVIEW OF PRECLINICAL DATA

5.2.1 STUDIES ON ANIMALS

Ayush-64 was tested on rats and mice for antimalarial activity. A set of five rats (150 to 200 g) and five mice (15 to 18 g) were infected with 10^6 *P. berghei*-infected donor red cells by intraperitoneal injection. Infection appeared by the third or fourth day of inoculation and was maintained by serial passages. Animals were given Ayush-64 when parasitaemia reached 1.5 to 2%. The dose given to rats was 3 g daily for 4 days, and to mice, 1.5 g daily for 5 days. Thin blood smears were prepared on alternate days and stained in Giemsa. All smears were checked for reduction in parasitaemia and compared with the untreated controls. Results revealed that there was neither any decrease in parasitaemia nor any increase in the survival time (CCRAS, 1987). The study concluded that Ayush-64 is not effective in clearing *P. berghei* infection in experimental animals.

A similar finding was reported by Kazim et al. (1991). The group studied the blood schizontocidal activity of Ayush-64 against four malaria parasites, viz., *P. berghei* and *Plasmodium yoelii nigeriensis* (resistant to chloroquine, mefloquine, and quinine) in Swiss mice and *Plasmodium cynomolgi* B and *Plasmodium knowlesi* in rhesus monkeys. Results revealed that that there was no difference in the parasitaemia and the mean survival time in any of the simian or rodent malaria parasite models.

Experimental studies with the active principles of *Alstonia scholaris* (active principle echitamine chloride), *Swertia chirata* (active principle swerchirin), and *Caesalpinea bonducella* (active principle β-caesalpin) were tested on albino rats infected with 10^6 *P. berghei* parasites intraperitoneally, and the smears were positive on the third or fourth day of inoculation. Parasite density was monitored daily and at 1% parasitaemia rats were administered the active principles at 1.6 mg/kg orally and 320 mcg/kg subcutaneously for 4 days in each case. Results revealed that echitamine chloride was effective at 320 mcg/kg subcutaneously, β-caesalpin was effective at 1.6 mg/kg orally, and swerchirin was effective at both levels and by both routes (CCRAS, 1987).

5.2.2 PHARMACOLOGICAL AND TOXICOLOGICAL STUDIES

Pharmacological and toxicological studies with Ayush-64 were carried out in the laboratories of the Central Council for Research in Ayurveda and Siddha (CCRAS, 1987). The appropriate authorities for pharmacological and toxicological studies approved the study protocol. The lethal dose 50% (LD50) of Ayush-64 given by oral route to mice and rats was 2 g/kg and >4 g/kg, respectively. Acute and subacute toxicity were carried out on adult male mice of uniform age and body weight between 150 and 200 g.

In acute toxicity tests, Ayush-64 was given orally to mice at a dose of 10 g/kg, and the treatment exhibited negative isotropic and chronotropic effects on the isolated frog heart. Treatment depressed the tone and amplitude of contraction of the isolated rabbit ileum. Ayush-64 was given orally to albino mice infected with *P. berghei* at doses of 100, 250, 500, and 750 mg/kg of body weight and did not exhibit any toxic effect in different tissues of the experimental animals, but showed significant anti-inflammatory effects.

In the subacute studies, male rats were given Ayush-64 500 mg/kg suspension in distilled water orally for 12 weeks. This exposure did not produce any significant difference in body weight or food and water intake. Blood glucose, blood urea, serum cholesterol, AST and ALT, and urine analysis matched normal controls. Heart, spleen, kidney, and adrenals remained normal, and there was no evidence of gastric ulceration. There was no change in the wet weight of various organs, except the liver, for which a significant increase was noted. There was a mild increase in the total white blood cell (WBC) count. Ayush-64 significantly decreased the polymorphs and increased the lymphocytes. Pathological studies revealed no effect in the kidney and liver of treated animals after microscopic examination.

In subacute toxicity studies Ayush-64 was found safe and was recommended for clinical trials. The acute toxicity studies in rats, mice, and rabbits failed to induce any mortality, even at high dosages (100 mg to 10 g/kg). Ayush-64 had no anaesthetic, anticonvulsant, and analgesic activity, and did not potentiate pentobarbitone sleep in mice. There was no effect on mammalian skeletal muscles, neuromuscular junction, or nerve conduction.

5.3 REVIEW OF CLINICAL TRIALS

5.3.1 TRIALS BY THE CCRAS

These trials were aimed to compare the efficacy of Ayush-64 vs. chloroquine in the treatment of malaria. The first trial in 1979 recruited 58 volunteers for this study. Thirty cases received supervised treatment with Ayush-64 (1500 mg on the first day and 750 mg daily for 3 days) and 28 cases with chloroquine (600 mg) and primaquine (30 mg on day 1 and 15 mg on day 2). Both treatments produced an equal response ($p > 0.05$).

The next trial consisted of vivax malaria cases (118 outpatients and 59 inpatients). Following the above-mentioned treatment schedule of Ayush-64 and chloroquine, all cases were treated with Ayush-64 or chloroquine plus primaquine (control) as mentioned above. Results revealed that in the outpatient department, 42 of 44 (95.4%) cases treated with Ayush-64 were successfully cured and 2 cases failed. In the chloroquine treated control group, all 53 cases (100%) were cured of malaria.

CCRAS carried out field trials with Ayush-64 for 5 to 6 years in different parts of India. The overall findings were that of the 1442 cases (*P. vivax* positive), the cure rate was 89.1%; in 2337 clinical malaria cases diagnosed by medical experts, the cure rate was 89.04%; and in 448 cases of clinical malaria diagnosed by the Ayurvedic experts, the cure rate was 91.96%.

Sharma et al. (1981) reported a clinical trial in 142 cases in the tribal pockets and backward areas of Madhya Pradesh, India. Fever cases were selected for the trial and subjected to thick and thin blood smear examination. Fifty-five cases were found parasite positive (27 *P. vivax*, 16 *P. falciparum*, and 12 mixed infection). Ayush-64 treatment was given to adults (two or three tablets) and children (one or two tablets) three times a day with milk or warm water. Treatment continued for 3 days and extended to 6 and 9 days after a blood test each time. During the period of treatment patients were advised not to take food and to take complete rest. Response to treatment was good in 45 cases (80%), 7 cases (13%) showed no response, and 3 cases dropped out of the study. In *P. vivax*, 19 (70%) and 5 cases were cured in the first and second follow-ups, respectively. In *P. falciparum*, 14 (90%) and 2 cases were cured in the first and second follow-ups, respectively. In 12 mixed infection cases, 9 (75%) cases were cured overall. However, none was cured after a 3-day treatment, only 50% were cured after a 6-day treatment, and three cases had no response even after a 9-day treatment. Ayush-64 was found more effective in the treatment of vivax malaria and required longer therapy to treat *P. falciparum* and mixed infection. The study found that the drug was more effective at the ring stage than at the other developing stages and killed the gametocytes of *P. falciparum*. It may be noted that there is no experimental proof in the literature of the activity of Ayush-64 on *P. falciparum* gametocytes. During the course of treatment no toxic or side effect of Ayush-64 was reported (CCRAS, 1987).

5.3.2 TRIALS BY CCRAS AND THE NMEP

These trials were carried out jointly by the CCRAS and the National Malaria Eradication Program (NMEP). Ayush-64 was tested in the treatment of *P. vivax* malaria in two primary health centers at Faruknagar and Sohna in the Gurgaon district, Haryana, from August 16, 1985 to October 10, 1986. Seventy vivax cases confirmed by blood smear examination were selected for the study, of which 13 dropped out. Fifty-seven patients (35 males and 22 females) completed the trial. They were given Ayush-64, two tablets of 500 mg each three times a day for 5 days, and there was no control. Urine samples were collected daily for 7 days (i.e., day 0 to day 7 of the treatment), and NMEP tested urine samples and confirmed the absence of chloroquine. Patients showing increased parasite count after administration of Ayush-64 were withdrawn and given chloroquine treatment.

Results of the clinical trials revealed that 37 patients (22 males and 15 females) were completely cleared of parasitaemia, i.e., 33 patients by day 5, and 2 each on days 6 and 7. There was no marked difference in the clearance of parasitaemia between both sexes. Thirteen cases showed persistence of parasitaemia after full treatment with Ayush-64. In seven cases, four showed decreased parasite count on day 1 in comparison to day 0 but exhibited an increasing trend on subsequent days, while two cases remained at the initial level of parasite count. Thus the investigators concluded that Ayush-64 provided relief in the treatment of Vishamajwara (malarial fever) in the field (CCRAS, 1987).

5.3.3 TRIALS BY MRC/NMEP AND CCRAS

The study was conducted by the Malaria Research Centre (MRC) in collaboration with the CCRAS and NMEP at the clinics in Delhi maintained by MRC and NMEP (renamed National Anti-Malaria Program, NAMP). This was a prospective comparative, open, and randomised trial. Proven cases of vivax malaria seeking treatment at the clinics were asked to volunteer for the study. The selected patients were 16 to 60 years of age, had asexual parasitaemia of <50,000/μl, with fever within the last 48 hours. Pregnant and lactating women and G6PD-deficient cases were excluded from the

TABLE 5.1
Curative Efficacy of Ayush-64 and Chloroquine

Parameter	Ayush-64	Chloroquine
Cases enrolled	54 (46♂ + 8♀)	50 (35♂ + 1♀)
Age	29.40 ± 9.04 years	30.46 ± 10.25 years
Parasitaemia on day 0	0.25 ± 0.26%	0.31 ± 0.32%
Dose of drug used	1 g (tds) for 5–7 days	1 g (base) over 3 days
Total cases completing 28 days	47	41
Cure rate	48.9%	100%
Recrudescence rate	17%	0
Parasite clearance time	3.16 ± 2.4 days	1.5 ± 0.5 days
Parasite clearance time	3.16 ± 2.4 days	1.5 ± 0.5 days
Clinical recovery	Slow	Rapid

Note: tds = three times a day.

Source: Valecha et al., 2000.

study. Also, patients were tested for chloroquine in urine and those found positive or informed to have taken chloroquine within 7 days prior to treatment were excluded. Patients were randomly allocated one of the two regimes; i.e., one group received a total of 1500 mg of chloroquine over 3 days followed by 15 mg of primaquine for 5 days as per the antimalaria drug policy of the NAMP, and the second group was given Ayush-64, two tablets of 500 mg each three times a day for 5 days. All cases were followed up for 28 days. Patients who did not clear parasitaemia were given Ayush-64 for two more days at the same dosage and were labeled as treatment failure if they remained positive for *P. vivax*.

The results of this study are given in Table 5.1. Fifty-four patients were enrolled in the Ayush-64 group and 50 in the chloroquine group. Seven cases in the Ayush-64 group dropped out, as they did not want to continue the treatment. In the chloroquine group nine cases dropped out as they became smear negative and were not willing to wait for a 28-day follow-up. Both treatment schedules were generally well tolerated, but the acceptability was poor and clinical recovery was slow in Ayush-64. In the Ayush-64 group 47 patients completed the 28-day follow-up. The cure rate was 48.9%, the recrudescence rate was 17%, and the parasite clearance time was 3.16 ± 2.4 days. Of 16 cases labeled as treatment failures, 3 had to be referred to the hospital due to deterioration of clinical condition, and they became negative after chloroquine treatment. In the chloroquine group 41 patients completed the 28-day follow-up. The cure rate was 100%, there was no recrudescence, and the parasite clearance time was 1.5 ± 0.5 days. Based on the clinical trials, Valecha et al. (2000) concluded that Ayush-64 does not have specific schizontocidal or gameto-cyticidal activity. It is not a primary drug for treatment of malaria, and in addition to increasing morbidity, slow parasite clearance may maintain transmission. Trials with *P. falciparum* were not undertaken due to poor drug efficacy and the possibility of complications/death in falciparum malaria.

5.4 PUBLIC HEALTH USE

In India, NAMP reports about 3 million parasite-positive cases (45 to 50% *P. falciparum*) and 500 to 1000 malarial deaths annually. Since most malaria cases are found in remote and inaccessible areas, the estimated cases vary between 20 and 25 million and deaths about 20,000 (Sharma, 1998). The tribal population of India is about 70 million; almost all of these people are settled in remote areas and have remained neglected from the main national stream. This tribal population accounts for 40% of total malaria and 67% of falciparum malaria incidence in India (Sharma, 1998; 1999). It may be noted that 75 to 80% of this population treat themselves using traditional Indian systems of medicine at the household level. Herbal drugs are low cost, indigenous, and highly accessible to poor and marginalised populations.

One and a half million practitioners of alternative systems of medicine use medicinal plants in preventive, promotive, and curative applications. There are about 460,000 registered practitioners of alternative medicine using plants in the codified streams. Further, there are 7843 registered pharmacies of the Indian System of Medicine and a number of unlicensed small-scale units. Ayush-64 is supplied to the Central Government Health Scheme beneficiaries and distributed by the state government dispensaries, e.g., Himachal Pradesh, Gujarat, Jammu and Kashmir, Madhya Pradesh, Maharashtra, Rajasthan, and the NAMP.

In the modern system of medicine malaria diagnosis and treatment have become highly specialised. Research has provided new nonmicroscopic diagnostic tools for use in the field such as the dipstick and parasite lactate dehydrogenase (pLDH) tests. Drugs are given depending on the species of *Plasmodium*, parasite development stage(s) to be attacked, and the status of drug resistance. Currently the World Health Organization (WHO) recommends drug combinations to stop or slow the progression of drug resistance, particularly in *P. falciparum*, a killer parasite. One major handicap in the modern system of medicine is their access and high cost of treatment. Predominantly, malaria is a disease of the poor and marginalised and of those settled in inaccessible areas. Herbal drugs would have an edge over the modern medicine, if these drugs can provide relief to the neglected sections of the society at an affordable cost.

Ayush-64 is marketed in a manner similar to that of modern medicine and therefore suffers from the disadvantage of distribution at the periphery. The drug is also not cheap compared to chloroquine, as the treatment cost of an adult with both treatments is Rs. 10 to 15 (U.S. $0.25 to 0.30) per case. Ayush-64 has poor antimalarial activity, and it is also not suitable in the treatment of *P. falciparum*, which causes 50% of cases and most malaria deaths. In *P. vivax*, the cure rate is 50% with 17% recrudescence; Ayush-64 has not been tested for its action on the hypnozoites. Clinical trials also revealed problems in compliance, and some cases became serious and had to be rushed to the hospital. The drug was obviously unsuitable for *P. falciparum* and drug-resistant malaria. The killing action on gametocytes is doubtful, as there is no experimental proof of its gametocytocidal activity. Clearly chloroquine has none of these disadvantages, with 100% cure in *P. vivax* and *P. falciparum* infections susceptible to chloroquine, and it kills gametocytes of *P. vivax*. Primaquine is given as an antirelapse drug in *P. vivax* and as a gametocytocidal drug in *P. falciparum*. In both cases it has no value in relieving the symptoms of clinical malaria.

5.5 SCOPE OF HERBAL DRUGS IN MALARIA CHEMOTHERAPY

Malaria has a history of the use of plant-based drugs that have saved humankind from many disasters. The natives of Peru discovered the first antimalarial compound, cinchona bark, and the bark was used in the treatment of intermittent fevers in the 16th century. In 1920 Pelletier and Caventou working in Paris isolated quinine from cinchona bark. Quinine is still one of the best medicines for the treatment of malaria (see Chapter 2). The second antimalarial drug is Qinghao (*Artemisia annua*), used in the treatment of fevers in China (see Chapter 3). The WHO supported research on *A. annua* and the active principle artemisinin was identified. Artemisinin formulations are used worldwide in the treatment of malaria, particularly for severe and drug-resistant cases (WHO, 1998). Both quinine and artemisinin are rapid-acting drugs with a short half-life, and therefore resistance against these drugs would be slow — there is no reported resistance against artemisinin compounds.

In India, *Azadirachta indica* (neem) is widely used in the treatment of diseases, including malaria (Sharma, 1993; see Chapter 6). It is noteworthy to mention that there are many important leads for antimalarial drug discovery; Guru et al. (1996) have reviewed the literature on the herbal remedies in the treatment of parasitic diseases and listed 81 plants with antimalarial activity. Sharma and Sharma (1998) reviewed the plants showing antiplasmodial activity from crude extracts to isolated compounds and listed the compounds in three major groups: alkaloids, terpines and quassinoids, and aromatic and miscellaneous compounds. They reported *in vitro* and *in vivo* activity of 231 crude plant extracts showing antiplasmodial activity against different species of plasmodia. The authors provide a list of alkaloids derived from 48 plants, terpines and quassinoids derived from 40 plants, and aromatic and miscellaneous compounds derived from 31 plants showing dose-related effects *in vivo* on *P. berghei* and *in vitro* on *P. falciparum*. Application of modern scientific techniques in herbal drug development holds the key to success in new compounds with antimalarial activity.

ACKNOWLEDGMENTS

I thank the Central Council for Research in Ayurveda and Siddha, Ministry of Health and Family Welfare, Government of India, New Delhi, for the consultation of its booklet titled *Ayush-64: A New Anti-Malarial Herbal Compound*, which was the main source of information. I acknowledge the kind assistance of and review of the manuscript by Dr. Neena Valecha, Malaria Research Centre, New Delhi, and Dr. N.K. Upadhyay, National Research Development Council, New Delhi, for useful discussions.

REFERENCES

CCRAS. (1987). *Ayush-64: A New Anti-Malarial Herbal Compound*. Central Council for Research in Ayurveda and Siddha, Ministry of Health and Family Welfare, Government of India, New Delhi, pp. 1–128.

Guru, P.Y., Singh, S.N., Chatterjee, R.K., Dhawan, B.N., and Kamboj, V.P. (1996). Traditional remedies in the management of parasitic diseases. In *Proceedings of the National Academy of Sciences, India, Section B — Biological Sciences*, Vol. 66, Sharma, V.P. and Kamboj, V.P., Eds. National Academy of Sciences, Allahabad, India, pp. 245–273 (special issue on parasitology).

Kazim, M., Puri, S.K., Dutta, G.P., and Narasimham, M.V.V.L. (1991). Evaluation of Ayush-64 for blood schizontocidal activity against rodent and simian malaria parasites. *Indian J. Malariol.*, 28, 255–258.

Sharma, V.P. (1993). Malaria control. In *Neem Research and Development*, Publication 3, Randhawa, N.S. and Parmar, B.S., Eds. Society of Pesticide Science, New Delhi, India, pp. 235–241.

Sharma, V.P. (1998). Fighting malaria in India. *Curr. Sci.*, 75, 1127–1140.

Sharma, V.P. (1999). Current scenario of malaria in India. *Parassitologia*, 41, 349–353.

Sharma, K.D., Kapoor, M.L., Vaidya, S.P., and Sharma, L.K. (1981). A clinical trial of "Ayush-64" (a coded antimalarial medicine) in cases of malaria. *J. Res. Ayurvedic Siddha*, 2, 309–326.

Sharma, P. and Sharma J.D. (1998). Plants showing antiplasmodial activity—from crude extracts to isolated compounds. *Indian J. Malariol.*, 35, 57–110.

Valecha, N., Devi, C.U., Joshi, H., Shahi, V.K., Sharma, V.P., and Lal, S. (2000). Comparative efficacy of Ayush-64 vs. chloroquine in vivax malaria. *Curr. Sci.*, 78, 1120–1122.

WHO. (1998). The Use of Artemisinin & Its Derivatives as Anti-Malarial Drugs, WHO/MAL/98.1086. Report of a joint CTD/DMP/TDR Informal Consultation, Geneva, June 10–12, 1998. Malaria Unit, Division of Control of Tropical Disease, World Health Organization, Geneva, pp. 1–33.

6 Neem (*Azadirachta indica*)

Merlin Willcox and Joanne Chamberlain

CONTENTS

6.1 INTRODUCTION

Azadirachta indica A. Juss. is a tree of the Mahogany family (Meliaceae) and is known by different names around the world. A synonymous Latin name is *Melia azadirachta* Linn. *Melia* is the Greek name for "ash" — and the name was given by Linnaeus because of the resemblance of the leaves to those of an ash tree.* In India, the tree is called neem (from the Sanskrit *nimba*, synonymous

---

* This is not to be confused with *Melia azedarach*, a different West Asian tree (Persian lilac).

FIGURE 6.1 Neem tree growing in the grounds of the Madurai temple, Tamil Nadu, India. (Copyright 2001, Gerry Bodeker.)

with *arishta*, which means "relieving sickness" — see Figure 6.1). The Spanish and Portuguese called it margosa, and in Swahili in East Africa it is known as mwarobaini, which means "of 40," reflecting the popular belief that the plant can cure 40 diseases (see Figure 6.2). The Latin name is derived from the Persian "azad darakht i hindi," which means "free tree of India" (National Research Council, 1992).

Neem is thought to have originated in Assam and Myanmar where it is common throughout the central dry zone and the Siwalik hills (National Research Council, 1992). However, the exact origin is uncertain and some authors suggest it is native to the dry forest of South and Southeast Asia, including Pakistan, Sri Lanka, Thailand, Malaysia, and Indonesia (Ahmed and Grainge, 1985). In these countries, neem has long been used in many ways, for example, twigs as toothbrushes, leaf teas as a tonic, and fresh leaves as an insect repellent. In the 19th century, neem was introduced by Indian immigrants to the Caribbean (Trinidad and Tobago, Jamaica, and Barbados), South America, and the South Pacific (National Research Council, 1992). The cultivation of neem spread to Africa in the early 20th century. The British governor of Ghana from 1919 to 1927 introduced neem seeds from India to Ghana, and from there it was planted over much of western Africa (National Research Council, 1992). Neem was introduced to Sudan in 1916, probably by a colonial forester, and then to several East African countries. Now it is well established in more than 30 countries worldwide.

6.2 PRECLINICAL STUDIES

6.2.1 History, Ethnobotany, and Ethnopharmacology

Neem is very widely used as a medicinal plant. A survey conducted in Niger found that the predominant use of neem was for medicinal purposes (Van der Burg and Haasane, 1990). Similarly,

Figure 6.2 Kikatiti Neem Project, near Arusha, Tanzania: board listing over 40 uses of Neem. (Copyright 1999, Merlin Willcox.)

a recent study by GTZ (Gesellschaft für Technische Zusammenarbeit) is the German government's overseas development agency. The same survey indicated that leaves were the most commonly used part of the tree for medicine. Childs et al. (1999) report that 87% of organisations responding to a postal questionnaire said that neem was used as a medicine by at least some farmers in the areas in which they worked. This figure was 93% where neem is native and 84% where it is grown as an exotic. One of the most common uses of neem is to treat malarial fevers, where a tea made with the leaves is drunk.

The first recorded use of neem as an antimalarial was in 1500 B.C., in the Ayurvedic text by Charaka. Table 6.1 lists some of the many classical Ayurvedic formulations containing neem for the treatment of vismama jvara (which corresponds to malarial fevers — see Chapter 13). There are many more formulations containing neem for the treatment of jvara (nonspecific fever), too numerous to list here. Furthermore, its use was noted by several European explorers. Sonnerat (1782), despite a condescending attitude toward Indian medicine, records that "for local fevers, they successfully use decoctions of powdered neem roots, which supplement our Cinchona; they claim that it is the same root, and that when fresh, it is better than that sent from Europe, which in the course of its journey has lost part of its strength."

The use of neem was also adopted by British colonists in India. A recipe for *Tinctura Azadirachta indicae* is given in the *Indian and Colonial Addendum to the British Pharmacopoeia* (General Council of Medical Education, 1901). It was prepared by macerating 100 g of grated neem bark in 1 l of 45% alcohol and was given in doses of 1.8 to 3.6 ml. This was used by some colonial doctors in India as an alternative to quinine (Deare et al., 1916). Waring (1897) writes:

>Ním bark is a valuable astringent tonic, and when dried and reduced to powder may be given in doses of one drachm [3.9 g] 3 or 4 times a day. A better form is a decoction prepared by boiling two ounces [56 g] of the bruised inner layer of the bark in a pint and a half [1.13 l] of water for a quarter of an hour, and straining whilst hot; of this, when cold, the dose is 2–3 oz [56–85 ml]. It, as well as the Powdered Bark, is a remedy of considerable value in *Ague or Intermittent Fever*; and in these cases it should be given every second hour previous to the time at which the attack is expected to return. It is chiefly adapted for mild, uncomplicated cases, especially in natives.

TABLE 6.1
Classical Ayurvedic Formulations for the Treatment of Visama Jvara (Malaria-Type Fever) Containing Neem

Formulation Name	Ingredients	Neem Part Used	Preparation	Ayurvedic Text	Date
Nimba Patola Kasaya	Nimba (*Azadirachta indica*), Patola (*Trichosanthes cucumerina*), Triphala (a group of three plants — *Terminalia chebula, Terminalia bellirica, Emblica officinalis*), Mrdvika (*Vitis vinifera*), Musta (*Cyperus rotundus*), Vatsaka (*Holorrhena pubescens*)	Stem bark	Decoction	Caraka samhita, Cikitsasthana, 3/201	1500 B.C.–400 A.D.
				Astanga samgraha, Cikitsa sthana, 1/61	500 A.D.
				Astanga hrdaya, Cikitsasthana, 1/49	600 A.D.
				Bhavaprakasa, Madhyama khanda, 8.1/747–749	1550 A.D.
Amrta Nimba Kvatha	Guduci (*Tinospora coridfolia*), Nimba, Dhatri (*Emblica officinalis*)	Stem bark	Decoction	Susruta Samhita, U, 39/213	1500 B.C.–500 A.D.
Pancatiktaka Ghrta	Vrsa (*Adhatoda vasica*), Nimba, Amrta (*Tinospora cordifolia*), Vyaghri (*Solanum virginianum*), Patola (*Trichosanthes cucumerina*)	Stem bark	Ghee preparation	Sarngadhara samhita, Madhyamakhanda, 9/91	1300 A.D.

Source: Database on Clinically Important Plants of Ayurveda — Version 1, Foundation for Revitalisation of Local Health Traditions, Bangalore, India.

Neem is still widely used in various parts of India as an antimalarial, as a dried powder most commonly of its leaves and bark (see Table 6.2). Various mixtures are used. Nadkarni (1976) reports that the bark decoction is mixed with a little black pepper (*Piper nigrum*) and chiretta (*Swertia chirata*) as a popular remedy for fevers. Singh Huidrom (1997) reports that neem bark powder is mixed with the plant juice of *Asparagus racemosus* and the leaf juice of *Oldenlandia corymbosa*, and honey, and this is taken thrice daily. Singh and Anwar Ali (1994), in their survey of five Indian states, report that neem leaves, stem bark, young roots, and fruits are ground and mixed with dried ginger and the dried powder of fruits of *Terminalia bellerica* (Gaertn) Roxb, *Terminalia chebula* Retz, and *Emblica officinalis* Gaertn. This mixture is taken with tepid water.

The use of neem also spread to other Asian countries. In Malaysia, neem sap was boiled together with young red leaves of *Vitis* spp. and resinous wood of *Aquilaria* spp.; the patient was advised to bathe in this mixture and to drink it for 3 days, in order to treat a "continued fever which is suppressed" (Gimlette and Burkill, 1930). In Cambodia, a decoction and a tincture of neem bark were used by traditional healers to treat malaria, and subsequently at colonial clinics in the 1920s, where they were sometimes used as a substitute for quinine (Crevost and Pételot, 1929). The recommended daily dose was 30 to 40 g of the yellow internal bark, powdered, macerated in 300 ml of cold water for 2 hours, filtered, and taken in three or four divided doses (Menaut, 1929). The

TABLE 6.2
Use of Neem in India for Malaria and Fever

Part Used	Preparation	Traditional Use	State	Reference
LF	Drp, 5 g bd with honey	Antimalarial	Orissa	Aminuddin and Subhan Khan, 1993
LF, SB, RT, FR	Mix, inf	Antipyretic	Bihar, Madhya Pradesh, Orissa, Rajasthan, Uttar Pradesh	Singh and Anwar Ali, 1994
SB powder	Mix, tds	Antipyretic	Manipur	Singh Huidrom, 1997
NR	NR	Antipyretic	Andhra Pradesh	Vedavathy and Rao, 1995
Fresh LF	Inf	Antimalarial	Rajasthan	Nargas and Trivedi, 1999
Fresh LF	Dec	Antimalarial	Uttar Pradesh	Singh et al., 2002
Dried LF	Fumes	Prophylactic	Rajasthan	Nargas and Trivedi, 1999

Note: LF = leaf; SB = stem bark; RT = root; FR = fruit; NR = not recorded; dec = decoction; drp = dry powder; inf = infusion; mix = mixture with other herbs; bd = twice a day; tds = three times a day.

Dutch also noticed the use of neem leaves for malaria in the East Indies (Heyne, 1927). In Sri Lanka, neem seed oil is one of ten herbal components of the remedy pastel tambuna, which is consumed two or three times daily for malarial fevers; however, this remedy has fallen into disuse following the introduction of Western antimalarials (Silva, 1991).

In Africa (Table 6.3), neem has also been integrated into local medicine. The plant is widely known and in common use by ordinary people, not just by traditional healers (Bodeker et al., 2001; Bugmann, 2000; Burford et al., 2000; Fourn et al., 2001; Nyamongo, 2002; Silva, 1991). Julvez et al. (1995) interviewed 114 families in urban and rural Niger, of whom 56% reported the use of neem decoctions for the treatment of malaria. Hirt and Lindsey (2000) recommend an infusion made by pouring 1 l of boiling water over 15 g of neem leaves and drinking 250 ml of this four times a day. However, the hot water and alcoholic extracts are used mainly in urban areas (Iwu, 1993), whereas the traditional medical practice is to make a cold-water extract. The leaves (and sometimes bark and roots) are left to macerate in cold water for up to several weeks. This brew is then dispensed cold, in a dose of one glass (c. 100 ml) two or three times daily for 2 to 3 days. Tablets made from neem bark and leaves have also been developed (Ibrahim et al., 1998).

Although complex mixtures of neem with other plants are not as common in Africa as in India, Ahorlu et al. (1997) report that other ingredients are sometimes added, such as lemon, lime, ginger, and camphor, whereas Hirt and Lindsey (2000) recommend drinking neem with lemon grass tea. Commonly, the neem leaves are boiled in water, and the patient sits over the bowl, covered with a blanket, inhaling the vapours until he is sweating freely; afterwards the patient drinks some of the decoction. Sometimes the patient may also bathe in the same decoction (Agyepong, 1992). This decoction can also be drunk prophylactically once a month (Iwu, 1993).

Neem is also traditionally used as an insect repellent. In Sri Lanka, 69% of households burn neem seeds and leaves in clay pots to repel mosquitoes in the evening (Konradsen et al., 1997). In India, it is believed that the fumes of dried neem leaves help to prevent malarial fevers (Nargas and Trivedi, 1999); this may be through their mosquito repellent action. In Kenya, Snow et al. (1992) found that 3% of households burned neem leaves to repel mosquitoes. Although there is no record of neem being used to kill mosquito larvae, it is used traditionally to protect stored cereals, and even books, from insect pests (National Research Council, 1992).

TABLE 6.3

Use of Neem in Africa for Malaria and Fever

Part Used	Preparation	Traditional Use	Country	Reference
LF	Inf	Antipyretic	Benin	Fourn et al., 2001
LF	Dec, inh, plt, bth, msg	Antimalarial	Burkina Faso	Bugmann, 2000
LF	Dec, inf	Antimalarial	Burkina Faso	Gbary et al., 1996
LF, RT	Dec, bth, mix	Antimalarial	Burkina Faso	Okrah et al., 2002
NR	NR	Antimalarial	Dem Rep Congo	Mulangu Binzambal, 1999
LF	Dec, inh	Antimalarial	Ghana	Agyepong, 1992
LF	Dec, inh, bth	Antimalarial	Ghana	Ahorlu et al., 1997
NR	NR	Antimalarial	Kenya	Mwenesi, 1994
LF, RT	NR	Antimalarial	Kenya	Munguti, 2000
LF, SB	Dec, bth	Antimalarial	Kenya	Nyamongo, 2002
LF, ST	Dec	Antimalarial, adjuvant	Madagascar	Ralantonirina, 1993
LF	NR	Antimalarial	Mali	Djimde et al., 1998
SB, RT	NR	Antimalarial	Nigeria	Iwu, 1982
LF	Dec, bth, mix	Antimalarial	Nigeria	Bhat et al., 1990; Fawole and Onadeko, 2001
LF, SB	NR	Antipyretic	Nigeria	Etkin and Ross, 1991
LF	Dec bd–tds, bth	Antipyretic	Niger	Julvez et al., 1995
LF	Dec, inf	Antimalarial	Sudan	El-Kamali and El-Khalifa, 1997
NR	NR	Antipyretic	Tanzania	Hausmann Muela et al., 2002
LF, RT	Dec, inh	Antimalarial	Tanzania	Burford et al., 2000
LF, SB, RT	NR	Antimalarial	Tanzania	Dery et al., 1999
LF, SB, RT	Dec, inf	Antimalarial	Tanzania	Gessler, 1995

Note: LF = leaf; SB = stem bark; SD = seeds; RT = root; FR = fruit; NR = not recorded; bth = bath; dec = decoction; inf = infusion; inh = inhalation; plt= poultice; mix = mixture with other herbs; msg = massage; bd = twice a day; tds = three times a day.

6.2.2 Insect Repellent and Insecticidal Activity

Sharma et al. (1993b) have demonstrated that heated cardboard mats impregnated with neem oil are as effective as chemically treated mats in repelling mosquitoes. The cost is estimated at U.S. $0.50 per room per year, compared to U.S. $25 per room per year for chemically treated mats. It has also been shown that when applied topically at 2% strength in a base of coconut oil, neem oil provides 100% protection against biting by all *Anopheles* species during a 12-hour period (Sharma et al., 1993a). In Kenya, neem oil is sold as an insect repellent and trials with neem oil-impregnated mosquito nets have been conducted (Knols, 2000). Neem oil is also used as a fuel, and when used in lamps, the oil has the advantage of repelling mosquitoes (Parmar, 1993). Further details about the repellent properties of neem are reviewed in Chapter 22.

Dhar et al. (1996) showed that neem oil and neem seed kernels inhibit the formation of mosquito eggs and oviposition in *An stephensi* and *An culicifacies*. A proprietary neem seed kernel extract with an azadirachtin content of 40% has been shown to inhibit moulting and kill second, third, and fourth instar larvae of *Aedes aegypti* (Boshitz and Grunewald, 1994). The lethal concentration 50% (LC_{50}) increases with the age of the larvae: it is 3.3 ppm for second instar, 4.8 ppm for third instar, and 8.4 ppm for fourth instar larvae. However, aqueous neem seed kernel extracts are less effective (Zebitz, 1984, 1986).

Interestingly, a waste by-product from neem oil refining, known as neem oil extractive, is also effective as a larvicide against *Culex fatigans* (Attri and Prasad, 1980). At 0.005% concentration, this product prevented 100% of larvae from emerging as adults and was nontoxic to fish and tadpoles (although concentrations of 0.01 to 0.02% were toxic to these animals).

6.2.3 ANTIMALARIAL ACTIVITY OF CRUDE EXTRACTS

6.2.3.1 *In Vitro* Antiplasmodial Activity

There have been many studies to determine the effectiveness of various neem extracts on *Plasmodium falciparum* in culture. Their results are summarised in Table 6.4, ranked according to the inhibitory concentration 50 (IC_{50}), the concentration required to inhibit growth of 50% of parasites. Some studies are not included in the table because an IC_{50} value was not derived. Udeinya (1993) found that an aqueous decoction of oven-dried leaves was ineffective against the ITG2F6 clone of *P. falciparum*, whereas a water–acetone extract completely inhibited parasite development at 20 µg/ml. Badam et al. (1987) also showed that ethanol extracts of leaves and seeds were effective *in vitro*, but used a different outcome measure (MED_{50} — the minimum dose at which 100% parasite clearance was observed in at least 50% of cultures). The MED_{50} of the ethanol leaf extract was 25 to 50 µg/ml for chloroquine-sensitive strains of *P. falciparum* and 75 µg/ml for a chloroquine-resistant strain.

There is a wide range of IC_{50} values for neem extracts, even prepared in the same way. Many factors may affect the IC_{50} value obtained. First, the method used to detect the effect on parasite growth may produce different results for exactly the same extract. Dhar et al. (1998) found that the IC_{50} was 4.6 times greater when parasite growth was measured using the radioactive hypoxanthine incorporation method than when it was measured using visual parasite counts on blood films. Benoit et al. (1996) found that the IC_{50} was lower if the parasites were incubated with the extract for 72 hours instead of the standard 24 hours.

Secondly, the freshness of the plant material is important. The most potent extracts were those prepared from fresh leaves. MacKinnon et al. (1997) obtained a 20-fold lesser IC_{50} using an ethanolic extract of fresh leaves than that obtained by Bray et al. (1990) using dried or frozen leaves. Thus active constituents could have been lost in the process of drying and freezing. However, Udeinya (1993) suggested that the activity of the water–acetone extract was not affected by sterilising in the autoclave at 125°C, although this was not proven by comparison with a nonautoclaved sample. Thirdly, the time and place of collection may also affect the yield of active constituents. Although most papers recorded the country of collection of the plant material, none recorded the exact site or season.

Fourthly, the method of extraction is important. Rochanakij et al. (1985) and Bray et al. (1990) found that ethanol extracts were much more effective than aqueous extracts. Benoit et al. (1996) report good IC_{50} values for aqueous decoctions and infusions prepared from dried stems and leaves, but this has not been replicated by other researchers.

Finally, the IC_{50} can differ according to the strain of *P. falciparum* used. Ofullah et al. (1995) report a 4.6-fold difference in the efficacy of the same preparation against four Kenyan strains of *P. falciparum*. However, other authors have found little difference in the efficacy of neem extracts against chloroquine-sensitive and chloroquine-resistant clones of *P. falciparum* (Badam et al., 1987; Benoit et al., 1996; MacKinnon et al., 1997).

Almost all *in vitro* experiments have concentrated on the effect of neem extracts on asexual, erythrocytic stages of *P. falciparum*. However, Dhar et al. (1998) showed that purified neem seed oil also has a gametocytocidal activity at 250 µg/ml, both on young and on mature gametocytes. This differs from the effect of the two known gametocytocidal drugs, artemisinin and primaquine, which are effective only against the immature stages. Thus neem seed oil has the potential to prevent transmission of malaria.

TABLE 6.4
IC$_{50}$ Values of Neem Extracts against *In Vitro* Cultures of *P. falciparum*

IC$_{50}$ (μg/ml)	Method	Plant Part	Collection Site	Extract	Falciparum Clone[a]	Reference
1.7	H	Dried leaves	Sudan	Methanol	Dd2	El Tahir et al., 1999
2.5	H	Fresh leaves	Togo	Ethanol	D6, W2	MacKinnon et al., 1997
5	V	Fresh leaves	Thailand	Ethanol decoction	K1	Rochanakij et al., 1985
5.8	H	Dried leaves	Sudan	Methanol	3D7	El Tahir et al., 1999
6.25	H	Stem and leaves, dried	Côte d'Ivoire	Aqueous decoction	F32	Benoit et al., 1996
7.29	H	Stem and leaves, dried	Côte d'Ivoire	Aqueous infusion and decoction	FcB1	Benoit et al., 1996
8.5	H	Dried stem bark	Sudan	Methanol	3D7	El Tahir et al., 1999
12.5	H	Stem and leaves, dried	Côte d'Ivoire	Aqueous infusion	F32	Benoit et al., 1996
16	V	Seeds	India	Centrifuged oil	NF54	Dhar et al., 1998
40	H	Dried stem bark	Sudan	Methanol	Dd2	El Tahir et al., 1999
50	H	Dried leaves	Nigeria	Ethanol	K1	Bray et al., 1990
50	H	Dried stem bark	Nigeria	Ethanol	K1	Bray et al., 1990
50	H	Frozen leaves	Ghana	Chloroform	K1	Bray et al., 1990
50–99	H	Leaves	Tanzania	Dichloromethane	K1	Weenen et al., 1990
75	H	Seeds	India	Centrifuged oil	NF54	Dhar et al., 1998
80	H	Fresh leaves	Kenya	Aqueous decoction	K67, S158	Kofi-Tsekpo et al., 1989
115	V	Fresh leaves	Thailand	Aqueous decoction	K1	Rochanakij et al., 1985
135	H	Leaves	Kenya	Aqueous	S104	Ofulla et al., 1995
241	H	Leaves	Kenya	Aqueous	Kil.9	Ofulla et al., 1995
100–499	H	Leaves	Tanzania	Petroleum ether	K1	Weenen et al., 1990
399	H	Leaves	Kenya	Aqueous	Ent 30	Ofulla et al., 1995
500	H	Frozen leaves	Ghana	Petroleum ether	K1	Bray et al., 1990
500	H	Frozen leaves	Ghana	Methanol	K1	Bray et al., 1990
625	H	Leaves	Kenya	Aqueous	Ent 22	Ofulla et al., 1995
NA	H	Frozen leaves	Ghana	Aqueous infusion or decoction	K1	Bray et al., 1990

Note: V = visual; H = hypoxanthine incorporation.

[a] Chloroquine-sensitive clones are 3D7, D6, F32, NF54, Ent 22, and Kil 9. Chloroquine-resistant clones are Dd2, FcB1, W2, Ent 30, and S104. A multi-drug-resistant clone is K1.

Etkin (1981) demonstrated that a cold leaf macerate had an oxidant effect *in vitro*. After a haemoglobin suspension had been incubated with this extract for 2 hours, 42% had been oxidised to methaemoglobin (cf. 9 and 7% for bark and root extracts, respectively). When reduced glutathione was incubated with the same extract for 2 hours *in vitro*, 95% was oxidised, whereas only 30 and 40% were oxidised by the bark and root extracts, respectively. This oxidising activity may be one mechanism of the antimalarial activity of *A. indica* (see Chapter 15).

6.2.3.2 *In Vivo* Antiplasmodial Activity

Tests of neem extracts in malaria-infected mice and chicks are uniformly disappointing (see Table 6.5). Ekanem (1978) conducted the most rigorous test and calculated that the ED_{50} (effective dose that inhibits 50% of parasite growth) was 1164 mg/kg, whereas the ED_{90} was calculated at 23,782 mg/kg. Extrapolating this to humans, the implication would be that a 60-kg adult would need to ingest almost 1.5 l of neem decoction daily in order to cure malaria. However, African patients have been observed to drink similar quantities of medicinal tea. Other studies did not calculate an ED_{50}, but few found any significant effect even at relatively high doses (Table 6.5; Tella, 1976). Makinde and Obih (1985) found that a mixture of plants with neem, mimicking a traditional recipe, had no effect on *Plasmodium berghei* in mice.

However, there are a number of caveats to consider. First, mice may not metabolise neem extracts in the same way as humans. Human metabolism may convert inactive compounds into active metabolites. Thus, trimethoprim is ineffective against murine malaria, but effective in humans.

TABLE 6.5
In Vivo **Efficacy of Neem Extracts**

Efficacy	Plant Part	Collection Site	Extract	Route	*Plasmodium* Species	Reference
No effect at 1100 mg/kg	Bark	NR	Aqueous	Oral	*P. gallinaceum*	Spencer et al., 1947
No effect up to 250 g/kg/day	Leaf	Nigeria	Decoction	Oral	*P. berghei*	Tella, 1977
No effect up to 8.5 g/kg/day	Leaf and bark	Nigeria	Aqueous decoction	Oral	*P. berghei and P. gallinaceum*	Odetola and Bassir, 1986
No effect up to 746 mg/kg/day	Fresh leaves	Thailand	Ethanol and aqueous decoctions	Oral and s-c	*P. berghei*	Rochanakij et al.,1985
41% suppression (dose unclear)	Leaves	Nigeria	Aqueous decoction	s-c	*P. berghei berghei*	Obih and Makinde, 1985
27% suppression at 15,000 mg/kg	Fresh leaves	Nigeria	Aqueous decoction	Oral	*P. yoelii nigeriensis*	Agomo et al., 1992
17% suppression at 500 mg/kg	Fresh leaves	Nigeria	Aqueous infusion	Oral	*P. berghei*	Abatan and Makinde, 1986
65% suppression at 500 mg/kg	Dried bark	India	Aqueous decoction	i-p	*P. berghei*	Devi et al., 2001
68% suppression at 800 mg/kg	Dried bark	Nigeria	Aqueous suspension	Oral	*P. yoelii nigeriensis*	Isah et al., 2003
79.6% suppression at 800 mg/kg	Dried leaves	Nigeria	Aqueous suspension	Oral	*P. yoelii nigeriensis*	Isah et al., 2003
95% suppression at 200 mg/kg	Fresh leaves	Nigeria	Freeze-dried concentrate of aqueous decoction	Oral	*P. yoelii nigeriensis*	Obaseki and Jegede-Fadunsin, 1986
ED_{50} =1164 mg/kg	Leaves	Nigeria	Aqueous decoction	s-c	*P. berghei*	Ekanem, 1978

Note: NRF = not recorded; s-c = subcutaneous; i-p = intraperitoneal.

Secondly, *P. berghei* may have different properties and sensitivities from *P. falciparum*. Thirdly, the plant extracts are only administered to mice for 4 days. Active constituents may have a different pharmacokinetic profile from chloroquine and may require more frequent doses for a longer period to maintain therapeutic plasma concentrations.

Obaseki and Jegede-Fadunsin (1986) showed that a concentrated aqueous extract (prepared by a decoction of 250 g of leaves in 300 ml of distilled water, filtration, concentration in a rotatory evaporator, and freeze-drying) was effective at a dose of 200 mg/kg. Interestingly, the efficacy peaked at this dose and declined at higher doses. Unfortunately, it seems that no attempt has been made to explain or replicate this finding.

Etkin and Ross (1991) used a different protocol, whereby infected erythrocytes were incubated for 2 hours with *A. indica* leaf extract, before intraperitoneal injection into mice. This resulted in complete absence of parasites at day 7, compared to 87% parasitaemia in controls.

The potential prophylactic use of neem extracts has also been investigated in mice. Obih and Makinde (1985) gave neem extracts to mice for 3 days, before then injecting them on the fourth day with *P. berghei*-infected erythrocytes. The stock solution suppressed 22% of parasite growth. Abatan and Makinde (1986) repeated a similar experiment, this time giving the extract orally for only 1 day. They found no prophylactic effect using this regimen. However, the experimental mode of transmission (intraperitoneal injection of infected erythrocytes) does not accurately reflect reality (small inoculum of sporozoites that first infect the liver).

6.2.3.3 Immunomodulatory Properties

The effects of neem on the immune system are complex and have not been fully elucidated; some evidence points to an immunostimulant effect, while other evidence suggests it can also act as an anti-inflammatory. The latter may be partly explained by inhibition of prostaglandin synthetase (Van der Nat et al., 1991; Okpanyi and Ezeukwu, 1981; Obaseki et al., 1985). However, Kroes et al. (1993) showed that in a classical Ayurvedic preparation for gout, nimba arishta, the anti-inflammatory activity is almost entirely attributable to ingredients other than neem.

Here we review the evidence for the immunostimulant effects of neem, and to what extent this may explain its perceived efficacy in the treatment of malaria. There is a small amount of evidence that neem may boost both the humoral and the cell-mediated immune responses to malarial infection, but it comes mostly from mice, which may have different immune responses from man, and in which neem has not been shown effective against malaria.

6.2.3.3.1 Humoral Immunity

Patients who produce specific antimalarial immunoglobulins M (IgM) and G (IgG) present with lower *p. falciparum* parasitaemia and are less likely to have recrudescent infections (Mayxay et al., 2001). IgG also protects newborns against malaria, following passive transfer across the placenta (Marsh, 1993). Ray et al. (1996) showed that mice fed on *A. indica* leaf extracts (in peanut oil) produced more IgM and IgG than controls in response to challenge with ovalbumin, and this effect was dose related. In rats and mice, production of antibodies in response to challenge with sheep red blood cells is increased after intraperitoneal injection of 100 mg/kg of aqueous neem bark or leaf extract, and immunosuppression due to stress was reduced (Sen et al., 1992; Njiro and Kofi-Tsekpo, 1999).

6.2.3.3.2 Cell-Mediated Immunity

Cytotoxic T cells and interferon-γ (IFN-γ) can eliminate intracellular plasmodia in hepatocytes and erythrocytes, and macrophages may also play a role (Plebanski and Hill, 2000). Neem seed oil injected intraperitoneally in mice induces the production of IFN-γ by spleen cells (Upadhyay et al., 1992). It also increases peritoneal leukocyte counts, enhances the phagocytic activity of macrophages, enhances the lymphocyte proliferative response to *in vitro* mitogen challenge, and enhances the cellular immune response to tetanus toxoid.

However, *in vitro*, a neem bark decoction reduced the phagocytic activity of neutrophils, but stimulated the production of migration inhibition factor (MIF) by human lymphocytes (Van der Nat et al., 1987). MIF attaches macrophages and monocytes to their site of action. Indeed, in mice, Ray et al. (1996) found that macrophage migration is inhibited by 100 mg/kg of neem leaf extract; however, in contradiction to this finding, Sen et al. (1992) found that aqueous neem leaf extract promotes leukocyte migration in rats.

6.2.3.4 Antipyretic Activity

The perceived efficacy of neem for the treatment of malaria may in part be due to an antipyretic effect. In rabbits made hyperpyretic by injection of lipopolysaccharide, a methanol extract of fresh neem leaves and bark given orally at a dose of 400 mg/kg reduced the temperature index by 15.7%, compared to 24% with aspirin and 30.6% with indomethacin (Okpanyi and Ezeukwu, 1981). Khattak et al. (1985) similarly showed that hexane, chloroform, and water extracts of neem leaves and twigs administered by gastric tube at 150 mg/kg reduced temperature in rabbits with a yeast-induced pyrexia. The hexane extract was as effective as aspirin. However, aqueous decoctions of neem bark and leaves have no significant antipyretic effect (Gujral et al., 1955; Odetola and Bassir, 1986). Neem oil and nimbidin have no antipyretic activity, although they prevented a secondary rise in rats injected subcutaneously with yeast. However, nimbidol is an effective antipyretic (Murthy and Sirsi, 1958).

6.2.4 CHEMICAL CONSTITUENTS AND THEIR ACTIVITY

A. indica contains at least 2 diterpenes, 7 triterpenes/steroids, 55 limonoids, and 16 polyphenolics, not to mention many proteins, polysaccharides, sulfides, fatty acids, and other miscellaneous compounds (Van der Nat et al., 1991). Very few of these have been investigated for their antimalarial or cytotoxic properties (see Table 6.6 and Table 6.7).

Azadirachtin is a powerful insect repellent and antifeedant compound (see Chapter 24). Its LC_{50} for *Aedes togoi* is 0.2 ppm, and this is not much different from a methanol extract, suggesting that there may be other related substances with a synergistic effect (Zebitz, 1986). There is no direct toxicity, only delayed lethality during development, due to blocking of the insect hormone ecdysone, which controls growth and moulting. Azadirachtin is mildly cytotoxic to *Drosophila* cells but not to mammalian cells (Cohen et al., 1996). It was found to have only mild effects on

TABLE 6.6
IC_{50} of Neem Compounds against *P. falciparum*

Alkaloid	IC_{50} (μmol/l)	Falciparum Clone	Reference
Gedunin	0.020	W2	MacKinnon et al., 1997
Gedunin	0.039	D6	MacKinnon et al., 1997
Gedunin	0.72	K1	Bray et al., 1990
Gedunin	1.7	FCMSU$_1$	Khalid et al., 1986
Nimbinin	0.77	K1	Bray et al., 1990
Nimbolide	1.74	K1	Bray et al., 1990
Nimbolide	2.0	K1	Rochanakij et al., 1985
Nimbin	50	K1	Bray et al., 1990
Salannin	50	K1	Bray et al., 1990
Meldenin	>50	K1	Bray et al., 1990

TABLE 6.7
Cytotoxic Activity of Neem Compounds

Alkaloid	Cell Line	IC_{50} $(\mu mol/l)_a$	Reference
Nimbolide	P-388	0.14	Kigodi et al., 1989
Nimbolide	KB	0.54	Kigodi et al., 1989
Nimbolide	HT-1080	0.67	Kigodi et al., 1989
Nimbolide	BC-1	0.84	Kigodi et al., 1989
Nimbolide	COL-2	0.88	Kigodi et al., 1989
Nimbolide	LU-1	0.90	Kigodi et al., 1989
Nimbolide	MEL-2	1.14	Kigodi et al., 1989
Nimbolide	143B.TK⁻	4.3	Cohen et al., 1996
Nimbolide	RAW 264.7	5.0	Cohen et al., 1996
Nimbolide	N1E-115	5.2	Cohen et al., 1996
Nimbolide	GPK	21.8	Bray et al., 1990
28-Deoxonimbolide	P-388	1.46	Kigodi et al., 1989
28-Deoxonimbolide	LU-1	1.86	Kigodi et al., 1989
28-Deoxonimbolide	HT-1080	2.30	Kigodi et al., 1989
28-Deoxonimbolide	KB	2.88	Kigodi et al., 1989
28-Deoxonimbolide	BC-1	3.63	Kigodi et al., 1989
28-Deoxonimbolide	COL-2	4.00	Kigodi et al., 1989
28-Deoxonimbolide	MEL-2	4.54	Kigodi et al., 1989
Gedunin	KB	4.8	MacKinnon et al., 1997
Gedunin	GPK	571	Bray et al., 1990
Nimbinin	GPK	15.3	Bray et al., 1990
Nimbinin	N1E-115, 143B.TK⁻	23–24	Cohen et al., 1996
Nimbinin	RAW 264.7	46	Cohen et al., 1996
Salannin	143B.TK⁻	89	Cohen et al., 1996
Salannin	N1E-115	133	Cohen et al., 1996
Nimbin	N1E-115, 143B.TK⁻	>200	Cohen et al., 1996
Deacetylnimbin	N1E-115, 143B.TK⁻	>200	Cohen et al., 1996
Azadirachtin	N1E-115, 143B.TK⁻	>200	Cohen et al., 1996

Note: BC-1 = breast cancer; COL-2 = colon cancer; GPK = guinea pig keratinocytes; KB = human epidermoid cancer; HT-1080 = human fibrosarcoma; LU-1 = lung cancer; MEL-2 = melanoma; N1E-115 = murine neuroblastoma cells; P-388 = human fibrosarcoma; RAW 264.7 = murine macrophages; 143B.TK⁻ = human osteosarcoma cells.

[a] In the interests of comparability, data quoted in μg/ml or other units in the original paper have been converted to μmol/l.

the electrophysiology of rat dorsal root ganglion neurons *in vitro*, causing a reversible reduction in the amplitude of the action potential and inhibition of voltage-activated K⁺ currents (Scott et al., 1999). Interestingly, azadirachtin blocks exflagellation (the development of the motile male gamete) of *P. berghei* (ED_{50} = 3.5 μ*M*) and of *P. falciparum* (fully effective at 100 μ*M*) *in vitro* (Jones et al., 1994).

Nimbolide was isolated by Rochanakij et al. (1985) and found to be active against chloroquine-resistant *P. falciparum in vitro*, but was inactive *in vivo* in *P. berghei*-infected mice. It also has strong cytotoxic effects on mammalian cells (Table 6.7). However, it is only toxic to mice when administered intravenously, with an LD_{50} of 24 mg/kg. When administered intragastrically, intra-muscularly, or subcutaneously, it is not toxic at doses up to 600 mg/kg (Glinsukon et al., 1986).

Gedunin, a limonoid similar to nimbolide, was isolated by Khalid et al. (1986, 1989) and was found to be active against *P. falciparum in vitro*. It was also cytotoxic to KB human epidermoid carcinoma cells. Gedunin was ineffective in treating mice infected with *P. berghei* (MacKinnon et al., 1997).

Nimbinin (otherwise known as epoxyazadiradione) has a strong *in vitro* activity against *P. falciparum*, with an IC_{50} of 0.77 μM. It also has cytotoxic effects, but at a much higher concentration: the IC_{50} is 23 to 24 μM (see Table 6.5 and Table 6.6).

Nimbin, and to a lesser extent nimbintin and nimbidinate, have a mild antipyretic activity. They prevent a yeast-induced temperature rise in rats, although they do not lower temperature to a normal level (Narayan, 1969). Nimbidin has antipyretic, hypoglycemic, and antiulcer properties and is nontoxic in doses up to 2 g/kg in mice and rats (Pillai et al., 1980; Pillai and Santhakumari, 1981, 1984a, 1984b).

All the above-mentioned compounds are limonoids. Iwu et al. (1986) isolated two flavonolglucosides (quercetin-3-rhamnoglucoside and quercetin-3-rhamnoside) as the main constituents of an aqueous extract prepared as follows: de-fatting with petroleum ether, methanol extraction of dry residue, and partition of methanol extract between water and chloroform. This aqueous extract was injected intraperitoneally into rats at a dose of 400 mg/kg daily for 4 days. After this time, the animals were sacrificed; the activity of the cytochrome C reductase in treated animals was reduced by 75%. Inhibition of this enzyme places oxidant stress on cells, to which parasites are especially susceptible, as they lack a pentose phosphate pathway and so cannot generate their own NADPH. Iwu et al. (1986) suggest that oxidative stress in infected cells caused by a lack of NADPH may explain the antimalarial activity of *A. indica*.

6.2.5 TOXICOLOGY

6.2.5.1 *In Vitro* Studies

The cytotoxicity of constituent compounds from neem has been reviewed above and in Table 6.7. The results vary widely between studies and according to the cell line tested. The cytotoxicity of nimbolide and nimbinin may seem of concern. However, crude neem extracts are much less cytotoxic (Table 6.8).

Genotoxicity was evaluated for an ethanolic neem extract developed as an insecticide (Margosan-O®). This was found to be nonmutagenic in five strains of *Salmonella typhimurium* (National Research Council, 1992). Similarly, Uwaifo (1984) found that nimbolide was not mutagenic in an Ames test on six strains of *S. typhimurium*.

Koga et al. (1987) found that neem seed oil inhibits the respiratory chain in mitochondria *in vitro*. This was more severe in the absence of carnitine, leading the authors to speculate that the toxicity of neem seed oil would be greatest in malnourished children.

TABLE 6.8
Cytotoxicity of Crude Neem Extracts

LD_{50} (μg/ml)	Plant Part	Collection Site	Extract	Cell Line	Reference
2000	Leaf	India	Ethanol	Chang liver	Badam et al., 1987
2500	Leaf	India	Ethanol	Vero	Badam et al., 1987
4000	Dry seed	India	Ethanol	Chang liver	Badam et al., 1987
5000	Dry seed	India	Ethanol	Vero	Badam et al., 1987

6.2.5.2 Acute Toxicity *In Vivo*

Sudanese farmers believe that neem is toxic to their livestock when they eat the leaves raw (Ali and Salih, 1982) and causes them to have a liver so bitter that it is inedible. Ali (1987) demonstrated this by feeding goats and guinea pigs with the fresh leaves of *A. indica* for 5 days, at a dose of 2000 mg/kg. All the goats and half of the guinea pigs died by the fifth day. At postmortem examination, the liver and kidneys were found to be fatty and hemorrhagic, the lungs were congested in the guinea pigs and hemorrhagic in the goats, the duodenum was hemorrhagic, the heart was flabby in some cases, and the brain was congested in some cases.

However, *A. indica* leaf decoction is less toxic: it has a maximum tolerated dose (MTD) per day of 25,000 mg/kg in mice and produces fewer acute toxic effects than some other medicinal plants or chloroquine (Agomo et al., 1992). The leaf decoction does not affect cell counts in the spleen, bone marrow, liver, or peritoneum, but slightly increases peripheral blood cell counts. However, a concentrated aqueous extract of neem leaves injected intraperitoneally in rats at a dose of 1.2 g/kg caused a 5- to 10-fold rise in serum 5'-nucleotidase activity, which was interpreted as demonstrating hepatobiliary toxicity; it is believed that neem metabolites are responsible for this toxicity, since the rise in enzyme activity is dampened by inhibitors of mixed-function oxidases (Komolafe et al., 1988; Obaseki et al., 1985). Oral, intraperitoneal, and intramuscular administration to mice of an aqueous suspension of dried neem leaves caused no deaths at a dose of 800 mg/kg, but was lethal at 1000 mg/kg, in an experiment by Isah et al. (2003). It is not clear why such a small dose increase caused an abrupt increase in mortality in this experiment. Variations in concentrations of active ingredients from one batch to the next may be one explanation.

Koley and Lal (1994) administered an ethanolic extract of neem leaves intravenously to anaesthetised rats, at doses of 100, 300, and 1000 mg/kg. Bradycardia, PR prolongation, and ventricular arrhythmias were observed at these three doses, respectively. There was also an immediate and sustained dose-dependent fall in blood pressure. In anaesthetised dogs given intravenous aqueous neem leaf decoction, Arigbabu and Don-Pedro (1971) showed an initial rise, followed by a protracted fall in blood pressure, and bradycardia. Similar findings were replicated in rabbits and guinea pigs, with the additional finding that intravenous aqueous neem leaf decoction had a weak antiarrhythmic activity (Thompson and Anderson, 1978). However, the conditions of these experiments were artificial and cannot be assumed to represent the effects of orally administered aqueous neem decoctions.

A standardised ethanolic extract of neem has been approved by the U.S. Environmental Protection Agency as an insecticide. This has undergone many toxicity tests (National Research Council, 1992). Toxic doses were not reached for ducks at up to 16 ml/kg or for rats at up to 5 ml/kg. There was low skin and eye irritation, no sensitisation, and no adverse immune response. A different methanol extract of neem leaves and bark was found to have an LD_{50} of 13 g/kg in mice (Okpanyi and Ezeukwu, 1981). Hexane and chloroform extracts of *A. indica* produced no side effects or mortality in rabbits at doses up to 1.6 g/kg (Khattak et al., 1989).

Neem seed oil is more toxic than the leaf decoction, although it has been speculated that the toxicity observed in some studies may be attributable to contaminants in industrially produced neem seed oil, such as aflatoxin (from fungi) or the similar but much more toxic seeds of *Melia azedarach* L. (Sinniah and Baskaran, 1981; Sinniah et al., 1982). Oil extracted from clean neem kernels in Germany produced no toxicity in rats at a dose up to 5 g/kg (National Research Council, 1992).

However, Ghandi et al. (1988), using industrially produced neem seed oil, found that the LD_{50} at 24 hours was 14 ml/kg in rats and 24 ml/kg in rabbits. Five milliliters per kilogram of the oil administered orally was well tolerated in rats, but from 10 ml/kg they started to act as if in a stuporous state, and most died within 48 hours. The dying animals developed labored breathing, tremor, and convulsions. At postmortem the lungs were found to be congested, with collapse of alveoli and hemorrhages into the air spaces. Levels of bilirubin and AST (aspartate transaminase) were slightly raised. The cause of death was thought to be shock secondary to severe hypoxia and clonic seizures. Intraperitoneal injection in rats of 5 ml/kg of neem seed oil produced a similar result, with the additional findings of reduced blood glucose, raised AST and ALT (alanine transaminase) levels, fatty infiltration of the liver, and mitochondrial degeneration on electron microscopy (Sinniah et al., 1985). Orally administered neem seed oil also produces hypoglycemia in rats and rabbits (Dixit et al., 1986; Pillai and Santhakumari, 1981).

Interestingly, a similar clinical picture has been observed in human infants treated with neem seed oil. Sinniah and Baskaran (1981) report a series of 13 babies ages 21 days to 2 years who had been given 5 to 30 ml of neem seed oil by their parents as a treatment for minor illnesses, such as vomiting, fever, or upper respiratory tract infections. Within $1/2$ to $4^1/2$ hours, the children developed drowsiness, tachypnea, and metabolic acidosis, followed by recurrent generalised seizures. Interestingly, hypoglycemia was not observed. Two of the babies died (one of these was also diagnosed with a streptococcal meningitis), one became developmentally delayed and suffered from recurrent fits, but the other 10 all made a good recovery. Acute hepatotoxicity with fatty infiltration was seen in one fatal case who had a liver biopsy. This pattern of illness has been likened to Reye's syndrome. Others have reported similar cases (Simpson and Lim, 1935; Lai et al., 1990).

6.2.5.3 Chronic Toxicity *In Vivo*

Ali (1987) administered fresh and dried *A. indica* leaves to goats for 56 days at doses of 50 and 200 mg/kg. From the third week, the goats developed signs of weakness and loss of appetite. As the experiment progressed, the animals' heart and respiratory rates slowed, and some developed tremors and ataxia. Body weight decreased by 23 to 38%. Levels of AST, sorbitol dehydrogenase, urea, creatinine, potassium, and cholesterol were increased, while blood glucose levels were decreased. No significant changes occurred in levels of sodium, chloride, or bilirubin. One third of the animals died before the end of the experiment. Postmortem findings were similar to those in the acute toxicity studies (see above), but also the spleen was found to be congested or hemorrhagic in some cases, and some animals had a hydropericardium.

Ibrahim et al. (1992) added 2 and 5% dried neem leaves to the diets of chicks over a period of 4 weeks, from the age of 7 days. This caused a reduction in weight gain, reduced efficiency of feed utilisation, and yellow discolouration of the combs and legs. Serum chemistry and histopathology revealed renal and hepatic toxicity.

However, a cold-water extract of neem leaves was administered to rats at a dose of up to 10 g/kg for up to 11 weeks, without affecting normal weight gain and without causing any apparent toxic effects other than infertility. This dose was, however, toxic to rabbits and guinea pigs, whose mortality was 90 and 75%, respectively, after 6 weeks of treatment (Sadre et al., 1984).

6.2.5.4 Reproductive Toxicology

The fertility of male rats treated with a cold-water extract of fresh neem leaves, at a daily dose of about 10 g/kg, was reduced to 33% of normal after 6 weeks, 16% of normal after 9 weeks, and the animals became infertile after 11 weeks (Sadre et al., 1984). The rats' testes were normal in weight and histopathological appearance, and sperm counts were normal; however, sperm motility was greatly reduced. Furthermore, goats eating fresh or dried neem leaves at a dose of 200 mg/kg daily for 56 days had smaller testes, and degeneration of the seminiferous tubules was found on histopathological examination (Ali, 1987).

The fertility of male rats was not affected by methanolic extracts of neem seeds at a dose of 0.1 ml of 1% solution for 8 days (Krause and Adami, 1984). Cohen et al. (1996) found that nimbolide has little or no effect on rabbit sperm *in vitro* with only a 20 to 30% reduction in motility and viability at 50 μM for 60 minutes.

Neem seed extract given orally at a dose of 0.6 ml for 3 days to female rats after 7 days of sperm-positive mating caused resorption of embryos in all cases, with no residual effect on fertility (Talwar et al., 1997). Thus neem oil has been proposed as an herbal morning-after pill.

6.2.5.5 Drug Interactions

Iwu et al. (1986) found that rats treated with neem extract for 4 days had a heavier liver, and that aniline hydroxylase activity and metabolism of phenobarbitone were enhanced, while the activity of NADPH cytochrome C reductase was reduced. Through these effects on hepatic metabolism, *A. indica* leaf extracts may increase or reduce plasma concentrations of other drugs that patients are taking at the same time. Furthermore, the increased oxidative stress could in theory lead to haemolytic anemia in patients with G6PD deficiency, as with the administration of primaquine, although this has not been demonstrated.

6.3 CLINICAL STUDIES

6.3.1 Clinical Pharmacology

Njoku et al. (2001) report a series of 25 patients with falciparum malaria treated with an aqueous extract of *A. indica*. Unfortunately, the only parameter reported is the patients' cholesterol levels, which were reduced compared to a non-malaria-infected control group, but not significantly different from those of malaria patients treated with chloroquine.

6.3.2 Safety and Tolerability

British and Indian doctors working in India in the early 20th century produced a report of their observations on the use of a tincture of *A. indica* bark for the treatment of malaria (Deare et al., 1916). There are reports of 54 cases, not all of whom had malaria confirmed on blood films. No serious side effects were reported, although it was mentioned that in some cases the tincture induced sweating, and in others that it had a laxative effect. Similarly, Crevost and Pételot (1929) report that a decoction and a tincture of neem bark were administered to many patients in French Cambodia, and no adverse effects were observed.

There are a few unpublished anecdotal reports of renal failure in Ghanaians who were drinking neem leaf teas as an antimalarial treatment (Okpanyi and Ezeukwu, 1981), and of toxicity following overconsumption of neem leaf extract in the Sudan in 1987, following a malaria epidemic and drug shortages (Ibrahim et al., 1992). Some traditional healers in Tanzania report that neem root and leaf infusions may cause palpitations and abortion (Gessler, 1995).

6.3.3 Efficacy

Twelve of the cases reported by Deare et al. (1916) had confirmed *Plasmodium vivax*, 10 had *P. falciparum*, and three had *Plasmodium malariae*. Of the falciparum cases, only three had parasite clearance by day 7 (and in one of these there was a recrudescence when the treatment was stopped), but seven improved clinically. Parasitaemia was monitored in four of the vivax cases; parasites were cleared in two of these, but recrudesced in both by day 8. Eight (66%) of the patients improved symptomatically. None of the patients with *P. malariae* improved clinically.

Of the 11 groups reporting cases, only 3 were of the opinion that the tincture of *A. indica* was effective for treating malaria. Six believed that it was ineffective, and two were undecided. All were unanimous that this preparation was less effective than quinine or *Cinchona* preparations.

Crevost and Pételot (1929) reported using neem bark decoctions and tinctures to treat chronic or recurrent malaria in Cambodia, sometimes as a substitute for quinine, with apparently good results. Unfortunately, there is no quantitative or parasitological data to substantiate these observations. An observational study of patients self-medicating with neem extracts is reported as showing that they were an effective treatment for malaria (Iwu, 1993).

6.4 BOTANY AND CULTIVATION

6.4.1 BOTANY

Neem (*A. indica* A. Juss. *Meliaceae*) is an attractive evergreen tree that can grow up to 20 m tall. Its spreading branches form rounded crowns up to 10 m in diameter. The short, usually straight trunk has a moderately thick, strongly furrowed bark that has a garlic-like odor and a bitter, astringent taste. Its leaves are imparipinate, 20 to 38 cm long, crowded near the branch end, oblique, lanceolate, and deeply and sharply serrate. Neem is rarely leafless and is usually in full foliage even during months of prolonged drought. Its small, white bisexual and staminate (functionally male) flowers are borne in axillary clusters. They have a honey-like scent and attract many bees, which act as pollinators. The fruit is a smooth, ellipsoidal drupe, 1 to 2 cm long, that is yellow when mature and comprises a sweet pulp enclosing a seed. The seed is composed of a shell and a kernel (sometimes two or three kernels), the latter having a high oil content. Neem will begin bearing fruit after 3 to 5 years, becomes fully productive in 10 years, and a mature tree can produce up to 50 kg of fruits annually (National Research Council, 1992; Tewari, 1992; Chandra, 1997; Ketkar and Ketkar, 1997; Gunasena and Marambe, 1998).

A. *indica* can be distinguished from the similar *M. azedarach*, whose seeds are toxic, by its leaves, flowers, and seeds. *M. azedarach* has two or three pinnate leaves with serrate, crenate, or dentate leaflets (Pennington, 1981). Its flowers are pink, whereas those of neem are white, and its seeds are round, whereas those of neem are long (Hirt and M'Pia, 2001).

6.4.2 HABITAT

Neem can grow in tropical and subtropical regions with semiarid to humid climates. Neem will typically experience a mean annual rainfall of 450 to 1200 mm, mean temperatures of 25 to 35°C, and grow at altitudes up to 800 m above sea level. The species is drought tolerant and thrives in many of the drier areas of the world. There is, therefore, considerable interest in neem as a means to prevent the spread of deserts and ameliorate desert environments, for example, in Saudi Arabia (Ahmed et al., 1989), sub-Saharan Africa (National Research Council, 1992), and western India (Gupta, 1994). Neem grows an all types of soils, including clay, saline, and alkaline soils, but does well on black cotton soils (National Research Council, 1992). It can tolerate dry, stony, shallow soil with a waterless subsoil, or in places where there is a hard calcareous or clay pan near the surface. Neem does not tolerate waterlogging, is fire resistant, and has a unique property of calcium mining that neutralises acidic soils (Gunasena and Marambe, 1998). The species is a light demander and when juvenile, neem will push up vigorously through scrubby vegetation. It is hardy but frost tender and does not withstand excessive cold, especially in the seedling and sapling stages (National Research Council, 1992; Chaturevedi, 1993). It coppices and pollards well and also produces root suckers.

6.4.3 PROPAGATION

6.4.3.1 Seed Harvesting

The fruits of neem are normally collected from the ground, de-pulped, and sown immediately because of the intermediate nature of neem seed and the risk of low viability (Nagaveni et al., 1987). Several authors therefore recommend that seeds should be collected from the tree when they are greenish yellow (rather than when they have fallen to the ground), are de-pulped, and dried under shade for two days (Nagaveni et al., 1987; Chaturevedi, 1993; Gunasena and Marambe,

1998). In this way, the risk from fermentation of the unopened cotyledons can be reduced, viability can be improved, and high germination percentages can be obtained for up to several months after harvest. Furthermore, if the seeds are to be used for production of oil, there is less risk of contamination by aflatoxins from mould.

6.4.3.2 Seed Storage

Neem seeds have been variously documented as orthodox, intermediate, or recalcitrant (Berjak et al., 1995; Bhardwaj and Chand, 1995). However, the species is generally regarded as intermediate, and viability can be extended through drying and ambient- to low-temperature storage conditions (Mohan et al., 1995; Singh et al., 1997). A 5- to 6-day drying period in shade conditions caused a drop in seed moisture content to 11% (Singh et al., 1997). Drying was found not to adversely affect germination rates, and seeds could be stored under ambient or refrigerated conditions (~10°C or less) for up to 6 months. Recent research has, however, shown a large variation in the lowest safe moisture content for different neem seed sources (C. Hansen, personal communication*). The minimum seed moisture content for two seed sources, one from Kenya and one from Myanmar, was 20%, whereas that for a second seed source from Kenya was 10%. These results suggest that there is considerable variation within neem for seed moisture content under storage conditions, and may explain why neem seed has been variously documented as orthodox, intermediate, and recalcitrant in the past.

6.4.3.2.1 Seed Germination

Fresh neem seed germinates 7 days after sowing, and germination is complete after 25 days (Singh et al., 1995). Bharathi et al. (1996) found that greenish yellow drupes had the highest viability and germination percentage when compared to both younger (green) and more mature (yellow) seeds. The seeds germinate readily without pretreatment, although soaking dried, de-pulped seeds for up to 24 hours has been found to improve germination (Hegde, 1993; Fagoonee, 1984).

6.4.3.2.2 Seedling Management

Neem seeds are commonly germinated in plastic bags in the nursery, although direct sowing is said to be successful where there is adequate rainfall (Benge, 1988; Vivekanandan, 1998). Seeds are best planted at a depth of 2.5 cm to avoid predation by rodents and birds (Bahuguna, 1997). Neem is responsive to VAM fungi;** hence, improved growth of seedlings in the nursery could be achieved through sowing in inoculated soils (Habte et al., 1993). Neem seedlings can be transplanted to a well-prepared site when 3 to 12 months old (Singh, 1982; Bahuguna, 1997). Once sown, it is important that neem seedlings receive adequate care. Seedlings are sensitive, like mature trees, to waterlogged and very dry soils (Gunasena and Marambe, 1998). Young neem trees cannot tolerate frost or excessive cold conditions (Chaturevedi, 1993; Hegde, 1993), but Bahuguna (1997) reported that neem seedlings are tolerant of shade. An application of mulch around the seedlings can help ensure that water is conserved and reduce weed buildup. Soil water conservation methods, i.e., planting in pits or trenches with an application of mulch, can be used in water-deficient areas (Gupta, 1994, 1995).

---

* Dr. Christian Hansen, DANIDA Forest Seed Centre, Denmark.
** Vesicular arbuscular mycorrhizal (VAM) fungi are beneficial fungi that penetrate and colonise the root of the plant, then send out filaments (hyphae) into the surrounding soil. The hyphae form a bridge that connects the plant root with large areas of soil and serves as a pipeline to funnel nutrients back to the plant. In return, the plant supplies the VAM fungi with carbon for their growth and energy needs.

6.5 PUBLIC HEALTH POTENTIAL

Neem is a promising candidate for malaria control programs because of its widespread distribution, ease of cultivation, and widespread use for the treatment of malaria. Not only could it provide treatment for malaria, but it also has well-established insect repellent and insecticidal properties.

The wealth of toxicological studies on different neem preparations is potentially confusing and off-putting. The aqueous decoction is basically safe in the acute setting, with a maximum tolerated dose of 25 g/kg, but ethanolic extracts have an LD_{50} of 13 g/kg, and neem seed oil is potentially more toxic, especially in young children. Long-term administration may cause renal and hepatic toxicity, infertility, and disturbed metabolism of other drugs in the liver, and therefore should also be avoided.

There is as yet insufficient evidence on the efficacy of neem preparations for the treatment of malaria. Some *in vitro* studies against *P. falciparum* are promising, but *in vivo* studies in mice are uniformly disappointing. There is some evidence for antipyretic and immunostimulatory effects, which may in part account for the perceived efficacy of neem for the treatment of febrile illnesses. Observational clinical studies are needed to resolve this issue, and if these are promising, formal clinical trials could be conducted, providing the preparation to be used has been proven to be safe.

ACKNOWLEDGMENTS

The authors thank Dr. Unnikrishnan for providing information from the Ayurvedic literature in Table 6.1 and Dr. Gerry Bodeker for editorial comments.

REFERENCES

Abatan, M.O. and Makinde, M.J. (1986). Screening *Azadirachta indica* and *Pisum sativum* for possible antimalarial activities. *J. Ethnopharmacol.*, 17, 85–93.

Agomo, P.U., Idigo, J.C., and Afolabi, B.M. (1992). "Antimalarial" medicinal plants and their impact on cell populations in various organs of mice. *Afr. J. Med. Med. Sci.*, 21, 39–46.

Agyepong, I.A. (1992). Malaria: ethnomedical perceptions and practice in an Adangbe farming community and implications for control. *Soc. Sci. Med.*, 35, 131–137.

Ahmed, S., Bamofleh, S., and Munshi, M. (1989). Cultivation of neem (*Azadirachta indica*, Meliaceae) in Saudi Arabia. *Econ. Bot.*, 43, 35–38.

Ahmed, S. and Grainge, M. (1985). Use of indigenous plant resources in rural development: potential of the neem tree. *Int. J. Dev. Technol.*, 3, 123–130.

Ahorlu, C.K., Dunyo, S.K., Afari, E.A., Koram, K.A., and Nkrumah, F.K. (1997). Malaria-related beliefs and behaviour in southern Ghana: implications for treatment, prevention and control. *Trop. Med. Int. Health*, 2, 488–499.

Ali, B.H. (1987). The toxicity of *Azadirachta indica* leaves in goats and pigs. *Vet. Hum. Toxicol.*, 29, 16–19.

Ali, B.H. and Salih, A.M.M. (1982). Suspected *Azadirachta indica* toxicity in a sheep. *Vet. Rec.*, 111, 494.

Aminuddin, R.D.G. and Subhan Khan, A. (1993). Treatment of malaria through herbal drugs from Orissa, India. *Fitoterapia*, 64, 545–548.

Arigbabu, S.O. and Don-Pedro, S.G. (1971). Studies on some pharmaceutical properties of Azadirachta indica or Baba Yaro. *Afr. J. Pharm. Pharm. Sci.*, 1, 181–184.

Attri, B.S. and Prasad, R. (1980). Neem oil extractive: an effective mosquito larvicide. *Indian J. Entomol.*, 42, 371–374.

Badam, L., Deolankar, R.P., Kulkarni, M.M., Nagsampgi, B.A., and Wagh, U.V. (1987). *In vitro* antimalarial activity of Neem (*Azadirachta indica* A. Juss) leaf and seed extracts. *Indian J. Malariol.*, 24, 111–117.

Bahuguna, V.K. (1997). Silviculture and management practices for cultivation of *Azadirachta indica* (neem). *Indian For.*, 123, 379–386.

Benge, M.D. (1988). Cultivation and propagation of the neem tree. In *Focus on Phytochemical Pesticides (1): The Neem Tree*, Jacobson, M., Ed. CRC Press Inc., Boca Raton, FL, pp. 2–17.

Benoit, F., Valentin, A., Pelissier, Y., et al. (1996). In vitro antimalarial activity of vegetal extracts used in West African traditional medicine. *Am. J. Trop. Med. Hyg.*, 54, 67–71.

Berjak, P., Campbell, G.K., Farrant, J.M., Omondi, O.W., and Pammenter, N.W. (1995). Responses of seeds of *Azadirachta indica* (neem) to short-term storage under ambient or chilled conditions. *Seed Sci. Technol.*, 23, 779–792.

Bharathi, A., Umarani, R., Karivaratharaju, T.V., Vanangamudi, K., and Manonmani, V. (1996). Effect of drupe maturity on seed germination and seedling vigour in neem. *J. Trop. For. Sci.*, 9, 147–150.

Bhardwaj, S.D. and Chand, G. (1995). Storage of neem seeds: potential and limitations for germplasm conservation. *Indian For.*, 121, 1009–1011.

Bhat, R.B., Etejere, E.O., and Oladipo, V.T. (1990). Ethnobotanical studies from Central Nigeria. *Econ. Bot.*, 44, 382–390.

Bodeker, G., Burford, G., Chamberlain, J., and Bhat, K.K.S. (2001). The underexploited medicinal potential of *Azadirachta indica* A. Juss. (Meliaceae) and *Acacia nilotica* (L.) Willd. ex Del. (Leguminosae) in sub-Saharan Africa: a case for a review of priorities. *Int. For. Rev.*, 3, 285–298.

Boshitz, C. and Grunewald, J. (1994). The effect of NeemAzal on *Aedes aegypti* (Diptera: Culicidae). *Appl. Parasitol.*, 35, 251–256.

Bray, D.H., Warhurst, D.C., Connolly, J.D., O'Neill, M.J., and Phillipson, J.D. (1990). Plants as sources of antimalarial drugs. Part 7. Activity of some species of Meliaceae plants and their constitutent limonoids. *Phytother. Res.*, 4, 29–35.

Bugmann, N. (2000). Le concept du paludisme, l'usage et l'efficacité *in vivo* de trois traitements traditionnels antipalustres dans la région de Dori, Burkina Faso. Doctoral thesis, Faculty of Medicine, University of Basel, Switzerland.

Burford, G.L., Rafiq, M.Y., Mollel, E.E., Ole Ngila, L., Tarimo, T., and Ole Lengisugi, N. (2000). Use of Plants for the Prevention and Treatment of Malaria among the Maasai and Wa-Arusha Peoples of Northern Tanzania. Unpublished document.

Chandra, V. (1997). Botany of neem. *Ann. For.*, 5, 182–188.

Chaturevedi, A.N. (1993). *Silviculture*. In *Neem Research and Development*, Randhawa, N.S. and Parmar, B.S., Eds., New Delhi, Indian Society of Pesticide Science, 38–48.

Childs, F.J., Chamberlain, J.R., and Harris, P.J.C. (1999). Current practices and potential for the development of the role of neem (*Azadirachta indica*) in crop protection among resource poor farmers: preliminary results from a postal survey. In *Commercialisation of Neem in Ghana, 19–21st October, Goethe Institute, Accra, Ghana*. GTZ Publications, Eschborn, Germany.

Cohen, E., Quistad, G.B., and Casida, J.E. (1996). Cytotoxicity of nimbolide, epoxyazadiadione and other limonoids from neem insecticide. *Life Sci.*, 58, 1075–1081.

Crevost, Ch. and Pételot, A. (1929). Catalogue des produits de l'Indochine (suite). *Plantes Méd. Bull. Econ. Indochine*, 32, 277–367 (296–299).

Deare, B.H., Mathew, C.M., Bahadur, R.C.L.B., Gage, A.T., and Carter, H.G. (1916). The Third Report of the Indigenous Drugs Committee. Superintendent Government Printing, Calcutta.

Dery, B.B., Otysina, R., and Ng'atigwa, C. (1999). *Indigenous Knowledge of Medicinal Trees and Setting Priorities for Their Domestication in Shinyanga Region, Tanzania*. International Centre for Research in Agroforestry, Nairobi.

Devi, C.U., Valecha, N., Atu, P.K., and Pillai, C.R. (2001). Antiplasmodial effect of three medicinal plants: a preliminary study. *Curr. Sci.*, 80, 917–919.

Dhar, R., Dawar, H., Garg, S., Basir, S.F., and Talwar, G.P. (1996). Effect of volatiles from neem and other natural products on gonotrophic cycle and oviposition of *Anopheles stephensi* and *An culicifacies* (Diptera: Culicidae). *J. Med. Entomol.*, 33, 195–201.

Dhar, R., Zhang, K., Talwar, G.P., Garg, S., and Kumar, N. (1998). Inhibition of the growth and development of asexual and sexual stages of drug-sensitive and resistant strains of the human malaria parasite *Plasmodium falciparum* by Neem (*Azadirachta indica*) fractions. *J. Ethnopharmacol.*, 61, 31–39.

Dixit, V.P., Sinha, R., and Tank, R. (1986). Effect of neem seed oil on the blood glucose concentration of normal and alloxan diabetic rats. *J. Ethnopharmacol.*, 17, 95–98.

Djimde, A.D., Plowe, C.V., Diop, S., Dicko, A., Wellems, T.E., and Doumbo, O. (1998). Use of antimalarial drugs in Mali: policy versus reality. *Am. J. Trop. Med. Hyg.*, 59, 376–379.

Ekanem, O.J. (1978). Has *Azadirachta indica* (Dongoyaro) any antimalarial activity? *Nig. Med. J.*, 8, 8–10.

El-Kamali, H.H. and El-Khalifa, K.F. (1997). Treatment of malaria through herbal drugs in the Central Sudan. *Fitoterapia*, 68, 527–528.

El Tahir, A., Satti, G.M.H., and Khalid, S.A. (1999). Antiplasmodial activity of selected Sudanese medicinal plants with emphasis on *Maytenus senegalensis* (Lam.) Exell. *J. Ethnopharmacol.*, 64, 227–233.

Etkin, N.L. (1981). A Hausa herbal pharmacopoeia: biomedical evaluation of commonly used plant medicines. *J. Ethnopharmacol.*, 4, 75–98.

Etkin, N.L. and Ross, P.J. (1991). Recasting malaria, medicine and meals: a perspective on disease adaptation. In *The Anthropology of Medicine*, 2nd ed., Romanucci-Ross, L., Moerman, D.E., and Tancredi, L.R., Eds. Bergin & Garvey, New York.

Fagoonee, I. (1984). Germination tests with neem seeds. In *Natural Pesticides from the Neem Tree and Other Tropical Plants*, Schmutterer, H. and Ascher, K.R.S., Eds. GTZ Press, Eschborn, Germany.

Fawole, O.I. and Onadeko, M.O. (2001). Knowledge and home management of malaria fever by mothers and care givers of under five children. *West Afr. J. Med.*, 20, 152–157.

Fourn, L., Sakou, G., and Zohoun, T. (2001). Utilisation des services de santé par les mères des enfants fébriles au sud du Bénin. *Santé Publique*, 13, 161–168.

Gbary, A.R., Sombié, I., Guiguemdé, T.R., and Guissoi, A.P. (1996). Connaissances et pratiques des tradipraticiens en matière de paludisme en milieu urbain de Ouagadougou et de Bobo-Dioulasso, Burkina Faso, Afrique de l'Ouest. *Santé Publique*, 8, 249–226.

The General Council of Medical Education and Registration of the United Kingdom. (1901). *Indian and Colonial Addendum to the British Pharmacopoeia 1898. Government of India Edition, 1901.* Spottiswoode & Co., Ltd., London, p. 47.

Gessler, M. (1995). The Antimalarial Potential of Medicinal Plants Traditionally Used in Tanzania, and Their Use in the Treatment of Malaria by Traditional Healers. Ph.D. dissertation, University of Basel, Switzerland.

Ghandi, M., Lal, R., Sankaranarayan, A., Banerjee, C.K., and Sharma, P.L. (1988). Acute toxicity of the oil from *Azadirachta indica* seed (neem oil). *J. Ethnopharmacol.*, 23, 39–51.

Gimlette, J.D. and Burkill, I.H. (1930). The medical book of Malayan medicine. *The Gardens' Bulletin*, Straits Settlements, 6, 347.

Glinsukon, T., Somjaree, R., Piyachaturawat, P., and Thebtaranonth, Y. (1986). Acute toxicity of nimbolide and nimbic acid in mice, rats and hamsters. *Toxicol. Lett.*, 30, 159–166.

Gujral, M.L., Kohli, R.P., Bhargava, K.P., and Saxena, P.N. (1955). Antipyretic activity of some indigenous drugs. *Indian J. Med. Res.*, 43, 89–94.

Gunasena, H.P.M. and Marambe, B. (1998). Neem in Sri Lanka: A Monograph. University of Peradeniya, Sri Lanka, 62 pp.

Gupta, G.N. (1994). Influence of rain-water harvesting and conservation practices on growth and biomass production of *Azadirachta indica* in the Indian desert. *For. Ecol. Manage.*, 70, 329–339.

Gupta, G.N. (1995). Rain-water management for tree planting in the Indian desert. *J. Arid Environ.*, 31, 219–235.

Habte, M., Muruleedhara, B.N., and Ikawa, H. (1993). Response of neem (*Azadirachta indica*) to soil P concentration and mycorrhizal concentration. *Arid Soil Res. Rehabil.*, 7, 327–333.

Hausmann Muela, S., Muela Ribera, J., Mushi, A.K., and Tanner, M. (2002). Medical syncretism with reference to malaria in a Tanzanian community. *Soc. Sci. Med.*, 55, 403–413.

Hegde, N.G. (1993). *Improving the Productivity of Neem Trees*. World Neem Conference, Bangalore, India, pp. 69–79.

Heyne, K. (1927). *De Nuttige Planten van Nederlandsch Indië*. Departement van Landbouw, Buitenzorg, Indonesia.

Hirt, H.M. and Lindsey, K. (2000). *Natural Medicine in the Tropics: Experiences*. Anamed, Winnenden, Germany.

Hirt, H.M. and M'Pia, B. (2001). *Natural Medicine in the Tropics*. Anamed, Winnenden, Germany.

Ibrahim, I.A., Khalid, S.A., Omer, S.A., and Adam, S.E.I. (1992). On the toxicology of *Azadirachta indica* leaves. *J. Ethnopharmacol.*, 35, 267–273.

Ibrahim, K.E., Isa, A.B., and Bangudu, A.B. (1998). Formulation of powdered *Azadirachta indica* A. Juss. (Meliaceae) bark and leaves into tablet dosage form. I. Powder characteristics and compact properties of tablets. *J. Pharm. Res. Dev.*, 3, 81–87.

Isah, A.B., Ibrahim, Y.K.E., and Iwalewa, E.O. (2003). Evaluation of the antimalarial properties and standardisation of tablets of *Azadirachta indica* (Meliaceae) in mice. *Phytother. Res.*, 17, 807–810.

Iwu, M.M. (1982). Perspectives of Igbo tribal ethnomedicine. *Ethnomedicine*, VII, 7–46.

Iwu, M.M. (1993). *Handbook of African Medicinal Plants*. CRC Press, Boca Raton, FL.

Iwu, M.M., Obedoa, O., and Anazodo, M. (1986). Biochemical mechanism of the antimalarial activity of Azadirachta indica leaf extract. *Pharmacol. Res. Commun.*, 18, 81–91.

Jones, I.W., Denholm, A.A., Ley, S.V., Lovell, H., Wood, A., and Sinden, R.E. (1994). Sexual development of malaria parasites is inhibited *in vitro* by the Neem extract Azadirachtin, and its semi-synthetic analogues. *FEMS Microbiol. Lett.*, 120, 267–274.

Julvez, J., Hamidine, M., Boubacar, Aa., Nouhou, A., and Alarou, A. (1995). Connaisances et pratiques face au paludisme. Enquête médicale en pays Songhay-Zarma (Niger). *Cahiers Santé*, 5, 307–313.

Ketkar, C.M. and Ketkar, M.S. (1997). Botany. In *Neem in Sustainable Agriculture*, Narwal, S.S., Tauro, P., and Bisla, S.S., Eds. Scientific Publishers, Jodhpur, India, pp. 1–12.

Khalid, S.A., Duddeck, H., and Gonzalez-Sierra, M. (1989). Isolation and characterisation of an antimalarial agent of the neem tree Azadirachta indica. *J. Nat. Prod.*, 52, 922–926.

Khalid, S.A., Farouk, A., Geary, T.G., and Jensen, J.B. (1986). Potential antimalarial candidates from African plants: an *in vitro* approach using *Plasmodium falciparum*. *J. Ethnopharmacol.*, 15, 201–209.

Khattak, S.G., Gilani, S.N., and Ikram, M. (1985). Antipyretic studies on some indigenous Pakistani medicinal plants. *J. Ethnopharmacol.*, 14, 45–51.

Kigodi, P.G.K., Blaskó, G., Thebtaranonth, Y., Pezzuto, J.M., and Cordell, G.A. (1989). Spectroscopic and biological investigation of nimbolide and 28-deoxonimbolide from Azadirachta indica. *J. Nat. Prod.*, 52, 1246–1251.

Knols, B. (2000). The Role of Neem in ICIPE's Malaria Project. Paper presented at the 8th Neem Workshop, ICIPE, Duduville, Nairobi, Kenya, May 2–5.

Kofi-Tsekpo, W.M., Runkunga, G.M., Aluoch, J.A., et al. (1989). A Preliminary Investigation of a Traditional Medicine (KRM 913) as a Potential Antimalarial and Antiparasitic Agent. Paper presented at the Proceedings of the KEMRI-KETRI Annual Medical Scientific Congress, Nairobi.

Koga, Y., Yoshida, I., Kimura, M., Yoshino, M., Yamashita, F., and Sinniah, D. (1987). Inhibition of mitochondrial functions by margosa oil: possible implications in the pathogenesis of Reye's syndrome. *Pediatr. Res.*, 22, 184–187.

Koley, K.M. and Lal, J. (1994). Pharmacological effects of *Azadirachta indica* (neem) leaf extract on the ECG and blood pressure of rat. *Indian J. Physiol. Pharmacol.*, 38, 223–225.

Komolafe, O.O., Anyabuike, C.P., and Obaseki, A.O. (1988). The possible role of mixed-function oxidases in the hepatobiliary toxicity of *Azadirachta indica*. *Fitoterapia*, 19, 109–113.

Konradsen, F., van der Hoek, W., Amerasinghe, P.H., Amerasinghe, F.P., and Fonseka, K.T. (1997). Household responses to malaria and their costs: a study from rural Sri Lanka. *Trans. R. Soc. Trop. Med. Hyg.*, 91, 127–130.

Krause, W. and Adami, M. (1984). Extracts of neem (*Azadirachta indica*) seed kernels do not inhibit spermatogenesis in the rat. In *Natural Pesticides from the Neem Tree and Other Tropical Plants*, Schmutterer, H. and Ascher, K.R.S., Eds. GTZ, Eschborn, Germany.

Kroes, B.H., Van den Berg, A.J.J., Labadie, R.P., Abeysekera, A.M., de Silva, K.T.D. (1993). Impact of the preparation process on immunomodulatory activities of the ayurvedic drug *Nimba arishta*. *Phytotherapy Res.*, 7, 35–40.

Lai, S.M., Lim, K.W., and Cheng, H.K. (1990). Margosa oil poisoning as a cause of toxic encephalopathy. *Singapore Med. J.*, 31, 463–465.

MacKinnon, S., Durst, T., Arnason, J.T., et al. (1997). Antimalarial activity of tropical Meliaceae extracts and gedunin derivatives. *J. Nat. Prod.*, 60, 336–341.

Makinde, J.M. and Obih, P.O. (1985). Lack of schizontocidal activity of three herbal decoctions on *Plasmodium berghei berghei* in mice. *Afr. J. Med. Med. Sci.*, 14, 54–58.

Marsh, K. (1993). Immunology of human malaria. In *Bruce Chwatt's Essential Malariology*, Gilles, H.M. and Warrel, D.A., Eds. Edward Arnold, London.

Mayxay, M., Chotivanich, K., Pukrittayakamee, S., Newton, P., Looareesuwan, S., White, N.J. (2001). Contribution of humoral immunity to the tereapeutic response in *falciparum malaria*. *Am. J. Trop. Med. Hyg.*, 65(6), 918–923.

Menaut, B. (1929). Matière médicale cambodgienne. Première partie: les drogues usuelles. *Bull. Econ. Indochine*, 32, 197–276.

Mohan, J., Chaudhari, L.D., and Jha, M. (1995). A note on increasing the viability of neem seed. *Indian For.*, 121, 1085–1086.

Moser, G. (1996). Status Report on Global Neem Usage. Pesticide Service Project, PN 86.2588.1. GTZ, Griesheim, Germany, 39 pp.

Mulangu Binzambal, O. (1999). Antimalarial plants used in the Democratic Republic of Congo. Poster presented at the First International Meeting of the Research Initiative on Traditional Antimalarials. Available at http://mim.nih.gov/english/partnerships/ritam_program.pdf.

Munguti, K. (2000). Indigenous knowledge in the management of malaria and visceral leishmaniasis among the Tugen of Kenya. *Indigenous Knowledge and Development Monitor*, 5. Available at http://www.nuffic.nl/ciran/ikdm/5-1/articles/mungutiart.htm.

Murthy, P.S. and Sirsi, M. (1958). Pharmacological studies on *Melia azadirachta*. *Indian J. Physiol. Pharmacol.*, 2, 456–461.

Mwenesi, H.A. (1994). Mothers' Definition and Treatment of Childhood Malaria on the Kenyan Coast, Geneva: WHO. TDR/SER/13.

Nadkarni, K.M. (1976). *Indian Materia Medica*. Sangam Books, London.

Nagaveni, H.C., Ananthapadmanabha, H.S., and Rai, S.N. (1987). Note on the extension of viability of *Azadirachta indica*. *Myforest*, 23, 245.

Narayan, D.S. (1969). Antipyretic effect of neem oil and its constituents. *Mediscope*, 12, 25–27.

Nargas, J. and Trivedi, P.C. (1999). Traditional and medicinal importance of *Azadirachta indica* Juss. in India. *J. Econ. Taxon. Bot.*, 23, 33–37.

National Research Council. (1992). *Neem: A Tree for Solving Global Problems*. National Academy Press, Washington, DC.

Njiro, S.M. and Kofi-Tsekpo, M.W. (1999). Effect of an aqueous extract of *Azadirachta indica* on the immune response in mice. *Onderstepoort J. Vet. Res.*, 66, 59–62.

Njoku, O.U., Alumanah, E.O., and Meremikwu, C.U. (2001). Effect of *Azadirachta indica* extract on plasma lipid levels in human malaria. *Boll. Chim. Farm.*, 140, 367–370.

Nyamongo, I.K. (2002). Health care switching behaviour of malaria patients in a Kenyan rural community. *Soc. Sci. Med.*, 54, 377–386.

Obaseki, A.O., Adeyi, O., and Anyabuike, C. (1985). Some serum enzyme levels as marks of possible acute effects of the aqueous extract of *Azadirachta indica* on membranes *in vivo*. *Fitoterapia*, 56, 111–115.

Obaseki, O. and Jegede-Fadunsin, H.A. (1986). The antimalarial activity of *Azadirachta indica*. *Fitoterapia*, 57, 247–251.

Obih, P.O. and Makinde, J.M. (1985). Effect of *Azadirachta indica* on *Plasmodium berghei berghei* in mice. *Afr. J. Med. Med. Sci.*, 14, 51–54.

Odetola, A.A. and Bassir, O. (1986). Evaluation of anti-malarial properties of some Nigerian medicinal plants. In *The State of Medicinal Plants Research in Nigeria*, Sofowora, A., Ed. Nigerian Society of Pharmacognosy, Ibadan.

Ofullah, A.V.O., Chege, G.M.M., Rukunga, G.M., Kiarie, F.K., Githure, J.I., and Kofi-Tsekpo, M.W. (1995). *In vitro* antimalarial activity of extract of *Albizia gummifera*, *Aspilia mossambicensis*, *Melia azedarach* and *Azadirachta indica* against *Plasmodium falciparum*. *Afr. J. Health Sci.*, 2, 312–314.

Okpanyi, S.N. and Ezeukwu, G.C. (1981). Anti-inflammatory and antipyretic activities of *Azadirachta indica*. *Planta Med.*, 41, 34–39.

Okrah, J., Traoré, C., Palé, A., Sommerfeld, J., and Müller, O. (2002). Community factors associated with malaria prevention by mosquito nets: an exploratory study in rural Burkina Faso. *Trop. Med. Int. Health*, 7, 240–248.

Parmar, B.S. (1993). Scope of botanical pesticides in integrated pest management. *J. Insect Sci.*, 6(1), 15–20.

Pennington, T.D. (1981). *Meliaceae*. Flora Neotropica Monograph 28. New York Botanical Garden, New York.

Pillai, N.R. and Santhakumari, G. (1981). Hypoglycaemic activity of *Melia azadirachta* Linn (Neem). *Indian J. Med. Res.*, 74, 931–933.

Pillai, N.R. and Santhakumari, G. (1984a). Toxicity studies on nimbidin, a potential antiulcer drug. *Planta Med.*, 50, 146–148.

Pillai, N.R. and Santhakumari, G. (1984b). Some pharmacological actions of "nimbidin" — a bitter principle of *Azadirachta indica* A Juss (neem). *Ancient Sci. Life*, IV, 88–95.

Pillai, N.R., Suganthan, D., and Santhakumari, G. (1980). Analgesic and antipyretic actions of nimbidin. *Bull. Med. Eth. Bot. Res.*, 1, 393–400.

Plebanski, M. and Hill, V.S. (2000). The immunology of malaria infection. *Curr. Opin. Immunol.*, 14, 437–441.

Ralantonirina, D. (1993). Aperçu sur les plantes médicinales dans le sud de Madagascar: Etude faite sur les adultes dans le périmètre de la Réserve spéciale de Beza-Mahafaly. Thèse pour l'obtention du doctorat en Médecine, Faculté de Médecine, Université d'Antananarivo, Madagascar.

Ray, A., Banerjee, B.D., and Sen, P. (1996). Modulation of humoral and cell-mediated immune responses by Azadirachta indica (neem) in mice. *Indian J. Exp. Biol.*, 34, 698–701.

Rochanakij, S., Thebtaranonth, Y., Yenjai, C., and Yuthavong, Y. (1985). Nimbolide, a constituent of *Azadirachta indica*, inhibits *Plasmodium falciparum* growth in culture. *Southeast Asian J. Trop. Med. Public Health*, 16, 66–72.

Sadre, N.L., Deshpande, V.Y., Mendulkar, K.N., and Nandal, D.H. (1984). Male antifertility activity of *Azadirachta indica* in different species. In *Natural Pesticides from the Neem Tree and Other Tropical Plants*, Schmutterer, H. and Ascher, K.R.S., Eds. GTZ, Eschborn, Germany.

Scott, R.H., O'Brien, K., Roberts, L., Mordue, W., and Mordue, J. (1999). Extracellular and intracellular actions of azadirachtin on the electrophysiological properties of cultured rat DRG neurones. *Comp. Biochem. Physiol. C*, 123, 85–93.

Sen, P., Mediratta, P.K., and Ray, A. (1992). Effects of *Azadirachta indica* A Juss on some biochemical, immunological and visceral parameters in normal and stressed rats. *Indian J. Exp. Biol.*, 30, 1170–1175.

Sharma, V.P., Ansari, M.A., and Razdan, R.K. (1993a). Mosquito repellent action of neem (*Azadirachta indica*) oil. *J. Am. Mosq. Control Assoc.*, 9, 359–360.

Sharma, V.P., Nagpal, B.N., and Srivastava, A. (1993b). Effectiveness of neem oil mats in repelling mosquitoes. *Trans. R. Soc. Trop. Med. Hyg.*, 87, 626.

Silva, K.T. (1991). Ayurveda, malaria and the indigenous herbal tradition in Sri Lanka. *Soc. Sci. Med.*, 33, 153–160.

Simpson, I.A. and Lim, E.C. (1935). Margosa fruit poisoning suspected to be cause of a case of fatal poisoning. *Malayan Med. J.*, 10, 138–139.

Singh, A.K., Raghubanshi, A.S., and Singh, J.S. (2002). Medical ethnobotany of the tribals of Sonaghati of Sonbhadra district, Uttar Pradesh, India. *J. Ethnopharmacol.*, 81, 31–41.

Singh, B.G., Mahadevan, N.P., Shanthi, K., Geetha, S., and Manimuthu, L. (1995). Multiple seedling development in neem (*Azadirachta indica*). *Indian For.*, 121, 1049–1052.

Singh, B.G., Mahadevan, N.P., Shanthi, K., Manimuthu, L., and Geetha, S. (1997). Effect of moisture content on the viability and storability of *Azadirachta indica* A. Juss (neem) seeds. *Indian For.*, 123, 631–636.

Singh, R.V. (1982). *Fodder Trees of India*. Oxford and IBH Publishing Co., New Delhi, pp. 77–84.

Singh, V.K. and Anwar Ali, Z. (1994). Folk medicines in primary health care: common plants used for the treatment of fevers in India. *Fitoterapia*, 65, 68–74.

Singh Huidrom, B.K. (1997). Studies on medico-botany of Meitei community in Manipur State, India (II). *Ad. Plant Sci.*, 10, 13–18.

Sinniah, D. and Baskaran, G. (1981). Margosa oil poisoning as a cause of Reye's syndrome. *Lancet*, 1, 487–489.

Sinniah, D., Baskaran, G., Looi, L.M., and Leong, K.L. (1982). Reye-like syndrome due to Margosa oil poisoning: report of a case with postmortem findings. *Am. J. Gastroenterol.*, 77, 158–161.

Sinniah, D., Schwartz, P.H., Mitchell, R.A., and Arcinue, E.L. (1985). Investigation of an animal model of a Reye-like syndrome caused by Margosa oil. *Pediatr. Res.*, 19, 1346–1355.

Snow, R.W., Peshu, N., Forster, D., Mwenesi, H., and Marsh, K. (1992). The role of shops in the treatment and prevention of childhood malaria on the coast of Kenya. *Trans. R. Soc. Trop. Med. Hyg.*, 86, 237–239.

Sonnerat. (1782). *Voyage aux Indes Orientales et à la Chine*, Vol. 1. Sonnerat, Paris.

Spencer, cf. Koniuszy, F.R., Rogers, E.F., et al. (1947). Survey of plants for antimalarial activity. *Lloydia*, 10, 145–174.

Talwar, G.P., Raghuvanshi, P., Misra, R., Mukherjee, S., and Shas, S. (1997). Plant immunomodulators for termination of unwanted pregnancy and for contraception and reproductive health. *Immunol. Cell Biol.*, 5, 190–192.

Tella, A. (1976). Studies on *Azadirachta indica* in malaria. *Br. J. Pharmacol.*, 58, 318P.

Tella, A. (1977). The effects of *Azadirachta indica* in acute *Plasmodium berghei* malaria. *Nig. Med. J.*, 7, 258–263.

Tewari, D.N. (1992). *Monograph on Neem (Azadirachta indica A. Juss.).* International Book Distributors, Dehra Dun, India, 279 pp.

Thompson, E.B. and Anderson, C.C. (1978). Cardiovascular effects of *Azadirachta indica* extract. *J. Pharm. Sci.*, 67, 1476–1478.

Udeinya, I.J. (1993). Anti-malarial activity of Nigerian neem leaves. *Trans. R. Soc. Trop. Med. Hyg.*, 87, 471.

Upadhyay, S.N., Dhawan, S., Garg, S., and Talwar, G.P. (1992). Immunomodulatory effects of neem (*Azadirachta indica*) oil. *Int. J. Immunopharmacol.*, 14, 1187–1193.

Uwaifo. (1984). The mutagenicities of seven coumarin derivatives and a furan derivative (nimbolide) isolated from three medicinal plants. *J. Toxicol. Env. Health*, 13, 521–530.

Van der Burg, G. and Haasane, K. (1990). *The Exploitation of Neem in Niger.* Ministere de l'Hydraulique et de l'Environnement, Niamey, Niger, 199 pp.

Van der Nat, J.M., Klerx, J.P.A.M., Van Dijk, H., De Silva, K.T.D., and Labadie, R.P. (1987). Immunomodulatory activity of an aqueous extract of *Azadirachta indica* stem bark. *J. Ethnopharmacol.*, 19, 125–131.

Van der Nat, J.M., van der sluis, W.G., de Silva, K.T.D., and Labadie, R.P. (1991). Ethnopharmacognostical survey of *Azadirachta indica* A. Juss (Meliaceae). *J. Ethnopharmacol.*, 35, 1–24.

Vedavathy, S. and Rao, D.N. (1995). Herbal folk medicine of Tirumala and Tirupati region of Chittor District, Andhtra Pradesh. *Fitoterapia*, 66, 167–171.

Vivekanandan, P. (1998). New tree, new system: neem production in South India. *Agrofor. Today*, 10, 12–14.

Waring, E.J. (1897). *Remarks on the Uses of Some of the Bazaar Medicines and Common Medicinal Plants of India.* J&A Churchill, London, p. 106.

Weenen, H., Nkunya, M.H.H., Bray, D.H., Mwasumbi, L.B., Kinabo, L.S., and Kilimali, V.A.E.B. (1990). Antimalarial activity of Tanzanian medicinal plants. *Planta Med.*, 56, 368–370.

Zebitz, C.P.W. (1984). Effect of some crude and azadirachtin-enriched neem (*Azadirachta indica*) seed kernel extracts on larvae of *Aedes aegypti*. *Entomol. Exp Appl.*, 35, 11–16.

Zebitz, C.P.W. (1986). Effects of three different neem seed kernel extracts and azadirachtin on larvae of different mosquito species. *Z. Angewandte Entomol.*, 102, 455–463.

7 Malarial-5: Development of an Antimalarial Phytomedicine in Mali

Drissa Diallo, Ababacar Maïga, Chiaka Diakité, and Merlin Willcox

CONTENTS

7.1 INTRODUCTION

In Mali, malaria is the third most important febrile illness (accounting for 12.6% of cases of fever) and is responsible for 3.7% of admissions to the medical wards at Bamako University Hospital. Cerebral malaria accounts for 51.7% of paediatric emergencies and causes 42% of child deaths; it is also the most important cause of mortality (13%) and morbidity (15.6%) in the whole population (Doumbia, 1997). All these data show that malaria remains a major public health problem.

In view of the increasing levels of drug resistance in *Plasmodium falciparum* in Mali (Plowe et al., 2001), intensive research is necessary, including in the field of traditional medicine. Herbal medicine is widely used in Mali for a variety of diseases, including malaria (Ancolio et al., 2002; Diallo et al., 1999; Djimde et al., 1998; Théra et al., 1998, 2000; Traore et al., 1993). In other

FIGURE 7.1 *C. occidentalis.*

countries this has led to the development of important antimalarials, as discussed in other chapters of this book.

Malarial was first formulated by Prof. Mamadou Koumaré, based on a recipe used in his family, and is now produced as a standardised phytomedicine by the Département de Médecine Tradition-nelle (DMT) of the Institut National de Recherche en Santé Publique (INRSP) of Mali. Malarial-5 is sold as a powder in 10-g sachets, one of which is boiled in water to make a decoction (Figure 7.1). The powder is a mixture of three plant species:

- Leaves of *Cassia occidentalis* L. (Caesalpinaceae): 64% in original formulation, now 62%
- Leaves of *Lippia chevalieri* Mold (Verbenaceae): 32%
- Flowerheads of *Spilanthes oleracea* L. (Compositae): 4% in original formulation, now 6%

These different plants are used separately in traditional medicine. Patients treated with Malarial feel symptomatically better from 48 hours after onset of treatment, and parasitaemia decreases, without the parasites being cleared totally. Here we will discuss preclinical and clinical studies, and the importance of Malarial in the Malian Malaria Control Program.

7.2 PRECLINICAL STUDIES

7.2.1 ETHNOBOTANY AND PHYTOCHEMISTRY

7.2.1.1 *Cassia (= Senna) occidentalis* L. (Caesalpinaceae)
      [Local bamanan name: Balambalan Kassa Go]

This shrub is widely distributed in the district of Bamako, Mali, and more widely in Africa, Asia, and Latin America. It is reputed for its antipyretic, diaphoretic, and diuretic properties (Adjanohoun et al., 1981; Hirt and M'Pia, 2001; Kirtikar and Basu, 1933). It is very widely used as an antimalarial

TABLE 7.1
Worldwide Use of *C. occidentalis* for Malaria and Fever

Part Used	Preparation	Traditional Use	Country	Reference
LF	NR	Antimalarial	Burkina Faso	Bugmann, 2000
RT	NR	Antimalarial	Brazil	Brandão et al., 1991, 1992
AP	NR	Antimalarial	Brazil	Milliken, 1997a
LF	Dec	Antimalarial	Brazil	Di Stasi et al., 1994
AP	NR	Antimalarial	Colombia	Milliken, 1997b
RB	NR	Antimalarial	Democratic Republic of the Congo	Tona et al., 2001
LF	Dec	Antipyretic	India	Dagar and Dagar, 1991
AP	Dec	Antimalarial	Madagascar	Rasoanaivo et al., 1992
RT, SB	Dec	Antimalarial	Madagascar	Rahantamalala, 2000
LF	Dec	Antimalarial	Mali	Koita, 1991
LF, RT, WP	Dec, jce	Antipyretic, antimalarial	Nicaragua	Coe and Anderson, 1996
LF, RT	NR	Antipyretic, antimalarial	Nigeria	Etkin and Ross, 1991
ST	Dec	Antimalarial	São Tomé and Principe	Madureira et al., 1998
LF	Inf	Antimalarial	Tanzania	Gessler et al., 1995
LF	Inh, bth	Antipyretic	Tanzania	Kokwaro, 1993
LF	NR	Antipyretic	Tanzania	Watt and Breyer-Brandwijk, 1962
AP	NR	Antimalarial	Venezuela	Milliken, 1997b

Note: AP = aerial part; LF = leaf; SB = stem bark; RB = root bark; RT = root; WP = whole plant; NR = not recorded; bth = bath; dec = decoction; inf = infusion; inh = inhalation; jce = juice.

and antipyretic throughout the tropics (Table 7.1). The leaves are alternate and even pinnate with five to eight pairs of oval, acuminate leaflets, characterised by a black gland at the base of the petiole (Figure 7.2). They measure 4.5 × 2.5 cm. The flowers are yellow, in axillary or terminal short clusters. The fruits are slightly arched pods, 15 cm long, containing 10 to 20 seeds.

Tests carried out by DMT on the leaves of *C. occidentalis* have revealed the presence of saponosides (foaming index = 200), condensed tannins and gallotannins, mucilages, and polyuronides in the aqueous extract; free anthraquinones in the chloroform extract; sterols and triterpenes in the ether extract; and cardiac glycosides in the ethanol extract. These tests have confirmed the absence of alkaloids. The dried leaves contain 3% water; 34.7% of the dried leaves are extractable with water. The 4% aqueous extract has a pH of 6.2.

7.2.1.2 *Lippia chevalieri* Mould (Verbenaceae)
[Local bamanan name: Nganniba]

This woody aromatic herb can grow up to 2.5 m tall (Figure 7.2). The leaves are in whorls of three. They are oblong or elliptical, with a cuneate base, an acuminate tip, and fine dentate margins. The white flowers are in terminal spikes, which are umbelliform, globular, or cylindrical.

In Central America, *L. chevalieri* is used for the treatment of influenza, bronchitis, cough, and asthma (Pascual et al., 2001). In Nigeria, an infusion of the plant is consumed for its sedative and relaxing properties (Pousset, 1989). In Senegal, it is used to flavor tea, and in Mali to treat fever (Traore, 1999; Kerharo and Adam, 1974).

Studies undertaken in the DMT laboratory on the leaves have revealed the presence of saponosides (foaming index = 250), condensed tannins and gallotannins, sugars and polyuronides in the aqueous extract; steroid glycosides in the ethanol extract; and sterols and triterpenes in the ether

FIGURE 7.2 *L. chevalieri.*

extract. The essential oil is yellow, less dense than water, with a refractive index of 1.5010 at 28°C, and it constitutes 1.5% of the leaves. There are also traces of a few alkaloids. Water content of the dried leaves is 7%; total ash is 21.1%; 24.5% of the powdered stems and leaves are water soluble. The pH of the aqueous extract is 8.1.

7.2.1.3 *Spilanthes oleracea* L. (Compositae) [Local bamanan name: Farimani]

This sprawling annual plant has a cylindrical stem (Figure 7.3). Its leaves are simple, lanceolate, and dentate. The limb/blade measures 2 to 4 cm × 10 mm. The yellow tubular flowers are grouped in small conical capitula, on long peduncles. Its uses in traditional medicine are summarised in Table 7.2.

The flowers, used as a phytomedicine, contain mainly sterols and triterpenes in the ether extract; condensed tannins, gallotannins, and mucilages in the aqueous extract; cardiac glycosides in the ethanol extract; alkaloids in an aqueous sulfuric acid extract; and spilanthol in a hexane extract (a local anaesthetic and larvicidal at 10 ppm for mosquito larvae). Water content is 3.8%; total ash is 10.7%; 24% of the dry flowerheads can be extracted with water. The pH of the decoction is 5.3.

7.2.1.4 Malarial

Phytochemical screening of Malarial carried out at the DMT revealed the following chemical constituents: anthracene derivatives, tannins, polyuronides, sterols, and triterpenes. Quality control

FIGURE 7.3 *S. oleracea.*

TABLE 7.2
Use of *S. oleracea* in Traditional Medicine

Plant Part	Mode of Administration	Use	Countries
Flowerheads alone	Chewing	Local anaesthetic for toothache	Africa, India
		To aid digestion	
	Infusion	Malaria	
	Decoction	Lotion for rheumatism	Philippines
Leaves alone		Diuretic and for renal stones	
	Topical (by rubbing)	Antipruritic	India
	Moist powder	Oral inflammation	Africa
Roots alone	Decoction	Lotion for scabies and psoriasis	Philippines
		Purgative	
Leaves and flowerheads	Chewing	For scurvy	Madagascar
Whole plants	Decoction	Against dysentery	Africa, Southeast Asia
		Antipyretic	

Source: Jellal, A. et al., (1998), *Le Spilanthes*, Etablissement National d'Enseignement Supérieur Agronomique (ENESA), Dijon, France.

is carried out by determination of foreign matter, of brown or yellow leaves, macroscopic examination of powders, thin-layer chromatography of extracts, and the consistent weight of the sachets.

Malarial is sold in packets containing 11 individual sachets of 10 g (see Figure 7.4). It is recommended for flu-like symptoms (joint pains, fever) and malaria. It is prepared by boiling one sachet of 10 g with a seedless slice of lemon in 500 ml of water for 10 minutes. The decoction is filtered and drunk tepid; sugar is added to taste. The adult dose is one sachet twice daily for 4 days, followed by one sachet once daily for 3 days. The paediatric dose is half of the adult dose.

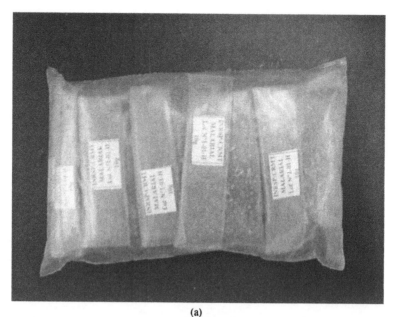

(a)

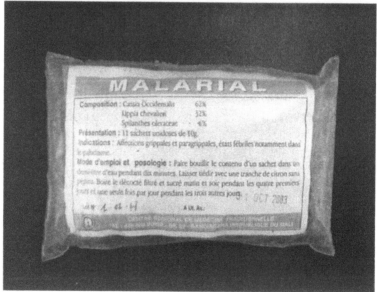

(b)

FIGURE 7.4 Sachets of Malarial.

7.2.2 TOXICITY

Acute toxicity has been determined in mice by oral administration of Malarial. Groups of five mice weighing 20 ± 2 g were given (1) the therapeutic dose (0.33 g/kg), (2) three times the therapeutic dose (0.99 g/kg), and (3) 90 times the therapeutic dose (30 g/kg). The mice were starved for 12 hours. The decoction was dissolved in water, and this was administered orally with a syringe and an esophageal probe. The animals were observed for 1 week, and during this time no adverse signs were seen. Postmortem examinations and histopathology were not performed. Tona et al. (2001) performed histopathological examinations of mice treated with ethanolic, dichloromethane, and lyophilised aqueous extracts of *C. occidentalis* root bark at a dose of 0.5 g/kg. No significant lesions were found. Cows and sheep have been fed large quantities of the leaves, with only transient gastrointestinal side effects (Watt and Breyer-Brandwijk, 1962).

7.2.3 PHARMACOLOGY

Each constituent of Malarial has a role. *C. occidentalis* powder and leaves are antimalarial: Iwalewa et al. (1990) found that a methanol and ether extract of dried, powdered leaves had an inhibitory concentration 50% (IC_{50}) of 0.59 μg/ml. *L. chevalieri* improves the taste of the medicine, and the flowerheads of *S. oleraceae* contain spilanthol, which is effective against *Plasmodium falciparum* (Jellal et al., 1998). The antimalarial activity of Malarial *in vitro* and *in vivo* was evaluated at the University of Marseille.

7.2.3.1 *In vitro* Antimalarial Activity

In order to evaluate the antimalarial activity of different extracts *in vitro*, inhibition of *P. falciparum* proliferation in continuous culture was determined according to the technique of Trager and Polonsky (1981). *In vitro* tests were done on two strains of *P. falciparum*:

FCC_2 (chloroquine sensitive, from Niger)
FZR (chloroquine resistant, from the Comores)

The strains were maintained in culture by the method of Jensen and Trager (1978).

Parasitised erythrocytes were suspended in the culture medium (RPMI 1640 Sigma + 10% human serum) with a predetermined concentration of the lyophilised water extract of Malarial-5 and of the part of each plant it is composed of. The lyophilised extracts were dissolved (stock solution: 2 mg/ml, then dilution by halves) in a culture medium of *P. falciparum* (RPMI 1640 Sigma + 10% human serum) sterilised on a Millipore filter (0.45 μ) to evaluate their antimalarial activity *in vitro*. Each concentration was tested three times. In each box, three negative controls were tested containing parasitised erythrocytes without Malarial in order to obtain the highest parasitaemia possible. Chloroquine was used as a positive control. After 48 hours of incubation, the parasitaemia in each box was quantified using thin-film microscopy (Gasquet et al., 1993).

Results were calculated as percent proliferation in comparison to the negative control. The IC_{50} values are given in Table 7.3. The antimalarial activity of each plant was similar on chloroquine-sensitive and chloroquine-resistant strains. The activity of the Malarial decoction was similar to that of *C. occidentalis*. *L. chevalieri* was twice as active, and *S. oleraceae* was three times as active as Malarial.

TABLE 7.3
IC_{50} Values of Different Extracts on Two Strains of *P. falciparum* (μg/ml)

Strain of *P. falciparum*	FCC_2	FZR
Malarial	600	470
C. occidentalis	660	580
L. chevalieri	300	380
S. oleraceae	180	200
Chloroquine	0.03	0.27

7.2.3.2 *In vivo* Antimalarial Activity

In vivo tests were carried out on mice (*Mus musculus* OF$_1$) infected with *Plasmodium berghei* according to the technique of Peters (1965). Mice were infected by intraperitoneal injection of 50 µl of blood containing 2×10^7 erythrocytes infected with *P. berghei*. The mice were maintained in the same conditions in isolation and fed with Totaliment.

Lyophilised extracts of Malarial were dissolved in water (20 mg/ml). Ten mice were treated with 200 mg orally for 5 days, starting on the day they were infected. Outcome measures were daily parasite counts on blood films from each mouse, and the survival time of treated mice compared to the control animals (Osdene, 1967).

The changes in parasitaemia in mice treated with decoctions of *C. occidentalis* leaves, *L. chevalieri* leaves, and *S. oleraceae* flowerheads were similar to those of the mice treated with Malarial. None of the treated mice were cured, but their parasitaemia was reduced. The treated mice survived 2 to 3 days longer than control mice. Chloroquine, given at 60 mg/kg orally for 5 days, cured the infected mice. Chloroquine-resistant strains were not resistant to Malarial (Gasquet et al., 1993).

7.3 CLINICAL STUDIES

Three clinical studies have been carried out to evaluate the safety and efficacy of Malarial. The first two studies were carried out at the Rural Health Research and Training Center in Sélingué (southern Mali), in collaboration with CREDES (Centre de Recherche d'Étude et de Documentation en Economie de la Santé), Terre des Hommes France, and Prof. Ogobara Doumbo. The third was conducted in Bamako.

7.3.1 RANDOMISED CONTROLLED TRIAL OF MALARIAL VS. CHLOROQUINE, 1987 (KOITA, 1989, 1991)

The study period was July to September 1987, when there was a high rate of malaria transmission. Patients presenting with malaria at the health center were screened for inclusion in the study. Inclusion criteria were as follows: ages 5 to 49, parasitaemia of 5000/mm^3 or above, a negative chloroquine urine test, and absence of severe malaria or concomitant illness. Patients were randomised (by tossing a coin) to Malarial or chloroquine.

Patients randomised to receive Malarial were shown how to prepare the decoction on the first day, and subsequently were asked to prepare the decoction at home. The 10-g sachet of Malarial powder was boiled in 500 ml of water for 15 minutes, adding sugar according to taste. Patients were asked to take this twice a day for the first 4 days, then once daily for the 3 remaining days. Patients in the chloroquine group were given chloroquine at a dose of 10 mg/kg/day for 3 days, followed by placebo capsules for 4 further days, so that their duration of treatment would also be 7 days.

Neither patients nor clinicians could be blinded to the treatment, since the Malarial was a decoction and chloroquine was in capsules. However, parasitaemia was assessed by a separate team, blinded to the treatment. Clinicians were blinded to the parasite counts. Patients were followed up for 3 weeks, with daily clinical examination and parasite counts.

There were 53 patients included, of which 36 were randomised to Malarial and 17 to chloroquine. Eighty-six percent of patients on Malarial completed the treatment to day 7, compared to 71% on chloroquine. This difference was not statistically significant (z = .06). Follow-up to day 21 was completed by 75% of the Malarial group and 59% of the chloroquine group.

Fever clearance was not significantly different between the two groups. There was a general trend toward better symptomatic improvement with Malarial (see Table 7.4), but this was not statistically significant. The same trend was observed on days 14 and 21 (Koita, 1989). However,

TABLE 7.4
Symptom Clearance by Day 7

Symptom	Clearance in Malarial Group (%)	Clearance in Chloroquine Group (%)
Fever	59	50
Headache	76	47
Shivering	63	80
Nausea	93	75
Vomiting	79	68

Source: Koita, 1991.

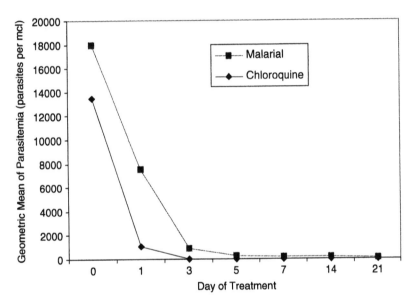

FIGURE 7.5 Parasitological response in first trial of Malarial vs. chloroquine. (From Koita, 1991.)

parasite clearance was better in the chloroquine group (see Figure 7.5). Mean parasitaemia on day 7 was 154 parasites per microliter in the Malarial group, compared to 3.4 per microliter in the chloroquine group. However, this is still a very significant reduction from the initial mean parasitaemia, which was 17,975 in the Malarial group and 13,414 in the chloroquine group.

Malarial was better tolerated than chloroquine. Three patients developed an allergy to chloroquine, causing them to abandon treatment on day 3, but there were no allergies reported in the Malarial group. One patient on Malarial developed constipation, compared to none in the chloroquine group.

7.3.2 COMPARISON OF MALARIAL WITH CHLOROQUINE, 1988 (GUINDO, 1988; KÉITA ET AL., 1990)

This study was conducted in the same area, with the same methods as the above study, but one year later, from July to October 1988. Inclusion criteria were the same, except that patients were required to have a temperature of 38°C or higher.

Fifty-three patients were included, 51% of them women and 49% men. Four villages took part in the study, with between 5 and 19 patients in each. There was no significant difference in the

TABLE 7.5
Baseline Characteristics of Patients

	Chloroquine	Malarial	Total
Number of patients	17	35	52
Mean age	8.55	9.10	8.92
Mean temperature	39.3	39.5	39.4
Mean parasitaemia	64,224	46,024	55,124

TABLE 7.6
Mean Parasitaemia from Days 1 to 21

	Chloroquine			Malarial			Total	
Day	Parasitaemia	N	% FU	Parasitaemia	N	% FU	N	% FU
0	64,224	17	100	46,024	35	100	52	100
1	16,797	13	76	44,646	27	77	40	77
3	35	11	65	22,010	32	91	43	83
5	0	10	59	8912	31	89	41	79
7	0	9	53	14,086	31	89	40	77
14	0	9	53	17,045	26	74	35	67
21	0	6	35	7533	19	54	25	48

Note: % FU = Percent of patients followed up.

mean ages, the mean temperature, or the mean parasitaemia of the two groups (Table 7.5). Parasitaemia at inclusion was higher than in the first study.

Follow-up of patients was a problem in both groups, more so in the chloroquine group from day 5 onwards (Table 7.6). This may be because all the patients treated with chloroquine cleared their parasites by day 5, whereas parasitaemia persisted in those followed up until day 21. The reduction in parasitaemia was slower for patients on Malarial, and their parasitaemia never went below 5000 parasites/mm^3 (Table 7.6, Figure 7.6).

The higher initial parasitaemia in this study may explain the poorer result than in the first study. Patients in this study were febrile at inclusion, with high parasite counts, so their immunity to malaria may have been less than that of the patients in the first study. It was felt that the amount of *S. oleraceae* (4%) present in this formulation of Malarial was insufficient for a truly effective schizonticidal activity. It was therefore decided to increase the amount of *S. oleraceae* in Malarial to 6%.

7.3.3 CLINICAL TRIAL OF MALARIAL WITH 6% *S. OLERACEAE* (DOUMBIA, 1997)

This was an observational cohort study on patients with uncomplicated malaria, conducted in Bamako from March to May 1995. Thirty patients were included, ages 5 and above, with a temperature of >37.5°C and a parasitaemia of >3000/μl *P. falciparum*. Patients with signs of severe disease or who had taken an antimalarial in the last 24 hours were excluded.

Adults were treated with one sachet of Malarial three times a day for 7 days, whereas children were given half this dose. Blood films were taken at 0, 3, and 7 days. There was no control group.

Fifteen male and 15 female patients were included. The mean age was 29, with a range of 8 to 60. The results are shown in Table 7.7. Parasitaemia declined and symptoms improved.

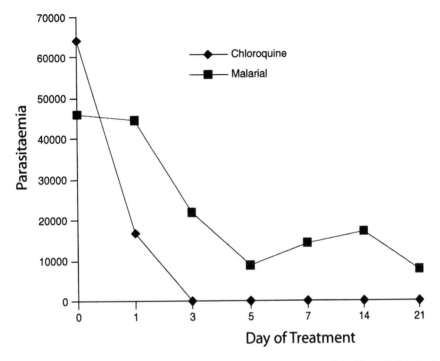

FIGURE 7.6 Parsaitological response in second trial of Malarial vs. chloroquine. (From Kéita, A., Doumbo, O., Koïta, N., Diallo, D., Guindo, M., and Traoré, A.K., (1990), *Méd. Traditionnelle Pharmacopée Bull. Liaison*, 4, 139–146.)

TABLE 7.7
Efficacy of Malarial with Enhanced *Spilanthes* Content

Day	Mean Parasitaemia	Mean Temperature (°C)
0	5465	39.3
5	1950	37.8
7	629	37.4

Source: Doumbia, S., (1997), *Etude des plantes antipaludiques au Mali*, Thèse Pharmacie, Bamako, Mali, 78 pp.

Parasitaemia at day 7 remained higher in patients ages 8 to 19 than in older patients. This suggests that patient immunity may be playing a role in clearing the parasites.

7.4 ROLE OF MALARIAL IN THE FIGHT AGAINST MALARIA IN MALI AND LESSONS FOR OTHER MALARIA-ENDEMIC COUNTRIES

The randomised controlled trials reported above show that chloroquine is significantly more effective than Malarial. However, the studies were conducted before the emergence of significant chloroquine resistance in Mali. By 1997, parasitological resistance levels had risen to 13 to 17%, causing treatment failure in 10 to 15% of patients (Plowe et al., 2001). These levels will probably rise as they have risen elsewhere in Africa. In these circumstances, Malarial may become more effective than chloroquine. Its efficacy in chloroquine-resistant malaria needs to be further evaluated. One disadvantage is that the volume of Malarial to be drunk is too high for young children; a more concentrated presentation would enable the administration of a correct therapeutic dose in the future.

Production and sale of the improved formulation of Malarial (6% *S. oleracea*) have been growing remarkably in Mali: 154 sachets in 1982, 892 in 1984, and 3840 in 2001. It is now included on the Malian National Formulary (Ministère de la Santé, 1998). The majority of the population use Malarial because it is culturally acceptable and readily available in community health centers. A course costs $0.95. In Mali, less than 20% of the population have ready access to Western medicine; there is a lack of qualified staff, and conventional pharmaceutical research is too costly. In these circumstances, Malarial is better than no treatment at all.

Malaria-endemic countries can use their botanical heritage to develop phytomedicines such as Malarial to treat patients affected by malaria, concentrating on the evaluation of safety and efficacy. As a result of the pharmacological studies and the first clinical trial of Malarial, a new formulation was adopted, containing 6% of *S. oleracea*. Two new studies are planned, one with this new formulation and one with a concentrated extract of *S. oleracea*.

ACKNOWLEDGMENTS

The authors thank Dr. Bertrand Graz and Dr. Jacques Falquet for their help in translating this article.

REFERENCES

Adjanohoun, J., Aké Assi, L., Floret, J.J., Guinko, S., Koumaré, M., Ahyi, A.M.R., and Raynal, J. (1981). *Médecine traditionnelle et Pharmacie. Contribution aux études ethnobotaniques et floristiques au Mali*. ACCT, Paris, 291 pp.

Ancolio, C., Azas, N., Mahiou, V., et al. (2002). Antimalarial activity of extracts and alkaloids isolated from six plants used in traditional medicine in Mali and Sao Tome. *Phytother. Res.*, 16, 646–649.

Brandão, M.G.L., Carvalho, L.H., and Krettli, A.U. (1991). Antimaláricos de uso popular na Amazônia. *Tome Ciência*, 13, 9–12.

Brandão, M.G.L., Grandi, T.S.M., Rocha, E.M.M., Sawyer, D.R., and Krettli, A.U. (1992). Survey of medicinal plants used as antimalarials in the Amazon. *J. Ethnopharmacol.*, 36, 175–182.

Bugmann, N. (2000). Le concept du paludisme, l'usage et l'efficacité *in vivo* de trois traitements traditionnels antipalustres dans la région de Dori, Burkina Faso. Doctoral thesis, Faculty of Medicine, University of Basel, Switzerland.

Coe, F.G. and Anderson, G.J. (1996). Ethnobotany of the Garífuna of eastern Nicaragua. *Econ. Bot.*, 50, 71–107.

Dagar, A.S. and Dagar, J.C. (1991). Plant folk medicines among the Nicobarese of Katchal Island, India. *Econ. Bot.*, 45, 114–115.

Diallo, D., Hveem, B., Mahmoud, M.A., Berge, G., Paulsen, B., and Maiga, A. (1999). An ethnobotanical survey of herbal drugs of Gourma district, Mali. *Pharm. Biol.*, 37, 80–91.

Di Stasi, L.C., Hiruma, C.A., Guimaraes, E.M., and Santos, C.M. (1994). Medicinal plants popularly used in Brazilian Amazon. *Fitoterapia*, 65, 529–540.

Djimde, A., Plowe, C.V., Diop, S., Dicko, A., Wellems, T., and Doumbo, O. (1998). Use of antimalarial drugs in Mali: policy versus reality. *Am. J. Trop. Med. Hyg.*, 59, 376–379.

Doumbia, S. (1997). *Etude des plantes antipaludiques au Mali*. Thèse Pharmacie, Bamako, Mali, 78 pp.

Etkin, N.L. and Ross, P.J. (1991). Recasting malaria, medicine and meals: a perspective on disease adaptation. In *The Anthropology of Medicine*, 2nd ed., Romanucci-Ross, L., Moerman, D.E., and Tancredi, L.R., Eds. Bergin & Garvey, New York.

Gasquet, M., Delmas, F., Timon-David, P., Keita, A., Guindo, M., Koita, N., Diallo, D., and Doumbo, O. (1993). Evaluation *in vitro* and *in vivo* of traditional antimalarial, "Malarial 5." *Fitoterapia*, 64, 423–426.

Gessler, M.C., Suya, D.E., Nkunya, M.H.H., et al. (1995). Traditional healers in Tanzania: the treatment of malaria with plant remedies. *J. Ethnopharmacol.*, 48, 131–144.

Guindo, M. (1988). Contribution à l'étude du traitement traditionnel du "Suma" (Paludisme). Thèse pour obtenir le grade de Docteur en Pharmacie, Ecole Nationale de Médecine et de Pharmacie du Mali.

Hirt, H.-M. and M'Pia, B. (2001). *Natural Medicine in the Tropics*. Anamed, Winnenden, Germany.

Iwalewa, E.O., Lege-Oguntoye, L., Rai, P.P., Iyaniwura, T.T., and Etkin, N.L. (1990). *In vitro* antimalarial activity of leaf extracts of *Cassia occidentalis* and *Guiera senegalensis* in *Plasmodium yoelii nigeriensis*. *West Afr. J. Pharmacol. Drug Res.*, 9 (Suppl.), 19–21.

Jellal, A., Lemerre, S., Michot, P., Oger, R., and Rabiller, P. (1998). *Le Spilanthes*. Etablissement National d'Enseignement Supérieur Agronomique (ENESA), Dijon, France.

Jensen, J.B. and Trager, W. (1978). *Plasmodium falciparum* in culture: establishment of additional strains. *Am. J. Trop. Med. Hyg.*, 27, 743–746.

Kéita, A., Doumbo, O., Koïta, N., Diallo, D., Guindo, M., and Traoré, A.K. (1990). Recherche expérimentale sur un anti-malarique traditionnel. "Etude préliminaire sur la faisabilité d'un protocole d'essai clinique." *Méd. Traditionnelle Pharmacopée Bull. Liaison*, 4, 139–146.

Kerharo, J. and Adam, J.G. (1974). *La Pharmacopée Sénégalaise Traditionelle. Plantes Médicinales et Toxiques*. Paris, Vigot Frères, p. 1011.

Kirtikar, K.R. and Basu, B.D. (1933). *Indian Medicinal Plants*, 2nd ed., Vol. II. Lalit Mohan Basu, Allahabab, India, 860 pp.

Kokwaro, J.O. (1993). *Medicinal Plants of East Africa*. Kenya Literature Bureau, Nairobi.

Koita, N. (1989). A Comparative Study of the Traditional Remedy "Suma-Kala" and Chloroquine as a Treatment for Malaria in Rural Area of Mali. Dissertation for M.Sc. CHDC, London School of Hygiene and Tropical Medicine, University of London.

Koita, N. (1991). A comparative study of the traditional remedy "Suma-Kala" and chloroquine as treatment for malaria in the rural areas. In *Proceedings of an International Conference of Experts from Developing Countries on Traditional Medicinal Plants, Arusha, Tanzania, February 18–23, 1990*, Mshigeni, K.E., Nkunya, M.H.H., Fupi, V., Mahunnah, R.L.A., and Mshiu, E., Eds. Dar Es Salaam University Press, Tanzania.

Madureira, M.C., Martins, A.P., Gomes, M., et al. (1998). Antimalarial Activity of Medicinal Plants from São Tomé and Príncipe Islands. Clone, Cure and Control: Tropical Health for the 21st Century. Liverpool School of Tropical Hygiene and Medicine, European Congress Office, Liverpool. Poster 64, abstract, p. 137.

Milliken, W. (1997a). Traditional antimalarial medicine in Roraima, Brazil. *Econ. Bot.*, 51, 212–237.

Milliken, W. (1997b). Malaria and antimalarial plants in Roraima, Brazil. *Trop. Doct.*, 27 (Suppl. 1), 20–25.

Ministère de la Santé, des Personnes Agées et de la Solidarité. (1998). *Formulaire Thérapeutique National*. Editions Donnaya, Bamako, Mali.

Osdene, T.S., Russel, P.B, and Rane, L. (1967). 2,4,7-Triamino-6-*ortho*-substituted Arylpteridines. A new series of potent antimalarial agents. *J. Med. Chem.*, 10, 431–434.

Pascual, M.E., Slowing, K., Carretero, E., Sánchez Mata, D., and Villar, A. (2001). *Lippia*: traditional uses, chemistry and pharmacology: a review. *J. Ethnopharmacol.*, 76, 201–214.

Peters, W. (1965). Competitive relationship between *Eperythrozoon coccoides* and *Plasmodium berghei* in the mouse. *Exp. Parasitol.*, 16, 158–166.

Plowe, C., Doumbo, O.K., Djimde, A., et al. (2001). Chloroquine treatment of uncomplicated *Plasmodium falciparum* malaria in Mali: parasitologic resistance versus therapeutic efficacy. *Am. J. Trop. Med. Hyg.*, 64, 242–246.

Pousset, J.L. (1989). *Plantes Médicinales Africaines. Ellipse*. Agence de Coopération Culturelle et Technique, Paris, pp. 102–103.

Rahantamalala, C. (2000). Contribution à l'inventaire et à l'évaluation en laboratoire de quelques plantes médicinales utilisées dans le traitement du paludisme. Thèse pour obtenir le grade de Docteur en Médecine, Faculté de Médecine, Université d'Antananarivo, Madagascar.

Rasoanaivo, P., Petitjean, A., Ratsimamanga-Urverg, S., and Rakoto-Ratsimamanga, A. (1992). Medicinal plants used to treat malaria in Madagascar. *J. Ethnopharmacol.*, 37, 117–127.

Théra, M.A., D'Alessandro, U., Thiéro, M., et al. (2000). Child malaria treatment practices among mothers in the district of Yanfolia, Sikasso region, Mali. *Trop. Med. Int. Health*, 5, 876–881.

Théra, M.A., Sissoko, M.S., Heuschkel, C., et al. (1998). Village Level Treatment of Presumptive Malaria: Experiences with the Training of Mothers and Traditional Healers as Resource Persons in the Region of Mopti, Mali. Clone, Cure and Control: Tropical Health for the 21st Century. Second European Congress on Tropical Medicine, Liverpool. Poster P72, p. 139.

Tona, L., Mesia, K., Ngimbi, N.P., et al. (2001). In vivo antimalarial Activity of *Cassia occidentalis*, *Morinda morindoides* and *Phyllanthus niruri*. *Ann. Trop. Med. Parasitol.*, 95, 47–57.

Trager, W. and Polonsky, J. (1981). Antimalarial activity of quassinoids against chloroquine-resistant *Plasmodium falciparum* in vitro. *Am. J. Trop. Med. Hyg.*, 30, 531–537.

Traore, F. (1999). Evaluation de l'activité antimalarique de *Glinus oppositofolius* (L.) ADC., *Nauclea latifolia* (SM), *Mitragyna inermis* (Willd) O. Kuntze, trois plantes utilisées en médecine traditionnelle au Mali. Thèse doctorat, Marseille, France, 285 pp.

Traore, S., Coulibaly, S.O., and Sidibe, M. (1993). Comportements et coûts liés au Paludisme chez les femmes des campements de pêcheurs dans la zone de Sélingué au Mali, TDR/SER/PRS/12. Institut National de Recherche en Santé Publique, Bamako, Mali.

Watt, J.M. and Breyer-Brandwijk, M. (1962). *The Medicinal and Poisonous Plants of Southern and Eastern Africa*. E&S Livingstone Ltd., Edinburgh.

8 *Cryptolepis sanguinolenta*

Jonathan Addae-Kyereme

CONTENTS

8.1 INTRODUCTION

Cryptolepis sanguinolenta Lindl. Schltr. (Periplocaceae) is a scrambling shrub native to West Africa (Figure 8.1). The genus *Cryptolepis* R.Br. includes about 20 species found mainly in tropical Africa and in Southeast Asia (Mendes and Jensen, 1984). In some literature, members of the family Periplocaceae are included in the family Asclepiadaceae. This explains why some texts assign *C. sanguinolenta* to the latter family.

Cryptolepis sinensis Lour. Merr. and *Cryptolepis buchanani* Roem. Shult. grow in parts of Southeast Asia. *C. sinensis* is used externally as a treatment for snake bites and scabies. The fruits and roots of *C. buchanani* are used to treat chills and edema (Ping-tao et al., 1995). Mendes and Jensen (1984) indicate that the aqueous extract of the roots of *Cryptolepis obtusa* N.E.Br. is used in Mozambique to relieve abdominal pains.

In Ghana *C. sanguinolenta* is found mainly on the slopes of the Akwapim and Kwahu mountain ranges. The extract of the roots of this plant has been used traditionally by some ethnic groups in the eastern region of Ghana for the treatment of fevers, urinary and upper respiratory tract infections, and thrush. It is also taken as a panacea (Boye, 2002) and a diuretic (Sam, 2002). The average adult dose prescribed by the herbalist is a cupful (about 150 ml) up to three times daily. More liberal amounts are taken when it serves as a panacea. This is thought to suggest that the extract is relatively safe. Collection of the roots of the wild, mature plant is done all year round (Boye, 2002). Some cultivation of the shrub is now done to meet an increased demand for the roots.

C. sanguinolenta was introduced to the Center for Scientific Research into Plant Medicine (CSRPM) in 1974. Decoctions of the roots of this plant have since then been used as an effective treatment for malaria and some bacterial infections (Boye and Ampofo, 1983). At the center, the decoction is prepared by boiling 3 parts of powdered root in 100 parts water (Sam, 2002).

FIGURE 8.1 *C. sanguinolenta.* (Copyright 2002, Jonathan Addae-Kyereme.)

8.2 BOTANICAL DESCRIPTION

According to Sarpong (1983), the leaves of *C. sanguinolenta* are thinly herbaceous and elliptic oblong to ovate or lanceolate in shape. Margins of the leaves are entire and apices curved and acuminate. They have symmetrical and obtuse or rounded bases. The texture of the leaf is described as glabrous and papery. It is odourless and has a slightly bitter taste. The flowers are small and yellow in colour (Figure 8.2). Seeds are small and flat (Amponsah et al., 2002). A transverse section through the stem reveals a narrow region of cork. The xylem is porous, has a yellow colour, and surrounds a central pith. The roots of the plant are tortuous and branched with a yellowish brown outer surface. A section through the root shows thin-walled rectangular cork cells containing yellowish brown material. A significant portion of the root is made up of thin-walled, irregularly shaped parenchymatous cells that are filled with starch grains (Sarpong, 1983).

8.3 CULTIVATION

C. sanguinolenta is propagated either by the seeds or by root cuttings (Amponsah et al., 2002). To facilitate harvesting of the roots, the plant is best grown on mounds or on raised ridges. Harvesting begins when the shrub is about a year old. Limiting root collection to about a third at a time ensures that plant growth is not disrupted.

8.4 PHYTOCHEMISTRY

Sarpong (1983) has determined that the leaves, stems, and roots of *C. sanguinolenta* contain approximately 1, 0.5, and 0.5% of alkaloids, respectively. The major alkaloid was found to be cryptolepine, an N-methyl derivative of the indoloquinoline compound quindoline (Figure 8.3). Cryptolepine was first isolated from *Cryptolepis triangularis* by Clinquart in 1929. Dwuma-Badu et al. (1978) subsequently isolated both cryptolepine and quindoline from the roots of *C. sanguinolenta*.

FIGURE 8.2 *C. sanguinolenta.* (Copyright 2002, Jonathan Addae-Kyereme.)

Other alkaloids that have been isolated from *C. sanguinolenta* include cryptospirolepine, a spirononacyclic compound (Tackie et al., 1993); cryptolepicarboline, a dimer of β-carboline and an indoloquinoline (Sharaf et al., 1994); and the indolobenzazepinone homocryptolepinone (Sharaf et al., 1995).

Paulo et al. (2000) have reported the presence in the leaves of *C. sanguinolenta* of cryptolepinoic acid, methyl cryptolepinoate, hydroxycryptolepine, and the indolobenzazepine alkaloid cryptoheptine. An earlier study had found hydroxycryptolepine and cryptoheptine also in the roots of *C. sanguinolenta* (Paulo et al., 1995). Neocryptolepine (an isomer of cryptolepine) and biscryptolepine (a cryptolepine dimer) have been isolated from the roots of *C. sanguinolenta* (Cimanga et al., 1996).

Fort and coworkers (1998) have isolated cryptolepinone from *C. sanguinolenta.* They concluded from its synthesis and from spectroscopic studies that it is the keto form of hydroxycryptolepine and further state that cryptolepinone is an oxidation artifact of cryptolepine extraction. Isocryptolepine is another isomer of cryptolepine that is found in *C. sanguinolenta* (Grellier et al., 1996).

Cryptosanguinolentine and cryptotackieine have been isolated from the roots of *C. sanguinolenta* and their structures elucidated by Sharaf et al. (1996a) (Figure 8.3). These alkaloids have structures identical to isocryptolepine and neocryptolepine, respectively. Cryptomisrine is one of the dimeric alkaloids from the roots of *C. sanguinolenta* (Sharaf et al., 1996b).

8.5 BIOLOGICAL ACTIVITY

Cryptolepine has been shown to lower the blood pressure and body temperature of anaesthetised dogs (Gellert et al., 1951). The compound was reported to have anti-inflammatory properties (Noamesi and Bamgbose, 1983). These workers demonstrated that cryptolepine reduced carrageenan-induced edema in the rat hind paw subplantar tissue. When given orally, the alkaloid significantly lowered glucose levels in the mouse model of diabetes. It was also shown to increase glucose uptake by 3T3-L1 cells (Luo et al., 1998).

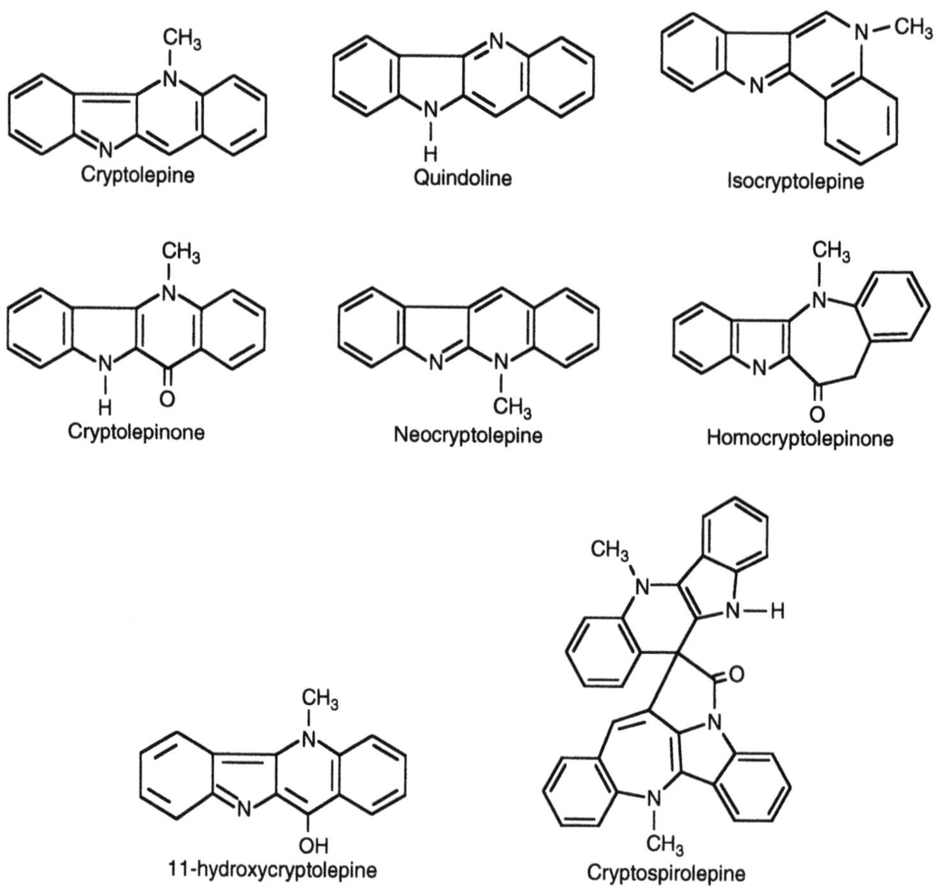

Figure 8.3 Some alkaloids isolated from *C. sanguinolenta*.

8.5.1 *IN VITRO* ANTIPLASMODIAL ACTIVITY

The most studied of the *C. sanguinolenta* alkaloids is cryptolepine. Wright et al. (2001), using the lactate dehydrogenase assay described by Makler et al. (1993), found that the alkaloid had potent *in vitro* activity ($IC_{50} = 0.44$ μM) against *Plasmodium falciparum* (multi-drug-resistant strain K1). In the same system, chloroquine diphosphate had comparable activity ($IC_{50} = 0.18 \mu M$). Against the chloroquine-sensitive strain HB3, the IC_{50} values for cryptolepine and chloroquine were 0.27 and 0.023 μM, respectively. However, quindoline was found to be inactive *in vitro* ($IC_{50} > 100$ μM) against K1 parasites. This was thought to be an indication that the N-methyl group in cryptolepine was necessary for antiplasmodial activity.

In an earlier *in vitro* study, IC_{50} values of 0.56 and 0.20 μM, respectively, were recorded for cryptolepine and chloroquine against chloroquine-resistant (FcB1) strains of *P. falciparum* (Grellier et al., 1996).

Cimanga et al. (1997) have also reported appreciable activity for 11-hydroxycryptolepine *in vitro*. Tests conducted recently, however, found this compound inactive *in vitro* (Wright, 2002).

The activities of various alkaloids extracted from *C. sanguinolenta* collected in Guinea Bissau have been screened *in vitro* against *P. falciparum*, strain K1, and the chloroqine-sensitive strain T996 (Paulo et al., 2000). Cryptolepine was found to be the most active (IC_{50} values of 0.23 and 0.059 μM against K1 and T996 strains, respectively). IC_{50} values of 0.26 μM (K1) and 0.019 μM (T996) were quoted for chloroquine. Other alkaloids tested were cryptoheptine (with IC_{50} values against K1 and T996 strains of 0.8 and 1.2 μM, respectively) and cryptolepinoic acid (no significant activity). Paulo and associates inferred from the results that cryptolepine was cross-resistant with chloroquine. In the report by Wright et al. (2001), however, this relationship was not evident.

Kirby et al. (1995) found cryptolepine as active against the K1 strains of *P. falciparum* as chloroquine ($IC_{50} = 0.11$ and $0.20\ \mu M$, respectively).

Eight naturally occurring anhydronium bases have been tested *in vitro* against the K1 strain of *P. falciparum* (Wright et al., 1996). Cryptolepine showed activity ($IC_{50} = 0.114\ \mu M$) similar to that of chloroquine ($IC_{50} = 0.20\ \mu M$).

C. sanguinolenta root extract was 1 of 20 plant extracts tested by Tona et al. (1999) against *P. falciparum in vitro*. At 6 μg/ml this extract was reported to inhibit more than 60% parasite growth.

8.5.2 *In vivo* Antimalarial Activity

Given orally at 50 mg/kg/day for 4 days to *Plasmodium berghei*-infected mice, cryptolepine suppressed parasitaemia by 80.5%. Chloroquine diphosphate (at 5 mg/kg/day) lowered parasitaemia by 93.5% in the same model (Wright et al., 1996). However, no effect was observed when the alkaloid was administered subcutaneously at 113 mg/kg/day (Kirby et al., 1995).

Grellier et al. (1996) found that a dose of 2.5 mg/kg/day for 4 days of cryptolepine given by the intraperitonial (i.p.) route reduced parasitaemia in *Plasmodium vinckei petteri*-infected mice by 80%. A similar level of activity was recorded for chloroquine administered at 1.5 mg/kg/day for 4 days by the same route.

However, when a group of *P. berghei*-infected mice were given 20 mg/kg/day of cryptolepine i.p., fatalities occurred on the second day (Wright et al., 2001).

8.6 TOXICOLOGICAL STUDIES

Ansa-Asamoah and Frempong-Asante (1983) studied the acute and subacute toxicity of cryptolepine in male albino mice. Two groups of mice were given daily subcutaneous injections of either 30 or 60 mg/kg of the alkaloid. When the animals were sacrificed after 2 weeks, no histological changes were detected in the liver and kidneys. After 6 weeks, extensive necrosis of liver cells was detected in both groups. While no obvious histological changes were observed in the kidneys of mice given the lower dose, widespread cortical cell damage was reported in the kidneys of those given the higher dose.

Cryptolepine has been shown to intercalate into DNA (Bonjean et al., 1998). The workers used DNAase I footprinting experiments to determine that the preferred site of interaction was with GC-rich sequences. Further quantitative DNA analysis led the group to conclude that the alkaloid extensively inhibited the synthesis of DNA. These findings are in agreement with those of Lisgarten et al. (2002), who established from competitive dialysis assays that the alkaloid selectively bound to CG-rich sequences containing nonalternating CC sites.

Further evidence of the cytotoxic action of cryptolepine has been reported by Dassonneville et al. (2000). The study was aimed at determining the mechanism underlying this activity of the alkaloid and its isomer neocryptolepine. They indicated that cryptolepine but not neocryptolepine caused cell death by inhibiting the action of poly (ADP-ribose) polymerase. Cryptolepine was found to be four times as cytotoxic as neocryptolepine.

Ansah and Gooderham (2001) have investigated the potential of aqueous extracts of *C. sanguinolenta* to cause cytotoxicity in Chinese hampster V79 cells. This was prompted by their concerns about the possible toxicity risks accompanying the human consumption of extracts of *C. sanguinolenta*. Using the alamar blue technique, trypan blue exclusion, and colony formation assay, these workers compared the growth inhibition abilities of cryptolepine and the aqueous extract of the roots of *C. sanguinolenta*. They concluded that the extract had a similar pattern of cytotoxicity to the alkaloid.

The distribution of 3H-cryptolepine after intravenous injection has been studied in whole-body autoradiography experiments in pigmented and albino mice and also in pregnant mice (Noamesi et al., 1991). Radioactivity measurements showed that the drug was extensively distributed in the body. However, no activity was recorded in the central nervous system. Relatively high and sustained counts were measured in the liver and adrenal medulla and in melanin-containing tissues of the eyes of both the adults and the fetuses. Activity was found to be significantly lower in the fetuses than in the mothers. Prolonged binding of the alkaloid to vital organs was thought to signal a potential toxicological risk after chronic use of the compound.

Phyto-Laria is a herbal tea preparation containing powdered roots of *C. sanguinolenta*. In a recent acute toxicity assessment of this product, up to 2 g/kg of body weight of the dry extract (corresponding to 1754-fold the human dose of the product) was given orally to groups of mice, rats, and rabbits. No adverse effects were observed in any of the rodents used in the study. No deaths occurred in any of the species of animals. Gross necropsies performed after 14 days on all the animals revealed no product-related organ damage. Some differences were observed in liver and kidney weights and in some aspects of the blood chemistry of animals given the maximum dose of 2 g/kg of body weight. It was noted that this dose was not likely to be reached in humans. The report concludes that no overt toxicity to the animals occurred at the dose levels of the extract tested (Sarpong, 2002).

8.7 CLINICAL STUDIES

Boye (1989) reported on a clinical study conducted at CSRPM on the antimalarial action of the aqueous extract of the roots of *C. sanguinolenta*. The work was an open, randomised trial that compared the efficacy of the extract with that of chloroquine in the treatment of malaria.

It was based on the World Health Organization (WHO) extended 7-day *in vivo* test. Patients used in the study were selected from the local population (semi-immune) and were between the ages of 16 and 60 years. They were all diagnosed with symptomatic uncomplicated falciparum malaria. Parasite levels were determined microscopically and ranged between 1000 and 100,000 per 8000 white cells. All the enrolled patients tested negative for chloroquine and sulfonamides in the urine.

The first group of 12 patients was treated with *Cryptolepis* extract (25 mg/kg of body weight three times daily). Ten others were given chloroquine (10 mg/kg on the first 2 days and 5 mg/kg on the third day). Parasite levels were determined each day for 7 days. Patients were monitored for signs of toxicity. Blood and urine samples were analysed weekly for indicators of any adverse drug effects.

An average parasite clearance time of 3.3 days was reported for the first group and 2.2 days for those treated with chloroquine. Patients treated with the extract were free of symptoms (headaches, bodily pains, and fever) in 36 hours. In the second group symptoms subsided after 48 hours. Fewer side effects were reported by patients taking the extract than those on chloroquine. Antipyretics were needed to control fever in patients given chloroquine but not in those taking the extract. No significant changes were detected in the blood and urine samples analysed. No cases of recrudescence were reported in any of the participants in the follow-up 28-day period.

It was concluded in the report that the activity of the aqueous extract of *C. sanguinolenta* was comparable with that of chloroquine and that the extract was safe.

8.8 CONCLUSION

Before 1974, the medicinal uses of *C. sanguinolenta* were limited to traditional practice. The introduction of the plant to CSRPM was followed in 1983 by an international conference held in Kumasi, Ghana, to assess its potential as an antimalarial. These developments led to an increased awareness of the usefulness of the plant. Presently *C. sanguinolenta* is incorporated into a number of herbal products such as Malaherb, Herbaquine, and Phyto-Laria on sale on the Ghanaian market for the treatment of malaria and fevers. Phyto-Laria is packaged for export. WHO is currently collaborating with CSRPM to develop a standardised form of the extract for use in health institutions in African countries (Armah, 2002). Clearly, the use of *C. sanguinolenta* as an antimalarial is increasing.

Some factors that may have contributed to the growth in the use of the shrub are the spread of chloroquine-resistant malaria parasites, the rather high cost of the newer antimalarials (a course of treatment with artesunate in Ghana costs the equivalent of about $4, while a 10-tablet course of treatment with chloroquine costs only about 10 cents), a preference of sections of the population for products of natural origin, and side effects (pruritus) experienced by some patients taking chloroquine. Efforts by WHO in the past decade to promote the use of traditional medicines in developing countries could also have been a contributory factor.

The studies cited in this review indicate that *C. sanguinolenta* contains alkaloids that have appreciable activity against malaria parasites *in vitro* and that the major alkaloid cryptolepine is active in rodent malaria models. Against chloroquine-sensitive organisms the alkaloids were generally less active than chloroquine.

The trial carried out in 1989 was not blinded (this would have been difficult to do because of the different treatment regimens). Blinding, however, would have minimised any potential biases in the interpretation of results. Also, the inferences drawn from this active control trial would have been enhanced if the results had been given a statistical treatment. A more appropriately designed study may therefore be needed to adequately evaluate the efficacy of the extract. Results of the studies listed do, however, suggest that the extract of *C. sanguinolenta* has some antimalarial activity, though it may not be as effective as chloroquine.

A point has been made that the drug requirements for treating malaria in semi-immune and nonimmune populations may possibly differ (Kirby, 1997). While the former group might benefit from local herbal extracts, the latter would need to be treated with more potent antimalarials.

Aqueous extracts of *C. sanguinolenta* have been used for several decades now without any evidence of significant harmful effects. Studies reviewed in this work indicate that it is safe. However, some researchers have expressed doubts about the continued use of the plant extract in the light of increasing reports of cytotoxicity due to some of the alkaloids of the plant. It is therefore imperative that investigations are undertaken into possible chronic toxicological risks that may be associated with its use. Data is needed on the absorption, metabolism, and excretion of the alkaloids of *C. sanguinolenta*. This information should give researchers an idea of what level of systemic exposure results from ingesting a given amount of the extract. Appropriate dose levels can then be determined for chronic toxicity studies in animals. False positive outcomes (toxicity elicited in animals but no adverse effects produced in humans) would be minimised. Such results can follow from studies in which higher doses are used in animals than are administered in humans. It is likely that the extensive organ damage reported in the test animals by Ansa-Asamoah and Asante-Frempong (1983) was a function of the rather high dose levels of cryptolepine used in the study. The amount of the alkaloid administered to the animals could explain the difference in outcomes of the studies by Grellier et al. (1996) and Wright et al. (2001). While 2.5 mg/kg/day of cryptolepine was not toxic to mice in the former work, fatalities occurred on the second day when a dose level of 20 mg/kg/day of the same compound was given by the same route in the latter study.

ACKNOWLEDGMENT

I am grateful to Dr. Colin W. Wright of the School of Pharmacy, University of Bradford, West Yorkshire, U.K., for reviewing this manuscript.

REFERENCES

Amponsah, K., Crensil, O.R., Odamtten, G.T., and Ofosuhene-Djan, W. (2002). *Manual for the Propagation and Cultivation of Medicinal Plants of Ghana*. Aburi Botanical Gardens, Aburi, Ghana, p. 22.

Ansa-Asamoah, R. and Frempong-Asante, J. (1983). Preliminary studies on the short term toxicity of cryptolepine in mice. In *Proceedings of the First International Seminar on Cryptolepine, July 27–30, University of Science and Technology, Kumasi, Ghana*, pp. 52–55.

Ansah, C. and Gooderham, N.J. (2001). The aqueous extract of *Cryptolepis sanguinolenta*, a West African herbal medicine, is cytotoxic. *Toxicology*, 168, 126–127.

Armah, G. (2002). Ministry to Present Patent Bill to House. *Daily Graphic* (Ghana), March 13, p. 32.

Bonjean, K., De Pauw-Gillet, M.C., Defresne, M.P., Colson, P., Houssier, C., Dassonneville, L., Bailly, C., Greimers, R., Wright, C.W., Quetin-Leclercq, J., Tits, M., and Angenot, L. (1998). The DNA intercalating alkaloid cryptolepine interferes with topoisomerase II and inhibits primary DNA synthesis in B16 melanoma cells. *Biochemistry*, 37, 5136–5146.

Boye, G.L. (1989). Studies on the antimalarial action of *Cryptolepis sanguinolenta* extract. In *Proceedings of International Symposium on East-West Medicine, Seoul, Korea*, pp. 242–251.

Boye, G.L. (2002). Private communication.

Boye, G.L. and Ampofo, O. (1983). Clinical uses of *C. sanguinolenta*. In *Proceedings of the First International Seminar on Cryptolepine, July 27–30, University of Science and Technology, Kumasi, Ghana*, pp. 37–40.

Cimanga, K., De Bruyne, T., Pieters, L., Claeys, M., and Vlietinck, A. (1996). New alkaloids from *Cryptolepis sanguinolenta*. *Tetrahedron Lett.*, 37, 1703–1705.

Cimanga, K., De Bruyne, T., Pieters, L., Vlietinck, A.J., and Turger, C.A. (1997). *In vitro* and *in vivo* antiplasmodial activity of cryptolepine and related alkaloids from *Cryptolepis sanguinolenta*. *J. Nat. Prod.*, 60, 688–691.

Clinquart, E. (1929). *Bull. Acad. R. Med. Belg.*, 9, 627. Cited in Dwuma-Badu, D. (1983). Review of the isolation and characterisation of the alkaloids of *Cryptolepis sanguinolenta*. In *Proceedings of the First International Seminar on Cryptolepine, July 27–30, University of Science and Technology, Kumasi, Ghana*, pp. 23–29.

Dassonneville, L., Lansiaux, A., Wattelet, A., Wattez, N., Mahieu, C., van Miert, S., Pieters, L., and Bailly, C. (2000). Cytotoxicity and cell cycle effects of the plant alkaloids cryptolepine and neocryptolepine: relation to drug induced apoptosis. *J. Pharmacol.*, 409, 9–18.

Dwuma-Badu, D., Ayim, J.S.K., Fiagbe, N.I.Y., Knapp, J.E., Schiff, P.L., Jr., and Slatkin, D.J. (1978). Constituents of West African medicinal plants XX. Quindoline from *Cryptolepis sanguinolenta*. *J. Pharm. Sci.*, 67, 433–434.

Fort, D.M., Litvak, J., Chen, J.L., Lu, Q., Phuan, P.W., Cooper, R., and Bierer, D.E. (1998). Isolation and unambiguous synthesis of cryptolepinone: an oxidation artifact of cryptolepine. *J. Nat. Prod.*, 61, 1528–1530.

Gellert, S., Raymond-Hamet, C.R., and Schlitterler, W. (1951). *Helv. Chim. Acta*, 34, 642. Cited in Boye, G.L. and Ampofo, O. (1983). Clinical uses of *C. sanguinolenta*. In *Proceedings of the First International Seminar on Cryptolepine, July 27–30, University of Science and Technology, Kumasi, Ghana*, pp. 37–40.

Grellier, P., Frappier, F., Trigalo, F., Ramiaramanana, L., Millerioux, V., Deharo, E., Bodo, B., Schreval, J., and Pousset, J.L. (1996). Antimalarial activity of alkaloids isolated from *Cryptolepis sanguinolenta*, cryptolepine and isocryptolepine. *Phytother. Res.*, 10, 317–321.

Kirby, G.C. (1997). Plants as a source of antimalarial drugs. *Trop. Doct.*, 27, 7–11.

Kirby, G.L., Noamesi, B.K., Paine, A., Warhurst, D.C., and Phillipson, J.D. (1995). *In vitro* and *in vivo* antimalarial activity of cryptolepine, a plant-derived indoloquinoline. *Phytother. Res.*, 9, 359–363.

Lisgarten, J.N., Coll, M., Portugal, J., Wright, C.W., and Aymami, J. (2002). The antimalarial and cytotoxic drug cryptolepine intercalates into DNA at cytosine-cytosine sites. *Nat. Struct. Biol.*, 9, 57–60.

Luo, J., Fort, D.M., Carlson, T.J., Noanesi, B.K., nii Amon-Kotei, D., King, S.R., Tsai, J., Quan, J., Hobensack, C., Lapresca, P., Waldeck, N., Mendez, C.D., Jolad, S.D., Bierer, D.E., and Reaven, G.M. (1998). *Cryptolepis sanguinolenta*: an ethnobotanical approach to drug discovery and the isolation of a potentially useful new antihyperglycaemic agent. *Diabetic Med.*, 15, 367–374.

Makler, M.T., Ries, J.M., Williams, J.A., Bancroft, J.E., Piper, R.C., Gibbins, B.L., and Hinrichs, D.J. (1993). Parasite lactate dehydrogenase assay as an assay for *Plasmodium falciparum* drug sensitivity. *Am. J. Trop. Med. Hyg.*, 48, 739–741.

Mendes, O. and Jensen, P.C.M. (1984). *Plantas medicinais, seu uso tradicional em Moçambique*, Vol. II. Instituto Natcional do Livro e do Disco, Lisboa. Cited in Paulo, A., Jimeno, M., Gomes, E.T., and Houghton, P.J. (2000). Steroidal alkaloids from *Cryptolepis obtuse*. *Phytochemistry*, 53, 417–422.

Noamesi, B.K. and Bamgbose, S.O.A. (1983). Cryptolepine-A pharmacological review. Clinical uses of *C. sanguinolenta*. In *Proceedings of the First International Seminar on Cryptolepine, July 27–30, University of Science and Technology, Kumasi, Ghana*, pp. 41–51.

Noamesi, B.K., Larson, B.S., Laryea, D.L., and Ullberg, S. (1991). Whole-body autoradiographic study on the distribution of 3H-cryptolepine in mice. *Arch. Int. Pharmacodyn. Ther.*, 313, 5–14.

Paulo, A., Gomes, E.T., and Houghton, P.J. (1995). New alkaloids from *Cryptolepis sanguinolenta*. *J. Nat. Prod.*, 58, 1485–1491.

Paulo, A., Gomes, E.T., Steele, J., Warhurst, D.C., and Houghton, P.J. (2000). Antiplasmodial activity of *Cryptolepis sanguinolenta* alkaloids from the leaves and roots. *Planta Med.*, 66, 30–34.

Ping-tao, L., Gilbert, M.G., and Stevens, W.D. (1995). Asclepiadaceae. Available at http://hua.huh. Harvard.edu/china/mss/volume16/Asclepiadaceae.published.pdf.

Sam, G.H. (2002). Informal consultation with staff of CSRPM.

Sarpong, K. (1983). Morphology and alkaloid distribution in *Cryptolepis sanguinolenta*. In *Proceedings of the First International Seminar on Cryptolepine, July 27–30, University of Science and Technology, Kumasi, Ghana*, pp. 15–22.

Sarpong, K. (2002). Private communication.

Sharaf, M.H.M., Schiff, P.L., Jr., Tackie, A.N., Boye, G.L., Phoebe, C.H., Howard, L., Meyer, C., Andrews, C.W., Minick, D., Johnson, L., Shockcor, J.P., Crouch, R.C., and Martin, G.E. (1994). Cryptolepicarboline. A Novel Indoloquinoline-β-Carboline Dimeric Alkaloid from *Cryptolepis sanguinolenta*. Presented at the International Research Congress of Natural Products, Halifax, Canada.

Sharaf, M.H.M., Schiff, P.L., Jr., Tackie, A.N., Phoebe, C.H., Jr., Davis, A.O., Andrew, C.W., Crouch, R.C., and Martin, G.E. (1995). Isolation and elucidation of the structure of homocryptolepinone. *J. Heterocyclic Chem.*, 32, 1631–1636.

Sharaf, M.H.M., Schiff, P.L., Jr., Tackie, A.N., Phoebe, C.H., Jr., and Martin, G.E. (1996a). Two new indoloquinoline alkaloids from *Cryptolepis sanguinolenta*: cryptosanguinolentine and cryptotackieine. *J. Heterocyclic Chem.*, 33, 239.

Sharaf, M.H.M., Schiff, P.L., Jr., Tackie, A.N., Phoebe, C.H., Jr., Johnson, R.L., Minick, D., Andrews, C.W., Crouch, R.C., and Martin, G.E. (1996b). The isolation and structure determination of Cryptomisrine, a novel indolo[3,2-b]quinoline dimeric alkaloid from *Cryptolepis sanguinolenta*. *J. Heterocyclic Chem.*, 33, 789.

Tackie, A.N., Boye, G.L., Sharaf, M.H.M., Schiff, P.L., Jr., Crouch, R.C., Spitzer, T.D., Johnson, R.L., Dunn, J., Minick, D., and Martin, G.E. (1993). *J. Nat. Prod.*, 56, 653. Cited in Sharaf, M.H.M., Schiff, P.L., Jr., Tackie, A.N., Phoebe, C.H., Jr., Davis, A.O., Andrew, C.W., Crouch, R.C., and Martin, G.E. (1995). Isolation and elucidation of the structure of homocryptolepinone. *J. Heterocyclic Chem.*, 32, 1631–1636.

Tona, L., Ngimbi, N.P., Tsakala, M., Mesia, K., Cimanga, K., Apers, S., De Bruyne, T., Pieters, L., Totte, J., and Vlietinck, A.J. (1999). Antimalarial activity of 20 crude extracts from nine African medicinal plants used in Kinshasa, Congo. *Ethnopharmacology*, 68, 193–203.

Wright, C.W. (2002). Private communication.

Wright, C.W., Addae-Kyereme, J., Breen, A.G., Brown, J.E., Cox, M.F., Croft, S.L., Gökçek, Y., Kendrick, H., Phillips, R.M., and Pollet, P.L. (2001). Synthesis and evaluation of cryptolepine analogues for their potential as new antimalarial agents. *J. Med. Chem.*, 44, 3187–3194.

Wright, C.W., Phillipson, J.D., Awe, S.O., Kirby, G.C., Warhurst, D.C., Quetin-Leclercq, J., and Angenot, L. (1996). Antimalarial activity of cryptolepine and some other anhydronium bases. *Phytother. Res.*, 10, 361–363.

9 Strychnos myrtoides: A Case Study of a Chemosensitising Medicinal Plant

David Ramanitrahasimbola, Jacques Ranaivoravo, Herintsoa Rafatro, Philippe Rasoanaivo, and Suzanne Ratsimamanga-Urverg

CONTENTS

9.1 INTRODUCTION

Madagascar has a very rich plant diversity with an unparalleled degree of endemicity and a wealth of ethnomedical heritage. More than 80% of the estimated 13,000 species are endemic to the island, and nearly 5000 species have ethnomedical uses (Rasoanaivo, 2000). These ethnobotanical data have been accumulated over several centuries in a dynamic process, the first written document dating back to the 1600s (Flacourt, 1642). In particular events such as epidemics that occurred in some regions at a given time, the population resorted to the methods of healing within its reach, and in certain ways has combined methods and therapeutic resources from traditional and modern medicine.

One of these events was the sudden recrudescence of malaria during the 1980s in the Highlands of Madagascar as one of the most devastating tropical diseases in the country. The severity of the infection is such that local populations believed it was a new disease, which they named *bemangovitra* (the disease of great shivering). A shortage of appropriate drugs due to the serious deterioration of the economic situation in Madagascar at that time, but also some cultural attachment to traditional healing, led the population back to the large-scale use of traditional herbal remedies. In our ethnobotanical fieldwork, we learned that local populations treated *bemangovitra* by means of self-medication with one or two tablets of chloroquine, a dose thought to promote chloroquine resistance, together with a decoction made from various plant species termed *chloroquine-adjuvants* (Rasoanaivo et al., 1992). To the best of our knowledge, this is the first time that healers in Madagascar have prescribed a medicinal plant in combination with a conventional medicine to treat malaria. Nearly 10 chloroquine-adjuvant plants were recorded and 5 of them investigated (Rasoanaivo et al., 1994). At this point, one of them, *Strychnos myrtoides*, has given promising results.

In this chapter, following a brief presentation of malaria epidemics in Madagascar and an outline of chemosensitising therapy in malaria, we wish to review all the scientific data that have enabled us to bring a standardised extract of *S. myrtoides* into a controlled double-blind randomised clinical trial. This work is a good illustration of a fruitful multidisciplinary collaboration between different stakeholders in our institute, namely, ethnobotanists, phytochemists, malariologists, and clinicians, and an effective cooperation with phytochemists in European laboratories, namely, the Instituto Supériore di Sanità, the Università degli Studi di Roma "La Sapienza" in Roma, and the Muséum National d'Histoire Naturelle in Paris. Their names have appeared in several joint publications cited in this case study.

9.2 MALARIA EPIDEMICS IN MADAGASCAR AND THEIR TREATMENT

Known first by the Malagasy word *tazo* (which generally means fever) and then *tazomoka* (fever of anopheles origin), malaria appears to have existed in Madagascar for several centuries (Flacourt, 1642). The first written document on the subject was a tragic story presented as a doctoral thesis at the Faculty of Medicine in Paris (Havet, 1827). Two brothers working at the Museum National d'Histoire Naturelle in Paris left their country in 1819 to visit Madagascar, known as a paradise for naturalists, and to meet King Radama I. Both brothers caught a strange disease on their way to Antananarivo. The younger brother miraculously recovered from the disease while the elder one died a few miles from Toamasina. Although the word *malaria* was not mentioned in the text, it is clear from the symptoms reported by the younger brother in his thesis and the discussion of its treatment that it probably was malaria.

One of the relevant characteristics of malaria in Madagascar is its sudden and unpredicted resurgence. The first reported malaria epidemic occurred in 1852 in the Highlands, and this was surprisingly accompanied by a massive epizootic that killed half of the zebus in Madagascar (Panou de Faymoreau, 1860). Ten years later, local populations and European residents in Madagascar began to hear rumours of a new disease called *ramanenjana*. This name did not convey any idea of its nature, and the accounts of it were so vague as to mystify rather than enlighten. This strange and inexplicable epidemic of dancing mania occurred first in the Southwest in February 1863 and reached Antananarivo where it began to be common in March. It appeared in the form of choreomania, in which dancing and singing are performed to an extraordinary and abnormal extent (Sibree, 1889). This mysterious disease is still the object of debate and controversy. In a work presented as a medical doctoral thesis in France, this epidemic was always observed to occur in the malarious period, at the time when the chief rice crop is ripening, and so was claimed to be a choreomania of malarial origin (Andrianjafy, 1902).

Another malaria epidemic occurred in 1878 in Ankazobe nearly 100 km from Antananarivo (Central Highlands of Madagascar) under the names *tazobe* (great malaria), *tazon'avaradrano* (malaria of the north part — with respect to the Ikopa River), or *aretin'olona* (someone's disease), killing thousands of people (Beauprez, 1901). One year later, another severe malaria epidemic took place in Fianarantsoa, South Highlands, under the local name *rapo-rapo* or *safo-tany* (Pearse, 1897). In 1903, there was a concomitant epidemic of malaria and influenza in Antananarivo, killing thousands of people (Fontoynont, 1903b). But the most severe malaria epidemic was probably that occurring in 1905–1906, again in the Central Highlands (Fontoynont, 1905a, 1905b), which is reported to have killed several tens of thousands of people. The next known tragic epidemic was the so-called *bemangovitra* disease that took place once more in the Central Highlands in the 1980s. The last malaria epidemic was reported to have appeared in 1994 in the south part of Madagascar (Champetier de Ribes, 1994).

It ensues that the Highlands of Madagascar have long been suffering from malaria epidemics. In these epidemics, it was reported that the treatment of choice was *Cinchona*, either as quinine sulfate or cinchona powder, but arsenic-based drugs (arrhenal, sodium cacodylate, methylarsinate, neo-treparsenan, and hectine) were also claimed to be efficient in the treatment of malaria (Fontoynont, 1903; Loiselet, 1912). A combination of quinine sulfate and arsenic-based drugs was said to be more efficient than one drug used alone (Loisel, 1885; Fontoynont, 1903a). The idea of quinine resistance when the drug was continuously used was mentioned, and a preliminary attempt to explain it was reported (Ravelomanantsoa, 1925). Except for the *bemangovitra* epidemic, it is a matter of regret that very little is documented in written form regarding the uses of medicinal plants to control malaria in other epidemic situations. The only document that substantially dealt with ethnomedical practices in the prevention and treatment of malaria was published by Ramisiray (1901) in his medical doctorate presented at the Faculty of Medicine in Paris.

9.3 CHEMOSENSITISING THERAPY IN MALARIA

The use of reversing agents to overcome drug resistance is a potential new treatment strategy in both malaria and cancer. One advantage of this approach in malaria is the possibility of prolonging the useful life of chloroquine, which is becoming less effective because of resistance. Chloroquine remains the drug of choice for the treatment of uncomplicated malaria in Madagascar because of its rapid onset of action, good tolerability, limited host toxicity, low cost, and versatility for both prophylactic and curative uses. On the other hand, extensive chloroquine monotherapy has been one of the main reasons behind the spread of resistance to this drug. The new second-line antimalarials, although effective, are out of reach of the family budget in most cases.

As a result, local populations rely on chloroquine to treat malaria, often by way of self-medication. Naturally, they use inexpensive local remedies to reinforce the activity of this antimalarial drug once it loses its effectiveness, and this behaviour is probably at the origin of chloroquine-adjuvant plants used by healers to this purpose. It is interesting to note that before the advent of rational drug combinations in malaria chemotherapy, this approach was already clinically used to treat quinine-resistant malaria in Madagascar as far back as 1880 in referring to combinations of quinine and arsenic-based drugs, and also empirically applied to enhance the action of chloroquine with chloroquine-adjuvants during the *bemangovitra* epidemic.

In the rational approach to resistance modulators, the pioneering work of Martin et al. (1987) attracted much attention in the 1990s. It was found that several calcium channel blockers, various tricyclic antidepressants, and tricyclic antihistamines reverse chloroquine resistance *in vitro* or *in vivo* (Rasoanaivo et al., 1996b). Laboratory studies, however, raised questions about the safety of this resistance-reversing therapy, which could potentiate more serious drug effects (Watt et al., 1990, 1993). Despite this, the potential of the tricyclic antidepressant desipramine and the tricyclic antihistamine cyproheptadine to improve the efficacy of chloroquine against chloroquine-resistant *Plasmodium falciparum* was then investigated in preliminary clinical trials (Björkman et al., 1990;

Warsame et al., 1992). Disappointingly, these clinical studies did not give any evidence for enhanced chloroquine efficacy through the use of the two chemosensitising agents in doses corresponding to the usual therapeutic range. Thus synthetic compounds that have a good pharmacological profile in experimental models may lack a useful application in man. This led to a decreased interest in the therapeutic value of the resistance-reversing therapy, and their use has been restricted to biochemical tools that may help understand the mechanism(s) of chloroquine resistance and its reversal. A renewed hope has come with the successful clinical trials of chloroquine in combination with chlorpheniramine in the treatment of acute uncomplicated malaria in Nigeria (Sowunmi et al., 1998), and it was even recently claimed that the resistance-reversing therapy is still a relevant approach for drug-resistant malaria (Schalkwyk et al., 2001).

At the inception of our research program on chemosensitising plants, very little was known about naturally occurring compounds that reverse chloroquine resistance in malaria, although relevant ethnobotanical data exist in the literature. We therefore started a research project aimed at evaluating the potential of the Malagasy flora to bring answers for the treatment of malaria through a rationally designed, interdisciplinary research team and a well-planned training program. We first focused our investigations on chloroquine-adjuvant plants that could enhance the efficacy of chloroquine. All this led to the isolation of various novel alkaloids that could restore chloroquine sensitivity to resistant strains of *Plasmodium* malaria (Rasoanaivo et al., 1994). Our interest was then focused on *S. myrtoides* for which preliminary promising results were obtained.

9.4 PRECLINICAL INVESTIGATION OF *STRYCHNOS MYRTOIDES*

9.4.1 ETHNOBOTANICAL BACKGROUND

All the information we obtained on the traditional uses of *Strychnos* species as antimalarials and chloroquine-adjuvants come from J.P. Abrahama, a highly experienced ethnobotanist and taxonomist. Not only was his knowledge of the trees of Madagascar exceptional, but also his knowledge of the traditional uses of plants as well as the significance of their vernacular names was remarkable. He tragically died in May 1996 as a result of a wound infection following an insect bite (Rasoanaivo, 1999). During a series of ethnobotanical field studies conducted in 1990–1992 with Abrahama in the eastern rainforests, one of us (P.R.) observed that he discretely used the infusion of stem scrapings of a particular plant when we went to the forests. Our colleague tactfully asked him about the purpose of this practice. Abrahama claimed that the infusion of this plant, known under the vernacular name *retendrika* and identified as *S. myrtoides*, protected him against malaria.

S. myrtoides grows in Ankarafantsika, in the west of Madagascar (see Figure 9.1). Assuming first that, like chloroquine, the infusion of this plant may have antimalarial activity with a long-lasting effect, we evaluated the *in vitro* and *in vivo* antiplasmodial activity of various extracts, but the results were rather disappointing since they all had a low activity. Many months later, our colleague tactfully made more inquiries into Abrahama's recipe. He maintained the use of *retendrika* as an efficient prophylactic remedy against malaria. But our colleague also learned that he had a personal recipe: he and his family used the infusion of stem scrapings of *S. myrtoides* in combination with one or two tablets of chloroquine as a curative treatment for chronic malaria. This subsequently guided our work into drug combination assessment and other developments.

9.4.2 CONFIRMATION OF THE ETHNOMEDICAL DATA: BIOASSAY-GUIDED FRACTIONATION

Assuming that alkaloids were responsible for the biological activity, we tested the crude alkaloid extract of *S. myrtoides* and found that they had a weak *in vitro* antiplasmodial activity, but when combined with chloroquine at dose levels much lower than those required for antiplasmodial effects, they markedly enhanced the *in vitro* and *in vivo* action of chloroquine (Figure 9.2). This confirmed

FIGURE 9.1 *S. myrtoides.* (Copyright 1997, Philippe Rasoanaivo.)

the validity of the traditional recipe in the experimental models. We then submitted the crude alkaloid extract to counter current distribution separation in a Craig apparatus, using chloroform as the stationary phase and a buffer solution at decreasing pH as the mobile phase. We isolated first the two major alkaloids in crystallised form, namely, strychnobrasiline and malagashanine (Figure 9.3), both of which had a chloroquine-potentiating effect *in vitro* (Rasoanaivo et al., 1994). We had previously isolated them from another Malagasy *Strychnos* species, *S. mostueoides*, but with an incorrect structure reported for malagashanine (Rasoanaivo et al., 1991).

While strychnobrasiline is a known compound, malagashanine with the revised structure turned out to be the parent compound of a novel group of *Strychnos* alkaloids with an unusual 3βH configuration, the N_b-C(21) seco-curan type, isolated from Malagasy *Strychnos* (Caira and Rasoanaivo, 1995).

9.4.3 Clinical Observational Study

After the successful confirmation of the traditional recipe and systemic toxicity studies, we submitted the infusion of *S. myrtoides* to a clinical observational study within the regulations regarding the evaluation of traditional medicine in Madagascar. This was carried out at the Service de Réanimation et de Toxicologie Clinique, HJRA Hospital in Antananarivo, under the framework of a medical doctoral thesis (Ranaivoravo, 1993). The methodology used in this study was to treat patients for which antimalarial treatment failure was observed upon admission to the hospital with infusion of 0.5 g of powdered stem barks or, alternatively, chloroquine together with the infusion if the quantity of antimalarial uptake was judged insufficient. Indeed, people generally treat malaria by self-medication with chloroquine, which is easily available; they only come to the hospital when treatment failure occurs. Twenty cases of antimalarial drug failure were recorded in the study. The most remarkable results from the study were the complete clearance of parasitaemia in two patients for whom treatment failure had been observed with several antimalarials, and in one patient with anaphylactic shock following an injection of quinine salts (Ramialiharisoa et al., 1994). Four patients with quinine intolerance were also successfully treated with the same therapeutic scheme. Regarding side effects, a few patients complained about slight fatigue (which is a symptom of malaria in any case), which was successfully treated with a nutraceutical rich in calcium and proteins (Masy calcium).

9.4.4 Phytochemical Studies

Historically, the investigation of Malagasy *Strychnos* species was initiated for a purely phytochemical purpose within a newly established collaboration with the Laboratory of Pharmaceutical

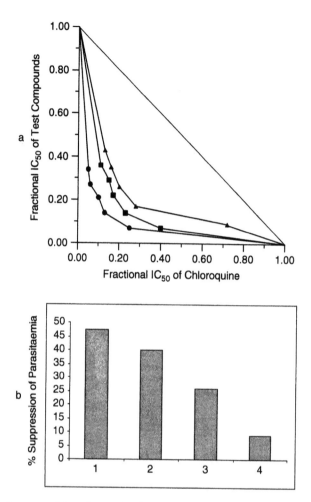

FIGURE 9.2 (a) Isobolograms of drug interaction between chloroquine and a crude alkaloid extract (▲) of *S. myrtoides*, malagashanine (■), and strychnobrasiline (●) against the chloroquine-resistant strain FCM29. The concave curve indicates a synergistic action; a convex curve would indicate an antagonistic effect, and a diagonal line would indicate a simple additive effect. (b) *In vivo* test on the crude alkaloid extract of *S. myrtoides* using the 4-day suppressive test of Peters: (1) controls, (2) mice treated with 100 mg/kg of extract, (3) mice treated with 0.75 mg/kg of chloroquine, (4) mice treated with 0.75 mg/kg of chloroquine + 100 mg/kg of extract.

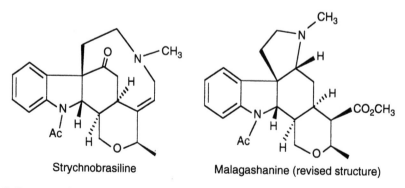

FIGURE 9.3 Structure of malagashanine and strychnobrasiline, major bioactive alkaloids of *S. myrtoides*.

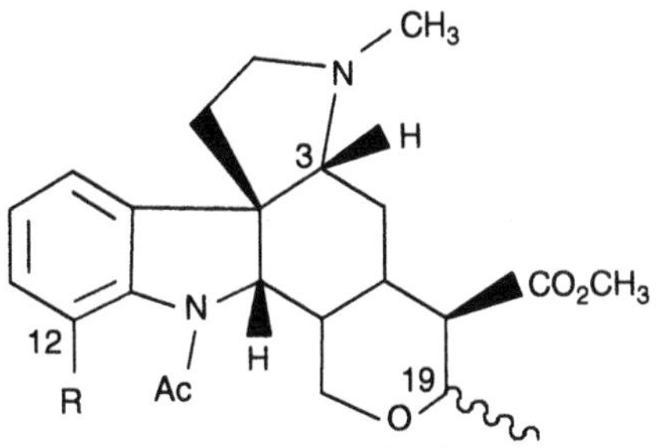

12-hydroxy-malagashanine
12-hydroxy-19-*épi*-malagashanine

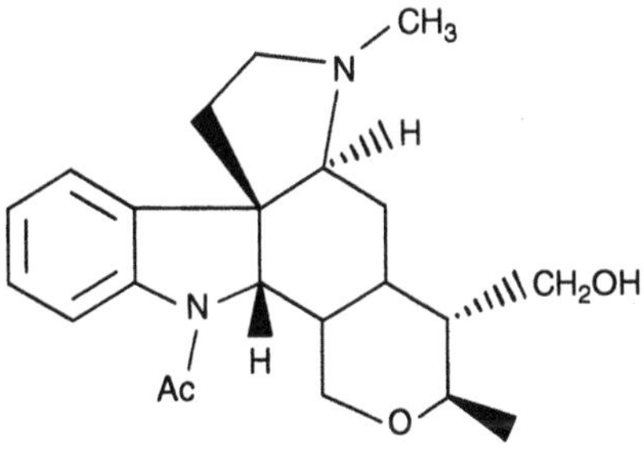

Malagashanol

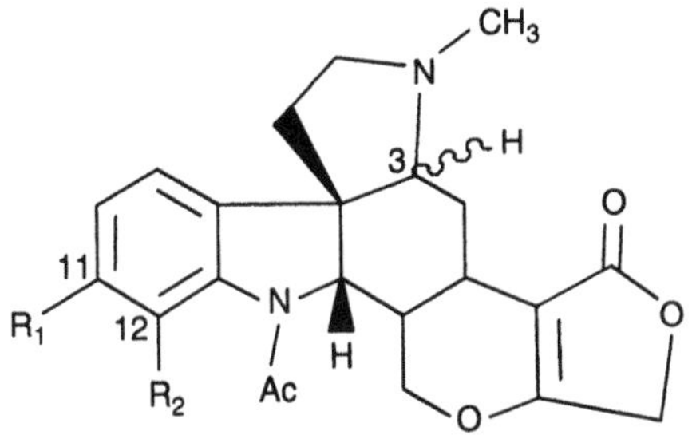

Myrtoidine series

R_1 = OCH$_3$, R_2 = H, 3βH: Myrtoidine
R_1 = OCH$_3$, R_2 = H, 3αH: 3-*epi*-myrtoidine
R_1 = R_2 = H, 3βH: 11-demethoxy-myrtoidine
R_1 = R_2 = H, 3αH: 3-*epi*-11-demethoxy-myrtoidine
R_1 = H, R_2 = OH, 3βH: 11-demethoxy-12-hydroxy-myrtoidine
R_1 = H, R_2 = OH, 3aH: 3-*epi*-11-demethoxy-12-hydroxy-myrtoidine

FIGURE 9.4 Structure of minor alkaloids of *S. myrtoides* and *S. diplotricha.*

Chemistry of the Instituto Supériore di Sanità of Roma. The team in this laboratory has acquired a long-standing experience in the phytochemical study of African and South American *Strychnos* species (Galeffi, 1980), and the main goal of the collaboration was to extend the work to the Malagasy *Strychnos*. There are 12 species available in Madagascar, of which 6 are endemic to the island, and this represents a good potential for phytochemical work (Leeuwenberg, 1984). *S. mostueoides* collected in the eastern rainforest of Madagascar, in the Ambatondrazaka region, was first investigated (Rasoanaivo et al., 1991). Eight alkaloids were isolated, of which the major alkaloid was by far strychnobrasiline; two others named malagashine and malagashanine were reported to be new. Particularly, the structure of malagashanine was established as a strychnobrasiline derivative. Later on, this structure was revised as a new subtype of *Strychnos* alkaloids, the N_b-C(21)-secocuran type (Rasoanaivo et al., 1996a), and completed by x-ray analysis (Caira and Rasoanaivo, 1995).

Meanwhile, our institute was moving with time by setting up a fully equipped section to carry out malaria chemotherapy research from plants traditionally used to treat malaria. The phytochemical investigation of the crude alkaloid content of *S. myrtoides* was thus coupled with biological work, and this led to the isolation of the two major alkaloids strychnobrasiline and malagashanine endowed with a chloroquine-potentiating effect as stated above (Rasoanaivo et al., 1994). The structure elucidation of all the minor alkaloids of *S. myrtoides* was then achieved, using mainly the two-dimensional nuclear magnetic resonance (2D-NMR) data (Martin et al., 1999).

Another *Strychnos* species, *S. diplotricha*, known under the same vernacular name of *retendrika* grows in the Northeast part of Madagascar in the Antalaha region. The phytochemical investigation

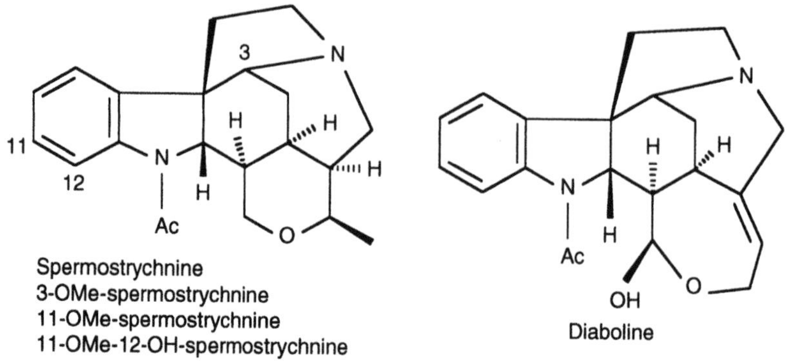

FIGURE 9.5 Structure of alkaloids isolated from *S. penthanta*.

of this species led to the isolation of strychnobrasiline and malagashanine as major alkaloids and several minor alkaloids structurally related to malagashanine (Rasoanaivo et al., 2001). Structures of all the minor alkaloids are presented in Figure 9.4.

Unexpectedly, all three Malagasy *Strychnos* species contain strychnobrasiline 1 and malagashanine 2 as major alkaloids. Minor alkaloids of the two species commonly named *retendrika* are related to the parent compound malagashanine. While malagashanine so far isolated has exclusively a 3βH configuration, minor alkaloids possess both 3αH and 3βH configurations.

The last *Strychnos* species we investigated was *S. penthanta*. Five known alkaloids were isolated, and their structures are shown in Figure 9.5 (Rasoanaivo et al., unpublished results).

9.4.5 PHARMACOLOGICAL STUDIES

Following our previous finding that the crude alkaloid extract of *S. myrtoides* markedly enhanced *in vivo* chloroquine action against *Plasmodium yoelii* N67, we evaluated the *in vivo* synergism between chloroquine and the individual major bioactive constituents strychnobrasiline and malagashanine. Using the classical 4-day suppressive test, malagashanine and strychnobrasiline, when tested individually at the dose level of 10 mg/kg, were devoid of any *in vivo* antimalarial effect, which is in agreement with the results of the *in vitro* test. In drug combination, malagashanine at the same dose level was found to markedly enhance the *in vivo* activity of chloroquine, while strychnobrasiline surprisingly was shown to lack such an effect when used at the same dose, in contrast with its chloroquine-enhancing activity *in vitro*. A dose of 50 mg/kg also failed to enhance the activity of chloroquine *in vivo* (Rafatro et al., 2000a). The exploration of higher doses was not justified as they approach a potentially toxic level. The comparatively high lipophilicity of malagashanine may play a key role in its *in vivo* biological activity since strychnobrasiline (which is predominantly water soluble at a physiological pH) failed to enhance *in vivo* chloroquine activity. Importantly, the *in vivo* chloroquine-enhancing activity of the crude alkaloid extract of *S. myrtoides* containing several minor alkaloids was much more pronounced than the effect of the individual major bioactive constituents. This gives support to the idea that crude extracts from herbal remedies may have better activity than the biologically active components, which may act synergistically in the extracts, thus justifying their use as phytomedicines.

To further investigate the reversal activity of malagashanine, we assessed its ability to enhance the action of 4-aminoquinolines, quinoline methanols, 9-aminoacridines, and phenanthrene methanol. We found in our study that the activities of quinine, mefloquine, pyronaridine, quinacrine, and even primaquine were significantly enhanced *in vitro* by malagashanine against the chloroquine-resistant FCM29/Cameroon strain. To the best of our knowledge, malagashanine is the only chemosensitiser acting on a broad range of antimalarials. It was also shown to display strong synergism when combined with halofantrine. We exposed the four different stages of the parasite life cycle in a sorbitol-synchronised culture to the test compounds, in order to know which of them were sensitive to malagashanine and its combination with chloroquine. Malagashanine was found to act specifically, like chloroquine alone, on the old trophozoite stage of the *P. falciparum* cycle when combined with chloroquine, which may preclude, at the low concentrations used, any significant action of malagashanine on a totally different target from the other test antimalarials (Rafatro et al., 2000a). Malagashanine and by extrapolation the crude alkaloid extract of *S. myrtoides* are therefore potential drugs for resistance reversal when used in combination with quinoline-containing drugs, which are by far the most widely used antimalarials.

9.4.6 TOXICOLOGICAL STUDIES

Prior to any use of malagashanine as an ingredient in a phytomedicine, we evaluated whether it also enhances the toxicity of chloroquine. First, the lethal dose 50% (LD_{50}) of malagashanine was determined as 400 mg/kg by the oral route and 62.5 mg/kg by intraperitoneal administration. Then, at a concentration much higher than the effective ones in chloroquine reversal activity, it was found that malagashanine did not affect the inherent cytotoxicity of chloroquine against KB and P-388 cancerous cell lines. Furthermore, malagashanine alone at concentration levels ranging from its inhibitory concentration 50% (IC_{50}) in the *in vitro* antiplasmodial activity down to one quarter did not affect the chronotropic and inotropic properties of the guinea pig isolated auricle muscles. Chloroquine alone at concentrations ranging from 100 to 800 n*M* had no chronotropic effect on the guinea pig isolated auricle muscles but showed a positive inotropic effect on the same organ in a dose-dependent manner. Malagashanine, when combined with chloroquine, did not significantly enhance the inherent chronotropic and inotropic effects of this antimalarial (Ramanitrahasimbola et al., 1999).

Based on these encouraging results, the cardiac effect of an infusion of *S. myrtoides* in combination with chloroquine was evaluated in healthy volunteers and malaria patients within the framework of a medical doctoral thesis (Raherimanana, 1999). The interpretation of the electro-cardiogram curves did not show any abnormality. No significant toxicity was observed.

9.4.7 PHARMACOKINETIC STUDIES

Pharmacokinetic investigations were deemed necessary to explain the biological activities of malagashanine. To this end, a reversed-phase high-performance liquid chromatography (HPLC) method was successfully developed for the quantitative analysis of malagashanine in rat plasma. The quantitation limit was 10 ng/ml. Both intraday and interday accuracy and precision data showed good reproducibility (Rafatro et al., 2000b). Then we investigated its pharmacokinetics and metabolites as a preliminary tool for the understanding of its biological activities. The pharmacokinetic parameters of malagashanine were thus determined using a single-dose administration in rat in three different routes (Rafatro et al., 2003). They are summarised in Table 9.1.

TABLE 9.1
Pharmacokinetic Parameters of Malagashanine after Single-Dose Administration to Male Rats by Different Routes

Parameters	Intravenous Bolus	Intravenous Infusion	Intraperitoneal Administration	Per os
Area under curve (μg·h/ml)	1.34 ± 0.07	2.11 ± 0.40	2.45 ± 0.31	8.36 ± 1.44
Half-life (h)	0.46 ± 0.06	0.52 ± 0.12	1.99 ± 0.58	1.81 ± 0.23
Clearance (l/h)	0.95 ± 0.03	0.87 ± 0.09		
Volume of distribution (l)	0.64 ± 0.04	0.71 ± 0.16		

Afterwards, the structure of two metabolites, N$_b$-demethyl-malagashanine and 5-hydroxy-mal-agashanine, isolated from rat urine and human liver microsomes was then elucidated by electrospray mass and tandem mass spectrometry and NMR spectroscopy (Rafatro et al., 2000c). The latter metabolite may be subject to a retro-Schiff ring-opening reaction followed by a dimerisation process leading to dimeric compounds as detected in the electrospray mass spectrum.

9.4.8 HEMISYNTHESIS OF MALAGASHANINE DERIVATIVES

In some cases, natural products may serve as starting materials for the hemisynthesis of useful derivatives. At this point, strychnobrasiline, which was present in large quantities in the stem barks of *S. myrtoides* but was lacking *in vivo* chemosensitising activity, was subjected to various chemical transformations. Thus, the oxidation of the N$_a$-deacetyl-strychnobrasiline with the metachloroper-benzoic acid (MCPBA) led to an unexpected Bayer–Villiger rearrangement leading to a new indole skeleton, as summarised in the scheme below (Trigalo et al., 2002).

Furthermore, the N$_a$-deacetyl-strychnobrasiline was successfully transformed into malagasha-nine derivatives with enhanced biological activities. They have been patented (Trigalo et al., 2002).

9.5 CLINICAL EVALUATION

We initiated a controlled, double-blind randomised clinical evaluation of a standardised alkaloid extract of *S. myrtoides* titrated at 20% malagashanine. It took place at a government-run outpatient clinic in the town of Ankazobe, located in the Northwest Central Highlands of Madagascar, nearly 100 km from the capital. The first reported malaria epidemic (*tazobe, tazon'avaradrano*) occurred in 1878 in this region. We chose this region because (1) a comparatively high level of chloroquine resistance was previously reported (Milijaona et al., 1998), and (2) it is readily accessible by car within 2 hours, which is important for the daily transportation of blood samples for *in vitro* chemosensitivity evaluation at Institut Malgache de Recherches Appliquées (IMRA). The main objective of the study was to assess the chloroquine resistance reversal activity of the standardised phytomedicine. We received a formal agreement of the ethical committee and an official approval of the Ministry of Health to conduct the study.

9.5.1 Preliminary Efficacy Evaluation of Chloroquine in Ankazobe

Prior to the clinical trial, we conducted a 2-month study (January 4 to March 7, 2001) on the clinical efficacy of chloroquine in Ankazobe. The main purpose was both to estimate the present level of resistance in the region and to learn about the feasibility of the project in terms of organisation and statistical requirements. Based on the previous World Health Organization (WHO) definitions (see Chapter 21), of 46 patients included, there were 26 S/RI, 2 RI early, 11 RI late, 3 RII, 2 RIII, and 2 lost to follow-up (Andriamanalimanana, 2001; Willcox et al., 2004).

9.5.2 Background of the Clinical Protocol

Basically, the clinical trial was conducted in two parallel groups: chloroquine + placebo vs. chloroquine + phytomedicine. The WHO guidelines for the assessment of uncomplicated malaria were used (WHO, 1996). The principal criterion for assessing treatment efficacy was adequate clinical response. After patient selection and consent, the investigators proceeded to a complete clinical examination and laboratory investigations that included:

- Body temperature
- A full blood count at days 0, 3, 7, 14, and 28
- Serum biochemistry at days 0, 3, 7, 14, and 28, including glucose, renal function, liver function, and lipids
- Dipstick urinalysis

At the inclusion time, blood samples were also taken for the determination of the parasite chemosensitivity using the isotopic semimicrotest method. Patients were followed up for at least 28 days. For those who could not come back to the health center to be followed up, every effort was made to meet them directly at their homes.

9.5.3 *In Vitro* Chemosensitivity Tests

Overall, 404 blood samples were collected from malarious patients enrolled in the clinical trial, of which 305 were at the time of inclusion and 99 when a treatment failure was observed. They were brought to IMRA at ambient temperature. Keeping aside those that did not fulfill the minimum requirements for *in vitro* tests (i.e., no parasite growth, low parasitaemia percentage, other *Plasmodium* species), 291 samples were tested for *in vitro* chemosensitivity, but 244 gave interpretable results in terms of IC_{50} and IC_{90} values. Results are summarised in Table 9.2.

Based on a previous assumption that an $IC_{50} > 100$ nM corresponds to chloroquine-resistant strains (Raharimalala et al., 1993), four isolates only were found to be resistant. We have tentatively proposed a classification as follows, taking into account the IC_{90} values:

$IC_{50} > 100$ nM: resistant isolates
$IC_{90} > 100$ nM: isolates with intermediate sensitivity, probably a mixture of sensitive and resistant strains
$IC_{90} < 100$ nM: truly sensitive isolates

In these conditions, 190 were truly sensitive, 58 had an intermediate sensitivity, and 4 were resistant in the overall study.

We selected at random 17 isolates and used them to assess drug interactions between chloroquine and the *Strychnos*-based phytomedicine. Isobolograms of drug interaction for the three most relevant results are reported in the Figure 9.6a.

A synergistic effect (concave curve) was observed with one isolate ($IC_{50} = 8.84$ µg/ml, $IC_{90} = 57.42$ µg/ml). An antagonistic effect was observed with one isolate ($IC_{50} = 13.97$ µg/ml,

TABLE 9.2
Results of the *In Vitro* Chemosensibility Evaluation of Isolates

	At Inclusion Time	When a Treatment Failure (or Reinfestation) Was Observed
$IC_{50} > 100$ nM	2	2
$IC_{90} > 100$ nM	35	15
$IC_{90} < 100$ nM	149	41

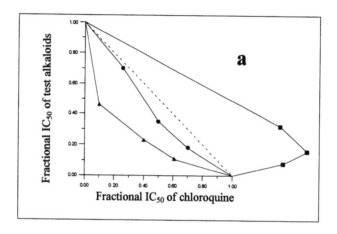

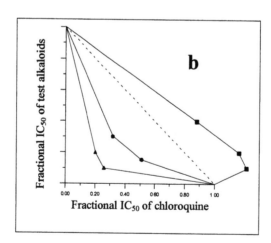

FIGURE 9.6 (a, left) *In vitro* responses of *Plasmodium falciparum* isolates determined as isobologrammes in drug interaction between chloroquine and a standardised extract of *Strychnos myrtoides*. (a) Isolates from three different patients are shown, to give an impression of the range of responses seen: (■) = patient 1, (●) = patient 2, (▲) = patient 3. (b) Isolates taken from a single patient: (■) at the first presentation, (●) = at the second presentation after one month, (▲) after one week of continuous culture of the isolate obtained at the second presentation.

TABLE 9.3
Monthly Distribution of Malaria Patients in the Clinical
Trial in Government-Run Outpatient Clinic in Ankazobe

Month of Inclusion	Number of Patients
March 2001	66
April 2001	98
May 2001	100
June 2001	36
July 2001	2
June 2002	71
Total	**373**

IC_{90} = 41.09 μg/ml), while a simple additive effect was obtained with the remaining isolates as exemplified by one curve following the diagonal (IC_{50} = 38.84 μg/ml, IC_{90} = 127.50 μg/ml). These results reflect the heterogeneity of the responses of the isolates to the chloroquine–phytomedicine combination, irrespective of the IC_{50} and IC_{90} values.

A case that deserves attention was a patient with a truly sensitive isolate (IC_{50} = 3.18 μg/ml, IC_{90} = 12.35 μg/ml) for which an antagonistic effect was observed in drug combination. He was successfully treated with chloroquine. One month later he went back to the health center for malaria symptoms. An isolate taken from this patient, although in the range of the truly sensitive ones (IC_{50} = 2.51 μg/ml, IC_{90} = 17.28 μg/ml), displayed a synergistic effect in drug combination. After 1 week of continuous culture, the same isolate showed both the same range of inhibitory concentration values (IC_{50} = 2.38 μg/ml, IC_{90} = 17.97 μg/ml) and synergistic effect (Figure 9.6b). He was successfully treated with sulphadoxine pyrimethamine at day 28. It turned out that, in this case, the *in vitro* responses did not correlate with clinical responses to drugs. Genetic factors of parasites may play an important role in the drug resistance irrespective of their chemosensitivity.

9.5.4 CLINICAL OUTCOMES

Overall, 373 patients were recruited, but 56 malaria cases were discontinued from the study because they did not fully comply with the protocol requirements. Table 9.3 summarises the monthly distribution of patients for the overall study.

The range of age extended from 6 to 60 years, with an average of 16.05. The statistical mode was 6 years, clearly indicating that children were most affected by malaria in Ankazobe. At the inclusion, 242 patients (64.9%) had a body temperature higher than 37.5°C, while 131 patients (35.1%) had a body temperature less than 37.5°C. Five cases of severe treatment failure were observed, of which three needed an urgent admission to the hospital. No malaria death was recorded during the study.

Table 9.4 gives an outline of the clinical outcomes, using either the adequate clinical response or the parasite clearance at day 14 as the main efficacy criteria.

Statistically (χ^2 = .0406, .5 < p < .9), there was no significant difference between the two treatments in terms of adequate clinical response or parasite clearance at day 14. In the overall study, the treatment failure rate was 11.5%.

9.5.5 DISCUSSION

In the controlled, double-blind randomised clinical trial, chloroquine + *S. myrtoides*-based chemo-sensitising phytomedicine failed to give positive results. One rapid explanation is that most isolates were sensitive, and the concerned phytomedicine was previously shown to lack any effect on

TABLE 9.4
Clinical Outcomes of the Clinical Trial in Terms of Adequate Clinical Response or Parasite Clearance

	Adequate Clinical Response at Day 14		Parasite Clearance at Day 14	
	Positive	Negative	Positive	Negative
Chloroquine + placebo	142	19	124	33
Chloroquine + phytomedicine	138	18	125	29
Total	**280**	**37**	**249**	**62**

sensitive strains or isolates. But in thinking it over, several lessons could be learned from this study when conducting future clinical trials.

All isolates were kept at ambient temperature during storage and transportation instead of 4°C. Further investigation is needed to clarify whether storage temperature and length of conservation may definitely influence the chemosensitivity of isolates.

Through our multicenter collaborations, we have observed that IC_{50} values for the same batch of extracts/compounds are subject to significant variations from one laboratory to another. Indeed, several factors, i.e., solvents used to dissolve the extracts (either dimethyl sulfoxide [DMSO] or methanol), parasitaemia, haematocrit, CO_2 concentration, stability of extracts, incubation time, and strains used, may affect the results. Standardising the *in vitro* chemosensitivity test is therefore a good step to achieve, and this is being done within the Interactive Networking of Francophonic Centers of Expertise on Antimalarial Drugs (Basco, unpublished data). Furthermore, delineating sensitive and resistant isolates with one fixed IC_{50} of 100 nM may be misleading and may yield erroneous conclusions. Indeed, IC_{50} was first used to determine the sensitivity of cloned strains of *P. falciparum*, and then extended to isolates for the same purpose. In fact, isolates are currently mixtures of strains with different levels of sensitivity. As a result, an isolate composed of 70% sensitive strains and 30% resistant strains will be ranked as a sensitive isolate on the basis of its IC_{50} value. More flexible data are needed, and this could be achieved through experiences gained in several related works. Polymerase chain reaction (PCR) may also be a helpful tool to assess this in the future.

In the overall study, 37 and 62 cases of clinical treatment failures were observed, respectively, depending on the principal efficacy criterion, adequate clinical response at day 14 or parasite clearance at day 14 (Table 9.4). On a case-by-case basis, 15 corresponded to isolates with intermediate sensitivity. This suggests that the concerned isolates might be a mixture of sensitive and resistant strains, and the treatment with chloroquine or chloroquine + phytomedicine would induce a selection of resistant strains that may account for the treatment failure. In these conditions, it is surprising to note that chloroquine + phytomedicine failed to give the expected results. One explanation is that malagashanine has a short half-life, and a 7-day treatment would be required to affect clinical outcome. Another explanation is that the decoction as it was successfully used in the clinical observational study may contain polar compounds that might contribute to the biological activity.

On the other hand, 41 cases of treatment failure based on parasite clearance at day 14 corresponded to truly sensitive strains, while four patients who carried resistant strains were successfully treated with chloroquine alone or chloroquine + phytomedicine. This strongly suggests that parasite resistance is not the only factor responsible for the treatment failure. Other parameters, particularly the bioavailability of chloroquine, may play an important role. The determination of chloroquine plasma concentration would be therefore necessary to clarify this point.

9.6 CONCLUSION

This case study illustrates a fruitful North–South collaboration based on the complementarities of expertise in a multidisciplinary work. To the best of our knowledge, this is the first time that a controlled, double-blind randomised clinical trial was done in Madagascar. The positive aspect of this study is the useful experience gained in conducting the clinical trial in very difficult conditions. The study raised several questions, and it is hoped that these would serve as a basis for further malaria clinical investigation in Madagascar for either phytomedicine or Western drugs.

ACKNOWLEDGMENTS

The authors acknowledge all those who helped and participated in the studies, including the laboratory staff and technicians at IMRA and in Ankazobe, namely, Bruno Sepiera Rakoto, Nantenaina Ranaivo, Emmanuel Rakotondrazafy, Rapiera Ombiasa, César Andriamalazavola, and Olivia Raherimalazanirainy; the medical students, namely, Radovola Dimbiniaina Andriamanalimanana, Domoina Lalaina Andrianasolo, Naivomahatratra Andriambelo, and Soloniaina Ramananjanahary; the medical staff in Ankazobe, namely, Dr. Lalao Radomarison, Dr. Pauline Farasoa, Dr. Soloniaina Raharimanana, Dr. Tahiana Randriamire, who collaborated with us, and Dr. Merlin Willcox; and, of course, all the patients who participated in the clinical trial. We thank the funding agencies who made this work possible: AUF under the convention ARC X/7.10.04/Palu95/IMRA.PR, and Fonds de Solidarité Prioritaire FSP 97008500, French Ministry of Foreign Affairs.

REFERENCES

Andriamanalimanana, R. (2001). Proposition de protocole d'évaluation de l'efficacité thérapeutique des antipaludiques pour le traitement du paludisme à *Plasmodium falciparum* non compliqué; réflexion à partir d'un test *in vivo*. Thèse de Doctorat en Médecine, Faculté de Médecine, Université d'Antananarivo, Madagascar.

Andrianjafy. (1902). Le ramanenjana à Madagascar: choréomanie d'origine palustre. Thèse de Doctorat en Médecine, Montpellier, France.

Beauprez. (1901). Informations: chronique du mois de juin (le froid et la mortalité des indigènes). *Rev. Madagascar*, 8, 593–599.

Björkman, A., Willcox, M., Kihamia, C.M., Mahikwano, L.F., Howard, P.A.P., Hakasson, A., and Warhurst, D. (1990). Field study of cyproheptadine/chloroquine synergism in falciparum malaria. *Lancet*, 336, 59–60.

Caira, M.R. and Rasoanaivo, P. (1995). X-ray analysis of malagashanine. *J. Chem. Crystallogr.*, 25, 725–729.

Champetier de Ribes, G., Ranaivoson, G., Rakotoherisoa, E., Rakotoson, J.D., and Andriamahefazafy, B. (1994). Une épidémie de paludisme dans le Sud de Madagascar. *Arch. Inst. Pasteur Madagascar*, 61, 66–69.

Flacourt, E. (1642). Histoire de la Grande Isle de Madagascar: 1642–1660. In *Collection des Ouvrages Anciens Concernant Madagascar*, Grandidier, A., Froidevaux, H., and Grandidier, G., Eds. Paris Union Coloniale, 1913, Tome VIII.

Fontoynont, M. (1903a). L'arrhénal dans le traitement des fièvres palustres à Tananarive. *Presse Méd.*, 72, 589–590.

Fontoynont, M. (1903b). Grippe et paludisme à Madagascar. *Presse Méd.*, 72, 637–638.

Fontoynont, M. (1905a). La nouvelle épidémie palustre deTananarive. *Bull. Acad. Malgache*, 334–335.

Fontoynont, M. (1905b). A propos des épidémies palustres actuelles sur les Hauts-Plateaux de Madagascar. *Rev. Méd. Hyg. Trop.*, 64–68.

Galeffi, C. (1980). New trends in the separation of active principles from plants. *J. Ethnopharmacol.*, 2, 127–134.

Havet, E.N. (1827). Dissertation sur une maladie qui règne à l'Ile de Madagascar: conseils hygiéniques à suivre pour l'éviter. Thèse de Doctorat en Médecine, Faculté de Médecine, Paris, No. 124.

Leeuwenberg, A.J.M. (1984). 167ème Famille, Loganiacées. In *Flore de Madagascar et des Comores*, Humbert, H., Ed. Muséum National d'Histoire Naturelle, Paris, pp. 70–104.

Loisel, L. (1885). Du traitement des fièvres paludéennes à Sainte-Marie de Madagascar: emploi simultané de la quinine et de l'arsenic, notamment en injection hypodermique dans les formes pernicieuses. Thèse de Doctorat en Médecine, Faculté de Médecine, Paris, No. 152.

Loiselet (1912). Traitement du paludisme par l'hectine. *Soc. Sc. Med. Madagascar*, 2, 5–9.

Martin, M.-T., Rasoanaivo, P., Palazzino, G., Galeffi, C., Nicoletti, M., Trigalo, F., and Frappier, F. (1999). Minor Nb-(C21) secocuran alkaloids of *Strychnos myrtoides*. *Phytochemistry*, 51, 479–486.

Martin, S.K., Oduola, A.M.J., and Milhous, W.K. (1987). Reversal of chloroquine resistance in *Plasmodium falciparum* by verapamil. *Science*, 235, 899–901.

Milijaona, R., Raharilmalala, L., Ramambanirina, L., Ranaivo, L.H., and Jambou, R. (1998). Chimiorésistance de *Plasmodium falciparum* sur les marges des hautes terres malgaches: perspectives pour le programme national de lutte. *Méd. Trop.*, 58, 261–265.

Panou de Faymoreau, A. (1860). Nossi-Bé: fièvres intermittentes. Thèse de Doctorat en Médecine, Faculté de Médecine, Paris, No. 77.

Pearse, J. (1897). A modern epidemic in the Betsileo province: the '*safo-tany*' or '*rapo-rapo*'. *Antananarivo Ann.*, XXI, 32–33.

Rafatro, H., Ramanitrahasimbola, D., Rasoanaivo, P., Ratsimamanga-Urverg, S., Rakoto-Ratsimamanga, A., and Frappier, F. (2000a). Reversal activity of the naturally-occurring chemosensitiser malagashanine in *Plasmodium* malaria. *Biochem. Pharmacol.*, 59, 1053–1061.

Rafatro, H., Rasoanaivo, P., Ratsimamanga-Urverg, S., Quetin-Leclercq, J., and Verbeeck, R.K. (2000b). HPLC assay of malagashanine in rat plasma and urine and its pharmacokinetic application. *J. Chromatogr. B*, 744, 121–127.

Rafatro, H., Rasoanaivo, P., and Verbeeck, R.K. (2003). Pharmacokinetic investigation of malagashanine following different single administration in rat. *Eur. J. Pharm. Sci.*, unpublished results.

Rafatro, H., Verbeeck, R.K., De Longhe, P.J.M., Rasoanaivo, P., Laurent, A., and Lhöest, G. (2000c). Isolation from rat urine and human liver microsomes, identification by electrospray and nanospray tandem mass spectrometry of new malagashanine metabolites. *J. Mass Spectrom.*, 35, 1112–1120.

Raharimalala, L., Lepers, J.P., Lepers-Rason, M.D., Rabartison, P., Ramambanirina, L., and Roux Jean. (1993). Aspects de la sensibilité de *Plasmodium falciparum* à la chloroquine à Madagascar de 1982 à 1993. *Arch. Inst. Pasteur Madagascar*, 60, 60–64.

Raherimanana, H. (1997). Action cardiaque de l'association chloroquine-retendrika sur des sujets paludéens et volontaires sains. Thèse de Médecine, Université d'Antananarivo, Madagascar, 117 pp.

Ramanitrahasimbola, D., Ratsimamanga-Urverg, S., Rasoanaivo, P., and Rakoto-Ratsimamanga, A. (1999). Effects of the naturally-occurring chemosensitiser malagashanine and its combination with chloroquine on KB and P388 cell lines and isolated auricle. *Phytomedicine*, 6, 331–334.

Ramialiharisoa, A., Ranaivoravo, J., Ratsimamanga-Urverg, S., Rasoanaivo, P., and Rakoto-Ratsimamanga, A. (1994). Evaluation en clinique humaine de l'action potentialisatrice d'une infusion de *Strychnos myrtoides* vis-à-vis d'antipaludéens. *Rév. Méd. Pharmacopée Africaines*, 8, 123–131.

Ramisiray, G. (1901). Croyances et pratiques médicales des Malgaches. Thèse de Doctorat en Médecine, Faculté de Médecine, Paris.

Ranaivoravo, J. (1993). Traitement du paludisme: évaluation en laboratoire (*in vitro* & *in vivo*) et en clinique humaine, de l'action potentialisatrice de la chloroquine d'une plante médicinale malgache. Thèse de Doctorat en Médecine, University of Antananarivo, Madagascar.

Rasoanaivo, P. (1999). Obituary: the Malagasy forester JP Abrahama. *J. Ethnopharmacol.*, 68, 1–2.

Rasoanaivo, P. (2000). Une banque de données sur les plantes médicinales de Madagascar, *Info-Essences*, 15, 5–6.

Rasoanaivo, P., Galeffi, C., De Vicente, Y., and Nicoletti, M. (1991). Malagashanine and malagashine, two alkaloids of *Strychnos mostueoides*. *Rev. Latinoam. Quim.*, 22/1, 32–34.

Rasoanaivo, P., Galeffi, C., Palazzino, G., and Nicoletti, M. (1996a). Revised structure of malagashanine: a new series of Nb,C(21)-secocuran alkaloids in *Strychnos myrtoides*. *Gaz. Chim. Ital.*, 126, 517–519.

Rasoanaivo, P., Palazzino, G., Galeffi, C., and Nicoletti, M. (2001). The co-occurrence of C(3) epimer Nb,C(21)-secocuran alkaloids in *Strychnos diplotricha* and *S. myrtoides*. *Phytochemistry*, 56/8, 863–867.

Rasoanaivo, P., Petitjean, A., Ratsimamanga-Urverg, S., and Rakoto-Ratsimamanga, A. (1992). Medicinal plants to treat malaria in Madagascar. *J. Ethnopharmacol.*, 37, 117–127.

Rasoanaivo, P., Ratsimamanga-Urverg, S., and Frappier, F. (1996b). Reversing agents in the treatment of drug-resistant malaria. *Curr. Med. Chem.*, 3, 1–10.

Rasoanaivo, P., Ratsimamanga-Urverg, S., Milijaona, R., Rafatro, H., Galeffi, C., and Nicoletti, M. (1994). *In vitro* and *in vivo* chloroquine potentiating action of *Strychnos myrtoides* alkaloids against chloroquine-resistant strain of *Plasmodium* malaria. *Planta Med.*, 60, 13–16.

Ravelomanantsoa, A. (1925). Ny tazo na paludisme. *Bull. Soc. Mutuelle Corps Méd. Malgache*, 185–187, 218–222, 301–305, 351–354, 355–358.

Schalkwyk, D.A., Walden, J.C., and Smith, P.J. (2001). Reversal of chloroquine resistance in *Plasmodium falciparum* using combination of chemosensitisers. *Antimicrob. Agents Chemother.*, 45, 3171–3174.

Sibree, J. (1889). The ramanenjana or dancing mania of Madagascar. *Antananarivo Ann.*, XIII, 19–27.

Sowunmi, A., Oduola, A.M.J., Ogundahunsi, O.A.T., and Salako, L.A. (1998). Enhancement of the antimalarial effect of chloroquine by chlorpheniramine in vivo. *Trop. Med. Int. Health*, 3, 177–183.

Trigalo, F., Frappier, F., Rasoanaivo, P., and Ratsimamanga-Urverg, S. Nouveaux composés dérivés d'alcaloïdes, leur procédé de préparation et leur utilisation pour la préparation de médicaments. Demande de Brevet Français 02/05740, déposée le 7/05/2002 au nom du CNRS.

Trigalo, F., Martin, M.-T., Blond, A., Rasolondratovo, B., Rasoanaivo, P., and Frappier, F. (2002). Oxydation of indolines to nitrones and new rearrangements in seco-curane type indole alkaloids. *Tetrahedron*, 58, 4555–4558.

Warsame, M., Wernsdorfer, W.H., and Björkman, A. (1992). Lack of effect of desipramine on the response to chloroquine of patients with chloroquine-resistant falciparum malaria. *Trans. R. Soc. Trop. Med. Hyg.*, 86, 235–236.

Watt, G., Long, G.W., Grogl, M., and Martin, S.K. (1990). Reversal of drug-resistant *falciparum* malaria by calcium antagonists: potential for host toxicity. *Trans. R. Soc. Trop. Med. Hyg.*, 84, 187–190.

Watt, G., Na-Nakorn, A., Batzeman, D.N., Plubha, N., Mothanaprakoon, P., Edstein, M., and Webster, H.K. (1993). Amplification of quinine cardiac effects by the resistance-reversing agent prochlorperazine in *falciparum* malaria. *Am. J. Trop. Med. Hyg.*, 49, 645–649.

WHO. (1996). Evaluation de l'efficacité thérapeutique des antipaludiques pour le traitement du paludisme à *P. falciparum* non compliqué dans les régions à transmission élevée, WHO/MAL/96.1077.

Willcox, M.L., Rakotondrazafy, E., Andriamanalimanana, R., Andrianasolo, D., and Rasoanaivo, P. (2004). Increasing chloroquine resistance in Ankazobe, Highlands of Madagascar. *Trans. Roy. Soc. Trop. Med. Hyg.*, 98(5), 311–314.

Part 3

Ethnomedical Research

10 Frequency of Use of Traditional Herbal Medicines for the Treatment and Prevention of Malaria: An Overview of the Literature

Merlin Willcox and Gerard Bodeker

CONTENTS

0-415-30112-2/04/$0.00+$1.50

10.1 INTRODUCTION

The first part of this book discussed the public health importance of traditional medicines and their potential role in malaria control programs; the second presented multidisciplinary literature reviews on some of the best known plant-based antimalarials. Now we turn to reviews of sociological, anthropological, and ethnomedical research on the worldwide use of traditional medicines for malaria.

It is often said that 80% of the world's population relies on traditional medicine for their primary health care (Bannerman et al., 1983). However, it is not clear what evidence there is for this statement (Bodeker, 2001), or how it relates specifically to the treatment of malaria. It is conceivable that, with widespread public health programs and the increasing availability of cheap modern antimalarials, the use of traditional medicines for malaria may have decreased. On the other hand, increasing levels of resistance to affordable drugs and the cost of more effective drugs may be driving patients back to traditional medicine.

Furthermore, it is possible that the use of traditional antimalarials varies according to the availability of safe and effective herbal remedies. If this were so, research programs could focus on localities where traditional antimalarials are widely used, to identify the remedies with the greatest perceived efficacy. These could be investigated more fully, and if shown to be clinically safe and effective, their use could be promoted in other areas where the relevant plants exist or could be cultivated.

Through a systematic overview of the literature on treatment seeking for malaria and usage of herbal antimalarials, this chapter aims to:

1. Analyse the frequency of use of plants for malaria in different areas and factors that affect this
2. Identify advantages and disadvantages of previous research methods and propose guidelines for the traditional medicine aspects of future studies of treatment seeking for malaria.
3. Consider priorities for sociological research on traditional herbal antimalarials

10.2 METHODS

A literature search was performed using the MEDLINE, CAB, SOCIOFILE, and EMBASE databases to find all articles published up to February 2003 referring to the frequency of use of traditional remedies for malaria (key words: malaria; traditional medicine; malaria, therapy; knowledge, attitudes, practice; self-medication; drug utilisation). References of relevant articles were also searched, and some journals were searched by hand, to identify as many relevant articles as possible. Experts were also consulted to supply unpublished papers.

Studies on the frequency of use of traditional plant-based treatments for malaria were identified and summarised (Table 10.1). Results from all these studies were combined (as a meta-analysis) to give an overall average frequency of use. Criteria for inclusion in this analysis were as follows:

1. Only studies that singled out traditional plant-based treatments were included. Many other forms of traditional medicine are used for malaria, including ingestion of old butter, massage, baths, soap, chanting, and exorcism (Adera, 2003; Julvez et al., 1995; Aikins et al., 1994). Therefore, studies that reported usage of undefined traditional medicine were excluded. Studies that reported undefined self-medication were also excluded, because this could have involved the use of herbs, tablets, a combination of both, or neither.

2. Only studies that singled out febrile illnesses or malaria were included. Many studies report global figures for frequency of use of traditional medicines for a range of conditions. These were not included unless there was a specific figure for fevers or malaria.

3. Only studies that reported a quantitative outcome (frequency of use) could be included in the meta-analysis. Many sociological and anthropological studies report a qualitative analysis of indigenous beliefs and practices, but not a quantitative outcome on the frequency of use of traditional medicines for malaria. Although these were excluded from the quantitative meta-analysis in Table 10.1, they were included in the qualitative analysis of factors affecting the use of traditional herbal medicines.

Additional criteria for exclusion from this analysis were as follows:

1. Studies referring only to consultation with traditional healers were excluded. Not all traditional healers necessarily use plant-based medicines, so the figure could overstate the importance of herbal medicine. Conversely, many people may self-medicate with traditional remedies without consulting a healer, so consultation rates with traditional healers may underestimate the importance of herbal remedies. Consultation rates with traditional healers for malaria have been summarised separately in Table 10.3 and are discussed later in the chapter.

2. Unobtainable unpublished papers quoted in reviews were excluded unless there was sufficient information given to establish that the study fulfilled the above criteria.

The studies were analysed to identify variables in the study design that may influence the reported frequency of use of plant-based treatments for malaria. On the basis of these variables, 10 criteria were developed for assessing the methodological relevance of studies to the objectives of this overview (see Box 10.1). Each included study was given a score out of 10 according to these criteria (see Table 10.2).

Studies of frequency of use of plant-based preventive measures were selected according to the same inclusion and exclusion criteria and summarised separately (Table 10.4 and Table 10.5). Similar criteria could be used to assess the design of these studies, substituting the phrase "preventive measures" for "treatments" in points 5 and 7 of Box 10.1. In this case, it is important for researchers to differentiate between traditional insect repellents, larvicides, and prophylactic herbal remedies.

10.3 RESULTS

10.3.1 Meta-Analysis of Treatment Seeking

One hundred and twenty-seven studies were found in the literature search described. Of these, 28 were included in the quantitative meta-analysis of treatment seeking (see Table 10.1). Ninety-nine studies were excluded for reasons explained above. Qualitative sociological and anthropological studies were reviewed and salient points have been included in the text below.

The proportion of patients using traditional herbal remedies to treat malaria varies widely from study to study. Meta-analysis of the figures in Table 10.1 reveals that of a total of 15,458 respondents, 3073 used herbal remedies for malaria — this yields an overall percentage of 19.9%. However, this figure is misleading, because the range is very wide, from 0 to 75%. The factors influencing this are discussed below.

TABLE 10.1
Frequency of Use of Plants to Treat Malaria

% Using Plants to Treat Malaria	Number Questioned	Subjects	Setting	Country	Design Score (see Table 10.2)	Reference
0%	216	Household heads (female)	Rural homes	Sri Lanka	4	Konradsen et al., 1997
1%	135	Residents	Rural homes	Ghana	3	Gardiner et al., 1984
1.9%	162	Heads of households	Urban homes	Ivory Coast	4	Dossou-Yovo et al., 2001
2%	346	Residents	Urban homes	Ghana	3	Gardiner et al., 1984
2%	100	Adults	Rural homes	Kenya	4	Karanja et al., 1999
2.5%	81	Mothers of children < 5 years	Rural homes	Tanzania	7	Hausmann Muela et al., 2002
3%	205	Children	Urban homes	Ghana	7	Agyepong and Manderson, 1994
3-9%	35	Adults	Rural homes	Kenya	6	Nyamongo, 2002
4%	91	Patients	Urban Western clinic	Burkina Faso	6	Bugmann, 2000
4%	99	Pregnant women with malaria	Rural homes	Uganda	5	Ndyomungyenyi et al., 1998
4.6%	421	Episodes of "homa" and headache in children	Rural homes	Kenya	8	Geissler et al., 2000
7%	118	Children	Urban and rural homes	Kenya	5	Mwenesi et al., 1995
10.2%	557	Caretakers of children <10 years	Urban and rural homes	Malawi	4	Slutsker et al., 1994
10.6%	3006	Parents of children aged 6 months–6 years with fever in last 2 weeks	Rural homes	Nigeria	6	Salako et al., 2001
10.7%	449	Households	Urban homes	Colombia	7	Lipowsky et al., 1992
11%	108	Episodes of fever	Rural homes	Kenya	9	Ruebush et al.,1995
11.1%	186	Caretakers of children aged 6 months–5 years with malaria in last 2 weeks	Urban clinics	Nigeria	5	Brieger et al., 2001
12%	460	Patients who had malaria in last 14 days	Urban and rural homes	Uganda	6	Nuwaha, 2002
15%	154	Caretakers of children <5 years with malaria in last 3 weeks	Urban and rural homes	Zambia	6	Baume et al., 2000
15.3%	463	Heads of households	Rural homes	Kenya	5	Munguti, 1998
16.8%	558	Episodes of malaria	Rural families	Sri Lanka	6	Jayawardene, 1993
19%	532	Adults	Rural homes	Mali	5	Djimde et al., 1998
20%	205	Adults	Urban homes	Ghana	7	Agyepong and Manderson, 1994

(continued)

TABLE 10.1 (CONTINUED)
Frequency of Use of Plants to Treat Malaria

% Using Plants to Treat Malaria	Number Questioned	Subjects	Setting	Country	Design Score (see Table 10.2)	Reference
20.4%	333	Students	Urban school	Sudan	3	Elzubier et al., 1997
25%	1935	Heads of households	Urban and rural homes	Nigeria	4	Ramakrishna et al., 1989
25.2%	483	Households	Rural homes	Colombia	7	Lipowsky et al., 1992
25.5	376	Mothers of children <5 years	Urban health centers	Nigeria	5	Fawole and Onadeko, 2001
26%	255	Children	Rural homes	Ghana	7	Agyepong and Manderson, 1994
32.2%	59	Caretakers of children <5 years	Urban and rural homes	Togo	4	Cook et al., 1999
38.4%	213	Patients	Urban and rural traditional clinics	Burkina Faso	6	Bugmann, 2000
42%	148	Educated, aged 13–80	Rural communities	Zanzibar	6	Alilio et al., 1998
42.1%	38	Teachers	Urban school	Sudan	3	Elzubier et al., 1997
45%	996	Adults	Rural	Gambia	5	Aikins et al., 1993
46%	274	Mothers of children <5 years with malaria in last 2 weeks	Rural homes	Benin	4	Fourn et al., 2001
52%	160	Noneducated, aged 13–80	Rural communities	Zanzibar	6	Alilio et al., 1998
55%	255	Adults	Rural homes	Ghana	7	Agyepong and Manderson, 1994
55.6%	114	Families	Urban homes	Niger	5	Julvez et al., 1995
74%	100	Households	Rural homes	Madagascar	3	Rahantamalala, 2000
75.1%	225	Women	Rural homes	Sudan	6	A/Rahman et al., 1995

10.3.2 STUDY DESIGN AND KEY CRITERIA

Many studies have sought to measure treatment seeking behaviour for malaria (McCombie, 1996). Unfortunately, they cannot easily be compared with one another because of greatly differing methodologies. The design of each study was assessed according to the key criteria listed in Box 10.1. Compliance with each of these criteria will be discussed in greater detail below.

10.3.2.1 Criterion 1: Study Setting

Some study populations consisted of patients at a hospital or clinic (Snow et al., 1992; Mnyika et al., 1995), which is not a representative sample, since only 8 to 25% of people with malaria visit health services (Brinkmann and Brinkmann, 1991). In order to overcome this problem, most researchers interviewed people in their homes. Ettling et al. (1989) interviewed both clinic attenders and villagers in their own homes. A traditional healer had been consulted by only 5% of clinic attenders, but by 12% of villagers. Bugmann (2000), in a study in Burkina Faso, found that 4% of

---

BOX 10.1: KEY CRITERIA FOR ASSESSING DESIGN OF INCLUDED STUDIES

1. Patients were studied in the community, not at a clinic (traditional or Western).
2. The ratio of men to women (and vice versa) did not exceed 60:40.
3. Interviewers were members of the local community.
4. Interviewers could not be perceived as having any bias towards either Western or traditional medicine.
5. A cohort was followed prospectively to see which treatments are actually used.
6. The illness being investigated was clearly defined (e.g., malaria).
7. Open questions were asked about what treatments were used.
8. Open questions were asked about the order of priority in which treatments were used.
9. Open questions were asked about treatments' perceived efficacy.
10. When traditional medicine or a visit to the traditional healer was mentioned, the researchers specified whether the treatment was herbal medicine or another traditional practice (and ideally, identified the herb(s)).

---

patients attending a Western clinic had previously taken traditional herbal medicines, compared to 38% of patients attending a traditional healer's clinic.

10.3.2.2 Criterion 2: Gender Balance

Many of the studies ensured that a majority of subjects were women. Women are usually those responsible for the treatment of children, so their treatment-seeking behaviour is particularly important. Gender differences in treatment seeking have been observed. In fishing communities in the Sélingué district of Mali, men are more likely to prefer traditional treatments than women (Traore et al., 1993), and more boys than girls believed in herbal medicine in surveys among the Luo in Kenya and in an urban secondary school in the Sudan (Geissler et al., 2000; Elzubier et al., 1997). However, others have suggested that women are less likely to be treated at modern facilities and are more likely to resort to traditional medicines (Tanner and Vlassoff, 1998). There are no concrete data on this. Many studies involved a majority of women, since they are the primary caretakers for their children.

10.3.2.3 Criteria 3 and 4: Choice of Interviewer

Very few studies explicitly stated that their interviewers were members of the local community and were unbiased towards either biomedical or traditional medicine. Yet subjects' perception of the interviewer may markedly influence their responses. People may feel it would be disrespectful to health professionals to deny use of drugs or to admit that they use traditional healers (Dabis, 1989). Some may simply wish to please the interviewers by saying that they use modern medicine, and others may even give this response in the hope of receiving aid in the form of medicines or health facilities (Traore et al., 1993).

For this reason, it is very important to use interviewers who are indifferent to the response given, or members of the local community who are known and trusted. Very few studies actually did this, one notable exception being that of A/Rahman et al. (1995) in Sudan; members of the village health committee interviewing women in their own villages found that about 75% use herbal remedies for malaria. Kengeya-Kayondo et al. (1994) found that in clinic interviews, drug use was more common than herb use, but focus group discussions and key informants revealed that use of herbs was the most common first treatment action.

TABLE 10.2
Assessment of Included Studies according to Key Criteria

Reference	Total Score	Key Criteria[a]									
		1	2	3	4	5	6	7	8	9	10
Agyepong and Manderson, 1994	7	+	+	?	+	−	+	+	+	−	+
Aikins et al., 1993	5	+	+	−	?	−	+	+	−	−	+
Alilio et al., 1998	6	+	+	+	+	−	−	+	−	−	+
A/Rahman et al., 1995	6	+	+	+	−	−	+	+	−	−	+
Baume et al., 2000	6	+	+	−	+	−	−	+	+	−	+
Brieger et al., 2001	5	−	+	+	−	−	+	+	−	−	+
Bugmann, 2000	6	−	+	−	−	−	+	+	+	+	+
Cook et al., 1999	4	+	+	−	−	−	−	+	+	−	−
Djimde et al., 1998	5	+	+	−	−	−	+	+	−	−	+
Dossou-Yavo et al., 2001	4	+	?	?	?	−	+	+	+.	−	−
Elzubier et al., 1997	3	+	−	−	−	−	+	−	−	−	+
Fawole and Onadeko, 2001	5	−	+	−	+	−	+	+	−	−	+
Fourn et al., 2001	4	+	+	?	?	−	−	−	−	+	+
Gardiner et al., 1984	3	−	+	−	−	−	+	−	−	−	+
Geissler et al., 2000	8	+	+	+	+	+	+	+	−	−	+
Jayawardene, 1993	6	+	+	−	?	+	+	+	+	−	−
Julvez et al., 1995	5	+	+	−	?	−	+	+	−	−	+
Karanja et al., 1999	4	+	+	−	−	−	−	+	−	−	+
Konradsen et al., 1997	4	+	+	−	−	−	+	+	−	−	−
Lipowsky et al., 1992	7	+	+	+	+	−	+	+	−	−	+
Munguti, 1998	5	+	?	?	?	−	+	+	+	−	+
Mwenesi et al., 1995	5	+	+	?	?	−	+	+	−	−	+
Nuwaha, 2002	6	+	+	−	−	−	+	+	−	+	+
Nyamongo, 2002	6	+	+	?	?	−	+	?	+	+	+
Rahantamalala, 2000	3	+	?	−	?	−	+	−	−	−	+
Ramakrishna et al., 1989	4	+	?	−	−	−	+	+	−	−	+
Ruebush et al., 1995	9	+	+	+	?	+	+	+	+	+	+
Slutsker et al., 1994	4	+	?	−	?	−	+	+	−	−	+
Total (28)		24	22	6	6	3	23	23	9	5	24

[a] See Box 10.1.

It may be difficult for researchers to adhere to these criteria as it may be difficult to find members of the community to act as interviewers; and it may also be difficult to find interviewers who are literate but do not have a bias towards Western medicine.

10.3.2.4   Criterion 5: Prospective Follow-Up

Most studies relied on recall of past behaviour or hypothetical actions in particular scenarios. Only three had a prospective design (Espino and Manderson, 2000; Jayawardene, 1993; Ruebush et al., 1995). A prospective study of actual behaviour is more costly and complicated to organise than a retrospective or hypothetical questionnaire.

However, theoretical actions often differ from actual behaviour patterns. Matthies (1998) found that although 74% of patients at a traditional healer's clinic claimed they would go to a formal health facility first, in fact 47% went to the traditional healer without seeking advice elsewhere, and only 16% first went to a formal health facility.

10.3.2.5 Criterion 6: The Definition of Malaria

Most of the studies in Table 10.1 relied on reports by patients or their caregivers, and when treating themselves with traditional remedies, almost all will have self-diagnosed. The relevance of this overview to malaria, as opposed to febrile illnesses in general, depends both on the local terms used by interviewers and on the accuracy of diagnoses by patients and their caregivers. Some studies were included in this overview although they looked at all febrile illnesses, because in the areas concerned, the majority of these febrile episodes would be compatible with malaria (Salako et al., 2001).

Malaria is a laboratory diagnosis, and without access to microscopy or other diagnostic tests, the diagnosis of mothers and clinicians alike will not be accurate in all cases. This problem can be surmounted by taking blood films from all patients concerned and measuring their temperatures; however, doing so will medicalise the interview and may influence the responses given. Diallo et al. (2001) compared maternal presumptive diagnoses with blood films and temperatures of 784 children under the age of five in Guinea. Although the terms used for malaria translated simply as "hot body," they found that the maternal diagnosis was 32% sensitive and 92% specific for malaria, with a positive predictive value of 37% and a negative predictive value of 90%. In this instance, mothers tended to underdiagnose malaria.

Many languages have no specific term for *malaria*, but have several words for different types of fever. In Nigeria, Ajaiyeoba et al. (1999) found that 21 types of fever were identified by local people. Many of these could be equated with different forms of malaria. Frequency of use of herbs and drugs varied according to type of fever. Similarly, Helitzer-Allen et al. (1993) were able to subdivide the term Malawian *malungo* into seven subcategories, each believed to have different causes, symptoms, and treatments.

In Swahili, several words denote different illnesses that overlap to a greater or lesser extent with a biomedical diagnosis of malaria. One of the most common illnesses reported is *homa*, a syndrome of fever and body pains, which is likely to include not only malaria, but also viral illnesses (Geissler et al., 2000; Winch et al., 1996). *Kibwengo* are spirits of the devil that may be encountered in the hot sun near large stones and trees; they cause headache, stomach pains, and fever, and can be worsened by modern medicine (Oberländer and Elverdan, 2000).

Complications of severe malaria may be seen as separate illnesses and treated differently. *Endwari ya inda* (splenomegaly) is usually treated with herbal medicines (Nyamongo, 2002). Similarly, *degedege* (childhood febrile convulsions) is believed to be caused by bad luck and to be treatable only by traditional medicine; injections are believed to be potentially fatal in this condition (Makemba et al., 1996; Matthies, 1998; Mwenesi et al., 1995; Oberländer and Elverdan, 2000; Tarimo et al., 2000; Winch et al., 1996). Interestingly, traditional medicine is preferred for the treatment of convulsions in many other ethnic groups, for example, in Madagascar, Malawi, Mali, and Zambia (Leon, 2002; Helitzer-Allen et al., 1993; Graz, unpublished; Baume et al., 2000).

10.3.2.6 Criterion 7: Study Questions

Wording of questionnaires is very important in determining the response. With regard to treatment seeking, it is particularly important to ask open questions about all treatment options, not only about drugs.

For example, frequency of herbal antimalarial usage in rural Ghanaian homes was reported at 1% by Gardiner et al. (1984), but at 55% by Agyepong and Manderson (1994). Gardiner et al.

interviewed selected individuals on their use of *drugs* and medicalised the interview by taking finger-prick blood samples. Agyepong and Manderson asked about *treatments* in general, using focus groups as well as individual informants.

Asking an open question alone is not enough. Some asked "How can malaria be treated?"; others, "What is the best way of treating malaria?"; others, "How would you treat malaria if you had it?"; and others, "How did you treat malaria last time you got it?" Questions like this last one, related to a specific episode of malaria, are likely to evoke a response more representative of the real situation. This could have been used as another criterion for assessing study design. Unfortunately, few papers specified the actual questions posed in their survey.

Focus group discussions and detailed interviews often reveal a more complex set of behaviours than questionnaires, including herbal medicines that may not otherwise be mentioned (Linhua et al., 1995). Ramakrishna et al. (1989) found that a general questionnaire underestimated the frequency of use of traditional herbal medicine, and that in-depth interviews revealed this to be the most common treatment for malaria.

10.3.2.7 Criterion 8: Sequence of Use

Use of traditional herbal remedies and use of pharmaceuticals are not mutually exclusive. Several studies have found that pharmaceuticals are often combined with herbs (Agyepong and Manderson, 1994; Bugmann, 2000; Gessler et al., 1995; Graz, unpublished; Jayawardene, 1993; Lipowsky et al., 1992; McCombie, 1996; Pagnoni et al., 1997). These may be taken together simultaneously, or as first- and second-line treatments.

Simultaneous use of drugs and herbs was reported by some traditional healers interviewed by Gessler et al. (1995) and Rasoanaivo et al. (1992); their belief is that combining the two modalities gives an additional effect. There is some laboratory evidence to substantiate this in Madagascar (see Chapter 9).

The results of most studies, reflected in the analysis in Table 10.1, assume that each episode of malaria is only treated once. However, this is not always the case. Many patients try two or more treatments before they feel they have been cured. Choice of first-line treatment differs from area to area. Sometimes, herbal remedies are given at home as the first-line treatment (Agyepong and Manderson, 1994; Bugmann, 2000; Klaver, 1993; Ruebush, 1995; Baume et al., 2000; Okrah et al., 2002), especially in mild cases of malaria (Miguel et al., 1998, 1999).

Sometimes, herbs are the second line after treatment with chemotherapy has failed (Dossou-Yovo et al., 2001; Théra et al., 1998; Hausmann Muela et al., 1998; Helitzer-Allen et al., 1993; Bitahwa et al., 1997; Bugmann, 2000; Utarini et al., 2003). Munguti (1998) found that in a rural area of Kenya, 7% used herbs as first choice of treatment, 17% as second choice, and 14% as third choice.

10.3.2.8 Criterion 9: Perceived Efficacy

Perceived efficacy is an important reason for the use of traditional herbal remedies and should be a key component of research on treatment seeking. It would seem logical that patients would not use a remedy unless they thought it was at least partly effective. On the other hand, some patients may use herbal remedies simply because they cannot afford pharmaceuticals. Adera (2003) found that Ethiopian patients chose traditional remedies for malaria primarily because of their greater accessibility (83%) and low cost (48%), but 7% of respondents believed that traditional remedies were more effective than modern medicine. Only 5 of the 28 studies included in this review (18%) asked respondents about their perception of efficacy of different treatments. Most of these interpreted their findings from a biomedical viewpoint.

In two studies where traditional medicines were perceived to be effective, this finding is interpreted in a negative light. Nuwaha (2002) writes that in the Mbarara region of Uganda, people continue to treat themselves with traditional medicine despite the availability of cheap modern antimalarials. Traditional medicines are perceived to be effective for malaria, and particularly for convulsions or splenomegaly, thus delaying attendance at biomedical facilities for severe malaria. An equally valid interpretation could be that traditional medicines are effective for uncomplicated malaria, and that this should be investigated, while encouraging use of modern medicine for severe malaria. Nyamongo (2002) suggests that the perceived efficacy of herbal medicines is spurious and related to concomitant treatment with over-the-counter pharmaceuticals. An alternative interpretation is that some herbal medicines may potentiate the effect of pharmaceuticals, and that concomitant use may help to overcome resistance. The interpretations chosen by these authors are consistent with their aims of improving compliance with biomedicine, with no consideration of the potential value of traditional medicines.

In the Dori region of Burkina Faso, Bugmann (2000) found that opinions differed between patients. Interestingly, some believed that plants effected a longer-term cure than pills, and that the side effects of diarrhoea and vomiting were actually beneficial in getting rid of the disease. However, fever clearance alone was sufficient to produce the perception of efficacy, whether or not it was associated with parasite clearance; thus, perceived efficacy does not necessarily imply therapeutic efficacy according to medical definitions.

Different tribes in the same country may have different traditional medicine systems, with different perceived efficacy for the treatment of malaria. For example, Luo communities in Kenya reported that modern medicine was more effective than their herbal remedies (Kawango, 1995; Ruebush et al., 1995). In contrast, the Gusii people in Kenya believe that their traditional medicine is effective in treating malaria (Sindiga, 1995).

Pagnoni et al. (1997) ran a project in rural Burkina Faso to make antimalarial drugs easily available. Although the proportion of people using tablets increased from 49 to 71%, the proportion using traditional remedies declined only from 70 to 54%, and the proportion of patients combining both treatments remained the same (25%). This may indicate that patients still perceived traditional remedies to be effective.

Perceived efficacy does not always determine choice of treatment. Although mothers of febrile children in South Benin perceive biomedicine to be effective for malaria, they often prefer to use traditional medicine because it is more affordable and easily available (Fourn et al., 2001). However, these mothers were not asked about the perceived efficacy of the traditional medicines.

10.3.2.9 Criterion 10: Definition of Traditional Medicine

A basic inclusion criterion of the overview was that the study should specifically report the use of traditional plant-based medicines, as distinct from other forms of traditional medicine. Nevertheless, while including measures of use of herbal medicines, a few of the studies also used other poorly defined categories that may or may not have included some herbal treatments (for example, popular or traditional medicine, home remedies, Ayurvedic or indigenous treatment). These made it difficult to quantify the total proportion of patients using herbal medicines for malaria.

10.3.3 EXTERNAL VARIABLES

Variables inherent to the study design can only explain part of the variation in frequency of use of herbal antimalarials. Other important factors are discussed below. These include the demographic characteristics of subjects and the availability of drugs and herbal antimalarials.

10.3.3.1 Study Location

Legislation in the country is an important factor. For example, in a study in Kenya in 1976, there was a widespread belief that any form of traditional medicine was forbidden by law, so respondents

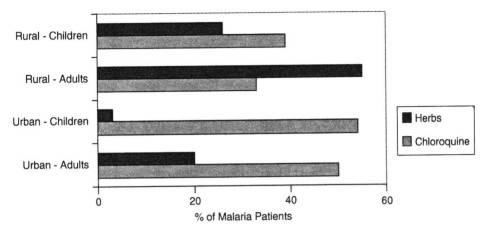

FIGURE 10.1 Use of herbs or chloroquine for the treatment of malaria in Ghana, according to age and location. (Agyepong and Manderson, 1994.)

did not wish to speak about it (Schulpen and Swinkels, 1980). This followed newspaper reports of "witch doctors" being prosecuted.

10.3.3.2   Demographics

Rates of use of herbal medicine vary according to the characteristics of the subjects. Because of the wide variety of methods used, different studies are not directly comparable. However, within-study differences are valid. Agyepong and Manderson (1994) stratified responses according to age (adults vs. children) and geography (urban vs. rural). Their findings are summarised in Figure 10.1. Herbs are more widely used in adults than in children, and more in rural than in urban areas. Frequency of use of herbs ranged from 55% of rural adults to 3% of urban children.

Such rural–urban differences have been replicated in some other studies (McCombie, 1996; Lipowsky et al., 1992) and are readily explained by greater availability of herbs and knowledge about them (or lesser availability of drugs) in rural areas. Lipowsky et al. (1992) found that 82.1% of rural households in Colombia knew at least one plant that could be used for the treatment of malaria, compared to 55.9% of urban households.

Luo children at a rural primary school in Kenya were able to identify a total of 70 herbal remedies, for all conditions (Geissler et al., 2000). Fifty-seven schoolchildren (ages 11 to 17) experienced 421 episodes of probable malaria, of which 49% were not treated, 40% were treated with assistance from adults, and 11% were self-treated. Eleven percent of self-treatments were with herbs, whereas 23% were with antimalarial drugs and 61% with painkillers. A total of 8.4% of assisted treatments were with herbal medicines (P.W. Geissler, personal communication).

Level of education may also influence treatments chosen. Usually it is reported that herbal medicines are more commonly used by uneducated people. However, the differences are not very great, and a substantial proportion of educated people still choose to use traditional medicines for malaria. For example, in a survey in Zanzibar, 52% of people with no formal education used herbal remedies for malaria, compared to 42% of those with at least primary education (Alilio et al., 1998). Even those who were educated consulted their grandparents, which may have led to high levels of use of traditional medicines. Fawole and Onadeko (2001) found that educated mothers were more likely to choose a formal health facility as their first port of call, but only slightly less likely to consult a traditional healer first. One survey found that education was positively associated with knowledge of herbal medicines for malaria: in an urban secondary school in Sudan, 42% of teachers, compared to 20% of students, believed that malaria could be treated with herbal medicine (Elzubier et al., 1997).

10.3.3.3 Availability of Drugs

Greater accessibility of traditional medicine was the most important reason for its use in Ethiopia, quoted by 83% of respondents in a recent survey (Adera, 2003). In Uganda, perceived absence of drugs at health centers is an important reason why patients do not attend for malaria (Ndyomugyenyi et al., 1998). In Malawi, shortages of chloroquine led health workers to prescribe insufficient doses (Helitzer-Allen et al., 1993). Even in a project training malaria extension workers to provide chloroquine to remote areas, one third of the workers had no chloroquine tablets for over 6 weeks (Théra et al., 1998). In many areas, distance to the health center is the limiting factor (Baume et al., 2000; Miguel et al., 1999; Okrah et al., 2002; Utarini et al., 2003). In these circumstances, the only available treatments are herbs.

10.3.3.4 Affordability of Drugs

Lack of money is a commonly quoted reason for malaria patients not seeking modern health care (Leon, 2002; Mugisha et al., 2002; Ndyomugyenyi et al., 1998; Théra et al., 2000). Even drugs provided at very low cost can be unaffordable to the very poor. For example, people were unable to afford chloroquine during a malaria epidemic in rural Malawi, and the drugs eventually had to be given away, or they would have reached their expiry date (Richardson, 1990, quoted in Foster, 1995). In Madagascar, some health professionals refuse to see patients unless they can pay; therefore, many poor patients have given up going to modern health facilities (Leon, 2002). Similar problems have been reported in Zambia (Baume et al., 2000). Low cost was the second most important reason for use of traditional medicine, quoted by 48% of respondents in the Ethiopian survey by Adera (2003).

10.3.3.5 Trust in the Doctor or Healer

In rural Gambia, a similar proportion of patients consulted traditional healers regardless of whether primary health facilities were available (De Francisco et al., 1994). The authors speculate that this may be due to the prevalence of traditional concepts of health and disease, as well as the direct accountability of traditional healers to their communities.

Asking people why they did not use modern health services, Abyan and Osman (1993) found that cost was the reason for only 50%, whereas lack of faith in doctors was the reason in the other 50%. In Burkina Faso, it has been observed that medical staff at the hospital are less trusted because they are often young, do not speak local languages, and are not welcoming to patients (Bugmann, 2000). Similarly, in Peru, some medical staff resent being sent to work in poor areas and are discourteous to patients, thus deterring them (Keme, 2000). Some health care professionals even scold parents for not looking after their children properly, which discourages them from attending — this has been reported in Madagascar and Zambia (Leon, 2002; Baume et al., 2000).

This distrust of modern medicine may in many cases be justified. Modern medical care is not always of high quality, and some studies have shown no decrease in child mortality in spite of increased treatment seeking at formal health facilities (Sodemann et al., 1997). Some dispensers prescribe the medicine that will maximise their profits rather than that which would be best for the patient (Bledsoe and Goubaud, 1985).

10.3.3.6 Availability of Medicinal Plants

Effective antimalarial herbs may not be available or known in all malarious areas. For example, in Mananjary (Madagascar) villagers reported that there were no local herbal remedies for malaria

(Leon, 2002). Even where plants exist, they may not be readily available. Indeed, in some areas people leave urban centers in order to gather plants that are believed to be more effective than drugs in treating malaria; this pattern of treatment seeking is common in the Maasai people of Tanzania (Burford et al., 2000). Other people may not have the opportunity to travel far for treatment, and this may reduce usage of botanical medicines, especially by city dwellers. Certain medicinal plants are becoming endangered or even extinct (FAO, 1997), and this will obviously reduce their use.

Herbal remedies are rarely made available on a large scale by public health systems. However, there are at least two exceptions to this rule. The first is Ayush-64, an Ayurvedic herbal antimalarial preparation, which was used during malaria epidemics in Rajasthan in 1994 and in Assam in 1996 (Bhatia, 1997). Over 3000 and 2200 cases were treated, respectively. The results were reportedly good, although information on this is scant (see Chapter 5).

Mali's national formulary (Ministère de la Santé, 1998) includes a page on Malarial, a herbal remedy composed of three local plants (see Chapter 7). The indication is for "febrile states linked to malaria," and there is a warning that parasites are not cleared completely. However, the medicine is contraindicated in children less than 15 years old. To our knowledge, this is the first traditional herbal remedy for malaria to be included in a national formulary.

10.3.4 TREATMENTS BY TRADITIONAL HEALERS

Most studies reporting consultations with traditional healers did not specify what treatment was given, so these were excluded from the quantitative meta-analysis in Table 10.1. There are many different types of healer, often specific to certain areas, and not all of them use herbal remedies (Traore et al., 1993; Miguel et al., 1998; Hausmann Muela et al., 2002). Baume (1998a) reports that when patients with malaria consulted traditional healers, they consulted only herbalists and "bone removers," not diviners or spiritualists. However, most studies did not specify this, and this finding may not be generalisable to other contexts. Other modes of treatment used by healers for malaria include herbal baths in Togo (Cook et al., 1999); scarification and application of herb ashes in Kenya (Nyamongo, 2002); massage and amulets in Madagascar (Leon, 2002); and prayers and sacrifices in Ivory Coast (Dossou-Yovo et al., 2001).

Some traditional healers even use modern pharmaceuticals, and some are formally trained to do so by health programs (Okanurak et al., 1992). In a low endemic area of the Philippines, traditional healers admitted that they did not know how to treat malaria and referred cases to a clinic (Espino et al., 1997). There is nothing to prevent Western drugs from becoming indigenised (Haak and Hardon, 1988), and this may happen even in the presence of traditional illness models (MacLachlan and Carr, 1994).

In general, the consultation rates reported in Table 10.3 are lower than the rates of herbal remedy usage reported in Table 10.1. In many societies, knowledge of herbal remedies is not confined to traditional healers, just as knowledge of drugs is not limited to doctors, and patients do not believe they need to seek external advice unless their initial treatment fails (Geissler et al., 2001; Leon, 2002). Mugisha et al. (2002) found that in rural Burkina Faso, 43% of malaria patients felt competent to treat themselves. Furthermore, certain demographic groups, such as children, may not visit the traditional healer on their own because the power difference is too wide (Geissler et al., 2000). Hausmann Muela et al. (1998) conducted a detailed anthropological investigation into beliefs about malaria near Ifakara, Tanzania. They discovered that although chemotherapy was the first choice, if this failed, people suspected witchcraft and went to see a traditional healer rather than trying a different medication.

However, there can be little doubt that in many developing countries traditional healers are consulted more frequently than doctors. Vongo (1999) reports that Zambia has 1 doctor per 20,000 population, 1 nurse per 2000, and 1 traditional healer per 200. In Kenya, the ratio of doctors to population has hardly increased in the last 40 years, partly because of rapid population growth;

TABLE 10.3
Consultation with Traditional Healers for Malaria

% Consulting Traditional Healer	Number Questioned	Subjects	Design	Setting	Country	Reference
0%	315	Households	Interviews	Rural, arid zone	Kenya	Mwabu, 1986
0.3%	376	Mothers	Hypothetical treatment	Urban and rural	Kenya	Snow et al., 1992
0.6%	177	Mothers	Interviews	Rural homes	Madagascar	Leon, 2002
1.1%	917	Mothers and caretakers	Interview	Urban clinic	Nigeria	Brieger et al., 2001
1.2%	81	Villagers	Interviews	Rural	Tanzania	Hausmann Muela et al., 1998
1.2%	162	Household heads	Interviews	Urban homes	Ivory Coast	Dossou-Yovo et al., 2001
1.7%	299	Household heads	Questionnaires	Rural homes	South Africa	Govere et al., 2000
2%	338	Households	Interviews by health workers	Urban homes	Senegal	Faye et al., 1996
2%	154	Caretakers of children under 5 years	Illness narratives	Rural and urban homes	Zambia	Baume, 1998b Baume et al., 2000
2.4%	1021	Patients	Prospective follow-up	Rural homes	Guatemala	Ruebush et al., 1992
3%	652	Guardians of children aged <5	Interviews	Rural homes	Tanzania	Nsimba et al., 2002
3%	254	Householders	Questionnaires	Rural homes	Ethiopia	Adera, 2003
3%	147	Mothers of children <10 years	Interviews	Rural homes	Kenya	Molyneux et al., 1999
3%	97	Malaria episodes	Prospective follow-up	Rural homes	Philippines	Espino and Manderson, 2000
3–13%	951	Caretakers of children <16 years	Interviews	Rural homes	Mali	Graz, unpublished
3–35%	3237	Cases of malaria	Interviews	Rural homes	Nepal	Mills, 1993
3.3%	394	Hospital patients	Prehospital treatment	Urban and rural	Kenya	Snow et al., 1992
5%	505	Children	Follow-up interviews	Rural	Togo	Deming et al., 1989
5.3%	532	Adults	Interviews	Rural homes	Mali	Djimde et al., 1998
5.5%	567	Villagers	Questionnaires	Rural homes	Tanzania	Winch et al., 1996
6%	236	Mothers of children <10 years	Interviews	Urban homes	Kenya	Molyneux et al., 1999

(continued)

TABLE 10.3 (CONTINUED)
Consultation with Traditional Healers for Malaria

% Consulting Traditional Healer	Number Questioned	Subjects	Design	Setting	Country	Reference
6.2%	97	Caretakers of children <5 years	Illness narratives	Rural homes	Kenya	Baume, 1998a
6.6%	557	Caretakers of children <10 years	Interviews	Rural and urban homes	Malawi	Slutsker et al., 1994
7.5%	200	Malaria patients	Interviews	Rural clinic	Thailand	Fungladda and Sornmani, 1986
9%	81	Women	Interviews	Rural homes	Nigeria	Okonofua et al., 1992
11.5%	949	Mothers	Questionnaire	Rural clinic	Nigeria	Ejezie et al., 1991
12%	199	Villagers	Interviews	Rural homes	Thailand	Ettling et al., 1989
13%	386	Malaria episodes in infants	Interviews	Rural homes	Malawi	Vaahtera et al., 2000
14.7%	523	Patients	Interviews	Urban hospital outpatients	Tanzania	Mnyika et al., 1995
20.1%	376	Mothers	Interviews	Urban health centers	Nigeria	Fawole and Onadeko, 2001
52.3%	200	Villagers	Interviews	Rural homes	India	Singh et al., 1998
56.8%	169	Villagers, fever with convulsions	Questionnaires	Rural homes	Tanzania	Winch et al., 1996
62%	158	Child deaths	Verbal autopsy	Rural villages	Gambia	De Francisco et al., 1994

57% of households must travel more than 4 km to the nearest health facility (Sindiga et al., 1995). Transport costs are usually cheaper when consulting a healer, as there is less distance to travel; the healer can often be paid in kind rather than in currency, and according to the patient's means rather than a fixed charge (Bugmann, 2000; Traore et al., 1993).

Matthies (1998) examined in detail the treatment-seeking patterns of patients attending the clinics of three traditional healers in Tanzania. She found that 47% of patients at one healer's clinic had gone directly there (and about 88% of patients at two other healers' clinics had gone directly there). However, when asked a theoretical question about treatment seeking, only 1.8% of patients said they would consult a traditional healer as a first step for malaria treatment, but 43.9% would consult as a second step, if their first method failed (usually visiting a formal health facility or treatment with chloroquine). Since this sample was taken at a traditional healer's clinic, it is likely to be biased. Bugmann (2000), also interviewing patients at a traditional healer's clinic in Burkina Faso, found that 70% had tried at least one other treatment before coming to the healer. Studies in

FIGURE 10.2 Patients with malaria waiting to be seen at a clinic run by traditional healers in Rukararwe, Bushenyi district, Uganda. (Copyright 1998, Merlin Willcox.)

Kenya and the Solomon Islands found that traditional healers were never consulted as the first treatment option (Baume, 1998a; Dulhunty et al., 2000).

This helps to explain the finding by De Francisco et al. (1994) that, of children dying of malaria, 62% had consulted a traditional healer. In fact, almost all of these had also consulted at least one health professional. There is no indication of the order of treatment seeking or of how many consulted a traditional healer alone. The high proportion consulting a traditional healer probably reflects the severity of the illness and the extent of different treatments sought. Similarly, Baume et al. (2000) report that although cases involving convulsions are more likely to be treated by traditional healers, 85% did receive modern care, and traditional treatments were most often used as a complementary rather than an alternative measure. In a systematic verbal autopsy study of child deaths in a defined birth cohort in Guinea-Bissau, 93% had been seen at a health center or hospital in the 2 weeks before their death (Sodemann et al., 1997). Over 30% of these children died of fever or malaria. This study demonstrated that contrary to popular belief, mothers sought formal care sooner in cases of fatal illness. In comparison with a previous study at the same site, appropriate care-seeking behaviour had increased, but child mortality had not decreased.

Another factor may be that convulsions are not always recognised as being caused by severe malaria and are believed to be best treated by traditional healers. Unfortunately, De Francisco et al. (1994) did not attempt to control for symptoms at presentation. Winch et al. (1996) found that up to 64% of people said they would see a traditional healer for degedege (fever with convulsions). In Mbarara, Uganda, 50% of interviewees believed that convulsions could not be treated by modern medicine, and one respondent even said that if a child gets convulsions in the hospital, he would bring a traditional healer to treat him there, or forcibly remove the child if this was not allowed (Nuwaha, 2002). Clearly, in such circumstances it will be difficult to improve the case management of cerebral malaria without involving traditional healers.

TABLE 10.4
Frequency of Use of Plants to Prevent Malaria

% Using Plants to Prevent Malaria	No. Questioned	Subjects	Design	Setting	Country	Reference
0%	~100	Selected community members	Focus group discussions	Urban community centers	Colombia	Nieto et al., 1999
0%	187	Households	Interviews	Rural village	Kenya	Ongore et al., 1989
2%	91	Clinic patients and key informants	Interviews	Rural village	Philippines	Espino et al., 1997
5%	1114	Villagers	Questionnaires	Rural villages	Mali	Traore et al., 1993
6.7%	951	Caretakers of children <16 years	Interviews	Rural homes	Mali	Graz, unpublished
10%	1935	Heads of households	Interviews	Urban and rural homes	Nigeria	Ramakrishna et al., 1989
12%	74	Women	Interviews	Rural homes	Nigeria	Okonofua et al., 1992
16.1%	807	Adult caregivers	Interviews	Rural homes	Zambia	Kaona et al., 2000
20.9%	376	Mothers of children <5 years with malaria	Interview	Urban health centers	Nigeria	Fawole and Onadeko, 2001
24%	103	Adults	Interview	Rural homes	Myanmar	Shein et al., 1998

10.3.5 Use of Medicinal Plants for the Prevention of Malaria

Few studies reported use of medicinal plants for the prevention of malaria (Table 10.4). However, the dichotomy between curative and preventive medicine is a biomedical concept that is not ubiquitous among populations at risk of malaria. The Luo of Kenya believe that illness is always present in the body and must be controlled continuously (Geissler et al., 2001). They do not view this as prevention. Food may also function as a prophylactic. The Mende of Sierra Leone increase their consumption of hot peppers in food during the cold season, and these are believed to combat fevers (Bledsoe and Goubaud, 1985). Etkin and Ross (1991) found that many food plants ingested during the malaria season in Nigeria had antimalarial properties, although they were not consciously being used as prophylactics. This inadvertent use of herbal prophylaxis may be one explanation for the finding that, despite open questions about prevention of malaria, several surveys had no responses mentioning herbal medicines (Klaver, 1993; Ongore et al., 1989; Nieto et al., 1999). Ramakrishna et al. (1989) found that 10% of people in rural Nigeria used an herbal tea (agbo) to prevent malaria. This was almost as common as bed net use (10.5%). Traore et al. (1993) found that although few people knew any methods of preventing malaria (12%), traditional plants were the common method quoted (38%). It is not clear whether these plants were ingested or burnt as repellents.

Herbal insect repellents are indeed used in some areas (Table 10.5). Interestingly, in two surveys, about twice as many people knew about burning repellent plants as actually used this method (Karanja et al., 1999; Ongore et al., 1989). Perhaps some people are deterred by the unpleasant smell of the smoke. In one survey in Sudan, about 70% of women used smoke to repel mosquitoes, and about 75% used vegetable oils applied on the skin (A/Rahman et al., 1995). However, it is not

TABLE 10.5
Use of Plants as Insect Repellents

% Using Plants to Repel Insects	No. Questioned	Subjects	Design	Setting	Country	Reference
0%	162	Household heads	Interviews	Urban homes	Ivory Coast	Dossou-Yovo et al., 2001
2%	409	Household heads	Interviews	Rural homes	Zimbabwe	Vundule and Mharakurwa, 1996
5%	299	Household heads	Questionnaires	Rural homes	South Africa	Govere et al., 2000
17%	187	Households	Interviews	Rural homes	Kenya	Ongore et al., 1989
18%	1531	Household heads	Interviews	Urban and rural homes	Malawi	Ziba et al., 1994
10–22%	100	Household women	Interviews	Rural homes	Kenya	Karanja et al., 1999
20.4%	226	Household heads	Interviews	Rural homes	Zimbabwe	Lukwa et al., 1999
32.2%	388	Mothers of children with malaria	Interviews	Hospital	Kenya	Snow et al., 1992
69%	216	Household heads (female)	Interviews	Rural homes	Sri Lanka	Konradsen et al., 1997
86%	161	Villagers	Interviews	Rural homes	Gambia	De Martin et al., 2001

stated whether the smoke was from special plants or from cow dung, which is also a common source of smoke for repelling mosquitoes.

Significantly, several studies asking open questions about methods of preventing malaria failed to elicit any mention of herbs. This implies that herbs are not used as widely for prevention as they are for treatment of malaria. However, some communities include prophylactic herbs in their diet without the conscious intention of preventing malaria (Etkin and Ross, 1991). Therefore, in certain communities, use of prophylactic herbs may be more common than estimated by the studies reviewed here.

10.4 DISCUSSION

Use of herbs for malaria varies by region, environment, and population subgroup. It certainly is not a homogenous 80% in developing countries. Traditional herbal antimalarials are more popular in some areas than in others. This is influenced by factors such as urbanisation, availability of plants and drugs, cultural familiarity, and perceived efficacy.

One may suppose that natural selection favors those who use effective treatments for potentially fatal diseases such as malaria. Therefore, in areas where no effective herbs exist, evolution will favor people who use modern drugs. The persistent use of herbs in areas where chemotherapy is available (and has been available for some time) suggests that herbs exist there that are perceived to have an intrinsic value in the treatment of malaria and that may in some cases prevent severe disease and thus improve survival.

Those communities where plant use is prevalent for the treatment or prevention of malaria should be the priority for future ethnobotanical, ethnopharmacological, and clinical cohort studies.

If certain herbal preparations are shown to be safe and effective, their use could be promoted on a larger scale by public health programs.

At every stage of this process, social science expertise is required. Of primary importance is to understand treatment-seeking behaviour — not just how many patients treat themselves with herbal medicines, but *why* they do so. Very few of the studies identified in this overview asked about the perceived efficacy of treatments. Yet when identifying communities for future research, it would be most useful to focus on those where patients use herbal remedies because they believe them to be effective, not just because they have no alternative.

Beyond the immediate scope of this chapter, good relationships established with communities by social scientists can lead to more detailed ethnobotanical studies (Chapter 14), as well as assessing safety (Chapter 18) and clinical efficacy (Chapters 20 and 21). Social scientists should form part of the multidisciplinary team designing and conducting each of these stages of research. If promoting cultivation and use of herbal remedies on a larger scale, it will be crucial to assess patients' willingness to cultivate, prepare, and use such remedies. It will also be crucial to understand operational processes related to changing and implementing national malaria control policies. At different stages, different social science disciplines will need to be involved: not only sociologists, but also anthropologists, health communicators, political scientists, and economists.

When designing malaria control programs, it is important to make a comprehensive assessment of local traditional herbal medicine. In the words of Etkin (1999),

> Top down, hierarchical approaches to malaria control do not work. The personnel of donor agencies, government ministries, and international health projects have more in common with one another than with local populations from whom they are culturally, politically and physically detached. Failure to deal with the behavioural dimension squanders the efficacy of both mosquito control technology and antimalarial pharmaceuticals.

It is widely recognised that there are few trained social scientists working on malaria, and that most sociobehavioural research on malaria control is undertaken using rapid assessment techniques by people with limited training in social science theory or methodology (Jones and Williams, 2002; Williams et al., 2002). The lack of consistency in methodology for studies of treatment-seeking behaviour for malaria has resulted in gaps in information and unclear findings, especially relating to the use of traditional herbal medicines. In a similar way that guidelines in the biological and clinical sciences are furthering the research agenda in this field, an agreed social science research framework would be of significant benefit in building on local knowledge as the basis for developing strategies for combating the current malaria epidemic.

The inclusion criteria for studies in this overview, as well as key criteria listed in Box 10.1, could form the basis for guidelines for the traditional medicine component of studies on treatment-seeking behaviour for malaria. These include using a prospective design, with local people as interviewers, open questions about treatments, and their perceived efficacy. It is important for studies to differentiate between different types of traditional medicine, and to include questions about the use of herbs for the prevention as well as for the treatment of malaria. It would be pertinent to find out more about the interplay between use of herbal antimalarials and use of pharmaceutical drugs, in which order they are used, and in what circumstances they are used together. The latter is particularly relevant, in view of the potential to use medicinal plants to reverse resistance to chloroquine and other antimalarial drugs.

This overview demonstrates that many treatment-seeking studies have only superficially acknowledged the use of traditional herbal medicines for malaria; many malaria control programs do not even do that. Yet it has also shown that traditional plant-based medicines are widely used for malaria in many communities. Increasing the quality and quantity of social science research on use of traditional herbal medicines is the first step in their evaluation. Social science expertise will be of continued importance in the next stages of evaluation, which are discussed in the following chapters.

ACKNOWLEDGMENTS

The authors thank Rosemary Byrd for providing some of the database searches. Holly Ann Williams and Carol Baume kindly provided some unpublished papers, as did Jocelyn Bruyere at TDR (WHO). Dr. Caroline Schulman, Dr. Holly Ann Williams, Dr. Elizabeth Hsu, and two anonymous reviewers kindly read the paper and made some helpful comments.

REFERENCES

Abyan, I.M. and Osman, A.A. (1993). Social and Behavioural Factors Affecting Malaria in Somalia, TDR/SER/PRS/11.

Adera, T.D. (2003). Beliefs and traditional treatment of malaria in Kishe settlement area, southwest Ethiopia. *Ethiop. Med. J.*, 41, 25–34.

Agyepong, I.A. and Manderson, L. (1994). The diagnosis and management of fever at household level in the Greater Accra Region, Ghana. *Acta Trop.*, 58, 317–330.

Aikins, M.K., Pickering, H., Alonso, P.L., D'Alessandro, U., Lindsay, S.W., Todd, J., and Greenwood, B.M. (1993). A malaria control trial using insecticide-treated bed nets and chemoprophylaxis in a rural area of the Gambia, West Africa. 4. Perceptions of the causes of malaria and of its treatment and prevention in the study area. *Trans. R. Soc. Trop. Med. Hyg.*, 87 (Suppl. 2), 25–30.

Aikins, M.K., Pickering, H., and Greenwood, B.M. (1994). Attitudes to malaria, traditional practices and bednets (mosquito nets) as vector control measures: a comparative study in five West African countries. *J. Trop. Med. Hyg.*, 97, 81–86.

Ajaiyeoba, E.O., Oladepo, O., Ebong, O.O., et al. (1999). Cultural Characterisation of Febrile Illnesses and the Use of Traditional Herbs in Southwestern Nigeria. Poster presentation at the First International Meeting of RITAM, Moshi, Tanzania. Available at http://mim.nih.gov/english/partnerships/ritam_program.pdf.

Alilio, M.S., Eversole, H., and Bammek, J. (1998). A KAP study on malaria in Zanzibar: implications for prevention and control. A study conducted for UNICEF sub-office Zanzibar. *Eval. Program. Plan.*, 21, 409–413.

A/Rahman, S.H., Mohamedani, A.A., Mirgani, E.M., and Ibrahim, A.M. (1995). Gender aspects and women's participation in the control and management of malaria in Central Sudan. *Soc. Sci. Med.*, 42, 1433–1446.

Bannerman, R.H., Burton, J., and Wen-Chieh, C. (1983). *Traditional Medicine and Health Care Coverage.* WHO, Geneva.

Baume, C. (1998a). Care Seeking for Fever and Convulsions in Bungoma District, Kenya: Implications for Malaria Programs, Final Report. BASICS Project, unpublished.

Baume, C. (1998b). Care-Seeking for Illnesses with Fever or Convulsions in Zambia, Final Report. BASICS Project, unpublished.

Baume, C., Helitzer, D., and Kachur, S.P. (2000). Patterns of care for childhood malaria in Zambia. *Soc. Sci. Med.*, 51, 1491–1503.

Bhatia, D. (1997). Role of Ayush-64 in malaria epidemic. *J. Res. Ayurveda Siddha*, XVIII, 71–76.

Bitahwa, N., Tumwesigye, O., Kabariime, P., Tayebwa, A.K.M., Tumwesigye, S., and Ogwal-Okeng, J.W. (1997). Herbal treatment of malaria: four case reports from the Rukararwe Partnership Workshop for Rural Development (Uganda). *Trop. Doct.*, 27(Suppl. 1), 17–19.

Bledsoe, C.H. and Goubaud, M.F. (1985). The reinterpretation of Western pharmaceuticals among the Mende of Sierra Leone. *Soc. Sci. Med.*, 21, 275–282.

Bodeker, G. (2001). Planning for cost-effective traditional health services. In *Traditional Medicine: Better Science, Policy and Services for Health Development.* WHO Kobe Centre for Health Development, Kobe, Japan, pp. 31–70.

Brieger, W.R., Sesay, H.R., Adesina, H., et al. (2001). Urban malaria treatment behaviour in the context of malaria transmission in Lagos, Nigeria. *Afr. J. Med. Med. Sci.*, 30 (Suppl.), 7–15.

Brinkmann, U. and Brinkmann, A. (1991). Malaria and health in Africa: the present situation and epidemiological trends. *Trop. Med. Parasitol.*, 42, 204–213.

Bugmann, N. (2000). Le concept du paludisme, l'usage et l'efficacité *in vivo* de trois traitements traditionnels antipalustres dans la région de Dori, Burkina Faso. Inaugural Doctoral Dissertation, Faculty of Medicine, University of Basel, Switzerland.

Burford, G., Rafiq, M.Y., Mollel, E.E., Ole Ngila, L., Tarimo, T., and Ole Lengisugi, N. (2000). Use of Plants for the Treatment of Malaria among the Maasai and Wa-Arusha Peoples of Northern Tanzania, unpublished paper.

Cook, J., Amevigbe, P.M., Crost, M., Gbetoglo, D., Tursz, A., and Assimadi, J.K. (1999). Le recours aux soins des enfants au Togo. *Rev. Epidémiol. Santé Publique*, 47, 2S93–2S113.

Dabis, F., Breman, J.G., Roisin, A.J., Haba, F., and ACSI-CCCD Team. (1989). Monitoring selective components of primary health care: methodology and community assessment of vaccination, diarrhoea, and malaria practices in Conakry, Guinea. *Bull. WHO*, 67, 675–684.

De Francisco, A., Schellenberg, J.A., Hall, A.J., Greenwood, A.M., Cham, K., and Greenwood, B.M. (1994). Comparison of mortality between villages with and without primary health care workers in Upper River Division, the Gambia. *J. Trop. Med. Hyg.*, 97, 69–74.

De Martin, S., Von Seidlein, L., Deen, J.L., Pinder, M., Walraven, G., and Greenwood, B. (2001). Community perceptions of a mass drug administration of an antimalarial drug combination in the Gambia. *Trop. Med. Int. Health*, 6, 442–448.

Deming, M.S., Gayibor, A., Murphy, K., Jones, T.S., and Karsa, T. (1989). Home treatment of febrile children with antimalarial drugs in Togo. *Bull. WHO*, 67, 695–700.

Diallo, A.B., De Serres, G., Béavogui, A.H., Lapointe, C., and Viens, P. (2001). Home care of malaria-infected children of less than 5 years of age in a rural area of the Republic of Guinea. *Bull. WHO*, 79, 28–32.

Djimde, A., Plowe, C.V., Diop, S., Dicko, A., Wellems, T., and Doumbo, O. (1998). Use of antimalarial drugs in Mali: policy versus reality. *Am. J. Trop. Med. Hyg.*, 59, 376–379.

Dossou-Yovo, J., Amalaman, K., and Carnevale, P. (2001). Itinéraires et pratiques thérapeutiques anti-paludiques chez les citadins de Bouake, Côte d'Ivoire. *Méd. Trop.*, 61, 495–499.

Dulhunty, J.M., Yohannes, K., Kourleoutov, C., et al. (2000). Malaria control in central Malaita, Solomon Islands. 2. Local perceptions of the disease and practices for its prevention. *Acta Trop.*, 75, 185–196.

Ejezie, G.C., Ezedinachi, E.N.U., Usanga, E.A., Gemade, E.I.I., Ikpatt, N.W., and Alaribe, A.A.A. (1991). Malaria and its treatment in rural villages of Aboh Mbaise, Imo State, Nigeria. *Acta Trop.*, 48, 17–24.

Elzubier, A.G., Ansari, E.H.H., El Nour, M.H., and Bella, H. (1997). Knowledge and misconceptions about malaria among secondary school students and teachers in Kassala, Eastern Sudan. *J. R. Soc. Health*, 117, 381–385.

Espino, F. and Manderson, L. (2000). Treatment seeking for malaria in Morong, Bataan, the Philippines. *Soc. Sci. Med.*, 50, 1309–1316.

Espino, F., Manderson, L., Acuin, C., Domingo, F., and Ventura, A. (1997). Perceptions of malaria in a low endemic area in the Philippines: transmission and prevention of disease. *Acta Trop.*, 63, 221–239.

Etkin, N.L. (1999). Ethnomedical Approaches to the Study of Indigenous Antimalarials. Presented at the First International Meeting of the Research Initiative for Traditional Antimalarials. Available at http://mim.nih.gov/english/partnerships/ritam_program.pdf.

Etkin, N.L. and Ross, P.J. (1991). Recasting malaria, medicine and meals: a perspective on disease adaptation. In *The Anthropology of Medicine*, 2nd ed., Romanucci-Ross, L., Moerman, D.E., and Tancredi, L.R., Eds. Bergin & Garvey, New York.

Ettling, M.B., Thimasarn, K., Krachaiklin, S., and Bualombai, P. (1989). Malaria clinics in Mae Sot, Thailand: factors affecting clinic attendance. *Southeast Asian J. Trop. Med. Public Health*, 20, 331–340.

FAO. (1997). Medicinal plants for forest conservation and health care. In *FAO Non-Wood Forest Series*, No. 11, Bodeker, G., Bhat, K.K.S., Burley, J., and Vantomme P., Eds. Food and Agriculture Organization of the United Nations, Rome.

Fawole, O.I. and Onadeko, M.O. (2001). Knowledge and home management of malaria by mothers and care givers of under five children. *West Afr. J. Med.*, 20, 152–157.

Faye, O., Faye, B., Dieng, B., et al. (1996). Soins informels dans le paludisme: déterminants de la demande et inventaire de l'offre. *Bull. Soc. Pathol. Exot.*, 89, 35–40.

Foster, S. (1995). Treatment of malaria outside the formal health services. *J. Trop. Med. Hyg.*, 98, 29–34.

Fourn, L., Sakou, G., and Zohoun, T. (2001). Utilisation des services de santé par les mères des enfants fébriles au sud du Bénin. *Santé Publique*, 13, 161–168.

Fungladda, W. and Sornmani, S. (1986). Health behaviour, treatment-seeking patterns and cost of treatment for patients visiting malaria clinics in Western Thailand. *Southeast Asian J. Trop. Med. Public Health*, 17, 379–385.

Gardiner, C., Biggar, R.J., Collins, W.E., and Nkrumah, F.K. (1984). Malaria in urban and rural areas of southern Ghana: a survey of parasitaemia, antibodies and antimalarial practices. *Bull. WHO*, 62, 607–613.

Geissler, P.W., Meinert, L., Prince, R., et al. (2001). Self-treatment by Kenyan and Ugandan schoolchildren and the need for school-based education. *Health Policy Plan.*, 16, 362–371.

Geissler, P.W., Nokes, K., Prince, R.J., Achienf'Odhiambo, R., Aagaard-Hansen, J., and Ouma, J.H. (2000). Children and medicines: self-treatment of common illnesses among Luo schoolchildren in western Kenya. *Soc. Sci. Med.*, 50, 1771–1783.

Gessler, M.C., Msuya, D.E., Nkunya, M.H.H., et al. (1995). Traditional healers in Tanzania: the treatment of malaria with plant remedies. *J. Ethnopharmacol.*, 48, 131–144.

Govere, J., Durrheim, D., La Grange, K., Mabuza, A., and Booman, M. (2000). Community knowledge and perceptions about malaria control and practices influencing malaria control in Mpumalanga province, South Africa. *South Afr. Med. J.*, 90, 611–616.

Graz, B. (Unpublished). Mali: prise en charge des enfants en cas de paludisme.

Haak, H. and Hardon, A.P. (1988). Indigenised pharmaceuticals in developing countries: widely used, widely neglected. *Lancet*, September 10, 620–621.

Hausmann Muela, S., Ribera, J.M., Mushi, A.K., and Tanner, M. (2002). Medical syncretism with reference to malaria in a Tanzanian community. *Soc. Sci. Med.*, 55, 403–413.

Hausmann Muela, S., Ribera, J.M., and Tanner, M. (1998). Fake malaria and hidden parasites: the ambiguity of malaria. *Anthropol. Med.*, 5, 43–61.

Helitzer-Allen, D.L., Kendall, C., and Wirima, J.J. (1993). The role of ethnographic research in malaria control: an example from Malawi. *Res. Sociol. Health Care*, 10, 269–286.

Jayawardene, R. (1993). Illness perception: social cost and coping strategies of malaria cases. *Soc. Sci. Med.*, 37(9), 1169–1176.

Jones, C. and Williams, H.A. (2002). Social sciences in malaria control. *Trends Parasitol.*, 18, 195–196.

Julvez, J., Hamidine, M., Boubacar, A., Nouhou, A., and Alarou, A. (1995). Connaissances et pratiques face au paludisme. Enquête médicale en pays Songhay-Zarma. *Cahiers Santé*, 5, 307–313.

Kaona, F., Siajunza, M.T., Manyando, C., Khondowe, S., and Ngoma, G.K. (2000). Utilisation of malarial drugs at a household level: results from a KAP study in Choma, Southern Province and Mporokoso, Northern Province of Zambia. *Cent. Afr. J. Med.*, 46, 268–270.

Karanja, D.M.S., Alaii, J., Abok, K., Adungo, N.I., Githeko, A.K., Seroney, I., Vulule, J.M., Odada, P., and Oloo, J.A. (1999). Knowledge and attitudes to malaria control and acceptability of permethrin impregnated sisal curtains. *East Afr. Med. J.*, 76, 42–46.

Kawango, E.A. (1995). Ethnomedical remedies and therapies in maternal and child health among the rural Luo. In *Traditional Medicine in Africa*, Sindiga, I., Nyaigotti-Chacha, C., and Kanunah, M.P., Eds. East African Education Publishers Ltd., Nairobi.

Keme, H. (2000). Villa El Salvador, Lima, Peru: A Case Study of Health Care under Neoliberalism. M.Phil. thesis in Latin American Studies, University of Oxford.

Kengeya-Kayondo, J.F., Seeley, J.A., Kajura-Bajenja, E., et al. (1994). Recognition, treatment seeking behaviour and perception of cause of malaria among rural women in Uganda. *Acta Trop.*, 58, 267–273.

Klaver, M. (1993). Malaria, Maize and Mangoes: A Descriptive Research into Local Perceptions of Malaria in South-West Uganda. Master's thesis, Subfaculty of Policy and Management in Health Care, Faculty of Medicine and Health Sciences, Erasmus University of Rotterdam, Netherlands.

Konradsen, F., van der Hoek, W., Amerasinghe, P.H., Amerasinghe, F.P., and Fonseka, K.T. (1997). Household responses to malaria and their costs: a study from rural Sri Lanka. *Trans. R. Soc. Trop. Med. Hyg.*, 91, 127–130.

Leon, T. (2002). *Listening to Women: Home Management of Childhood Fever in a Cyclone Affected Area of Madagascar*. John Snow, Inc., Antananarivo, Madagascar.

Linhua, T., Manderson, L., Da, D., et al. (1995). Social aspects of malaria in Heping, Hainan. *Acta Trop.*, 59, 41–53.

Lipowsky, R., Kroeger, A., and Vazquez, M.L. (1992). Sociomedical aspects of malaria control in Colombia. *Soc. Sci. Med.*, 34, 625–637.

Lukwa, N., Nyazema, N.Z., and Curtis, cf. Mwaiko, G.L. and Chandiwana, S.K. (1999). People's perceptions about malaria transmission and control using mosquito repellent plants in a locality in Zimbabwe. *Cent. Afr. J. Med.*, 45, 64–68.

MacLachlan, M. and Carr, S.C. (1994). From dissonance to tolerance: toward managing tropical health in tropical countries. *Psychol. Dev. Soc.*, 6, 119–129.

Makemba, A.M., Winch, P.J., Makame, V.M., et al. (1996). Treatment practices for degedege, a locally recognised febrile illness, and implications for strategies to decrease mortality from severe malaria in Bagamoyo district, Tanzania. *Trop. Med. Int. Health*, 1, 305–313.

Matthies, F. (1998). Traditional Herbal Antimalarials: Their Role and Their Effects in the Treatment of Malaria Patients in Rural Tanzania. Ph.D. dissertation, University of Basel, Switzerland.

McCombie, S.C. (1996). Treatment seeking for malaria: a review of recent research. *Soc. Sci. Med.*, 43, 933–945.

Miguel, C.A., Manderson, L., and Lansang, M.A. (1998). Patterns of treatment for malaria in Tayabas, the Philippines: implications for control. *Trop. Med. Int. Health*, 3, 413–421.

Miguel, C.A., Tallo, V.L., Manderson, L., and Lansang, M.A. (1999). Local knowledge and treatment of malaria in Agusan del Sur, the Philippines. *Soc. Sci. Med.*, 48, 607–618.

Mills, A. (1993). The household costs of malaria in Nepal. *Trop. Med. Parasitol.*, 44, 9–13.

Ministère de la Santé, des Personnes Agées et de la Solidarité. (1998). *Formulaire Thérapeutique National.* Editions Donnaya, Bamako, Mali.

Mnyika, K.S., Killewo, J.Z.J., and Kabalimu, T.K. (1995). Self-medication with antimalarial drugs in Dar es Salaam, Tanzania. *Trop. Geogr. Med.*, 47, 32–34.

Molyneux, C.S., Mung'ala-Odera, V., Harpham, T., and Snow, R.W. (1999). Maternal responses to childhood fevers: a comparison of rural and urban residents in coastal Kenya. *Trop. Med. Int. Health*, 4, 836–845.

Mugisha, F., Kouyate, B., Gbangou, A., and Sauerborn, R. (2002). Examining out-of-pocket expenditure on health care in Nouna, Burkina Faso: implications for health policy. *Trop. Med. Int. Health*, 7, 187–196.

Munguti, K.J. (1998). Community perceptions and treatment seeking for malaria in Baringo District, Kenya: implications for disease control. *East Afr. Med. J.*, 75, 687–691.

Mwabu, G.M. (1986). Health care decisions at the household level: results of a rural health survey in Kenya. *Soc. Sci. Med.*, 22, 315–319.

Mwenesi, H.A., Harpham, T., Marsh, K., and Snow, R.W. (1995). Perceptions of symptoms of severe childhood malaria among mijikenda and Luo residents of coastal Kenya. *J. Biosoc. Sci.*, 27, 235–244.

Mwenesi, H., Harpham, T., and Snow, R.W. (1995). Child malaria treatment practices among mothers in Kenya. *Soc. Sci. Med.*, 40, 1271–1277.

Ndyomugyenyi, R., Neema, S., Magnussen, P. (1998). The use of formal and informal services for antenatal care and malaria treatment in rural Uganda. *Health Policy and Planning*, 13(1), 94–102.

Nieto, T., Méndez, F., and Carrasquilla, G. (1999). Knowledge, beliefs and practices relevant for malaria control in an endemic urban area of the Colombian Pacific. *Soc. Sci. Med.*, 49, 601–609.

Nsimba, S.E.D., Massele, A.Y., Eriksen, J., et al. (2002). Case management of malaria in under-fives in a Tanzanian district. *Trop. Med. Int. Health*, 7, 201–212.

Nuwaha, F. (2002). People's perception of malaria in Mbarara, Uganda. *Trop. Med. Int. Health*, 7, 462–470.

Nyamongo, I.K. (2002). Health care switching behaviour of malaria patients in a Kenyan rural community. *Soc. Sci. Med.*, 54, 377–386.

Oberländer, L. and Elverdan, B. (2000). Malaria in the United Republic of Tanzania: cultural considerations and health-seeking behaviour. *Bull. WHO*, 78, 1352–1357.

Okanurak, K., Sornmani, S., and Chitprarop, U. (1992). The Impact of Folk Healers on the Performance of Malaria Volunteers in Thailand, TDR/SER/PRS/10.

Okonofua, F.E., Feyisetan, B.J., Davies-Adetugbo, A., and Sanusi, Y.O. (1992). Influence of socioeconomic factors on the treatment and prevention of malaria in pregnant and non-pregnant adolescent girls in Nigeria. *J. Trop. Med. Hyg.*, 95, 309–315.

Okrah, J., Traoré, C., Palé, A., Sommerfeld, J., and Müller, O. (2002). Community factors associated with malaria prevention by mosquito nets: an exploratory study in rural Burkina Faso. *Trop. Med. Int. Health*, 7, 240–248.

Ongore, D., Amunvi, F.K., Knight, R., and Minawa, A. (1989). A study of knowledge, attitudes and practices (KAP) of a rural community on malaria and the mosquito vector. *East Afr. Med. J.*, 66, 79–90.

Pagnoni, F., Convelbo, N., Tiendrebeogo, J., Cousens, S., and Esposito, F. (1997). A community-based programme to provide prompt and adequate treatment of presumptive malaria in children. *Trans. R. Soc. Trop. Med. Hyg.*, 91, 512–517.

Rahantamalala, C. (2000). Contribution à l'inventaire et à l'évaluation en laboratoire de quelques plantes médicinales utilisées dans le traitement du paludisme. Thèse pour obtenir le grade de Docteur en Médecine, Faculté de Médecine, Université d'Antananarivo, Madagascar.

Ramakrishna, J., Brieger, W.R., and Adeniyi, J.D. (1989). Treatment of malaria and febrile convulsions: an educational diagnosis of Yoruba beliefs. *Int. Q. Commun. Health Educ.*, 9, 305–319.

Rasoanaivo, P., Petitjean, A., Ratsimamanga-Urverg, S., and Rakoto-Ratsimamanga, A. (1992). Medicinal plants to treat malaria in Madagascar. *J. Ethnopharmacol.*, 37, 117–127.

Ruebush, T.K., Kern, M.K., Campbell, C.C., and Oloo, A.J. (1995). Self-treatment of malaria in a rural area of western Kenya. *Bull. WHO*, 73, 229–236.

Ruebush, T.K., Zeissig, R., Klein, R.E., and Godoy, H.A. (1992). Community participation in malaria surveillance and treatment. II. Evaluation of the volunteer collaborator network of Guatemala. *Am. J. Trop. Med. Hyg.*, 46, 261–271.

Salako, L.A., Brieger, W.R., Afolabi, B.M., et al. (2001). Treatment of childhood fevers and other illnesses in three rural Nigerian communities. *J. Trop. Paediatr.*, 47, 230–238.

Schulpen, T.W.J. and Swinkels, W.J.A.M. (1980). Machakos project studies: agents affecting health of mother and child in a rural area of Kenya. XIX. The utilisation of health services in a rural area of Kenya. *Trop. Geogr. Med.*, 32, 340–349.

Shein, H., Sein, T.T., Soe, S., Aung, T., Win, N., and Aye, K.S. (1998). The level of knowledge, attitude and practice in relation to malaria in Oo-do village, Myanmar. *Southeast Asian J. Trop. Med. Public Health*, 29, 546–549.

Sindiga, I. (1995). Traditional medicine in Africa: an introduction. In *Traditional Medicine in Africa*, Sindiga, I., Nyaigotti-Chacha, C., and Kanunah, M.P., Eds. East African Education Publishers Ltd., Nairobi.

Singh, N., Singh, M.P., Saxena, A., Sharma, V.P., and Kalra, N.L. (1998). Knowledge, attitude, beliefs and practices (KABP) study related to malaria and intervention strategies in ethnic tribals of Mandla (Madhya Pradesh). *Curr. Sci.*, 75, 1386–1390.

Slutsker, L., Chitsulo, L., Macheso, A., and Steketee, R.W. (1994). Treatment of malaria fever episodes among children in Malawi: results of a KAP survey. *Trop. Med. Parasitol.*, 45, 61–64.

Snow, R.W., Peshu, N., Forster, D., Mwenesi, H., and Marsh, K. (1992). The role of shops in the treatment and prevention of childhood malaria on the coast of Kenya. *Trans. R. Soc. Trop. Med. Hyg.*, 86, 237–239.

Sodemann, M., Jakobsen, M.S., Mølbak, K., Alvarenga, I.C., and Aaby, P. (1997). High mortality despite good care-seeking behaviour: a community study of childhood deaths in Guinea-Bissau. *Bull. WHO*, 75, 205–212.

Tanner, M. and Vlassoff, C. (1998). Treatment-seeking behaviour for malaria: a typology based on endemicity and gender. *Soc. Sci. Med.*, 46, 523–532.

Tarimo, D.S., Lwihula, G.K., Minjas, J.N., and Bygbjerg, I.C. (2000). Mothers' perceptions and knowledge on childhood malaria in the holendemic Kibaha district, Tanzania: implications for malaria control and the IMCI strategy. *Trop. Med. Int. Health*, 5, 179–184.

Théra, M.A., D'Alessandro, U., Thiéro, M., et al. (2000). Child malaria treatment practices among mothers in the district of Yanfolia, Sikasso region, Mali. *Trop. Med. Int. Health*, 5, 876–881.

Théra, M.A., Sissoko, M.S., Heuschkel, C., et al. (1998). Village Level Treatment of Presumptive Malaria: Experiences with the Training of Mothers and Traditional Healers as Resource Persons in the Region of Mopti, Mali. Clone, Cure and Control: Tropical Health for the 21st Century. Second European Congress on Tropical Medicine, Liverpool. Poster P72, p. 139.

Traore, S., Coulibaly, S.O., and Sidibe, M. (1993). Comportements et coûts liés au Paludisme chez les femmes des campements de pêcheurs dans la zone de Sélingué au Mali, TDR/SER/PRS/12. Institut National de Recherche en Santé Publique, Bamako, Mali.

Utarini, A., Winkvist, A., and Ulfa, F.M. (2003). Rapid assessment procedures of malaria in low endemic countries: community perceptions in Jepara district, Indonesia. *Soc. Sci. Med.*, 56, 701–712.

Vaahtera, M., Kulmala, T., Maleta, K., Cullinan, T., Salin, M.L., and Ashorn, P. (2000). Epidemiology and predictors of infant morbidity in rural Malawi. *Paediatr. Perinatal Epidemiol.*, 14, 363–371.

Vongo, R. (1999). Traditional Medicine on Antimalarials. Paper presented at the First International Meeting of the Research Initiative for Traditional Antimalarials. Available at http://mim.nih.gov/english/partnerships/ritam_program.pdf.

Vundule, C. and Mharakurwa, S. (1996). Knowledge, practices and perceptions about malaria in rural communities of Zimbabwe: relevance to malaria control. *Bull. WHO*, 74, 55–60.

Williams, H.A., Jones, C., Alilio, M., et al. (2002). The contribution of social science research to malaria prevention and control. *Bull. WHO*, 80, 251–252.

Winch, P.J., Makemba, A.M., Kamazima, S.R., et al. (1996). Local terminology for febrile illnesses in Bagamoyo district, Tanzania and its impact on the design of a community-based malaria control programme. *Soc. Sci. Med.*, 42, 1057–1067.

Ziba, C., Slutsker, L., Chitsulo, L., and Steketee, R.W. (1994). Use of malaria prevention measures in Malawian households. *Trop. Med. Parasitol.*, 45, 70–73.

11 An Overview of Ethnobotanical Studies on Plants Used for the Treatment of Malaria

Merlin Willcox, Gemma Burford, and Gerard Bodeker

CONTENTS

11.1 BACKGROUND

As the previous chapter has shown, plants are widely used for the treatment of malaria throughout the Tropics. The questions to be addressed here are which plant species are used, how widely they are used, which should be prioritised for future research, and which should be prioritised for conservation measures. To date, there has been no systematic overview of the literature to assess which plant species are the most important candidates for further research and conservation.

Natural selection probably favors the behaviour of self-medication with effective remedies. Malaria is a potentially fatal disease and has shaped human evolution through the selection of mutations such as the sickle-cell trait, G6PD deficiency, and thalassemia (Miller, 1999). These mutations in themselves are disadvantageous, and in homozygotes may prove fatal, but the protection they afford against malaria has been sufficient for them to spread in the human population. Therefore, it would not be surprising if evolution also favored behavioural patterns that included the use of effective antimalarial medicinal plants, even if these sometimes have unwanted adverse effects. Other primates have also been observed to self-medicate with antiparasitic plants (Huffman et al., 1996; Koshimizu et al., 1994).

It can further be hypothesised that the plant species in most common use today for the treatment of malaria are among the safest and most effective. Traditional healers constantly experiment with different plant remedies in an effort to find the safest and most effective preparations, and this knowledge is transmitted from generation to generation, either through the oral tradition (as in Africa) or through written manuscripts (as in traditional Chinese medicine and Ayurveda). For a remedy to be in common use, it needs to be not only safe and effective, but also widely available and easy to prepare. These are also important characteristics for a plant remedy to be used as part of a malaria control program.

Bioprospectors, in search of new natural products, often screen plant extracts at random on a large scale (Mateo, 2000). However, important plants may be deemed inactive as a result of insufficient tests (see Chapter 16). Random screening yields few active extracts, compared to testing plants recommended by traditional healers (Brandão et al., 1985; Carvalho et al., 1991; Krettli et al., 2001). Leaman et al. (1995) have shown that plants widely used as antimalarials by traditional healers are significantly more active *in vitro* against *Plasmodium falciparum* than plants that are not widely used, or not used at all, for the treatment of malaria.

Paradoxically, effectiveness as a medicine may in the modern world cause a plant species to become endangered (Marshall, 1998). Plants have evolved over millennia to produce antiparasitic and other compounds for their own self-defense. Today, habitat destruction and overexploitation for whatever reason (including medicinal uses) threaten many plant species. There has been until now no published data on how many potentially antimalarial plants are under threat.

In this chapter, therefore, we set out systematically to examine the ethnobotanical literature, in order to identify the range and frequency of use of different plant species as antimalarials, and to discover which herbal remedies are the most widely used. We also investigated which of these species are threatened from a conservation viewpoint.

11.2 METHODS

An attempt was made to identify as many primary ethnobotanical studies as possible from malaria-endemic countries. MEDLINE, EMBASE, CAB, and SOCIOFILE databases were searched systematically using the terms *traditional medicine* and *malaria*, and references to more studies were identified in these papers. Selected journals were searched by hand. Researchers and experts were also contacted to supply copies of unpublished work. Studies were included if they contained firsthand reports of the medicinal uses of plants. Literature reviews were consulted, but where possible the primary sources were found.

Using Microsoft Access, a database is being constructed of all remedies and plant species reportedly used for the treatment and prophylaxis of malaria, fevers (including febrile convulsions), and splenomegaly. For each remedy, data recorded included the plant species and part(s) used; method of preparation; route of administration; dose, frequency, and duration; contraindications; and traditional use(s). Conservation information was found by consulting the 2002 *IUCN Red Data Book* online (http://www.redlist.org).

Each species was assigned an IVmal (importance value for the treatment of malaria) according to how widely its use was reported. This scoring system was first developed for use at a local level by Leaman et al. (1995), with values 1 to 4. We have extended this scoring system to apply at an international level by adding values 5 to 8 (see definitions in Table 11.1).

11.3 RESULTS

Ninety-four original ethnobotanical publications have been included to date, concerning 33 tropical countries. Overall, 1277 species from 160 families have been reported for the treatment of malaria and fevers. This data set is still incomplete, as there are many more studies that have yet to be included in the database.

Eight hundred and forty-nine of the listed species were quoted by only a single study, which did not give enough information for an IVmal to be allocated — but it must be 3 or less. The number of species in each category of IVmal are given in Table 11.1. There were 11 species used as antimalarials or antipyretics in all three tropical continents (IVmal = 8; see Table 11.2), and 47 used in two continents (IVmal = 7; see Table 11.3 and Figures 11.1 and 11.2). A few of these have already been discussed in more detail in Part 2 of this book.

TABLE 11.1
Number of Species according to IVmal

IVmal	No. Species	Definition
?	849	Insufficient data
1	95	Reported once in a single ethnobotanical survey
2	30	Reported twice in one community
3	6	Reported at least three times in one community
4	42	Reported in more than one community
5	91	Reported in more than one survey, in the same country
6	106	Reported in more than one country, in the same continent
7	47	Reported in two continents
8	11	Reported in three continents

TABLE 11.2
Species Used for Fevers or Malaria in Three Continents
(IVmal = 8)

Family	Species
Anacardiaceae	*Mangifera indica* L.
Annonaceae	*Annona muricata* L.
Crassulaceae	*Kalanchoe* (= *Bryophyllum*) *pinnata* Lam
Cucurbitaceae	*Momordica charantia* L.
Euphorbiaceae	*Jatropha curcas* L.
	Ricinus communis L.
Leguminosae	*Senna* (= *Cassia*) *occidentalis* (L.) Link[a]
	Senna (= *Cassia*) *tora* L.
Malvaceae	*Sida rhombifolia* L.
Menispermaceae	*Cissampelos pareira* L.
Zingiberaceae	*Zingiber officinale* Roscoe[b]

[a] An ingredient in Malarial (see Chapter 7).
[b] An ingredient of the traditional preparation of Changshan (see Chapter 4).

The vast majority (1213) of the species were not recorded in the *IUCN Red Data Book*. However, 5 species were listed as endangered, 13 as vulnerable, and 3 as near threatened (see Table 11.4). The endangered species are listed in Table 11.5, and the vulnerable species in Table 11.6.

11.4 DISCUSSION

This is the first published overview of the global ethnobotanical literature regarding traditional medicinal plants for fever and malaria. There still remain many studies to be added to it, and the number of species will certainly increase as the database expands. Milliken (1997) performed a more localised review of antimalarial and antipyretic plants in Latin America and found that just in this continent, 956 species of 569 genera and 140 families were cited in the literature as being used for fevers or malaria. Oliveira et al. (2002) reviewed the Brazilian ethnomedical literature and identified 197 species used for the treatment of malaria and fevers in Brazil. They quantified the importance of each species by assigning them a score according to citation frequency and the

TABLE 11.3
Species Used for Fever or Malaria in Two Continents
(IVmal = 7)

Family	Species
Anacardiaceae	*Spondias mombin* L.
Asclepiadaceae	*Calotropis procera* (Ait) R.Br.
Bixaceae	*Bixa orellana* L.
Caricaceae	*Carica papaya* L.[a]
Chenopodiaceae	*Chenopodium ambrosioides* L.
Compositae	*Ageratum conyzoides* L.
	Elephantopus scaber L.
	Tagetes erecta L.
	Tinthonia diversifolia (Hemsl) A Gray
Euphorbiaceae	*Euphorbia hirta* L.
	Phyllanthus amarus Schum & Thonn
	Phyllanthus niruri L.
Labiatae	*Lantana camara* L.[b]
	Leonotis nepetaefolia R.Br.
	Ocimum americanum L.
	Ocimum basilicum L.
	Ocimum canum Sims
	Ocimum gratissimum L.
Leguminosae	*Albizia lebbek* Benth
	Caesalpinia bonducella Fleming
	Cassia fistula L.
	Erythrina indica Lamk
	Prosopis juliflora (Sw) DC
	Senna alata (L.) Roxb
	Senna obtusifolia L.
	Tamarindus indica L.
Malvaceae	*Gossypium barbadense* L.
Meliaceae	*Azadirachta indica* A Juss[c]
	Cedrela odorata L.
	Melia azedarach L.
Moraceae	*Moringa oleifera* Lam
Myrtaceae	*Eucalyptus citriodora* Hook
	Psidium guajava L.
Oxalidaceae	*Biophytum sensitivum* (L.) DC
Plantaginaceae	*Plantago major* L.
Plumbaginaceae	*Plumbago zeylanica* L.
Ranunculaceae	*Nigella sativa* L.
Rhamnaceae	*Zizyphus mauritiana* Roxb
Rubiaceae	*Cinchona officinalis* L.[d]
Rutaceae	*Citrus aurantiifolia* (Christm) Swingle
Scrophulariaceae	*Scoparia dulcis* L.
Solanaceae	*Capsicum annuum* var *frutescens* L.
	Solanum nigrum L.
Umbelliferae	*Cuminum cyminum* L.
Zingiberaceae	*Curcuma longa* L.

[a] See Figure 11.1.
[b] See Figure 11.2.
[c] See Chapter 6.
[d] See Chapter 2.

Figure 11.1 Pawpaw (*Carica papaya*): leaf and root preparations are widely used in Africa and Brazil for the treatment of malaria. (Copyright 2000, Merlin Willcox.)

weight of each reference (greatest for primary ethnobotanical studies in malarious areas). There was good agreement with the IVmal scores in this review, in that the two highest scoring species were *Senna occidentalis* and *Momordica charantia*, which were both assigned an IVmal of 8. Further research is needed to find the best method for quantifying the potential importance of plants and prioritising species for future research.

IVmal is only one such method. It has only been tested to a limited extent, and for values only up to 4 (Leaman et al., 1995). The next step would be to compare IVmal with antimalarial efficacy (for example, inhibitory concentration 50% (IC_{50})) of species that have been investigated. However, if plants with a high IVmal are found to be ineffective against *P. falciparum in vitro*, this may be a clue to other mechanisms of antimalarial activity or symptom control. Some, for example, might act synergistically with components of other plants or with synthetic antimalarials — as in the case of *Strychnos* sp., which reverses resistance to chloroquine (see Chapter 9). Others might have immunostimulant or antipyretic properties, or may act against the hepatic stages of malaria (see Chapters 16 and 17).

One drawback of assigning an IVmal to a plant species is that a single species may be prepared and used in many different ways. The efficacy and safety may differ according to which part of the plant is used and how it is prepared. Therefore, a more useful concept would be the IVmal of

Figure 11.2 *Lantana camara*, often regarded as a weed in the Tropics, is prepared as a leaf or root infusion for the treatment of malaria and fevers in Guyana and several African countries. (Copyright 2000, Merlin Willcox.)

a particular remedy, which may consist of one or more parts of one or more plant species. Our database of ethnobotanical studies is currently being expanded to include this information. However, some studies did not quote the plant part used or its method of preparation, and very few gave information about precise doses and length of treatment. It will be important to gather such information in future ethnobotanical studies.

Another disadvantage of IVmal is that it depends partly on the geographical distribution of a plant — if it only occurs in one country, it cannot score above 5. Therefore, species with an IVmal of 5 or above should be considered as a priority for research. IVmal also depends on good quantity and quality of ethnobotanical studies. If a plant has not been identified in more than one study, it cannot score greater than 4. There are many areas where not even a single ethnobotanical study has been undertaken.

Further ethnobotanical research in such areas is a priority. Traditional medicines are being forgotten as some traditional healers die with no successors to their knowledge; fewer young people are interested in this as modern medicine is promoted. In cultures with no written tradition, healers take their knowledge with them to the grave, unless they have apprentices. Even in societies where the dependence on specialist healers is relatively low, as treatments for common illnesses are well known in the community as a whole, recent studies have shown an alarming decline in ethnobotanical knowledge with successive generations (Florey, 2001; Zent, 2001).

Furthermore, some of the important plant species are also endangered and may be lost before they have even been evaluated. Good conservation data are as important as ethnobotanical data, both to safeguard the survival of these species and to assess the feasibility of using a herbal remedy on a larger scale. If a particular plant species is to be promoted as an antimalarial in public health programs, its availability and ease of cultivation will be crucial. There is little point in researching the safety and efficacy of a plant that is not widely available and cannot be cultivated. However, most of the antimalarial plants identified did not appear in the International Union for Conservation of Nature and Natural Resources (IUCN) database. Although most of these are probably not at high risk, more species may turn out to be endangered or at risk when they are assessed. There is a need for conservation data on more of these species.

Some plants that are not recorded as rare on a global scale may be locally threatened as a result of collection for medicinal use or deforestation for other purposes. Conservation measures are needed to protect these too. For example, collectors in Nepal have found that *Swertia chirayita* is becoming more difficult to collect as demand has increased (Olsen, 1998). This plant is widely used in India to treat several ailments, including fevers, and is a component of the antimalarial Ayush-64 (see Chapter 5). In Tanzania, *Acacia* species that are common at the international and national levels are thought to be at risk in certain localised areas, due to the high demand for timber, charcoal, and wood fuel (Ole-Lengisugi and Mziray, 1996).

Species known to be at risk should be prioritised for conservation. There needs to be further research on what methods of conservation are effective. Harvesting of some plant parts is more destructive than harvesting of others. Evaluating the efficacy of different parts of a species may yield data to support the use of a part that can be harvested nondestructively (for example, leaves or fruits). Some antimalarial plants are currently harvested by uprooting the whole plant (e.g., *Cassia occidentalis*, *Calotropis procera*), while others have a ring of bark removed (e.g., *Khaya senegalensis*, *Warburgia ugandensis*), which will kill growing trees (Ochieng' Obado and Odera, 1995; Cunningham, 1997). Even in areas where unwritten community laws preclude ring barking and other such destructive practices, commercial collectors — for whom the survival of individual trees is of less importance than the total harvest — may ignore these restrictions (Pergola and Burford, 2001).

Traditional mechanisms for the conservation of specific areas of forest still exist in some societies. They are often linked to religious belief and ritual practice, as in the case of sacred groves in India (Gadgil and Vartak, 1976) and Zimbabwe (Byers et al., 2001). The reserved areas may also have an explicit role in health care: an example is the use of forest sites for the healing ritual of *olpul* by Maasai in Tanzania and Kenya (see Figure 11.3). Cutting of trees in these so-called bush hospitals is forbidden (Burford, 2002). Yet in all of the above examples, the influence of the traditional law is in decline, due to socioeconomic, cultural, and religious change. Community education initiatives, through indigenous peoples' organisations, can raise awareness of the importance of medicinal plant conservation, or revive interest in the issue in areas where it was previously acknowledged.

Additionally, cultivation of medicinal plants can be encouraged, rather than harvesting from the wild (Cunningham, 1996). At a local level, domestic or community-owned herbal gardens can be started to cultivate priority species and provide a sustainable source of herbal medicines (Majumdar, 1991). Traditional healers can also start medicinal plant gardens for their own use, and to educate the local community about conservation (Figure 11.4 and Figure 11.5). Some medicinal plants, such as *Prunus africana*, can also be grown as a cash crop by farmers (Okoko, 2000). However, some of the species most at risk are slow growing (e.g., *Warburgia ugandensis*) and sold at a low price, making cultivation less attractive (Cunningham, 1997). This illustrates the importance of combining both *in situ* and *ex situ* conservation strategies and of determining and addressing the underlying reasons for threats to individual species.

FIGURE 11.3 Maasai warrior oversees the forest area used for the healing ritual of *olpul*. Tradition forbids felling of trees in this area. (Copyright 2002, Bob Webzell.)

The validity of ethnobotanical data, as well as the success of conservation programs, depends on the quality of the methods used and the depth of understanding of local ethnomedical systems. The following chapters present the contrasting approaches of British colonial doctors (Chapter 12) and Ayurvedic practitioners (Chapter 13) in India. Standards to ensure the quality of ethnobotanical research are proposed by Etkin et al. (Chapter 14). Good ethnobotanical data are the key to further research on plant-based antimalarials, by guiding laboratory and clinical evaluation (see Chapters 16 and 21).

If humanity is to retain the full wealth of biodiversity as a bank of potential medicines for current and future diseases, ethnobotany and conservation are of utmost importance. Ethnobotanical studies assist conservation by revealing the potential usefulness of many plant species. However, wider knowledge of the medicinal properties of plants can lead to their overexploitation. Therefore, studies of conservation status and measures should be conducted in parallel. Priorities for research in these disciplines are summarised below.

BOX 11.1: PRIORITIES FOR FUTURE RESEARCH

1. Expanding the database to include more studies and more detail about individual remedies
2. Conducting high-quality ethnobotanical studies in uncharted areas
3. Evaluating IVmal and other methods to prioritise species and remedies for future research
4. Investigating the safety and efficacy of remedies and species with high IVmal
5. Assessing conservation status of more antimalarial species
6. Evaluating and implementing conservation measures for all species, especially those at risk

---

FIGURE 11.4 Medicinal plant garden at traditional healers' clinic in Rukararwe, Bushenyi district, Uganda. (Copyright 1998, Merlin Willcox.)

ACKNOWLEDGMENTS

The authors are extremely grateful to Mr. Ian Edwards, of the IT training department, Oxford Radcliffe Hospitals, for his very generous and effective help in designing the database that was used as the basis for this paper. We also thank Bob Webzell for permission to reproduce his photograph here.

FIGURE 11.5 *Prunus africana*, a vulnerable species with several medicinal uses, including treatment of fevers, growing in the medicinal plant garden at Rukararwe. (Copyright 1998, Merlin Willcox.)

REFERENCES

Brandão, M.G.L., Botelho, M.D.G.A., and Krettli, A.U. (1985). Quimioterapia experimental antimalárica com produtos naturais. I. Uma Abordagem mais racional? *Ciên. Cult.*, 37, 1152–1163.

Burford, G. (2002). Linking Healthcare and Natural Resource Management: The Ritual of *olpul* among Ilkisongo Maasai in Monduli District, Tanzania. Dissertation submitted for the degree of Master of Science (M.Sc.) in Environmental Anthropology, University of Kent, Canterbury.

Byers, B.A., Cunliffe, R.N., and Hudak, A.T. (2001). Linking the conservation of culture and nature: a case study of sacred forests in Zimbabwe. *Hum. Ecol.*, 29, 187–218.

Carvalho, L.H., Brandão, M.G.L., Santos-Filho, D., Lopes, J.L.C., and Krettli, A.U. (1991). Antimalarial activity of crude extracts from Brazilian plants studied *in vivo* in *Plasmodium berghei*-infected mice and *in vitro* against *Plasmodium falciparum* in culture. *Braz. J. Med. Biol. Res.*, 24, 1113–1123.

Cunningham, A.B. (1996). *People, Park and Plant Use*. UNESCO, Paris.

Cunningham, A.B. (1997). Medicinal plants for forest conservation and health care. In *Medicinal Plants for Forest Conservation and Healthcare*, Bodeker, G., Bhat, K.K.S., Burley, J., and Vantomme, P., Eds. FAO, Rome.

Florey, M. (2001). Threats to indigenous knowledge: a case study from eastern Indonesia. In *On Biocultural Diversity: Linking Language, Knowledge and the Environment*, Maffi, L., Ed. Smithsonian Institution Press, Washington, pp. 325–342.

Gadgil, M. and Vartak, V.D. (1976). Sacred groves of the western Ghats in India. *Econ. Bot.*, 30, 152–160.

Huffman, M.A., Koshimizu, K., and Ohigashi, H. (1996). Ethnobotany and zoopharmacognosy of *Vernonia amygdalina*, a medicinal plant used by humans and chimpanzees. In *Compositae: Biology & Utilisation. Proceedings of the International Compositae Conference, Kew, 1994*, Vol. 2, Caligari, P.D.S. and Hind, D.J.N., Eds. Royal Botanical Gardens, Kew, U.K., pp. 351–360.

Koshimizu, K., Ohigashi, H., and Huffman, M.A. (1994). Use of *Vernonia amygdalina* by Wild chimpanzee: possible roles of its bitter and related constituents. *Physiol. Behav.*, 56, 1209–1216.

Krettli, A.U., Andrade-Neto, V.F., Brandão, M.G.L., and Ferrari, W.M.S. (2001). The search for new antimalarial drugs from plants to treat fever and malaria or plants randomly selected: a review. *Mem. Inst. Oswaldo Cruz.*, 96, 1033–1042.

Leaman, D.J., Arnason, J.T., Yusuf, R., et al. (1995). Malaria remedies of the Kenyah of the Apo Kayan, East Kalimantan, Indonesian Borneo: a quantitative assessment of local consensus as an indication of biological efficacy. *J. Ethnopharmacol.*, 49, 1–16.

Majumdar, A. (1991). Tribal medicines and their role in community herbal gardens. In *Recent Advances in Medicinal, Aromatic and Spice Crops*, Vol. 1, Raychaudhuri, S.P., Ed. Today and Tomorrow's Printers and Publishers, New Delhi, pp. 241–247.

Marshall, N. (1998). *Searching for a Cure: Conservation of Medicinal Wildlife Resources in East and Southern Africa*. TRAFFIC International, Cambridge, U.K.

Mateo, N. (2000). Bioprospecting and conservation in Costa Rica. In *Responding to Bioprospecting*, Svarstad, H. and Dhillion, S.S., Eds. Spartacus Forlag AS, Oslo.

Miller, L.H. (1999). Evolution of the human genome under selective pressure from malaria: applications for control. *Parassitologia*, 41, 77–82.

Milliken, W. (1997). *Plants for Malaria, Plants for Fever. Medicinal Species in Latin America: A Bibliographic Survey*. Royal Botanical Gardens, Kew, U.K.

Ochieng' Obado, E.A. and Odera, J.A. (1995). Management of medicinal plant resources in Nyanza. In *Traditional Medicine in Africa*, Sindiga, I., Nyaigotti-Chacha, C., and Kanunah, M.P., Eds. East African Education Publishers Ltd., Nairobi, Kenya.

Okoko, T. (2000). Scientists Take Shortcut to Save Africa's Medicinal Tree. *Environment News Service*, October 31.

Ole-Lengisugi, N. and Mziray, W. (1996). The Role of Indigenous Knowledge in Sustainable Ecology and Ethnobotanical Practices among Pastoral Maasai: Olkonerei-Simanjiro Experience. Paper presented at the 5th International Congress of Ethnobiology in association with the Festival of Living Traditions: Old Ways to the Future, Kenyatta International Conference Centre, Nairobi, Kenya, September 2–8, 1996.

Oliveira, F.Q., Brandão, M.G.L., Junqueira, R.G., Stehmann, J.R., and Krettli, A.U. (2002). Plants used to treat fevers and malaria in Brazil. Poster B192 of the 50th Annual Congress of the Society for Medicinal Plant Research. *Rev. Fitoterapia*, 2, 310.

Olsen, C.S. (1998). The trade in medicinal and aromatic plants from central Nepal to northern India. *Econ. Bot.*, 52, 279–292.

Pergola, T. and Burford, G. (2001). Plant Utilisation Scoping Project, Monduli District: Final Report. UNDP/GEF/NEMC East African Cross Border Biodiversity Project, Arusha.

Zent, S. (2001). Acculturation and ethnobotanical knowledge loss among the Piaroa of Venezuela: demonstration of a quantitative method for the empirical study of traditional environmental knowledge change. In *On Biocultural Diversity: Linking Language, Knowledge and the Environment*, Maffi, L., Ed. Smithsonian Institution Press, Washington, pp. 190–211.

12 The Discovery of Indigenous Febrifuges in the British East Indies, c. 1700–1820

Mark Harrison

The discovery in the 17th century of the febrifuge qualities of cinchona bark — then known as Peruvian or Jesuits' bark — was an important event in the history of medicine (Brown, 1698; Honigsbaum, 2001; Jarcho, 1993; see Chapter 2). The bark of the tree *Cinchona officinalis* L. was, as is well known, the source of quinine — the first truly effective antimalarial drug. It also paved the way for other so-called specific remedies that challenged the Galenic orthodoxy that fevers, like other diseases, should be treated by restoring the body to a healthy equilibrium. The bark came, gradually, to displace the most common methods of fever therapy like blood letting, which dominated European medicine until the middle of the 18th century. However, it also prompted a search for other specific medicines and for different varieties of febrifuge, varieties that occurred outside the domains of the Spanish–American empire. The motive for this search was partly economic, in that the trading companies of the various European powers wanted to end Spain's profitable monopoly of the cinchona trade. But there was an important medical rationale too: the importance in the Western medical tradition of the principle of locality — the notion that diseases of particular places were best treated by medicines available nearby.

In this short survey, I will examine some of the discoveries made by the British in their East Indian colonies. These discoveries are interesting historically because they represent a forgotten chapter in the history of febrifuges. While much has been written about cinchona and quinine, practically nothing is known about the other febrifuges native to the East Indies and that were widely used by locals and European colonists. In their quest for alternatives to cinchona, medical practitioners depended heavily upon local knowledge, which was generally considered as reliable, if not more so, as scientific trials conducted at home. Although some of the East Indian febrifuges were discussed in learned journals, most appear to have been used and evaluated entirely within the locality in which they originated. The East India Company explicitly encouraged its medical practitioners to rely on local drugs because this reduced the costs of importation (Arnold, 1998). But it was also believed that indigenous plants were more likely to cure fevers of the same locality than imported drugs. This was a vestige of the old belief in divine providence: that where a disease occurred, its remedy was to be found nearby. It also reflected the then common view that all diseases were peculiar to time and place — the product of peculiar environmental and atmospheric conditions. This view of fever remained strong until the late 19th century, especially among medical practitioners working in the tropics (Arnold, 1996; Harrison, 1999).

The use of local plants was also recommended because of continuing doubts over the efficacy and safety of cinchona. Although its therapeutic properties were widely recognised by the mid-18th century, there remained some uncertainty over the effectiveness of the drug and how it should be administered. It was also widely known that patients treated with bark could experience unpleasant side effects, including nausea and constipation (Maehle, 1999). In view of these reservations, some British practitioners warned against slavish dependency upon the bark, condemning it as "that idol nostrum of the Faculty" and as a "systematic deceiver of the world." William Stevenson of Newark (1781), for example, believed that the bark of his native oak was far more effective than a remedy imported from South America.

Although there was a good deal of interest in species of cinchona indigenous to British colonies — like those in the West Indian islands (Lindsay, 1794; Ryan, 1782) — the bark failed to dominate medical practice in India, where the East India Company had established trading posts that would later become the centers of colonial power. In India and other outposts of the East India Company, there were no native species of cinchona and various local febrifuges were commonly used alongside or instead of the bark. One specific remedy for fever used by the company's servants during the 17th century was the so-called Malabar Rattle-broom, which was reported in the Royal Society's *Philosophical Transactions* in 1698. The informant was a Mr. Samuel Brown, an East India Company surgeon at Fort St. George, Madras, who had sent details of the therapeutic powers of this and other plants to a fellow of the society. The seeds of this finger-leaved plant (possibly the Malabar cardamom, *Elettaria cardamomum*) were made into a decoction that, when introduced into a bath, was said to cure fevers, especially tedious Tertian agues. This and the other medicinal plants mentioned by Brown were all collected within 12 miles of the fort. Some had already been mentioned in the *Hortus Malabaricus*: a compendium of plants from Southwest India, compiled by the Dutchman Van Rheede; others had apparently been recommended for use by local brahmans, who also gave details of the ways in which they should be used (Brown, 1698). This reliance on local knowledge was a well-established feature of European medicine in the East Indies at that time, and despite growing differences between Western and indigenous medical traditions, interest in local medicines remained strong throughout the 18th century (Harrison, 2001).

Some of the cures recommended for fevers in the East Indies were better known as spices and condiments. Pepper, ginger, and other spices had been used for some time as remedies for stomach complaints and East India Company surgeons were recommended to carry them for this purpose in their medicine chests (Harrison, 2001, p. 48). In the early 17th century, they were also being recommended for the treatment of agues and other fevers. Again, this knowledge was learned from practitioners of Ayurveda and other indigenous medical traditions. A Jesuit priest, whose remarks were reported to the Royal Society in 1713, declared that he had "seen them [practitioners of Ayurvedic medicine] cure Fevers which begin with a shivering Fit, by giving the Patient three large Pills made of Ginger, black Cumin and long Pepper. For Tertian Agues, they give the Person for three Days together three Spoonfuls of the Juice of the *Tecorium* [*sic*] or Great Germander, with a little Salt and Ginger" (Papin, 1713). The priest appears to have been referring to *Teucrium ciiamaedrys*, which is still used in India as a specific remedy for gout, rheumatism, and diseases of the spleen and other organs. Long pepper (*Piper longum*), ginger (*Zingiber officinale*), and cumin (*Nigella sativa*) are also found in modern antimalarial preparations in India, as well as in Africa and Central America (Kurrup et al., 1979; Shankar and Venuvopal, 1999; Sharma, 1999; Singh and Anwar Ali, 1994).

Later in the 18th century, as the East India Company began to acquire territory and carry out botanical surveys in India, knowledge of indigenous medical plants was gathered more systematically, and this knowledge was centralised in the medical departments of the company's presidencies in Madras, Bombay, and Bengal (Desmond, 1992). Those who worked as company botanists invariably had medical training, and even though they were searching for plants that might be grown and traded by the company, the usefulness of plants in relieving the miseries of humankind remained a guiding inspiration. Many risked death or, certainly, illness in pursuance of their work, though some also found considerable fame. One of these was William Roxburgh, who joined the Madras Medical Service in 1776 and who later became superintendent of the company's botanical gardens at Samalkot (1781–1793) and Calcutta (1793–1813). During his time as a botanist in India, Roxburgh corresponded with many other botanists and medical practitioners. One letter, received by Roxburgh from a British Moravian missionary working in the Danish colony of Tranquebar, mentions that a febrifuge identified by Roxburgh was being used to treat fevers in the local hospital (John, 1792). The febrifuge mentioned was what Roxburgh termed *Swietenia febrifuga*, a species of the Indian redwood or mahogany tree, known to locals by its Sanskrit name, *Seymida*. The plant is now more commonly referred to as *Soymida febrifuga*, following the Sanskrit. During the early 1790s, Roxburgh subjected the bark of this tree to a series of experiments comparing its efficacy with that of the Peruvian bark. He concluded that the *Swietenia* bark was not only equal to that of cinchona but superior, at least in treating the fevers of India. He prepared the bark as a powder and mixed it with cold water, using a dose of between 20 and 60 grains per day. Some of the doctors with whom Roxburgh was in contact at Tranquebar also tried the remedy, though with more mixed success, and via Danish and German missionary connections, the bark and the seeds of the tree were sent to various botanists and doctors in Europe (Roxburgh, 1792).

Another Indian febrifuge that was commonly known — though apparently little used outside India — was the Spikenard, or *Nardus indica*, known locally as *Terankus* in the northwest provinces of India, where the British first came to know of it. The eminent naval physician Gilbert Blane described the plant in an article for the London medical journal *Medical Facts and Observations* in 1791, so it must have become quite widely known. Blane came across the plant near Lucknow while on a hunting expedition with the Nawab of Oudh. "It is called by the natives *Terankus*," he wrote, "which means literally, in the Hindu language, 'fever-restrainer.'" He explained that "They infuse about a drachm of it in half a pint of hot water, with a small quantity of black pepper. This infusion serves for one dose, and is repeated three times a day. It is esteemed a powerful medicine in all kinds of fevers, whether continued or intermittent. I have not made trial of it myself, but shall certainly take the first opportunity of doing so" (Blane, 1791). He thought the plant sufficiently important to warrant sending a dried specimen to Sir Joseph Banks, the director of Kew Gardens and the hub of an ambitious attempt to collect and classify useful plants from around the world (Drayton, 2000; Gascoigne, 1998). Blane's confidence in the *Nardus* was underpinned by his belief that it had an ancient provenance, corresponding to the plant described in Arrian's history of the expedition of Alexander the Great to India; it was apparently used by Hippocrates and Galen. Various species of *Nardostachys*, as the plant came to be known, were still in use as febrifuges at the end of the 19th century, although apparently rarely (Dey, 1896). Nor is there much evidence of its use as a febrifuge today.

It is interesting to note that Indian names were still commonly used to designate plants that had medicinal properties. Within a few decades, this practice began to die out, as the arrogance of the British in India and their confidence in the power of Western science increased. But, in the late 18th and early 19th centuries, partly under the influence of Orientalists (some of whom were surgeons), there was more tolerance and even enthusiasm for the use of Indian names (Harrison, 2001). Francis Hamilton — well known for his botanical and other surveys of India — told the Royal Society of Edinburgh in 1824 that "I prefer using the Sanscrit names, both as being more scientific and as being more likely to remain permanent; for, after a lapse of many ages, they continue to be known to all Hindus of learning" (Hamilton, 1824). Like most other Indian botanists, Hamilton was heavily dependent on local expertise, and his preference for indigenous terminology reflected this.

The same can be said of James Johnson, a naval surgeon working in India at the beginning of the 19th century. Although he spent only a few years in India, Johnson's *The Influence of Tropical Climates on European Constitutions* (1815) was the chief work on medicine in tropical climates in the first half of the 19th century, and it went through several editions, the last being published posthumously in 1856. In this work, Johnson devoted a great deal of attention to the intermittent and remittent fevers that he regularly encountered in Bengal and wrote much about their causes and treatment. Although he saw himself in the vanguard of medical progress, he was still willing, like many other surgeons in India, to learn from indigenous practitioners. He thus informed his readers that Ayurvedic practitioners in Bengal used the Catcaranja nut (*Caesalpinia bonducella*) in the final stages of treating fever. The kernel of the nut was pounded into a paste, with three or four corns of pepper, and taken from three to five times a day, in conjunction with a decoction of Cherettah (probably either *Swertia chirata* or *Swertia purpurescens*). "The kernel is intensely bitter, and possessed of … the febrifuge powers of Peruvian bark in a very high degree," he wrote, adding that it also had "a manifest advantage over the latter; for, instead of producing any constipating effects in the bowels, it … proves mildly laxative." This quality, he noted, made the febrifuge well suited to the climate of Bengal, which, he observed, tended to produce constipation in Europeans. Johnson reported that the remedy had been found so successful that it had been "adopted by many European practitioners" and believed that it would "probably, at no distant period, supersede the bark" (Johnson, 1815). The "fever" or "physic" nut, as it was sometimes known, remained in use throughout the 19th century (Dey, 1896). It is still commonly used as a febrifuge and antimalarial in India today (Kurrup et al., 1979; Gupta, 1981; Sharma, 1999; Singh and Anwar Ali, 1994).

Had the active ingredients of cinchona not been extracted during the 1830s, it might well have been replaced by indigenous febrifuges, just as Roxburgh and Johnson speculated. As it turned out, quinine came to dominate malaria therapy in India for the next century to the virtual exclusion of local remedies, among practitioners of Western medicine at least. But among practitioners of Indian medical systems, such as Ayurveda and Siddha, and as folk remedies, many of the substances mentioned above remained in common use.

However, practitioners of Western medicine did not entirely lose interest in indigenous drugs. After the posthumous publication of William Roxburgh's *Flora Indica* in 1820 and 1824, a number of East India Company surgeons such as John Forbes Royle and W.B. O'Shaughnessy attempted to provide scientific accounts of the medical properties of Indian plants. They were soon joined by Indian doctors trained in Western medicine at such institutions as the Calcutta Medical College, which was founded in 1835. Several classic accounts of the pharmacopoeia of India, written by Indians, took their place alongside those of Royle, O'Shaughnessy, and E.J. Waring, who opened up a new era in Indian material medica with the publication of his comprehensive volume *Pharmacopoeia of India* in 1868. Like Waring, the leading Indian writers were keen to strip Indian material medica of its superstitious associations and to base it on a sound knowledge of chemistry. One such was the Calcutta graduate K.L. Dey, whose classic work *Indigenous Drugs of India* was published in 1867, going to a second, enlarged edition in 1896. By this time, the profession of medicine in India had become more inclined toward nationalism, though still of a moderate kind. Authors such as Dey fully acknowledged their debts to the British pioneers but took great pride in what they had achieved in the field of medicine. Such was their interest in indigenous materia medica that a whole session of the Indian Medical Congress of 1894 was devoted to it. Dey and colleagues looked ahead to a time when India would have the confidence to draw on its own medical traditions and become largely self-sufficient in the production of its own remedies (Dey, 1896; Harrison, 2001). It is interesting to note that *Caesalpinia bonducella* and *Swertia chirata* are key components of the modern antimalarial Ayush-64, which has been used in malaria control programs in the 20th century (see Chapter 5). However, although British colonial doctors took an interest in the local plants used to treat fevers, most did not understand the Ayurvedic system of medicine from which they were taken, and which will be discussed further in Chapter 13.

REFERENCES

Arnold, D. (1998). Introduction. In *Imperial Medicine and Indigenous Societies*. Manchester University Press, Manchester, U.K., pp. 1–26.

Arnold, D., Ed. (1996). *Warm Climates and Western Medicine*. Rodopi Press, Amsterdam.

Blane, G. (1791). Account of the *Nardus indica*, or Spikenard. *Med. Facts Obs.*, 1, 153–163.

Brown, S. (1698). An account of some Indian plants, etc, with their names, descriptions and vertues [*sic*]; communicated in a letter to Mr. James Pettiver, apothecary and fellow of the Royal Society. *Philos. Trans.*, 20, 324.

Desmond, R. (1992). *The European Discovery of the Indian Flora*. Oxford University Press, New Delhi.

Dey, L.V. (1896). *Indigenous Drugs of India*. Thacker, Spink & Co., Calcutta.

Drayton, R. (2000). *Nature's Government: Science, Imperial Britain and the 'Improvement' of the World*. Yale University Press, New Haven.

Gascoigne, J. (1998). *Science in the Service of Empire: Joseph Banks, the British State and the Uses of Science in the Age of Revolution*. Cambridge University Press, Cambridge, U.K.

Gupta, S.P. (1981). Native medicinal uses of plants by the Asurs of Netarhat Plateau (Bihar). In *Glimpses of Indian Ethnobotany*, Jain, S.K., Ed. Oxford & IBH Publishing Co., New Delhi, pp. 218–231.

Hamilton, F. (1824). Some notices concerning the plants of various parts on India, and concerning the Sanscrit names of those regions. *Trans. R. Soc. Edinburgh*, 10, 171.

Harrison, M. (1999). *Climates and Constitutions: Health, Race, Environment and British Imperialism in India 1600–1850*. Oxford University Press, New Delhi.

Harrison, M. (2001). Medicine and orientalism: perspectives on Europe's encounter with Indian medical systems. In *Health, Medicine and Empire: Perspectives on Colonial India*, Pati, B. and Harrison, M., Eds. Orient Longman, Hyderabad, India, pp. 37–87.

Honigsbaum, M. (2001). *The Fever Trail: The Hunt for the Cure of Malaria*. Macmillan, London.

Jarcho, S. (1993). *Quinine's Predecessor: Francesco Torti and the Early History of Cinchona*. Johns Hopkins University Press, Baltimore.

John, S. Letter to W. Roxburgh, August 17, 1792. Roxburgh Papers, MSS Eur D 809. Oriental and India Office Collections, British Library.

Johnson, J. (1815). *The Influence of Tropical Climates on European Constitutions*. J. Callow, London.

Kurrup, P.N.V., Ramadas, V.N.K., and Joshi, P. (1979). *Hand Book of Medicinal Plants*. CCRAS, New Delhi.

Lindsay, J. (1794). An account of the *Quassia polygama*, or bitter-wood of Jamaica; and of the *Cinchona brachycarpa*, a new species of Jesuit's bark found in the same island. *Trans. R. Soc. Edinburgh*, 3, 205–214.

Maehle, A.H. (1999). *Drugs on Trial: Experimental Pharmacology and Therapeutic Innovation in the Eighteenth Century*. Rodopi Press, Amsterdam.

Papin, Fr. (1713). A letter from Father Papin to Father Gobien, containing some observations upon the mechanic arts and physick of the Indian. Taken from the ninth volume of letters of the missionary Jesuits. *Philos. Trans.*, 28, 225–230.

Roxburgh, W. (1792). A Botanical Description of a New Species of Swietenia (Mahogany) with Experiments and Observations on the Powers of Its Bark Comparing Them with Those of Peruvian Bark for Which It Is Proposed as a Succedaneum. Royal College of Physicians of Edinburgh, MS, 1792.

Ryan, D. (1782). Further remarks on the method of treating the remitting fevers of the West Indies; with observations on the best means of preserving health in Jamaica. *Lond. Med. J.*, 2, 66–67.

Shankar, D. and Venugopal, S.N. (1999). Understanding of Malaria in Ayurveda and Strategies for Local Production of Herbal Anti-Malarials. Paper presented at the First International Meeting of the Research Initiative on Traditional Antimalarials. Available at http://mim.nih.gov/english/partnerships/ritam _program.pdf.

Sharma, K.D. (1999). An Antimalarial Herbal Preparation. Paper presented at the First International Meeting of the Research Initiative on Traditional Antimalarials. Available at http://mim.nih.gov/english/partnerships/ritam_program.pdf.

Singh, V.K. and Anwar Ali, Z. (1994). Folk medicines in primary health care: common plants used for the treatment of fevers in India. *Fitoterapia*, 65, 68–74.

Stevenson, W. (1781). Cases in medicine. *Lond. Med. J.*, 2, 182.

13 The Ayurvedic Perspective on Malaria

P.M. Unnikrishnan, S.N. Venugopal, Sarika D'Souza, and Darshan Shankar

CONTENTS

13.1 INTRODUCTION

In spite of massive investments in research and eradication programs, malaria still continues to be a major public health problem. The management of malaria by the conventional biomedical strategy has well-known limitations due to changing patterns of resistance to antimalarials and pesticides and the increased spread of *Plasmodium falciparum* infection posing new challenges (Narayan, 2002). In this context, this chapter aims to make two points: (1) The indigenous medical knowledge systems of India and perhaps of other countries have had their own understanding of malarial fevers and found local solutions before modern medical science arrived at its own. (2) Many current research projects for finding antimalarials from traditional medicine are inadequate because they do not incorporate in their design the etiology, classification, and management schemes of traditional medical knowledge.

13.2 TRADITIONAL MEDICINE IN INDIA

Traditional medicine in India includes codified systems like Ayurveda, Siddha, Unani, and Tibetan medicine and the noncodified oral traditions. Codified systems are grounded in a theory of physiological functioning, disease etiology, and clinical practice. They have formal traditions of training and possess a vast array of written documents recording the *materia medica*, specialised subjects related to medicine and surgery, clinical procedures, and medical ethics. The noncodified or folk traditions, such as those represented by bonesetters, birth attendants, pediatric specialists, veterinary healers, poison healers, and healers specialised in specific diseases like jaundice, eye diseases, gastrointestinal diseases, etc., have been transferred as oral traditions for generations through a people-to-people process. Another feature of folk traditions is that they are ethnic community and ecosystem specific, and thus embody tremendous geocultural diversity. Folk medicine also includes what is popularly known as grandmothers' remedies, the household knowledge about primary health care, different health food recipes, seasonal health regimens and health customs, rituals, etc. Traditional medicine thus comprises a holistic approach to healthy living through drugs, diet, and lifestyle (Shankar, 1992; Shankar and Manohar, 1995; Hafeel and Shankar, 1999).

In this article we largely base our discussions on the Ayurvedic system of medicine, because our studies so far have been confined to the Ayurvedic literature. This chapter therefore does not take into account the presence of antimalarial treatments in Siddha, Unani, or Tibetan medicine, as well as in folk traditions, all of which are worthy of separate review articles.

13.3 MALARIAL FEVERS IN AYURVEDA

The term *malaria* is not known in Ayurveda, and the etiological explanation in terms of transmission, the presence of plasmodia, and their progress through asexual to sexual stages is not described in Ayurveda. Nevertheless, there is evidence to show that the malaria type of fever was carefully studied and understood. The view taken in this chapter is that an understanding of these approaches, concepts, and principles holds significant potential for the contribution to malaria control and more generally to wider health care strategies (Venugopal and Shankar, 2001).

From the period of the *Atharva Veda* (1500 B.C.), there are descriptions of malarial types of fever. *Takman*, for instance, is the term used for a type of fever attended with trembling, rigor, headache, debility, and cough ending in pallor and yellowness. It was endemic to particular regions in the Indian subcontinent like *Munjavan, Mahavrsa, Gandhara, Anga*, and *Magadha*. The epidemics mostly occurred in summer, the rainy season, and autumn. There were fever types such as *anyedyu* (quotidian), *trtiyaka* (tertian), and *sadandi* (remittent). In severe types the patient often was reported to have suffered from delirium and died. From the descriptions it is assumed that the mortality rate was high. There were medicinal preparations and a number of drugs mentioned for the management of the condition (Sharma, 1992).

Classical texts of Ayurveda describe what seems to be malaria in the following way:

Suksma suksma tarasyesu doso raktadi margesu sanairalpa cirena yat kramoyam tena vicchhinna santapo laksyate jvara visamo visamarambha kriya kalo anusangavan

— *Astanga samgraha, Jvara nidana, 69*

Vitiated dosa* spreads slowly in minute channels of blood and other orifices and does not spread to all parts of the body simultaneously but only to predominant parts. This is the manner in which fever is seen with interrupted or irregular onset and relief. It is called visama jvara because of the irregular onset, action, time and reappearance.

Of the eight main types of fever explained in Ayurvedic medical texts, there is a certain category of fever referred to as *visama jvara* (Caraka samhita, *cikitsa sthana*, 3/55–75; Susruta samhita, *Uttara tantra*, 39/51–58; Astanga hrdaya, *nidana sthana*, 1/56–57). This type of fever is characterised by irregular onset, action, and recurrence. It is this category of fever that can be correlated to malaria. The symptoms exhibited by patients suffering from this kind of fever vary according to its stages. Ayurvedic theory describes these stages as being due to the involvement of particular body tissues, viz., *rasa* (primary nutrient formed from the food), *rakta* (blood), *mamsa* (muscles), *meda* (adipose tissue), *asthi* (bone), *majja* (marrow), and *sukra* (reproductive tissues).

---

* According to Ayurveda there are three biological principles in the body: *vata, pitta,* and *kapha.* These govern all physiological as well as pathological functions in the body. A state of equilibrium of these three is considered healthy. A deranged or vitiated state leads to disease.

For instance, the classical texts like Caraka samhita, Susruta samhita, and Astanga hrdaya speak of *santata*, a type of fever characterised by continuous fever for 7 days or more and symptoms such as heaviness of the body, nausea, debility, vomiting, loss of appetite, delusion, giddiness, delirium, rashes, and thirst. The *satata* type presents twice a day (within 24 hours). The *anyedyu* type of fever is characterised by remission after 24 hours with symptoms of pain in the calves, thirst, high temperature, burning sensation, convulsions of hands/legs, and trembling. The *trtiyaka* form of fever is characterised by remission on every third day and symptoms like low back pain, profuse sweating, thirst, fainting, delirium, vomiting, and loss of appetite. These can be correlated with the modern understanding of malaria (see Table 13.1). The Ayurvedic understanding of this fever also anticipates mental symptoms such as confusion, intolerance, restlessness, and fear that may arise in acute and chronic fever of high intensity, i.e., in a fever that persists for more than 21 days. The *caturthaka* type has an interval of 2 days and presents on every fourth day.

The treatment of the different fevers described in the Ayurvedic medical texts is not uniform but needs to be varied according to the stage and specific tissues involved and the presentation of particular symptoms (see Table 13.2). In addition to treating the symptoms, certain other treatments like *panca karma*,* a set of body purification, balancing, and rejuvenation methods, are prescribed (see Table 13.2 and Table 13.3). There are also important dietary rules prescribed for certain conditions of the fever.

Preventive measures for malarial-type fever, which is known to occur in India during the monsoon, also figure in Ayurvedic texts. These include herbal fumigation using herbs like *Azadirachta indica* (*nimba*), *Acorus calamus* (*vaca*), *Brassica campestris* (*sarsapa*), *Terminalia chebula* (*haritaki*), and *Saussurea lappa* (*kustha*) and oral medication involving *Azadirachta indica* (*nimba*), *Albizia lebbeck* (*sirisa*), and *Alstonia scholaris* (*saptaparna*) (Shankar et al., 2002).

13.4 PROBLEMS IN THE CURRENT TRADITIONAL MEDICINE RESEARCH IN MALARIA

It would seem evident from the chapters in this book that different cultural regions of the world have had their own understanding of malarial fevers and have equipped themselves to combat the disease in localised ways. This is clearly evidenced by the fact that the two globally most important antimalarial drugs have had their origin in traditional medical knowledge drawn from two far corners of the world, Peru (see Chapter 2) and China (see Chapter 3). However, when it comes to searching for further solutions for combating the disease, the contemporary relevance of the traditional knowledge is often overlooked. It is clear from the textual descriptions in Ayurveda that malaria has been described in detail and has been managed by traditional medicine healers in India in the past. Due to various reasons, like lack of political patronage for traditional medicine, overdependency on the biomedical drugs, lack of interest among young people to continue the tradition, and depletion of natural resources, at present these are not widely being practiced in the country by traditional physicians, although this knowledge may have much to offer to current research.

The search for local solutions and the empirical evaluation of traditional practices and formulations are becoming important in view of the resurgence of malaria. Searching for antimalarials from traditional medicine has largely been through the ethnobotanical, pharmacological approach from a biomedical paradigm. It is our contention here that this approach fails to recognise the comprehensive nature of traditional medicine's concept of health. Some of the problems of the current research can be summarised as follows:

---

* *Panca karma* is a purification and rejuvenation treatment in Ayurveda that comprises five types of methods such as emesis, purgation, two different types of medicated enemas, and nasal medication.

TABLE 13.1

Various Presentations of *Visama Jvara*

SN	Types of Malarial Fevers	Equivalent Sanskrit Name	Involvement of Body Elements (*dhatus*)	Onset	Symptoms
1	Continuous	*Santata*	*Rasa*[a] *Rakta* (blood)	Continuous for 7, 10, or 12 days for *vata*, *pitta*, and *kapha*[b] types, respectively (remission may happen after this period with a gap of 2 days)	If *rasa* is involved, heaviness of body, nausea, debility, vomiting, loss of appetite, and timidity occur. If *rakta* is involved, hemoptysis, burning sensation, delusion, vomiting, giddiness, delirium, rashes, and thirst will be present
2	—	*Satata*	—	Appears twice in 24 hours	—
3	Double quotidian	*Anyedyu*	*Mamsa* (muscle)	Once in 24 hours	Pain in the calves, thirst, increased urination, high temperature, burning sensation, convulsions of hands/legs, or trembling
4	Quotidian	*Trtiyaka*	*Medas* (adipose tissues)	On third day	Catching type of lower back pain, or pain in the head; profuse sweating, thirst, fainting, delirium, vomiting, foul smell, loss of appetite, exhaustion, intolerance
5	Tertian	*Caturtthaka*	*Asthi* (bone), *majja* (bone marrow)	On every fourth day	Pain in the calf muscles or pain in the head. In *asthi*, splitting pain in the bones, moaning, breathing difficulty, vomiting, diarrhoea, and convulsions. In *majja*, fainting, hiccups, cough, cold, vomiting, deep breath, burning sensation inside body, cutting pain in vital parts like heart, etc.
6	Tertian (variety)	*Caturtthaka viparyaya*	—	On every second and third day in a 4-day course	—

Note: SN = serial number.

[a] Primary nutrient formed from the food.

[b] According to Ayurveda, *vata*, *pitta*, and *kapha* are the three humoral principles in the body on which all the physiological as well as pathological processes are understood. In this the word *vata* is derived from the Sanskrit term *va*, which means to move or destruct. *Vata* is involved in functions such as transportation, excretion, movements, respiration, and degeneration of the body. *Pitta* derives from *tap*, which means to heat or burn. *Pitta* is responsible for metabolic processes, maintenance of body temperature, vision, intelligence, complexion, etc. *Kapha* derives from *ka* and *pha*, which means that which nourishes. *Kapha* is responsible for stability, lubrication, and compactness in the body.

Source: FRLHT database; http://www.frlht-india.org. FRLHT, 2002 and 2003.

TABLE 13.2
Treatment Measures for *Visama Jvara*

SN	Type	Body Tissue	Treatment	Specific Formulations
1	*Santata* remittent	*Rasa, rakta*	*Panca karma* treatments	*Indrayava* (seeds of *Holarrhena pubescens*), *Patola* leaves (*Trichosanthes cucumerina*), *Katurohini* (*Picrorhiza scrophulariiflora*)
2	*Satata*	—	—	*Patola* (*Trichosanthes cucumerina*), *sariba* (*Hemidesmus indicus*), *musta* (*Cyperus rotundus*), *patha* (*Cyclea peltata*), *katurohini* (*Picrorhiza scrophulariiflora*)
3	*Anyedyu* quotidian	*Mamsa*	*Panca karma* treatments Fasting	*Nimba* (*Azadirachta indica*), *patola* (*Trichosanthes cucumerina*), *Triphala* (*Phyllanthus emblica, Terminalia chebula*, and *Terminalia bellirica*), *mrdvika* (*Vitis vinifera*), *musta* (*Cyperus rotundus*), *kutaja* (*Holarrhena pubescens*)
4	*Trtiyaka* tertian	*Medas*	—	*Kiratatikta* (*Andrographis paniculata*), *guduci* (*Tinospora cordifolia*), *candana* (*Santalum album*), *sunthi* (*Zingiber officinale*)
5	*Caturtthaka* quartan	*Asthi, majja*	*Panca karma* treatments	*Guduci* (*Tinospora cordifolia*), *amalaki* (*Phyllanthus emblica*), *musta* (*Cyperus rotundus*)

Note: SN = serial number.

Source: FRLHT database; http://www.frlht-india.org. FRLHT, 2002 and 2003.

1. Biomedical approaches do not recognise the conceptual and theoretical aspects (epistemological) of traditional medicine and thus fail to establish reliable correlation between malaria and an appropriate traditional medicine entity. Therefore, the most suitable traditional management of malarial fevers is not identified and subsequently tested for its efficacy. The differential management approaches of traditional medicine remain unidentified.
2. Ethnobotanical approaches involve anthropological, sociological designs, which are not sensitive to the indigenous culture and knowledge systems. For example, if there is traditional advice to collect, north-facing roots of a plant, or a prescription to collect a plant during a particular constellation, or if it were advised to pray to a plant and seek its permission before uprooting it, such prescriptions may be considered to be irrelevant by the ethnobotanical approach and ignored.
3. The research focus has been largely looking for active ingredients, biomolecules, or novel structures from medicinal plants for currently understood targets and thus seeks solutions similar to those already known, and this focus ignores the probability of new modes of action and alternative management strategies.
4. Lack of proper correlation between traditional/indigenous medicinal plant nomenclature and botanical nomenclature, so that there can be errors about the identity of the plant itself.
5. Bypassing subtle aspects of traditional medicine such as principles of drug formulation design, quality standards for collection, and processing techniques of medicinal plants.

TABLE 13.3
Treatment Rules

SN	Methods	Brief Explanation	Methods of Treatment
1	Purification measures	Remove toxic materials from the body through five main procedures known as *panca karma*	If *vata* vitiation is more, enema of nonunctuous herbs If *pitta* vitiation is more, purgation If *kapha* vitiation is more, emesis is recommended
2	Palliative measures	Give external and internal medicines in different forms against vitiation of *dosas* in body elements	In case of *vata* predominant fever, treat with various types of ghee or with milk and ghee If *pitta* vitiation is more, treat with various bitter and cold drugs If *kapha* vitiation is more, treat with non-oily diet, astringent, and hot drugs
4	Psychosomatic measures	Various types of malarial fevers may result in mental symptoms in chronic stages	Treatment methods recommended for mental disorders are to be selected
5	Preventive measures	Abstain from causative factors, prepare the body elements against the disease, and seek environmental sanitation	Follow certain local practices, for example, *sirisa* (*Albizia lebbeck*), chewing neem bark, etc.; use of neem and other drugs for fumigation

Note: SN = serial number.

Source: FRLHT database; http://www.frlht-india.org. FRLHT, 2002 and 2003.

An epistemological understanding of traditional medicine implies the development of a new cross-cultural research methodology. One of the key problems in the context of cross-cultural medical research is how to correlate malaria or, for that matter, any other disease entity with the correct traditional medicine entity. The textual descriptions of fevers in the classical Ayurvedic literature do not match with the disease presentation today. To bridge this gap, it would be fruitful if the malaria existing today were studied using the diagnostic principles of traditional medicine. This is a challenging task. It is not only essential to correlate the symptoms of malaria as it presents today with Ayurvedic symptoms, but also to understand the pathogenesis according to Ayurveda so that the point of convergence or difference of opinions between the different systems can be explicitly stated. Ayurveda understands disease as a process. Identification of this process is as important as the exercise of interpretation of symptoms. In Ayurveda such a study will have to include *panca laksana nidana* (fivefold diagnosis): *nidana* (etiology), *purvarupa* (initial signs and symptoms), *rupa* (symptoms), *upasaya* (therapeutic diagnosis), and *samprapti* (complete manifestation). Treatment could be *nidana parivarjana* (removal of the cause) or *prakrti vighata* (altering the internal environment). Once disease and its management is well understood, it will help in designing proper research protocols and mutually agreeable end points (Unnikrishnan, 2002).

For this purpose exhaustive literature studies need to be initiated, and it would also be beneficial to consult eminent practitioners in the field and to seek their understanding of a particular condition and its management. There is also a need for critical clinical documentation of malaria patients according to Ayurvedic parameters. Understanding and documenting diseases according to the Ayurvedic framework will help build Ayurvedic diagnostics and in turn contribute to rejuvenating clinical research. Fresh field documentation of diseases according to the understanding of traditional medicine is essential to fill the chronological gaps and understand newly emerging conditions.

It is important to note that traditional medicine advocates a management strategy and not merely a set of herbal formulations. For example, treatment of *jvara* (fever) in different stages starts by a method of *langhana* (fasting), then goes on to *sadanga pana* (medicated drinks), then to decoctions, to *ksirapaka* (medicated milk preparations), *ghrta* (preparations with clarified butter), *leha* (linctuses), etc. This sequenced management described in the classical Ayurvedic literature is necessary to completely root out the condition and establish health. But when research is done only on the effects of isolated entities such as one formulation or a single plant, it is not sensitive to the etiological understanding or the stage, types, or form of medicine explained in Ayurveda (Unnikrishnan, 2002).

A shortcut approach to malaria is reflected in the research strategy of the Central Council for Research in Ayurveda and Siddha in India, which produced a formulation in the 1970s called Ayush-64 (see Chapter 5). The research strategy (under pressure from the biomedicine collaborators) did not take into account the differential diagnosis and the comprehensive treatment approach of Ayurvedic science. The formulation showed only 49% clinical efficacy in a recent clinical study (Shankar et al., 2002; see also Chapter 5). Resistance to active ingredient-based antimalarials from natural sources has been experienced in the case of quinine.

In the area of plant studies, studies on the concept of formulation, drug design, and mode of action are essential. Designing evaluation methods based on Ayurvedic parameters and proper cross-cultural studies in selecting drugs for research are also essential. By ignoring the basic concepts of traditional medicine, the research will inevitably result in incomplete understanding of disease management.

There are also a number of subtle features in traditional medicine practice related to the quality of plants, which need to be taken into account in designing research. Ayurvedic classical texts detail specific habitats for collection of plants, specific seasons, stages of maturity of the plants, and methods of collecting plants in order to manifest their best qualities. There are also certain quality parameters prescribed for processing and storing drugs.

A great challenge therefore exists for standardising methodology for research on traditional medicine in a manner that captures the theoretical underpinnings of the traditional system.

13.5 PARTICIPATORY RESEARCH

The Foundation for Revitalisation of Local Health Traditions (FRLHT) is a nongovernmental organisation in Bangalore, India, dedicated to revitalising traditional medical knowledge and the conservation of medicinal resources. For finding out best solutions to public health problems, FRLHT has been carrying out a program called documentation and participatory rapid assessment of local health practices followed by clinical trials. The rapid assessment is carried out through community-level workshops. In these workshops the knowledgeable households, village healers, and physicians from various systems of medicines such as Ayurveda, Siddha, Unani, and biomedicine meet together and review selected local health practice. The medical systems represented use clinical data from their own classical texts in Ayurveda, Siddha, Unani, and pharmacology to back up their comments. This rapid assessment method is applied to promote effective local remedies. The workshop provides a platform for the community and people from the professional medical community to interact. It increases the community's confidence and legitimacy in its own practices (Hafeel et al., 2001) and also provides maximum ethnographic data.

The Research Initiative on Traditional Antimalarial Methods (RITAM)–India program was started by the foundation in collaboration with two other agencies — Community Health Cell, Bangalore, and BAIF Institute for Rural Development, Tiptur, in South India. The activities carried out so far are on the following lines. An initial study was done to document the presentation of malaria on Ayurvedic parameters in a selected location. Around 90 patients diagnosed as having malaria by blood smear tests were also documented in a comprehensive case sheet prepared with Ayurvedic parameters during two malaria seasons. These data were analysed to get a picture of how traditional medicine would perceive malaria. These data were then correlated with the information existing in important traditional medical texts to arrive at a standard correlation of biomedical entities and the Ayurvedic understanding of the condition. Following this, a package of treatment is selected from the Ayurvedic medical literature through an exhaustive referencing process. Around 170 plant drugs were listed belonging to 115 formulations from 20 classical Ayurvedic texts before finalising a package for the management of malaria in the selected location (D'Souza, 2002). After standardising this package, it will be tested during the pilot study in a select number of patients.

Apart from this, the local healers who are treating malaria are being visited and the details of their malaria management are being documented. In the pilot phase the folk healers' treatments will be recorded with blood smear tests before and after their treatment period. The objective is to follow the local healers' treatment without any intervention. Following this, further studies on both malaria management packages (folk and Ayurvedic) will be carried out.

In one early finding, patients of malaria affected in the selected location have been observed to fall mainly into a category of fever discussed in Ayurveda called *trtiyaka jvara* (D'Souza, 2002). This is characterised by remission on every third day and symptoms like low back pain, profuse sweating, thirst, fainting, delirium, vomiting, and loss of appetite. The Ayurvedic understanding of this fever also anticipates mental symptoms such as confusion, intolerance, restlessness, and fear that may arise in acute and chronic fever of high intensity (i.e., a fever that persists for more than 21 days). It is thus assumed that a general malaria remedy may not be effective for the condition in the selected geographical location, and that the remedy will have to be specific to *trtiyaka jvara*.

It would be constructive also to develop a suitably designed field and laboratory research program on malaria in collaboration with biomedicine so that the entire medical community can be satisfied with the antimalarial potential of traditional herbal drugs, including the *panca karma* techniques and dietary advice. The focus of research should be on success in management of malaria and not on efficacy of a particular drug.

In countries where malaria has been managed by traditional means, local traditional medicine councils should be given responsibility and should be encouraged to implement Roll Back Malaria programs, based on traditional medical systems. Given the recent upsurge in the spread of malaria, it is prudent to harness all available means, including a fresh look at the experience and knowledge of traditional systems of medicines.

13.6 CONCLUSION

Here we have tried to give an overview of the elaborate understanding of malaria in traditional medicine citing the example of Ayurveda. It is to be understood that the theoretical framework and worldview of traditional medicine is different from that of biomedicine. Due to various sociopolitical reasons, there is an erosion in these practices and there is a knowledge-rich, practice-poor situation today. Collaborative research with biomedicine is desirable, but the research designs have to be sensitive to the issues relating to the differences in the foundational framework of two different systems. A diversity of worldviews on health care is essential for development and progress.

REFERENCES

Athvale, A.D., Ed. (1980). *Vrddha Vagbhata Viracita Astanga Samgraha with Indu Vyakhya*. Pune, Ayurvidya mudranalaya, *Jvara nidana* section, verse 69.
D'Souza, S. (2002). *Report on RITAM Progam*, unpublished. FRLHT, Bangalore.
FRLHT (2002). Clinically Important Plants of Ayurveda (CD-ROM). Bangalore: Foundation for the Revitalisation of Local Health Traditions.
Hafeel, A. and Shankar, D. (February 1999). Revitalising Indigenous Health Practices. *Compas News Letter*, No. 1. COMPAS, Leusden.

Hafeel, A., Suma, T., Unnikrishnan, P.M., and Shankar, D. (March 2001). Participatory Rapid Assessment of Local Health Traditions. *Compas News Letter*, No. 4. COMPAS, Leusden.

Narayan, R. (2002). Beyond biomedicine: the challenge of socio-epidemiological research. In *Trends in Malaria and Vaccine Research: The Current Indian Scenario*, Raghunath, D. and Nayak, R., Eds. Tata McGraw-Hill Publishing Company Ltd., New Delhi.

Paradkar, H., Ed. (1995). *Srimad Vagbhata Viracita Astanga Hrdaya*. Krishnadas Academy, Nidana sthana, Varanasi, 1/56–57.

Shankar, D. (1992). Indigenous health services: the state of the art. In *State of India's Health*, Mukhopadhyay, A., Ed. Voluntary Health Association of India, New Delhi.

Shankar, D. and Manohar, R. (1995). Ayurvedic medicine today: Ayurveda at the cross roads. In *Oriental Medicine: An Illustrated Guide to the Asian Arts of Healing*, Alphen, J.V. and Aris, A., Eds. Serindia Publications, London.

Shankar, D., Unnikrishnan, P.M., and Venugopal, S.N. (2002). *An Urgent Need to Harness Ayurveda's Creativity for Management of Malaria in Satabdi Smaranika*. Arya Vaidya Sala, Kottakkal.

Sharma, P.V., Ed. (1983). *Caraka Samhita: Agnivesa's Treatise Refined by Caraka and Redacted by Drdhabala*. Chowkhambha Orientalia, Cikitsa sthana, Varanasi, 3/55–75.

Sharma, P.V., Ed. (1992). *History of Medicine in India*. Indian National Science Academy, New Delhi.

Unnikrishnan, P.M. (2002). Validation of Ayurveda: limitation of current cross cultural medical approaches. In *Ayurveda at the Crossroads of Care and Cure: Proceedings of the Indo-European Seminar on Ayurveda*, Arrabida, Lisbon, November 2001.

Venugopal, S.N. and Shankar, D. (2001). The indigenous understanding and management of malaria. In *Amruth*. FRLHT, Bangalore.

Yadavji, T. and Narayan, R.A. (1992). *Susruta Samhita of Susruta*. Choukhambha Orientalia, *Uttara tantra*, Varanasi, 39/51–58.

14 Guidelines for Ethnobotanical Studies on Traditional Antimalarials

Nina L. Etkin, Maria do Céu de Madureira, and Gemma Burford

CONTENTS

14.1 INTRODUCTION

The continued reliance of contemporary indigenous populations on plant medicines warrants close scrutiny of those species for pharmacological action, adverse interactions (e.g., with pharmaceuticals), and long-term toxicity.* In view of the growing rate of drug and pesticide resistance among plasmodia and anophelines, respectively, these plants may eventually contribute as well to antimalarial drug discovery and the development of more effective mosquito repellents and antibreeding

---

* Plants containing toxins that manifest over the short term typically do not become established in pharmacopoeia, because their toxic effects are apparent.

agents. Over the shorter term, these plants can also continue to play a role in primary health care and insect management, especially where pharmaceuticals and other biomedical technology is not available on a sustainable basis. We argue that the ethnopharmacological/ethnobotanical* approach is the best means to explore these plants.

Ethnopharmacology is predicated on the principle that indigenous uses of plants offer the strongest clues to the biological activities of those species. This is confounded by the fact that some plants are selected as well, or instead, for colour, location of growth, ritual purposes, and other ideational criteria. Nonetheless, chemosensory perception and empirical observation of therapeutic outcome guide much indigenous plant use, and numerous studies have demonstrated that these are strongly predictive of pharmacological potential. Indeed, among drug discovery techniques, the ethnobotanical approach has been the most productive, generating higher rates of positive laboratory and clinical results than specimens collected randomly or by some other criterion of selection (e.g., Carvalho et al., 1991; Moerman, 1996).

The field techniques used by ethnopharmacologists have evolved rapidly and substantially over the last 10 to 15 years as methodologies matured from the generation of catalogs that merely enumerated plants used for particular conditions or listed medicinal applications for particular species. The adoption of more sophisticated research plans reflects the recognition that inventories do little to characterise the complex circumstances in which people select, process, combine, prepare, administer, dose, and judge the efficacy of medicines. It reflects as well advances in phylogenetics, plant ecology, and phytochemistry, including how allelochemicals (secondary plant metabolites) are variably distributed among families, genera, species, populations, individuals, and even tissues of the same plant. The genetic basis of plant activity is further influenced by such ecological factors as UV radiation, rainfall, soil composition, presence and proximity of other (plant and animal) species, and anthropogenic activities such as land and water management. This translates into significant variation over time (even over 24 hours) and space (including very short distances). Consequently, how and why some people use a particular plant can differ from the application of that same species by someone else, including within the same village.

The foregoing underscores how important it is to discern as many details of use and application as possible. This is accomplished through careful ethnographic study, for which ethnopharmacology relies heavily on the traditions of anthropological (ethnographic) field research in which a suite of methods is triangulated — used serially or in parallel. This combination of methods is not intended to test the veracity or consensus of study participants, but to elicit information through a variety of formats to assure reliability — i.e., that the *researcher* "gets it right." Examples of the application of such methods for identification of prophylactic and therapeutic antimalarials include Etkin (1997), Etkin and Ross (1997), and Ross et al. (1996).

This chapter outlines a set of practical guidelines for ethnopharmacological research that researchers can use to identify prophylactic and therapeutic antimalarial plants in both indigenous human and veterinary medicine. It also incorporates suggestions for low-cost activities that can be carried out at the community level, with the aim of translating research directly into improvements in the local management of malaria or the sustainability of medicinal plant utilisation.

14.2 ANTHROPOLOGICAL (ETHNOGRAPHIC) FIELD METHODS

Although there is no right length of time for conducting research, for anthropologists in a field site for the first time, 12 months has become a standard — a year of close, sustained, in-residence observation and interaction, speaking the local language(s). Where long-term study is not possible, a series of shorter-duration rapid ethnobotanical assessments may be substituted and include the participation of resident study participants in the collection of data (e.g., Hoddinott, 1993; Martin,

---

* For purposes of this chapter, the terms *ethnopharmacology* and *ethnobotany* are used interchangeably.

1995; Scrimshaw and Gleason, 1992). Inclusion of community members helps to assure the capture of seasonal, sociocultural, epidemiological, and other continuities and discontinuities that influence plant selection and use.

Readers of this volume are more likely to be interested in rapid assessment, rather than conventional in-depth ethnography, and will want to take advantage of these characteristics of rapid assessment techniques, modified from Trotter and Schensul (1998) and specified for antimalarial plants:

- Narrowly circumscribed to malaria or to certain species such as those reviewed in Part 2 of this volume
- Problem-oriented, intended to help decision makers develop programs or policy — e.g., about incorporating certain antimalarial plants into primary health care
- Participatory, including local partners such as healers and government health aids
- Involve small sample size
- Conduct rapid sampling of representative sectors such as healers, mothers, and other household members responsible for home or self-care
- Focus on cultural patterning rather than intracultural complexity, to discern commonly used plants and patterns of collection, preparation, and administration

14.2.1 ETHICS AND INFORMED CONSENT

Indigenous people's control of their cultural and biological resources eroded significantly during the colonial and neocolonial eras. In the last 20 years researchers, international entities, national governments, indigenous people's organisations, and self-appointed nongovernmental organisations have paid increasing attention to the protection of intellectual property rights (IPRs) and sharing both benefits and knowledge. For ethnobotanists, IPRs have become an increasingly politicised issue, in large part because the pharmaceutical industry has an increasing presence in natural products research: the potential profit from drug discovery has heightened tensions, although, paradoxically, little has been gained to date. Professional bodies such as the American Anthropological Association, the Society for Economic Botany, and the International Society for Ethnobiology have elaborate guidelines that help to ensure ethical conduct and appropriate compensation. Research institutions have increasingly formalised these concerns, and funding agencies require the prior informed consent (PIC) of research participants. In view of growing concern about IPRs and the difficulty of creating transnationally and transculturally appropriate agreements, legislation currently in place to protect IPRs is complex, contentious, and inadequate.

Still, individual research teams that include local participants can create context-specific, culturally appropriate PIC agreements that build local capacity and ensure compensation. Minimally, official and unofficial permissions should be secured; source of funding, research objectives, and methods should be identified; healers and other indigenous peoples with whom one works should share colleague status; the nature of the research should be low impact and low risk; copies of interview, photograph, and voucher data should be deposited in the host country (including the village or other specific locus of research); and research results should be returned to the community (Fluehr-Lobban, 1998; Martin, 1995). It is also important that the nature and schedule of compensation be determined by the indigenous people from whom information and botanical materials originate (Kloppenburg and Balick, 1996), and that contracts and other agreements remain dynamic, subject to modification as the circumstances and outcomes of the research change.

Typically it is difficult to establish exactly who/whether anyone is qualified to represent the community: providing financial benefits at the level of formal village government or other authority may exaggerate existing power asymmetries and social divisions (Eghenter, 2000). This can be partially overcome by redefining benefit sharing as a long-term investment in community health and well-being, through the funding of locally managed projects aimed at documenting, developing, and sustaining the traditional health sector. Such interventions might, for example, use low-technology methods to improve the safety, efficacy, palatability, convenience, and availability of traditional preparations.

14.2.2 CONSERVATION OF CULTURAL AND BIOLOGICAL DIVERSITY

Conservation of cultural and biological diversity is closely related to issues of ethics. As the number of threatened plant species increases, ethnobotanists should help to conserve species that local populations regard to be important, including medicines. This will help to reverse the current trend in which conservable species are identified by outsiders who are culturally and politically detached from the threatened environments (Etkin, 2002). Thus, a primary goal of ethnobotany should be to study plants with an aim that *local* people will gain from the endeavour, or at least that we will gain knowledge of their particular circumstances from the study of their medicines. A primary objective of ethnopharmacology should not be bioprospecting, which in many circumstances should be construed as biopiracy (Svarstad and Dhillion, 2000).

Ethnobotanists can contribute to global conservation efforts by thorough documentation, including identification of endangered species, and sharing information at local, national, and regional levels. Conservation does not necessarily require an ambitious, large-scale project: it may begin by simply empowering families to cultivate two or three priority species in domestic or communal gardens. The Green Belt Movement of Kenya (Seabrook, 1993) is a good example of the mobilisation of women at the grassroots level to promote the cultivation of indigenous trees.

Another important issue is the maintenance of indigenous ethnobotanical knowledge itself. In some communities, such as the Piaroa of Venezuela (Zent, 2001), researchers have documented a significant loss of ethnobotanical information from one generation to the next. Causative factors include economic, ecological, and cultural change; rural–urban migration; and the active suppression of medicinal plant knowledge by religious institutions. The presence of foreign researchers is often enough, in itself, to trigger a revival of interest in health care traditions that were previously dismissed as primitive and backward, but sustaining that interest among young people after the team's departure may require a more formalised program of recognition and accreditation. The Tanzanian cultural association Aang Serian is developing a community college project aimed at bridging the gap between formal classroom-based education and the acquisition of ethnobotanical knowledge from elders (Burford and Ole-Ngila, in preparation).

14.2.3 PARTICIPANT OBSERVATION/OBSERVATION OF PARTICIPANTS

Participant observation involves the researchers in the study community getting close to people, hanging around, learning who's who, and generally familiarising themselves with the substance and rhythm of daily life in a particular community. Learning the community and its resources on this experiential level begins with introduction to the study community and continues through all phases of research. Researchers do not seek corroboration of what study participants say they do, but instead a substantiation that the researcher has understood. In the case of accelerated ethnobotanical assessments, the researcher apprehends community life in less detail, but participant observation is still an important element. For ethnopharmacological research, especially relevant contexts for participant observation consist of ordinary occasions of illness prevention and treatment, including home or self-treatment, healing ceremonies, accompanying plant collectors and medicine preparers, and discussions about health. These experiences illustrate action and process that, through repeated observation, help the researcher develop the depth of insight that may not arise during abstract conversation and interviews (Bernard, 2002; Dewalt et al., 1998; Etkin, 1993).

14.2.4 Key Respondents (Informants)

Key respondents or consultants are individuals with whom the researcher easily establishes rapport and who understand his or her objectives and are interested in participating in the research. Key respondents mediate between the researchers and the community, but are not necessarily representative of it. They need not be specialists in healing, plants, or anything else. Especially when these key individuals conceptualise cultural data in the ethnobotanist's frame of reference, they play important tutorial roles in the identification of relevant topics, instruction about the appropriate way to pose questions, design of survey instruments, and collection and interpretation of data. Selecting key respondents is case and context specific and should be an early, but not first, step in the research process. Researchers should allow time first for some preliminary participant observation (Bernard, 2002; Etkin, 1993).

14.2.5 Focus Groups

Focus groups are exploratory sessions into which, typically, 6 to 12 participants are recruited to discuss a particular theme. They generate qualitative data that complement interview, survey, and questionnaire data, and in fact may be convened to discuss questions for interviews and surveys — e.g., Are the questions naïve? Is the instrument too long? Focus group methodology includes factorial design (Bernard, 2002), for example, two types of healer and two genders (or levels of formal education, etc.) require four focus groups. The objective is not an orderly consensus but a group approach that creates a dynamic in which the range and diversity of opinions are revealed, at the same time that idiosyncratic views do not obscure patterns (Etkin, 1993). Examples of focus groups that are helpful for ethnobotanical research include generalist healers, midwives, mothers and others who administer home care, and people who are knowledgeable about plants.

14.2.6 Semi- and Unstructured Interviews and Questionnaires

Interview type varies along a continuum of structure. *Unstructured interviews* resemble more an exploratory conversation than a formal technique and consist of open-ended questions that allow study participants to elaborate to whatever level of detail they want and to shape the direction of further discussion. Such an interview might begin with questions such as: What causes malaria? Who gets it? What plants treat it? *or* What is this plant used for? Which other symptoms does it treat?

As the researcher interacts with the growing database, patterns are revealed that can form the nucleus of *semistructured interviews*. For example, the 16 or so antimalarial plants or the 5 kinds of fever that were distinguished by participants in unstructured interviews become the lists of stimuli to which each respondent is exposed. The stimuli commonly are open-ended questions, but can be photographs (e.g., of plants or disease symptoms), word lists, or something else that elicits responses that can be compared (Bernard, 2002; Etkin, 1993).

Free listing employs a kind of semistructured interview in which a respondent is asked, for example, to list all the antimalarial plants he knows, or all the kinds of fever that occur, or all the foods that can treat malaria. Having generated a list, a respondent is then asked to prioritise and otherwise compare items and to identify criteria for inclusion and exclusion. The software package ANTHROPAC 4® (Borgatti, 1996) enables the researcher to identify plants with high salience and significant agreement among respondents. It also allows for comparison between individual respondents with respect to their degree of similarity to the cultural norm, and assessment of the respective relevance of factors such as gender, age, and geographical locality in the listing of antimalarial species.

The *frame elicitation* method employs sentences with this structure:

Artemisia can treat X.
 Response type: all fevers, some fevers, malaria and yellow fever, mild malaria
Malaria is treated by X.
 Response type: *Artemisia, Diospyros mespiliformis, Azadirachta indica*
X is a kind of Y: *Azadirachta indica* is a malaria medicine, *Momordica charantia* is a general fever medicine.

In a *triad exercise* a study participant is shown three items and asked to choose the one that does not fit. For example, with the triads (periodic fever, haemoglobinuria, rash) and (*Artemisia annua, Azadirachta indica, Allium sativum*) the researcher presents as the first two items malaria symptoms and antimalarial plants, respectively. It is not necessary, or even expected, that all respondents will group symptoms or assign plants in the same way. Like other methods, the triad exercise serves more to generate discussion and data than to seek consensus.

In *pile sorts* a respondent is presented with a collection (e.g., of plants, photographs, words written on slips of paper) and asked to group items that belong together (Bernard, 2002). This method can be extended into multiple sorts from which the researcher eventually can create taxonomic trees and networks that characterise such cultural domains as botanicals, medicinal plants, food plants, malaria preventives, and healers.

The *questionnaire* is the most structured of the interview and other interactive methods and includes a fixed list of questions with a very limited number of possible responses. For ethnobotanical research, such questions might include a plant name and instructions to respond "does" or "does not" treat (malaria, other fevers, diarrhoea).

14.2.7 PRIORITISING METHODS

Participant observation, open-ended interviews, and free listing are recommended as standard first-tier methods that allow maximum flexibility to learn about a domain (antimalarial plants, for example) and to identify research objectives and communication style for the community of research participants. The other methods outlined are best launched on the strength of more familiarity with the community and some preliminary knowledge about local medicines and plants. Which of these second-tier methods work best varies from one community to another, depending on local language structure, communication patterns, facility with visual images, and so on. Clearly, not all studies use all available methods, but the standard rule is that the more diverse one's methodology, the more reliable and rich the data.

14.2.8 QUANTITATIVE DATA

As for the other elicitation methods outlined above, the outcomes of free listing, triad tests, pile sorts, and questionnaires become the nucleus of continued discussions, and all except the unstructured interview yield quantifiable data. The most common quantification of ethnobotanical data estimates the local significance of a plant, or a disease, or a resource management practice based simply on the percentage of respondents who use that plant, identify that disease, or engage that practice. Other indices require more specific data. A *saliency index* ranks plants that are commonly used. The *use value* of a species calculates the overall utility of a species based on participants' identification of all plants and their uses in a particular area or transect. The *preference ranking* index assigns a mean numerical value to species to reflect participants' perceptions of the plant's significance (Martin, 1995), as determined by asking respondents to rank species by such criteria as antimalarial potential, personal preference, and availability (Box 14.1).

BOX 14.1: PREFERENCE RANKING OF ANTIMALARIAL PLANTS WITH REFERENCE TO EFFICACY[a,b]

Species	Respondents					Total Score	Rank
	A	B	C	D	E		
Acacia arabica Del	2	1	3	1	3	10	2
Azadirachta indica A. Juss.	3	4	4	3	4	18	4
Cassia tora L.	1	2	2	2	1	8	1
Guiera senegalensis JF Gmel	4	3	1	1	2	11	3

[a] Most effective = 4; least effective = 1.
[b] Respondent D ranks two plants as equally and most effective.

BOX 14.2: DIRECT MATRIX RANKING, SUMMED ACROSS RESPONDENTS A TO E[a]

	Acacia arabica	*Azadirachta indica*	*Cassia tora*	*Guiera senegalensis*
Efficacy	2	4	1	3
Availability	2	1	3	4
Food value	3	1	4	2
Total score	7	6	8	9
Rank	2	1	3	4

[a] 4 = most effective/available/nutritious; 1 = least effective/available/nutritious.

Preference ranking can be further complicated by combining two or more criteria in *direct matrix ranking* (Martin, 1995). The example in Box 14.2 combines the dimensions antimalarial efficacy, availability, and food value.

This type of ranking can serve as a standard that facilitates comparison among studies of antimalarial plants. Moreover, it better approximates the nuanced and complex nature of real-life decisions in the selection of medicinal plants.

14.2.9 DISCOURSE, NARRATIVE, AND TEXT ANALYSIS

Discourse-centered research methods pay attention to the use of language in social contexts. The challenge is to capture natural discourse, rather than conversations significantly influenced, or elicited, by the researchers. Ethnographers of an earlier generation referred to the same methods as the ethnography of speaking/communication. For purposes of this chapter, discourse methods would center on listening (and recording or taking notes) while study participants talk about particular plant species, about malaria and how to treat it, how particular species are selected, and so on. Similarly, *narrative* analysis occurs in the context of listening to someone tell (or sing, or act) a story and overlaps *oral history*. Both discourse and narrative methods are used in participant observation. Field research on antimalarial plant medicines would elicit narratives of "the great malaria epidemic of 1980," "the famous healer who treated so many fever cases," and so on. And, similarly again, *text* analysis uses written accounts — indigenous literature (e.g., medicinal formularies, dictionaries, letters, novels) in which plants, fevers, healers, etc., are included. Data generated by these (and other) qualitative methods can be managed electronically (e.g., using The Ethnograph®) and lend themselves to quantitative analysis (Bernard, 2002; Bernard and Ryan, 1998; Farnell and Graham, 1998).

14.2.10 How to Select Study Participants, What Do You Want to Know?

Significant intracultural variation exists in knowledge and practice, even in small communities that appear on superficial examination to be culturally homogeneous and socially undifferentiated. Traditional healers, shamans, and other specialists are valuable sources of information and have been the focus of attention of most ethnobotanical research. However, disease prevention and treatment are ongoing community activities, which often begin with home or self-treatment. A good research strategy extends beyond the specialists to explore the scope of knowledge about plants and malaria. Ideally, one selects a random — or at least representative — sample of adults who span the range of income, education, religion, occupation, language, and other demographics (Cotton, 1996; Etkin, 1993; Martin, 1995). These laypeople are an invaluable resource whose knowledge derives from interacting with plants in the context of farming, caring for children, hunting, cooking, religious rituals, home care, etc. These minor specialists, or nonspecialists, may be more readily available than healers and less likely to worry about revealing the substance of their professional activities.

As discussed above, it is important to learn about antimalarial plants in as much detail as possible. Interviews can be initially organised around a list of plants, for each of which study participants are asked to describe physical attributes, availability, and medicinal and other uses. Similarly, following a list of malaria symptoms, the researcher asks the respondent to describe commonly used medicines, including the criteria used in the selection of that species, source, preparation, additional constituents, approximate dose and schedule of administration, therapeutic objectives, how efficacy is judged, any adverse effects or contraindications, alternatives for circumstances in which that medicine is not available or does not produce the expected results, and the relative merits of different antimalarial species.

The issue of therapeutic objectives merits further explanation. Researchers need to step outside the biomedical paradigm, which centers on symptom resolution. In virtually all medical systems, "getting better" is indeed the ultimate goal, but therapy (like illness) is a process, and in native medical systems there typically exist more proximate (earlier) therapeutic objectives. Thus, researchers need to accommodate medical ideologies in which the objectives may be the expulsion of spirits, darkening of skin to mark transition from one phase of illness to the next, or the occurrence of rash or vomiting, which evidence that the disease agent leaves the body. To understand traditional antimalarial plants in a way that informs bioscience requires that extensive interviews and other methods identify those plants that are designated to reduce the physiological signs of malaria — fever and hepatosplenomegaly (Etkin, 1993, 1997; Etkin and Ross, 1997). Researchers also should pay attention to all kinds of information relevant to disease prevention, such as foods made with plants used for the treatment of malaria, medicines taken regularly to strengthen the body, fumigations, use of botanical insecticides and insect repellents, and so on.

Another important aspect is documenting the safety of the medicinal plants, based on their long-term use. According to the World Health Organization (2000), and as a basic rule, documentation of a long duration of use should be taken into consideration when assessing the safety of a botanical preparation, and could form the basis of the risk assessment. Thus, when there are no detailed toxicological studies, researchers should also report how long a plant species has been in use, the disorders and symptoms it is used to treat, and the number of users in the region of the study. If long-term traditional use (10 years or more) cannot be documented, or there is concern about safety, toxicity data will need to be obtained.

14.3 BOTANICAL FIELD METHODS

Qualitative and quantitative data generated by ethnographic/anthropological methods are analysed in conjunction with conventional botanical methods.

14.3.1 COLLECTION FORMS

Collection forms are a convenient way to collect and represent data in an abbreviated and standardised format (Elisabetsky et al., 1996).

The collector is the person who conducts the ethnobotanical research (interviews, collects the plants, and gathers the information) and should be designated for each plant.

It is important to know all the details of study participants who show the plants and provide information on its uses. Age, birthplace, profession, education level, etc., can influence how an individual apprehends, uses, and judges the efficacy of plants. The training, skills, and experience of traditional medical practitioners and nonspecialists should be taken into account.

All local names of the plant must be included on the form. Each collector gives a number for each plant collected, in order to organise the plant material and to connect each species collected to the form that contains the information about the plant. The same collection number must be written on both the form and the label attached to the plant samples. A simple description of plant type should be written (herb, vine, shrub, tree), its height, colour and odor of its leaves, flowers, fruit, and sap. Information should identify the collection site (date, place, abundance, ecological conditions, etc.) and specify whether the specimen is wild or cultivated. Notation should be made as well about the conservation status of the plant: Is it endangered, vulnerable, at low risk, or not evaluated (IUCN, 2000)? If the same plant is collected several times, in different places or at different times, an additional form and distinct collection number will be assigned.

Each study participant should be asked about the plants used to treat or prevent malaria or fevers. For each plant the following descriptors should be recorded: part(s) of plant used; sap or other excretions; whether used fresh or dried; quantity to be used and mode of preparation (e.g., macerated, powdered); any other plants or materials added to the medicine; and means of administration, including dose, schedule, duration of treatment, and preventive or therapeutic objectives.

The collection form (Box 14.3) will help organise the information and is a way to direct the research on each plant. These forms complement but do not replace interviews, participant observation, and the other methods outlined above.

BOX 14.3: COLLECTION FORM

Collector: name, address

Person Interviewed: name, address, and other demographic details

Collected Plant: local name(s), collection numbers, description, conservation status

Traditional Use: indications, preparation, administration

14.3.2 VOUCHER SPECIMENS

Preparation of herbarium voucher specimens is essential because the identification of plant species (to family, genus, species, and authority) by trained taxonomists is the critical link between bioscientific and indigenous knowledge. Vernacular names vary widely from one location to another, even within the same village, and change over time. Thus, one cannot reliably extrapolate from identifications published or otherwise reported by other researchers.

A good voucher specimen consists of a dried, pressed section of a plant containing as many reproductive (flower, fruit) and vegetative structures as possible. For herbaceous species, whole plants (including root) can be collected. For trees and shrubs, pieces of root, trunk, and stem barks should be included. Plant parts that are difficult to press (such as large fruit) can be preserved in alcohol. Photographs of the plant before and after cutting add an important dimension.

If samples cannot be dried on the same day, a wet method should be used to prepare voucher specimens: samples are labeled and pressed in newspaper, stacked together, pressed flat, and tied tightly with nylon cord; samples are preserved by pouring diluted ethanol (50:50 v/v) into and through the specimens, until the newspaper pages are well soaked; samples are placed in a plastic bag, removing as much air as possible and sealing with adhesive tape. With this procedure the plant samples do not decompose and can be conserved for a long time until they can be sent to an herbarium where they will be dried in a routine manner.

Researchers should record growth form, habitat, geographic coordinates, specific location, topography, and soil type (Alexiades, 1996; Elisabetsky et al., 1996; Martin, 1995; WHO, 2000).

It is advisable to collect at least three sets of duplicates, leaving one or two voucher sets in-country, depositing another in the researcher's institution, and reserving another to donate as a professional courtesy to an additional herbarium that might assist with troublesome identifications. Wherever suitable facilities are available, a community herbarium established at a village school or health center can be a valuable tool for maintaining local involvement with ethnobotany over the longer term. This is particularly important in areas where medicinal plant knowledge is con-centrated among older generations, with youth exhibiting a loss of interest or confidence in traditional health systems. Training one or two young research assistants as specimen collectors, and encouraging them to accompany the principal investigators to the national or regional herbar-ium, can further enhance enthusiasm.

14.3.3 COLLECTING FOR PHYTOCHEMICAL STUDIES

Preliminary screening of plants can be conducted in the field to test for the presence of, for example, alkaloids, steroids, terpenoids, and saponins. These techniques are not precise and yield limited findings. Another drawback of field screening is that it takes significant time away from other aspects of ethnobotanical inquiry and compromises the quality of ethnographic information col-lected. That problem can be addressed by increasing the size of the field team, but field screening is not likely to be useful in the search for antimalarial plants. Alternatively, samples can be sent from the field to a laboratory, preferably in-country (for screening studies one should collect approximately 1 kg dry weight for each sample).

After screen results for antimalarial activity are analysed, bulk collection can proceed for promising species. Bulk plant materials typically are dried, but may be sent to the laboratory fresh, frozen, or preserved or extracted in alcohol. All collections must be accompanied by carefully documented vouchers. For phytochemical studies one generally needs larger samples — e.g., 10 to 100 kg dry weight. Rare species should never be collected in bulk, and all harvesting should be sustainable (Daly and Beck, 1996; Martin, 1995). Everywhere, and at virtually all local and national levels, formal permissions are required for collecting plants for vouchers, phytochemical testing, and export.

The World Health Organization (2000) guidelines for evaluation of indigenous plant medicines advance this principle. On the one hand, the methodologies should guarantee the safety and efficacy of plant medicines and traditional procedure-based therapies. On the other hand, the investigation of medicinal plants should not become an obstacle to the application and development of traditional medicine. This is a complex issue that has concerned national health authorities and scientists in recent years.

14.4 LITERATURE REVIEW

Conventional literature searches yield enormous amounts of information. Researchers should review the botanical, ethnopharmacological, and ethnographic literature for their region in advance of entering the field, and can begin to conduct more in-depth literature review as soon as the medicinal plants in their study have been taxonomically identified. One objective is not to repeat phytochemical explorations already conducted by other research teams. Minimally, researchers should consult the NAPRALERT File, Dr. Duke's Phytochemical and Ethnobotanical Databases, and the Center for International Ethnomedicinal Education and Research (CIEER).

14.5 MULTIDISCIPLINARY TEAMS

Given the diversity of expertise required to conduct ethnopharmacological research, fieldwork teams should be multidisciplinary, involving anthropologists, botanists, pharmacologists, physicians, traditional healers, and other local counterparts who are or can be trained to the research endeavour and who can sustain the research after the out-of-country researchers have departed.

14.6 RESULTS

Research results should be made available to all members of the research team, study participants, national health authorities, and other scientists. Factoring in IPR conventions, the findings should be presented in varied formats to assure accessibility — minimally as a document in the local and one transnational (e.g., English) or regional (e.g., Hausa, Spanish, Portuguese) language. Some nongovernmental organisations, such as ENDA-Caribe in the Caribbean and the Foundation for the Revitalisation of Local Health Traditions in India, are also committed to providing research feedback in vernacular languages in the form of pamphlets, audiovisual materials, community meetings, and performance arts such as music and theater.

If funding is available, perhaps as part of benefit-sharing contracts, the provision of feedback can be taken a step further by sharing the findings of literature reviews on important plant species. Local people — especially those with some formal education and a tendency to dismiss traditional medicine as outdated — may be surprised and pleased to learn that scientists in industrialised countries have already demonstrated antimalarial or antimicrobial activity in "grandmother's remedies." Conversely, if a particular medicine has been demonstrated to have long-term adverse effects, such as carcinogenicity or hepatotoxicity, imparting this information may help to reduce the burden of chronic illness for later generations.

If traditional healers and other local people do not agree to the immediate or long-range publication of research, this should not hinder research. In that case, the results can be made available on a confidential basis to approved collaborators. For example, the findings can serve as the basis for preliminary clinical research and can be formatted electronically to permit data manipulation. Information for each identified species should be presented in a standardised format, as depicted in Box 14.4.

14.7 SUMMARY

Conventional (long-duration) ethnographic field methods can be streamlined to a rapid assessment format that is narrowly circumscribed to malaria, participatory, problem oriented, pattern focused, and representative of the study community. Standard first-tier ethnographic methods favor flexibility and provide a foundation for other approaches that elicit more specific, in-depth information on antimalarial medicines. Quantitative and qualitative analyses can be effectively integrated. Comparison among diverse studies of antimalarial plants is advanced by using a standard such as a direct matrix ranking that combines use value and preference. Box 14.1 summarises the minimal content for forms used in conjunction with botanical collections. Box 14.2 identifies a useful format for the presentation of data. Conservation, IPRs, and other issues that overlap issues of ethical and cultural sensitivity must be attended throughout all phases of research. Wherever possible, they should also be continued after the end of the formal study through the empowerment of community members to carry out their own ethnobotanical work and plant cultivation activities.

BOX 14.4: FORMAT FOR DATA PRESENTATION

Botanical Name (genus and species, including authority and family)
Synonyms (most common and well-established)
Voucher Number (and the herbarium where specimen was deposited)
Vernacular Name(s)
Geographic Distribution (locally, general range, worldwide)
Plant Part(s) used, and how (specific medical application, food, etc.)
Preparation and Administration of each medicine (including combination, dose, schedule)
Ethnoveterinary Medicine applications
Contraindications and precautions (e.g., do not use during pregnancy)
Previously Reported Medicinal Uses
Isolated Constituents
Pharmacological Effects
Clinical Findings
Toxicological Data

REFERENCES

Alexiades, M.N. (1996). Standard techniques for collecting and preparing herbarium specimens. In *Selected Guidelines for Ethnobotanical Research: A Field Manual*, Alexiades, M., Ed. New York Botanical Garden, Bronx, NY, pp. 99–126.

Bernard, H.R. (2002). *Research Methods in Anthropology: Qualitative and Quantitative Approaches*, 3rd ed. Alta Mira Press, Walnut Creek, CA.

Bernard, H.R. and Ryan, G.W. (1998). Text analysis: qualitative and quantitative methods. In *Handbook of Methods in Cultural Anthropology*, Bernard, H.R., Ed. Alta Mira Press, Walnut Creek, CA, pp. 595–646.

Borgatti, S. (1996). ANTHROPAC 4. Analytic Technologies, Natick, MA.

Carvalho, L.H., Brandão, M.G.L., Santos-Filho, D., Lopes, J.L.C., and Krettli, A.U. (1991). Antimalarial activity of crude extracts from Brazilian plants studied *in vivo* in *Plasmodium berghei*-infected mice and *in vitro* against *Plasmodium falciparum* in culture. *Braz. J. Med. Biol Res.*, 24, 1113–1123.

Center for International Ethnomedicinal Education and Research (CIEER). (2002). http://www.cieer.org. Accessed May 30.

Cotton, C.M. (1996). *Ethnobotany: Principles and Applications*. Wiley, Chichester, U.K.

Daly, D.C. and Beck, H.T. (1996). Collecting bulk specimens: methods and environmental precautions. In *Selected Guidelines for Ethnobotanical Research: A Field Manual*, Alexiades, M., Ed. New York Botanical Garden, Bronx, NY, pp. 147–164.

Dewalt, K.M., Dewalt, B.R., and Wayland, C. (1998). Participant observation. In *Handbook of Methods in Cultural Anthropology*, Bernard, M., Ed. Alta Mira Press, Walnut Creek, CA, pp. 259–299.

Dr. Duke's Phytochemical and Ethnobotanical Databases. (2002). http://www.ars-grin.gov/duke/plants.html. Accessed May 30.

Eghenter, C. (2000). What is *tana ulen* good for? Considerations on indigenous forest management, conservation and research in the interior of Indonesian Borneo. *Hum. Ecol.*, 28, 331–357.

Elisabetsky, E., Trajber, R., and Ming, L.C. (1996). Appendix: manual for plant collections. In *Medicinal Resources of the Tropical Forest: Biodiversity and Its Importance to Human Health*, Balick, M.J., Elisabetsky, E., and Laird, S.A., Eds. Columbia University Press, New York, pp. 409–420.

ENDA-Caribe with International Development Research Center of Canada. (2003). TRAMIL: Program of Applied Research to Popular Medicine in the Caribbean. http://funredes.org/endacaribe/traducciones/tramil.html. Accessed January 21.

Etkin, N.L. (1993). Anthropological methods in ethnopharmacology. *J. Ethnopharmacol.*, 38, 93–104.

Etkin, N.L. (1997). Antimalarial plants used by Hausa in northern Nigeria. *Trop. Doct.*, 27, 12–16.

Etkin, N.L. (2002). Local knowledge of biotic diversity and its conservation in rural Hausaland, northern Nigeria. *Econ. Bot.*, 56, 73–88.

Etkin, N.L. and Ross, P.J. (1997). Malaria, medicine and meals: a biobehavioural perspective. In *The Anthropology of Medicine*, 3rd ed., Romanucci-Ross, L., Moerman, D.E., and Tancredi, L.R., Eds. Praeger Publishers, New York, pp. 169–209.

Farnell, B. and Graham, L.R. (1998). Discourse-centered methods. In *Handbook of Methods in Cultural Anthropology*, Bernard, H.R., Ed. Alta Mira Press, Walnut Creek, CA, pp. 411–457.

Fluehr-Lobban, C. (1998). Ethics. In *Handbook of Methods in Cultural Anthropology*, Bernard, H.R., Ed. Alta Mira Press, Walnut Creek, CA, pp. 173–202.

Foundation for the Revitalisation of Local Health Traditions. (2003). http://www.frlht-india.org. Accessed January 20.

Hoddinott, J. (1993). Fieldwork under time constraints. In *Fieldwork in Developing Countries*, Devereux, S. and Hoddinott, J., Eds. Lynne Reinner, Boulder, CO, pp. 73–85.

International Union for Conservation of Nature and Natural Resources (IUCN). (2000). Redlist of Threatened Species. http://www.redlist.org. Accessed May 30.

Kloppenburg, J.R. and Balick, M.J. (1996). Property rights and genetic resources: a framework for analysis. In *Medicinal Resources of the Tropical Forest: Biodiversity and Its Importance to Human Health*, Balick, M.J., Elisabetsky, E., and Laird, S.A., Eds. Columbia University Press, New York, pp. 174–190.

Martin, G.J. (1995). *Ethnobotany: A Methods Manual*. Chapman & Hall, London.

Moerman, D.E. (1996). An analysis of the food plants and drug plants of native North America. *J. Ethnopharmacol.*, 52, 1–22.

NAPRALERT File (NAtural PRoducts ALERT). (2002). http://www.cas.org/ONLINE/DBSS/napralertss.html. Accessed May 30.

Ross, P.J., Etkin, N.L., and Muazzamu, I. (1996). A changing Hausa diet. *Med. Anthropol.*, 17, 143–163.

Scrimshaw, N.S. and Gleason, G.R. (1992). *Rapid Assessment Procedures: Qualitative Methodologies for Planning and Evaluation of Health Related Programmes*, Scrimshaw, N.S. and Gleason, G.R., Eds. International Nutrition Foundation for Developing Countries, Boston, MA.

Seabrook, J. (1993). *Pioneers of Change: Experiments in Creating a Humane Society.* New Society Publishers, Philadelphia, pp. 49–51.

Svarstad, H. and Dhillion, S.S. (2000). *Responding to Bioprospecting: From Biodiversity in the South to Medicines in the North*, Svarstad, H. and Dhillion, S.S., Eds. Spartacus Forlag, Oslo.

Trotter, R.T. and Schensul, J.J. (1998). Methods in applied anthropology. In *Handbook of Methods in Cultural Anthropology*, Russell Bernard, H., Ed. Alta Mira Press, Walnut Creek, CA, pp. 691–735.

World Health Organization. (2000). *General Guidelines for Methodologies on Research and Evaluation of Traditional Medicine.* WHO, Geneva.

Zent, S. (2001). Acculturation and ethnobotanical knowledge loss among the Piaroa of Venezuela. In *On Biocultural Diversity: Linking Language, Knowledge and the Environment*, Maffi, L., Ed. Smithsonian Institution Press, Washington, DC, pp. 190–211.

Part 4

Laboratory Research

15 Pharmacological Properties of the Active Constituents of Some Traditional Herbal Antimalarials

Colin W. Wright

CONTENTS

15.1 INTRODUCTION

Many of the plant species used as antimalarials in traditional medicine have been subjected to laboratory investigations in an attempt to provide evidence to support their clinical use. Often, these studies have been limited to determining the activities of crude extracts of the plants against malaria parasites *in vitro* or *in vivo*, although in many cases the compounds responsible for the antiplasmodial effects have been isolated, identified, and assessed for their antimalarial activities (for recent reviews see Camacho Corona et al., 2000; Wright, 2002). In contrast, very few antimalarial plants and their active constituents have undergone pharmacological studies to investigate their antimalarial modes of action, the mechanisms by which toxic effects may occur, and their pharmacokinetic properties, but this information is required if traditional antimalarials or compounds derived from them are to be employed for the effective and safe treatment of malaria. The purpose of this chapter is to review the pharmacological properties of the antimalarial constituents of a number of important traditional antimalarial plants that have been investigated in more detail.

15.2 COMPOUNDS INTERACTING WITH HEME

Heme **1** plays a crucial role in the antimalarial mode of action of several of the most important natural product-derived antimalarial drugs, including quinine **2** and artemisinin **10**. Malaria parasites growing in red blood cells utilise haemoglobin as a source of most, but not all, of the amino acids they require for growth and development (Rosenthal and Meshnick, 1998). The protein (globin) part of the haemoglobin molecule is hydrolysed by a number of proteases present in the parasite acidic food vacuole, leaving heme, the iron-containing core of the molecule, behind. Free heme is toxic to the parasites (Orjih et al., 1981) and therefore must be removed or detoxified. In humans, heme is degraded by the enzyme heme oxidase, but malaria parasites lack enzymes for this purpose, so instead heme is detoxified by converting it into nontoxic malaria pigment, also known as hemozoin. Initially it was thought that this process required an enzyme, heme polymerase (Slater and Cerami, 1992), but it has now been established that hemozoin formation may take place spontaneously in the acidic parasite food vacuole (Dorn et al., 1995). The structure of hemozoin appears to be identical to that of β-hematin, in which heme molecules are linked together in a polymer such that the propionic acid side chain of one molecule is coordinated onto the iron of the next (Bohle et al., 1997), although a more recent study concluded that β-hematin consists of heme dimers linked by hydrogen bonds to form long chains (Pagola et al., 2000). The blood schizontocidal actions of quinine (and related compounds) and of artemisinin and its derivatives require the presence of free heme (and hence they do not kill parasites in liver cells where no haemoglobin digestion takes place), but the mechanisms involved are very different; these are illustrated in Figure 15.1 and discussed in the following sections.

15.2.1 QUININE AND OTHER CINCHONA ALKALOIDS

The bark of trees belonging to the genus *Cinchona*, native to South America, has been used to provide effective treatment for malaria for over 300 years, and its use has been discussed in Chapter 2. Quinine **2**, a quinoline alkaloid, is the major antimalarial compound present in *Cinchona*, although some other constituents, including cinchonidine **3**, cinchonine **6**, and quinidine **5**, also have antimalarial properties. Using the quinine molecule as a template, a number of synthetic quinoline antimalarials such as chloroquine have been developed.

The quinoline antimalarials appear to act by binding to free heme released following the digestion of haemoglobin, thus preventing its conversion to hemozoin (β-hematin) (Figure 15.1). The drug–heme complex is believed to be toxic to the malaria parasite. Compounds may be tested for their ability to inhibit the formation of β-hematin by means of a simple assay that does not require malaria parasites (Egan et al., 1994). In this test, heme (as hemin chloride) is dissolved in sodium hydroxide, neutralised with hydrochloric acid, and then incubated and stirred in acetate buffer (4.5 M, pH 5.0) at 60°C for 30 minutes. The cooled mixture is then filtered and the Fourier-transformed infrared (FTIR) spectrum of the dried precipitate recorded. The presence of two characteristic peaks at 1660 and 1210 cm^{-1} in the FTIR spectrum confirms the formation of β-hematin (hemozoin). Compounds may be tested by adding 3 equivalents (with respect to hemin) to the reaction mixture, and inhibition of β-hematin formation is shown by the absence of the above characteristic peaks in the FTIR spectrum. Using this method, the *Cinchona* alkaloids, quinine (Egan et al., 1994), cinchonine, and cinchonidine (Wright, unpublished results), have been shown to block β-hematin formation. However, 9-*epi*-quinine **4** (also found in *Cinchona* as a minor constituent) failed to prevent β-hematin formation, and this alkaloid is also inactive against malaria

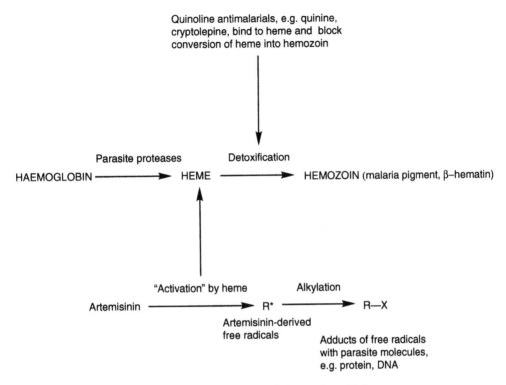

FIGURE 15.1 Modes of action of antimalarial compounds interacting with heme.

parasites (Egan et al., 1994), thus demonstrating that the method is selective. Synthetic antimalarials based on quinine such as amodiaquine, chloroquine, quinacrine (mepacrine), and mefloquine also bind to hemin and inhibit the reaction, but importantly, primaquine (an 8-aminoquinoline) acts only on liver-stage malaria parasites and does not inhibit β-hematin formation.

The complex of quinine with hemin is thought to be toxic to the parasite and may damage the parasite membrane by lipid peroxidation (Sugioka and Suzuki, 1991) or by a reduced glutathione-dependent release of non-heme iron (Atamna and Ginsburg, 1995). However, the interaction of quinine with hemozoin appears to be different from that of chloroquine since the clumping of hemozoin granules by chloroquine in *Plasmodium berghei*-infected erythrocytes is antagonised and reversed by quinine (Peters, 1987). Quinine was found to be three- to fourfold less potent than chloroquine as an inhibitor of β-hematin formation in a cell-free microassay but was of similar potency to mefloquine (Parapini et al., 2000). The selective toxicity of the quinoline antimalarials to malaria parasites may also be partly due to their basic nature, which facilitates their accumulation in the acidic food vacuole of the parasite (Krogstad and De, 1998). It is also important to note that the antimalarial mode of action of quinine may also involve other mechanisms such as blocking the transport of ions into the parasite (Aceti et al., 1990) and inhibition of the production of tumor necrosis factor (TNF), which may be involved in the pathogenesis of cerebral malaria (Gantner et al., 1995).

15.2.2 CRYPTOLEPINE

A decoction of the roots of the West African climbing shrub *Cryptolepis sanguinolenta* (Asclepi-adaceae/Periplocaceae) is used in traditional medicine to treat malaria. A number of indoloquinoline alkaloids are present in this species, of which cryptolepine **7** is the major one. The ethnobotany and phytochemistry of this species and the pharmacology of cryptolepine have been reviewed in Chapter 8.

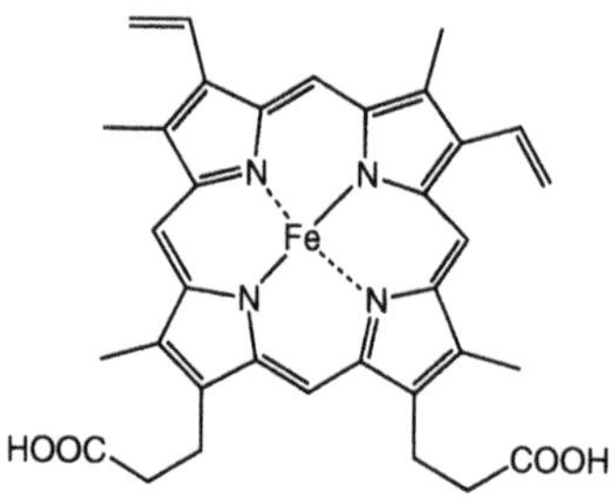

1 heme

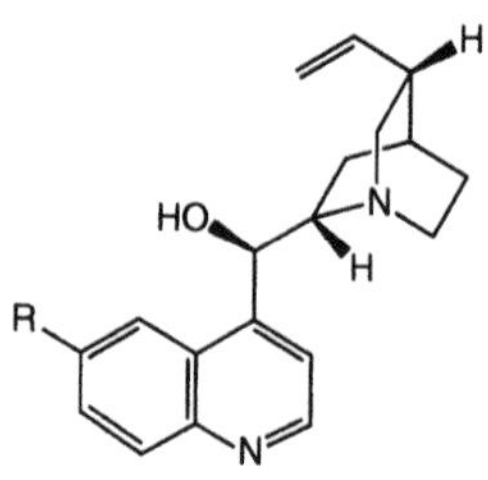

2 R = H;   quinine
3 R = CH₃O; cinchonidine

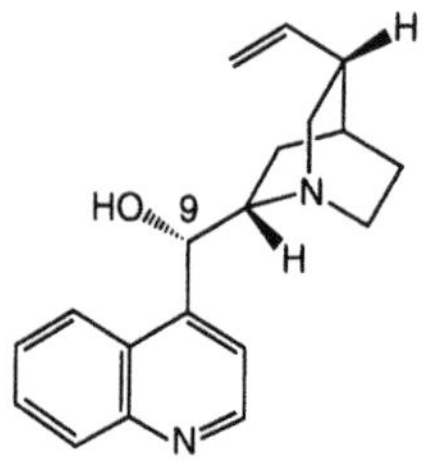

4 9-*epi*-quinine

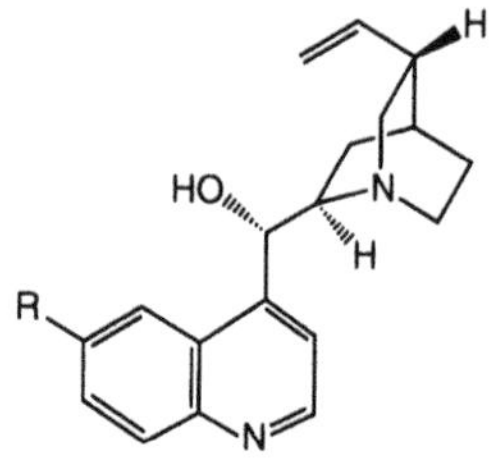

5 R = H;   quinidine
6 R = CH₃O; cinchonine

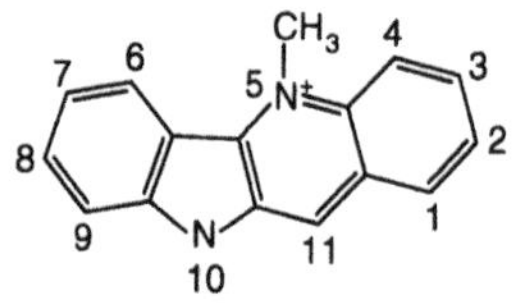

7 cryptolepine

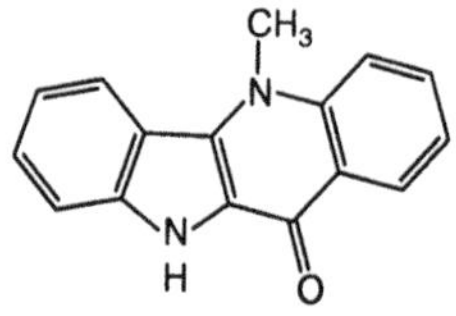

8 cryptolepine-11-one

Cryptolepine has potent *in vitro* antiplasmodial activity against both chloroquine-sensitive and chloroquine-resistant *Plasmodium falciparum* (Wright et al., 2001), but it is also moderately cytotoxic. Cytotoxicity is associated with the ability of the alkaloid to intercalate into DNA and inhibit topoisomerase II as well as DNA synthesis (Bonjean et al., 1998). Recently, it has been shown that cryptolepine intercalates into DNA in a unique way as it preferentially targets nonalternating GC sequences (Lisgarten et al., 2002). Further investigations of the cytotoxic mode of action of cryptolepine using cell cycle analysis and measurements of caspase activities have shown that it induces apoptosis in leukemia (HL60) cells, but degradation of DNA (a late effect in apoptosis) was not observed; mitochondrial function was also disrupted with the release of cytochrome *c* (Dassoneville et al., 2000). Cryptolepine was found to be only twofold less toxic to HL60 cells resistant to mitoxantrone (HL60/MX2 cells) than to mitoxantrone-sensitive cells. This suggests that inhibition of topoisomerase II may not be the main target in the cytotoxic action of this alkaloid, and it is suggested that a number of mechanisms are involved (Dassoneville et al., 2000). However, the antiplasmodial action of cryptolepine appears to be due (at least in part) to a quinine-like action since it inhibits β-hematin formation (see Figure 15.1), although it is possible that interactions with DNA could contribute to the antimalarial action of cryptolepine (Wright et al., 2001). Using fluorescence microscopy, Arzel et al. (2001) found that cryptolepine was localised in *P. falciparum* in structures that *could* correspond to the parasite nuclei, but further studies are needed to confirm this.

A number of cryptolepine analogues have been synthesised in an attempt to reduce the cytotoxicity while retaining or enhancing the antiplasmodial activity (Wright et al., 2001). Compounds lacking the 5-*N*-methyl group of cryptolepine (e.g., quindoline) were inactive against malaria parasites. Loss of this group greatly affects the charge distribution on the molecule (cryptolepine

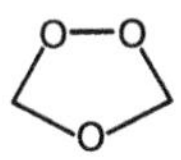

9 1,2,4-trioxane ring

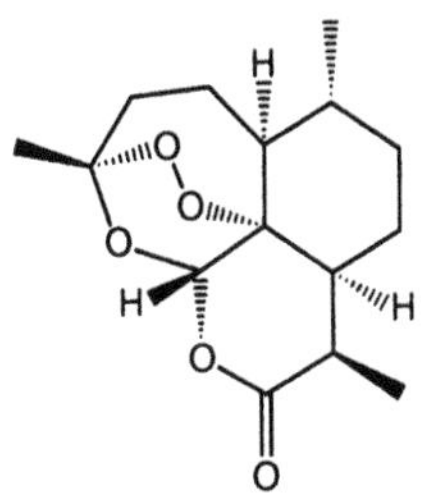

10 artemisinin

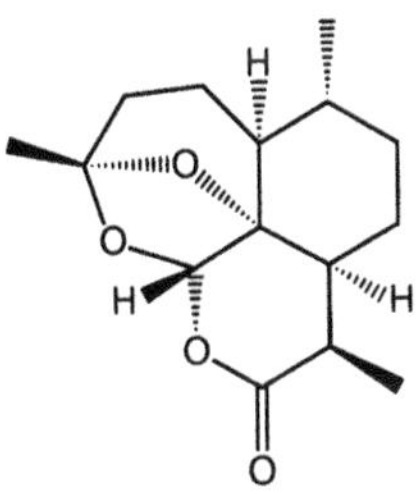

11 deoxyartemisinin

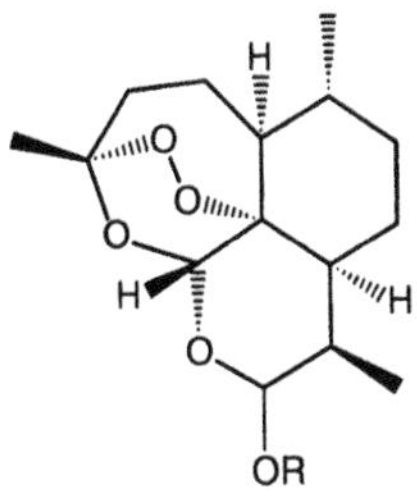

12 R = H;   dihydroartemisinin
13 R = COCH$_2$COONa; sodium artesunate
14 R = CH$_3$;   artemether

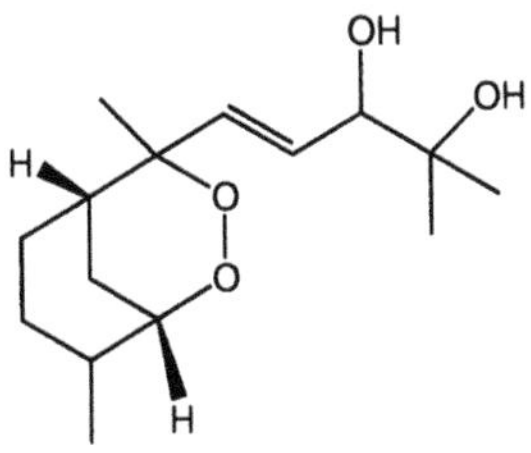

15 yingzhaosu A

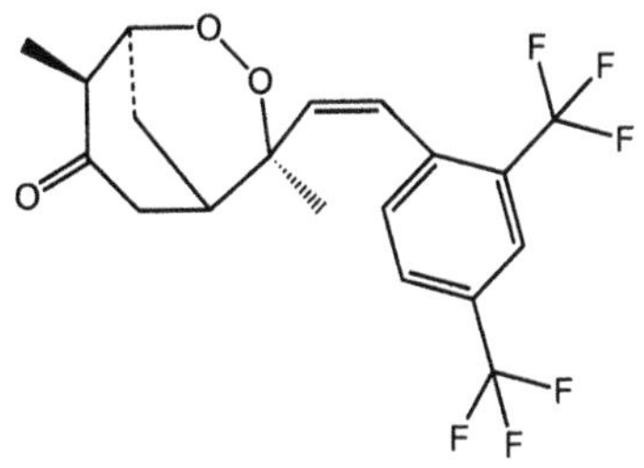

16 arteflene

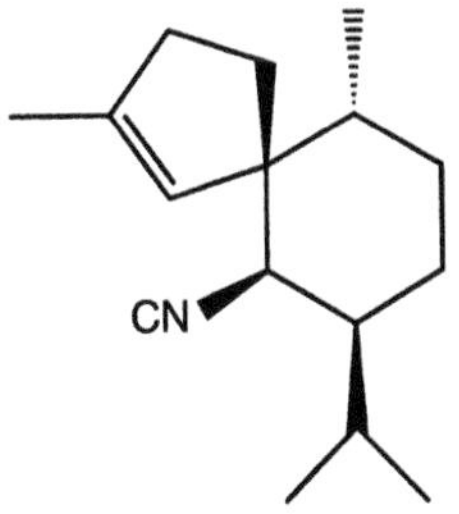

17 axisonitrile-3

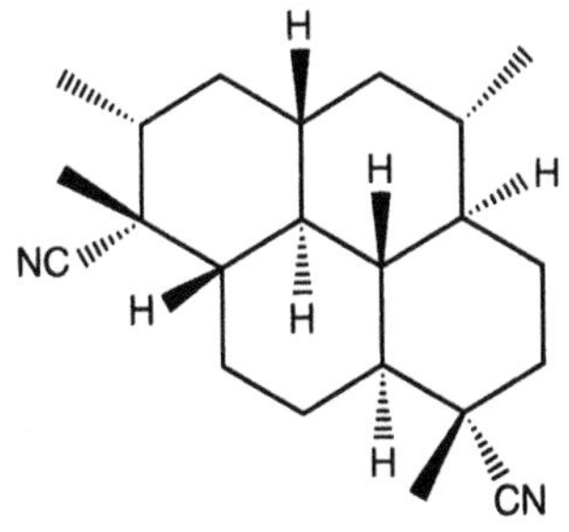

18 diisocyanoadociane

is an anhydronium base), and this may be important for its biological activity. The 7-nitro-, 7,9-dinitro-, and 7-*N*-acetyl- derivatives were of similar antiplasmodial activity to the parent while 9-nitrocryptolepine was 10-fold less potent. Improved antiplasmodial activities were seen with the 2-bromo- and 7-bromo- analogues, and this prompted the synthesis of 2,7-dibromocryptolepine, which was ninefold more potent than cryptolepine. Interestingly, although 11-chlorocryptolepine was twice as potent as cryptolepine, a number of dihalogenated derivatives that were 11-chloro-substituted, including 2-bromo, 11-chlorocryptolepine, were found to have weak antiplasmodial activities.

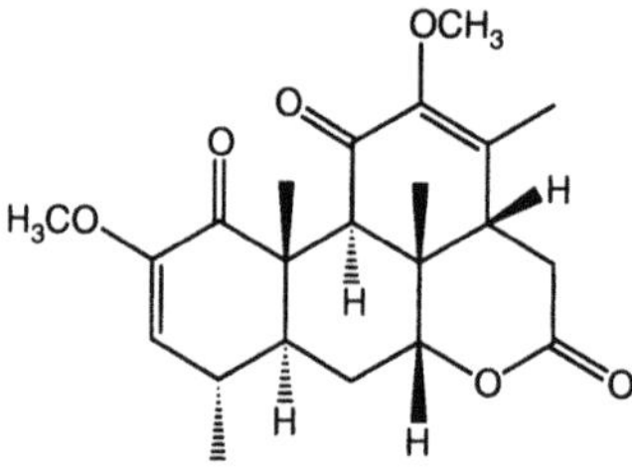

19 quassin

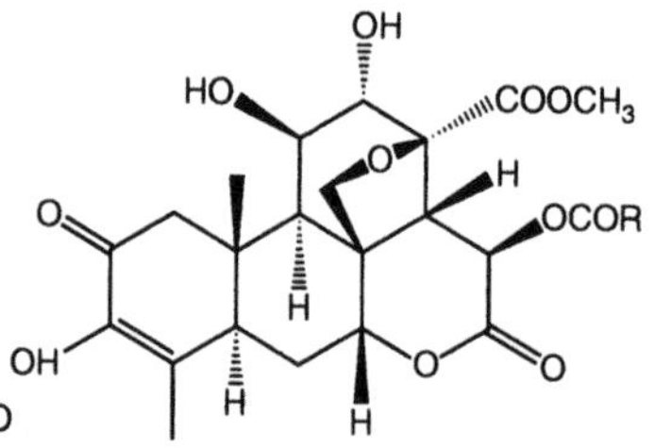

20 R = CH=C(CH$_3$)$_2$;   brusatol
21 R = CH=C(CH$_3$)CH(CH$_3$)$_2$; bruceantin

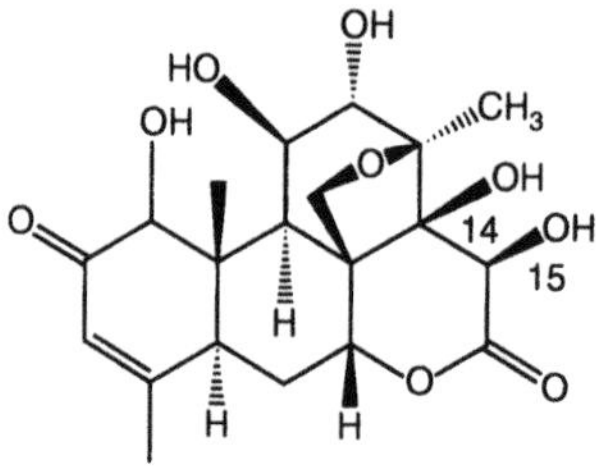

22 bruceine D

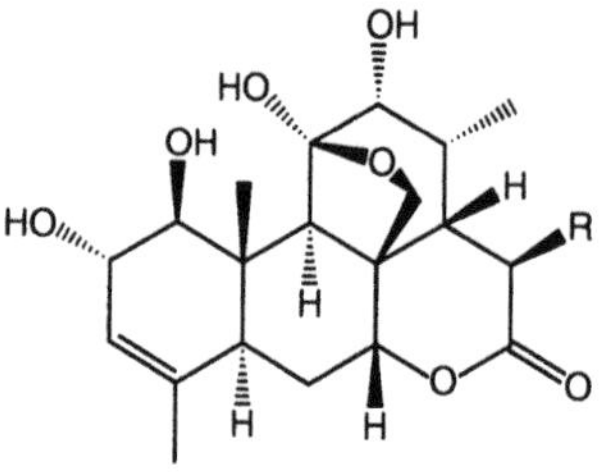

23 R = H;   chapparin
24 R = OH; glaucarubol

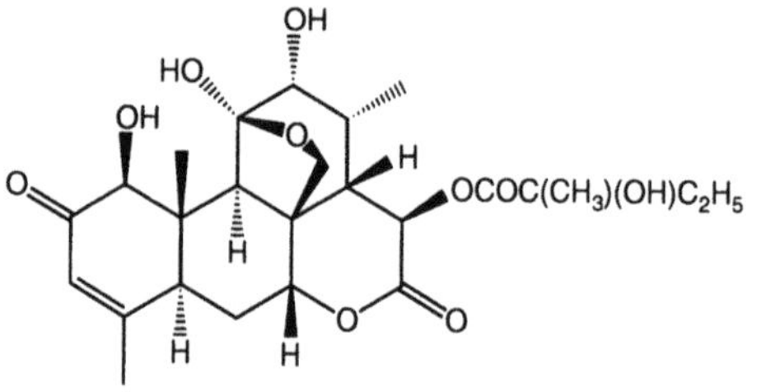

25 glaucarubinone

A comparison of the antiplasmodial activities of the active cryptolepine derivatives against chloroquine-sensitive and chloroquine-resistant strains of *P. falciparum* showed that with the possible exception of 7-nitrocryptolepine there was no cross-resistance with chloroquine. Like cryptolepine, several of the active compounds were also found to inhibit β-hematin formation, but in contrast to cryptolepine, they were less able to interact with DNA as judged by thermodenaturation studies (ΔTm values), although there was little reduction in their *in vitro* cytotoxities. When tested against *P. berghei* in mice by intraperitoneal injection using Peters' 4-day suppressive test, 2,7-dibromocryptolepine reduced parasitaemia by 89% compared to infected controls at a dose of 12.5 mg kg^{-1} day^{-1} with no apparent toxicity to the mice. The 2-bromo, 7-bromo-, 7-nitro-, 9-nitro-, and 7-N-acetyl- derivatives had less effect on parasitaemia but were not toxic to the mice at 20 mg kg^{-1} day^{-1}. In contrast, cryptolepine and the 11-chloro- analogue were toxic to the mice at 25 mg kg^{-1} day^{-1} (C.W. Wright et al., 2001). Little is known concerning the absorption and metabolism of cryptolepine. In a previous study in mice infected with *P. berghei*, oral cryptolepine at a dose of 50 mg kg^{-1} day^{-1} was not toxic to the mice and parasitaemia was suppressed by 80% compared to infected controls (Wright et al., 1996). The lack of toxicity suggests slow or incomplete absorption or metabolism to inactive metabolites. Recent work has shown that cryptolepine may be oxidised to cryptolepine-11-one 8 (tautomeric with 11-hydroxycryptolepine) by rabbit or guinea pig liver aldehyde oxidase (Stell and Wright, unpublished results), and this compound has also been found as an oxidation artifact during the isolation of cryptolepine from *C. sanguinolenta* (Fort et al., 1998). Cryptolepine-11-one has little *in vitro* antiplasmodial activity, and it is possible that

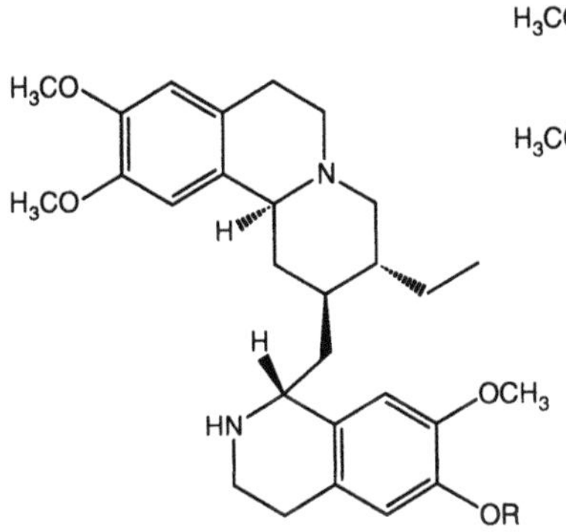

26 R = H;   cephaeline
27 R = CH₃;   emetine

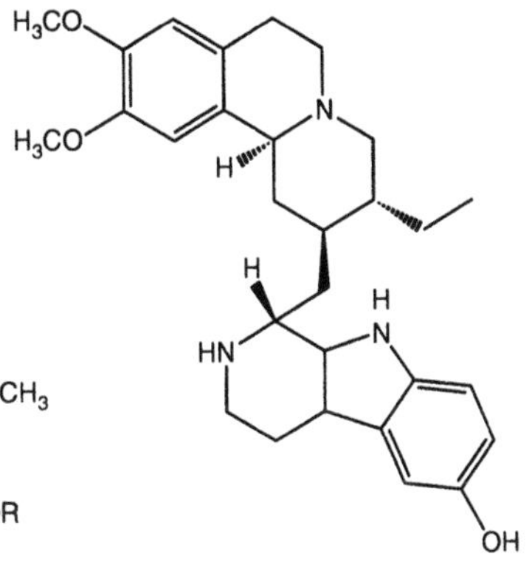

28 tubulosine

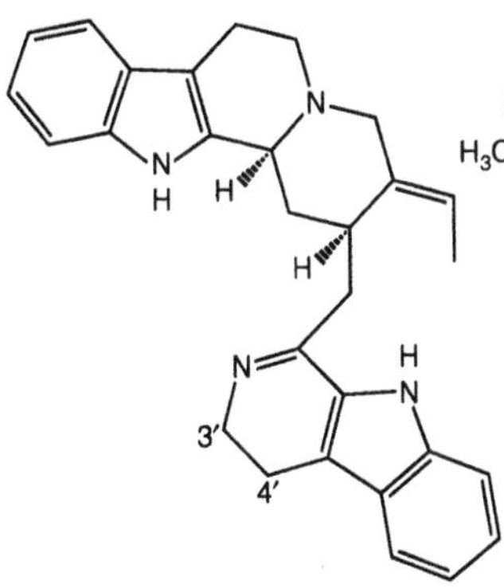

29 3′, 4′ - dihydrousambarensine

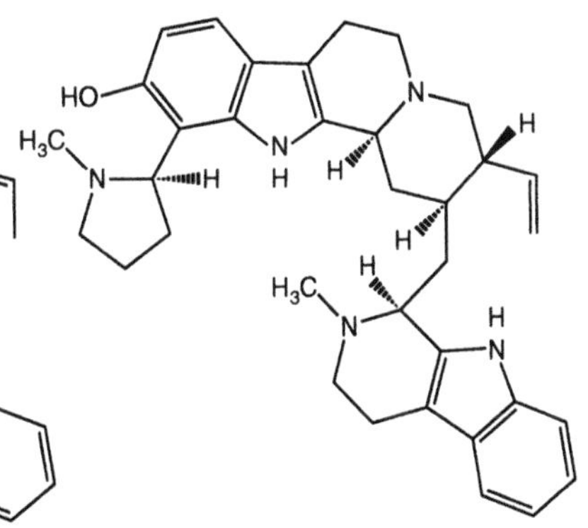

30 strychnopentamine

cryptolepine may be metabolically deactivated in man, but further studies are needed to confirm this. In contrast, 2,7-dibromocryptolepine does not appear to be metabolised by aldehyde oxidase (Stell and Wright, unpublished results). The distribution of cryptolepine in mice has been studied using autoradiography of mice treated with [³H]-cryptolepine (Noamesi et al., 1991). Radioactivity was distributed in most tissues except for the central nervous system with relatively high and long lasting levels in the liver and organs with rapid cell proliferation, but the most prolonged retention was found in the adrenal medulla and in the melanin-containing tissues of the eye. This raises the possibility that cryptolepine may be oculotoxic, and it will be important for studies to be carried out not only on the latter but also on cryptolepine analogues, which have potential use as antimalarial agents.

The studies reviewed above indicate that cryptolepine analogues have been prepared that have reduced *in vivo* toxicity compared to cryptolepine, and in particular, 2,7-dibromocryptolepine has promising *in vivo* antimalarial activity. Further work on cryptolepine analogues as leads to new antimalarial agents is warranted.

15.2.3 ARTEMISININ

For well over a millenium, the Chinese herb Qing Hao (*Artemisia annua*) has been used traditionally in China for the treatment of fevers, but it was only in the late 1960s that Chinese researchers discovered that ether extracts of *A. annua* cured malaria in mice (Klayman, 1985; see Chapter 3). The active principle, artemisinin **10**, isolated in 1972, is a remarkable compound by virtue of its unusual chemical structure and its potent and rapid antimalarial activity against multi-drug-resistant malaria parasites. Artemisinin is a sesquiterpene lactone containing an endoperoxide group that

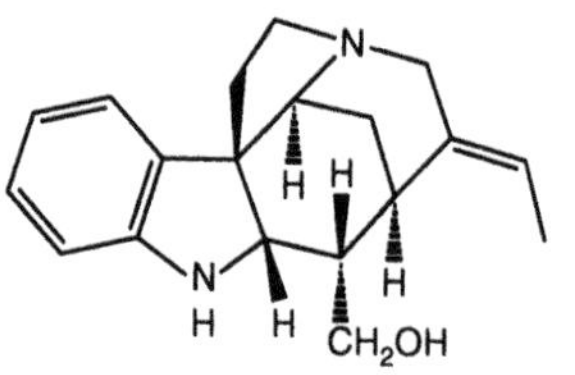

31 retuline

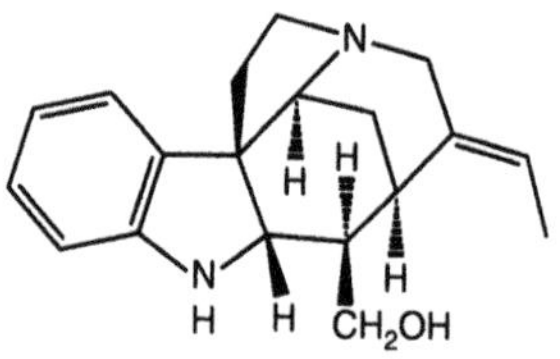

32 isoretuline

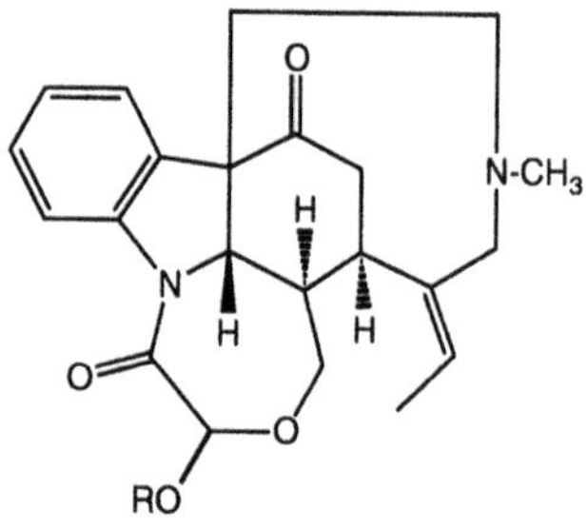

33 R = H;   holstiline
34 R = CH₃; holstiline

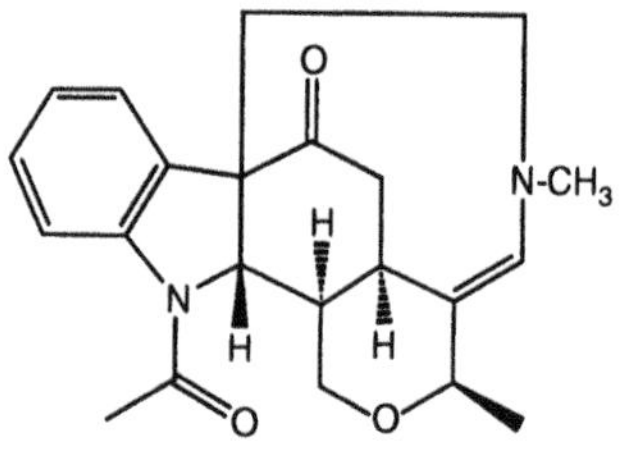

35 strychnobrasiline

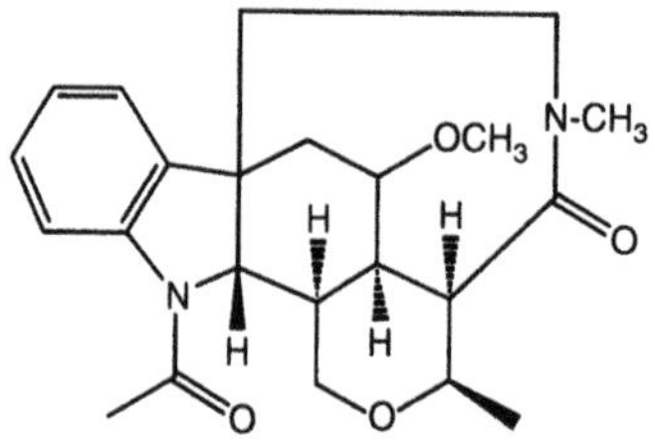

36 malagashanine

bridges an oxygen-containing seven-membered ring. Looking at the structure it can be seen that the two oxygens of the peroxide group and the oxygen of the seven-membered ring are present in a six-membered 1,2,4-trioxane ring **9**, so that artemisinin and its derivatives are often referred to as trioxane antimalarials. In contrast to the antimalarials of the quinine group, which, being alkaloids, contain nitrogen and are basic, artemisinin is a neutral, nonpolar compound with a totally different type of structure, and it is therefore not surprising that it has a unique mode of action against malaria parasites.

The effectiveness of the quinoline antimalarials is due in part to their basicity that allows them to accumulate in the acidic food vacuole of malaria parasites. Artemisinin, as a neutral compound, cannot take advantage of this (ion-trapping) mechanism, although artemisinin derivatives that include an amine function have been shown to have enhanced antimalarial activity (Hindley et al., 2002). However, recent evidence suggests that the partition of artemisinin into parasitised red blood cells may be facilitated by a carrier-mediated mechanism (Vyas et al., 2002). The process was shown to be rapid, saturable, temperature dependent, and irreversible.

There is now considerable experimental evidence that suggests that the antiplasmodial effects of artemisinin and also those of its derivatives are due to a two-step process. In the first, the interaction of artemisinin with heme, the residue remaining following the digestion of haemoglobin by the malaria parasite, gives rise to highly reactive free radicals. The second step involves the reaction of the artemisinin-derived free radicals with parasite molecules, thus disrupting normal metabolic processes and ending in parasite death (Figure 15.1). While there is general agreement with respect to this overall mechanism of action, the details are less clear. The experimental evidence

37 lapachol   38 lapinone

39 atovaquone

40 febrifugine

in support of the above has been reviewed (Wright and Warhurst, 2002) and a brief summary of the main points is presented below.

The importance of the endoperoxide moiety in the artemisinin molecule is clearly demonstrated by the lack of antiplasmodial activity of deoxyartemisinin 11, a constituent of *A. annua* that has a single oxygen in place of the peroxide group of artemisinin. Since it is well known that peroxides react with iron to give free radicals (Halliwell and Gutteridge, 1989), the interaction of artemisinin with iron and iron-containing molecules was investigated (Meshnick, 1994). Of particular signifi-cance is the observation that artemisinin does not react with the haemoglobin of uninfected red blood cells, but does react with heme (Hong et al., 1994). This explains the selectivity of artemisinin for parasite-infected red blood cells and why the drug kills the blood stages of malaria parasites while having no effect on parasites in the liver cells where haemoglobin digestion does not take place. The above is supported by cyclic voltammetry studies in which the reduction of the peroxide group of artemisinin in the presence of heme was examined (Chen et al., 1998). These studies also suggest that artemisinin preferentially reacts with heme containing Fe(II) iron rather than Fe(III). Molecular modeling studies have provided further support for the interaction between heme and the endoperoxide group of artemisinin (Shukla et al., 1995).

Chloroquine-resistant *P. berghei* does not have visible hemozoin and is also resistant to artem-isinin, suggesting that haemoglobin breakdown is essential for its action (Peters et al., 1986). The observation that the action of artemisinin is antagonised by iron chelators that chelate heme iron (Meshnick et al., 1993) and by chloroquine that binds to heme (Chou et al., 1980) also indicates a pivotal role for heme. Electron microscopy has revealed that when parasites are treated with artemisinin derivatives, damage to the parasite food vacuoles is seen first (Li et al., 1981). As artemisinin was also found to damage mitochondrial membranes, Zhao et al. (1986) investigated the possibility that inhibition of the mitochondrial heme-containing enzyme cytochrome oxidase may be important to the mode of action of artemisinin. However, the doses required to inhibit the enzyme *in vitro* and *in vivo* were found to be considerably higher than the doses required for the antimalarial effects of artemisinin.

Much work has been carried out with the aim of identifying the highly reactive products that are believed to result following the interaction of artemisinin derivatives with heme iron. As a result of the use of different iron compounds, solvents, and reaction conditions, apparently disparate results have been obtained by a number of research groups. However, Wu et al. (1998) have proposed a "united mechanistic framework for the Fe(II) induced cleavage of artemisinin and its derivatives" in an attempt to rationalise published results. Their proposed reaction scheme provides possible pathways by which the activation of artemisinin could lead to highly reactive molecules, including carbon-centered free radicals, high-valent Fe(IV)=O species, and alkylating agents that could contribute to the death of parasites, but it must be noted that many of the reactions reported involved the use of organic solvents and iron compounds other than heme, so that it is difficult to know how closely the scheme mimics what actually occurs in the malaria parasite.

Exactly how the reactive products resulting from the interaction between artemisinin and heme iron kill the parasite is unknown. In one study (Asawamahasakda et al., 1994b) red blood cells infected with *P. falciparum* were incubated with $[10-^3H]$-dihydroartemisinin; a number of nonabundant parasite proteins were selectively alkylated, but no alkylation of proteins was observed in uninfected red blood cells. One of the proteins has been identified as the translationally controlled tumor protein homologue (TCTP) of *P. falciparum* (Bhisutthibhan et al., 1998). Binding was found to be dependent on the presence of heme (FeIII), and a single cysteine residue in TCTP has an important role in binding the drug, but the function of this protein in *P. falciparum* is not known. These experiments show that parasite proteins may be alkylated by artemisinin, but this does not prove that this is a mechanism by which parasite death occurs. Damage to DNA has also been shown to occur when DNA is incubated in the presence of artemisinin and ferrous sulfate (Wu et al., 1996).

15.2.3.1 Interactions of Artemisinin with Heme and Hemozoin

It has been suggested that artemisinin may bind to heme or hemozoin forming a toxic complex and inhibit hemozoin formation. Artemisinin–hemozoin adducts were formed when radiolabeled artemisinin was incubated with infected erythrocytes or isolated hemozoin (Hong et al., 1994), but the adducts were not active against malaria parasites *in vitro* (Meshnick et al., 1991). Berman and Adams (1997) have shown that the ability of heme (FeIII) to oxidise membrane lipids is enhanced by artemisinin, and they have proposed a model in which the drug binds irreversibly to heme, thus preventing its conversion to hemozoin and promoting heme-catalysed oxidation of the vacuolar membrane by molecular oxygen, leading to vacuole rupture and parasite autodigestion. Recently, Robert et al. (2002) have shown that in the presence of artemisinin and glutathione, iron (III)–protoporphyrin IX (heme containing ferric iron) is alkylated by an artemisinin-derived C4-centered radical, but how alkylation of heme might result in parasite death is not clear. Alkylation of heme could possibly prevent hemozoin biosynthesis or facilitate the release of iron, but attempts to provide evidence in support of these hypotheses have been unsuccessful (Meshnick, 1998). As noted above, artemisinin appears to react preferentially with heme containing Fe(II) iron rather than with heme (FeIII), and glutathione may have a role in maintaining heme iron as Fe(II). Another interesting observation is that glutathione has been shown to be alkylated by artemisinin via a carbon-centered radical in the presence of Fe(II/III) (Wang and Wu, 2000). On the other hand, the antimalarial action of artemisinin is antagonised by glutathione and experimental evidence suggests that artemisinin may react with reduced glutathione in the presence of glutathione transferase (Mukanganyama et al., 2001), presumably resulting in the inactivation of artemisinin.

Conflicting results were obtained when the effect of artemisinin on hemozoin formation was examined; Orjih (1996) reported that artemisinin inhibited hemozoin formation in parasitised erythrocytes, but Asawamahasakda et al. (1994a) reported that artemisinin did not inhibit hemozoin formation. More recently, Tripathi et al. (2001) reported that artemisinin inhibited hemozoin formation in plasma from mice infected with *Plasmodium yoelii*.

In addition to their remarkable effectiveness in the treatment of malaria, there is evidence that artemisinin derivatives are able to prevent malaria transmission. Although these agents do not kill mature gametocytes, the early stages (I to III) are inhibited and the asexual forms from which they develop are killed (Kumar and Zheng, 1990). Other studies (Chen et al., 1994; Tripathi et al., 1996) have confirmed these findings, and there is also evidence from the field showing that the incidence of *P. falciparum* malaria declined in areas of Southeast Asia when artemisinin derivatives were introduced (Price et al., 1996).

15.2.3.2 Toxicity of Artemisinin

Although artemisinin and its analogues have been found to have very low toxicity in clinical use, two areas have been highlighted as giving cause for concern. First, in a study in rodents artemisinin was found to cause fetal resorption, although no mutagenicity or teratogenicity was observed (Qinghaosu Antimalarial Coordinating Research Group, 1979). The few reports to date of the use of artemisinin in pregnancy have not shown evidence of abnormalities in the children of mothers who received artemisinin or its analogues in the second or third trimester (Wilairatana and Looareesuwan, 2002). In a prospective study of 461 pregnant women (including 44 in the first trimester) treated for falciparum malaria with artesunate **13**, or artemether **14**, no evidence of adverse effects was seen (McGready et al., 2001). These drugs are recommended for the treatment of severe malaria in the second and third trimesters of pregnancy, but in the first trimester the potential benefits must be weighed against the fact that there is limited information on their safety.

Second, toxicity tests in animals revealed that artemisinin derivatives caused a delayed, dose-dependent neurotoxicity with symptoms of gait disturbance, loss of pain response and spinal, brainstem, and eye reflexes, and electrocardiogram abnormalities. Lesions were found especially in the pons and medulla in animals receiving doses comparable to those used clinically in man (Brewer et al., 1994). The delay in toxicity is suggested to be due to the accumulation of the major metabolite, dihydroartemisinin **12**, that has a longer elimination half-life than the parent compounds and accumulates with repetitive dosing. Studies have shown that the auditory system may be particularly sensitive to low doses of artemisinin analogues and that, at least in rats, auditory dysfunction could be an early sign of neurotoxicity. Although there has been one report of a patient with ataxia and slurred speech following treatment with an artemisinin derivative (artesunate) (Miller and Panosian, 1997), there is no definite proof of a causal relationship in man; the above symptoms may have been due to other causes, including malaria itself (Wilairatana and Looareesuwan, 2002). No evidence of significant neurotoxicity (including auditory effects) was found in a case control study of patients treated with two or more courses of artemether or artesunate within a period of 3 years (van Vugt et al., 2000).

15.2.3.3 Mechanisms of Neurotoxicity

Artemisinin and its analogues are cytotoxic to neuronal and glial cells in culture (Fishwick et al., 1995). Since the brain contains a high level of iron, this raises the possibility that neurotoxicity may involve mechanisms similar to those responsible for the antiplasmodial activity of artemisinin discussed above. In support of this, it has been shown that the toxicity of neuronal cells to artemisinin derivatives is markedly enhanced by the presence of heme (FeIII) (Smith et al., 1997), and as hemozoin is present in the brains of patients with cerebral malaria, this mechanism may possibly have clinical relevance. When mouse neuroblastoma cells were incubated with labeled artemisinin, several neuronal proteins were alkylated but less efficiently than *P. falciparum* proteins incubated under identical conditions (Kamchonwongpaisan et al., 1997). Further studies by Fishwick et al.

(1998a) provided evidence in support of protein alkylation, and another study (Fishwick et al., 1998b) suggested that mitochondria and other membranous organelles may be disrupted by artemisinin derivatives. The effects of artemisinin on neuronal brainstem cell cultures have been compared with those on cortical neuronal and astrocyte cell cultures (Schmuck et al., 2002). As expected from the pattern of neurotoxicity seen in animal studies, the brainstem cell cultures were more sensitive to artemisinin, but effects were also seen in the other cell types. Neurodegeneration was associated with effects on the cytoskeleton, mitochondrial function, and oxidative stress.

15.2.3.4 Pharmacokinetics and Metabolism

An understanding of the absorption, distribution, and metabolism of artemisinin is essential in order to be able to design treatment regimens that are effective and avoid the important problems of recrudescence (see below) and the possible development of malaria parasites resistant to artemisinin. This applies to the use of artemisinin itself as well as to crude preparations made from *A. annua* herb. Pharmacokinetic studies on artemisinin and its derivatives are difficult as the artemisinin molecule does not have a chromophore and therefore cannot be detected by standard methods utilising UV absorption (Woerdenbag and Pras, 2002). Another problem is that artemisinin may be converted to active but nondetectable derivatives in whole blood so that blood levels of the drug may not be related to therapeutic effectiveness (Muhia et al., 1994). Artemisinin is rapidly metabolised in the body to dihydroartemisinin (artemorin) **12**, a metabolite that has greater *in vitro* antiplasmodial activity than the parent compound (Lee and Hufford, 1990). The latter is then converted to the glucuronides of both 12α- and 12β-dihydroartemisinin; the 12α-isomer is inactive against plasmodia and the 12β-isomer is about 20-fold less active than dihydroartemisinin (Ramu and Baker, 1995).

Although artemisinin is difficult to formulate on account of its poor solubility in water and oil, the pharmacokinetics of oral, intramuscular (IM), and rectal administration have been studied (Titulaer et al., 1990). The drug was rapidly and well absorbed orally (peak levels of 260 ng/ml after 1 hour, elimination half-time of 1.9 hours) with a bioavailability that was 32% of that found with an oil suspension administered by IM injection (peak levels of 209 ng/ml after 3.4 hours, elimination half-time of 7.4 hours), and animal studies have shown that extensive first-pass metabolism occurs; IM absorption from an aqueous suspension is poor. Rectal absorption of artemisinin was much slower with peak levels (170 ng/ml) achieved after 11.3 hours, while the elimination half-time was 4.1 hours (Shen, 1989, cited in White, 1994).

The main problem experienced with the clinical use of artemisinin (and other artemisinin derivatives) is that of recrudescence where parasites are apparently cleared from the blood following drug treatment but reemerge some time later (usually about 2 weeks) after treatment has been completed. This is not due to drug resistance since the parasites remain drug sensitive, but to a failure of treatment to completely eradicate parasites. The reasons for this phenomenon are not well understood, but as it is more commonly seen with oral therapy, limited bioavailability has been suggested as a possible cause, but the rapid elimination of the drug may be important.

The artemisinin ester derivative artesunate is rapidly metabolised to dihydroartemisinin and has a short (45 minutes) elimination half-life (Yang et al., 1986), and recrudescence is a major problem. However, recrudescence appears to be much less of a problem with derivatives such as artemether and arteether that are metabolised more slowly and have longer elimination half-lives (especially after IM injection). Recrudescence may occur in some patients who have received 7 days treatment with an artemisinin derivative that, according to theoretical models, should be sufficient to eradicate all parasites (White, 1998). A more recent study has confirmed the frequent occurrence of recrudescence with artemisinin and also showed that extending treatment from 5 to 7 days does not reduce the rate of recrudescence (Giao et al., 2001). Possible explanations include the suggestion that the release of liver stages (not killed by artemisinin) is not continuous and may be delayed until drug treatment has ceased (Murphy et al., 1990). Although artemisinin derivatives are more rapidly acting and have a broader window of activity than other antimalarials against the various blood stages, some parasites may not be killed because they have not entered the growth phase or because they are damaged but not killed, thus delaying their development perhaps because there is insufficient heme to induce destruction by artemisinin (White, 1998). Another possible factor is that the metabolism of artemisinin may be autoinduced, leading to declining levels with subsequent doses (Alin et al., 1996; Gupta et al., 2001).

15.2.3.5 Drug Interactions

Artemisinin has been shown to potentiate the effect of mefloquine and tetracyline against *P. falciparum in vitro* as well as in mice infected with *P. berghei*, but antagonism was seen with antifolates, including sulfadoxine (Chawira et al., 1987). Chloroquine antagonises the action of artemisinin *in vitro*, but this is not seen *in vivo* in mice infected with *P. berghei* (Chawira et al., 1986). Doxycycline has been shown to potentiate the action of artemisinin against fresh isolates of *P. falciparum* from Thailand (Sponer et al., 2002). A number of methoxylated flavones have been shown to potentiate the action of artemisinin against *P. falciparum in vitro*, and thus it is possible that flavonoids present in *A. annua* may be important when the herb is used for malaria treatment (Elford et al., 1987). Recently, Bilia et al. (2002) reported that some flavonols isolated from *A. annua* enhanced the formation of a complex of artemisinin and hemin when incubated in dimethylsulfoxide. In a clinical trial in China, malaria patients were treated with capsules containing the dried ethanolic extract of *A. annua*, but although parasites were cleared initially, recrudescence rates were high, suggesting that the flavonoid constituents did not increase the ability of artemisinin to cure malaria infections (Yao-De et al., 1992).

Artemisinin derivatives such as arteether that are metabolised by liver (cytochrome P450) enzymes have the potential to interact with other drugs such as mefloquine and halofantrine by inhibiting their metabolism and vice versa (Grace et al., 1998). Grapefruit juice that inhibits the cytochrome P450 isozyme CYP3A4 has been shown to increase the oral bioavailability of artemether (van Agtmael et al., 1999; reviewed by Wright, 2002).

15.2.3.6 Malaria Parasite Resistance to Artemisinin

Malaria parasites resistant to chloroquine and other antimalarials are now widespread, and in parts of Southeast Asia the artemisinin derivatives are the only reliable antimalarial agents. In view of this, the development of parasites resistant to the latter would be a very serious matter. In the laboratory, artemisinin-resistant strains of *P. falciparum* have been produced by exposing parasites to mutagenic agents (Inselburg, 1985), and more significantly, *P. yoelii* resistant to artemisinin was selected in mice as a result of drug pressure (Chawira et al., 1986). Artemisinin-resistant *P. yoelii* (strain ART) were found to accumulate 43% less drug *in vitro* than the sensitive strain NS, and artemisinin appeared to react with the same parasite proteins in both strains. The artemisinin target protein, translationally controlled tumor protein, did not differ between the strains. No DNA sequence difference was found, but the resistant strain was found to express 2.5-fold more protein than the sensitive strain, and it appears that artemisinin resistance in *P. yoelii* is multifactorial (Walker et al., 2000). In 1992, Oduola et al. reported the isolation of four strains of *P. falciparum* from malaria patients in Africa that exhibited a transient 7- to 14-fold decrease in sensitivity to artemisinin *in vitro* during their adaptation to continuous culture, and in 1994, a highly resistant strain of *P. falciparum* was isolated from a traveller to Africa (Gay et al., 1994). However, in other studies (e.g., Alin, 1997), no parasites isolated from malaria patients have been found to be resistant to artemisinin, nor has cross-resistance with other antimalarial drugs been observed. More recently it has been shown that mutations in the pfmdr1 gene of *P. falciparum* that encodes for the P-glycoprotein homologue 1 protein can confer resistance to mefloquine, quinine, and halofantrine (Reed et al., 2000; Durasingh et al., 2000). While chloroquine-resistant parasites are generally more sensitive to artemisinin than are chloroquine-sensitive strains, resistance to mefloquine and halo-fantrine is associated with a decreased response to artemisinin. When short courses of artemisinin and its derivatives are used, it is recommended that treatment is followed by a second antimalarial agent in order to prevent recrudescence (Wilairatana and Looareesuwan, 2002), and this may also help to prevent or at least delay the development of malaria parasites resistant to artemisinin. Studies in mice have shown that a combination of artemisinin with mefloquine can impede to a significant degree, but by no means completely, the selection of resistance to compounds in both *P. berghei* and *P. yoelii* ssp. NS (Peters and Robinson, 2000). It is possible that resistance may be more likely to develop when *A. annua* herb or extracts are used to treat malaria, especially if the preparations used contain inadequate amounts of artemisinin. Further studies are needed to assess the effectiveness of such preparations and to assess the risk of artemisinin-resistant parasites emerging.

15.2.4 MISCELLANEOUS NATURAL PRODUCTS THAT INTERACT WITH HEME

Like *A. annua*, the Chinese herb Ying Zhao, *Artabotrys unciatus* (Annonaceae), has been used traditionally for the treatment of malaria, and this species contains the terpenoid peroxides yingzhaosu A **15** and C (Zhang et al., 1988). Arteflene **16**, a synthetic antimalarial derived from yingzhaosu A, was evaluated in Gabonese children for the treatment of falciparum malaria (Radloff et al., 1996). Treatment with a single dose of 25 mg/kg was not effective in eradicating parasites, and this was shown to be due to recrudescence rather than reinfection. More worrying, an indication of both R2 and R3 resistance development was seen in 8 of the 20 patients treated. Because of these problems, the development of arteflene as an antimalarial has been discontinued, but there is interest in other analogues (Posner, 1998). Although rare in nature, a number of other peroxide-containing natural products have been isolated, particularly from marine sources, and these may be worthy of investigation as antimalarial agents (reviewed by Newman et al., 2000).

A number of terpene isonitrile compounds isolated from marine sponges have been shown to have *in vitro* antiplasmodial activity, and these appear to act like the quinoline antimalarials by inhibiting the detoxification of heme (A.D. Wright et al., 2001). In addition, two of the most potent compounds, axisonitrile-3 **17** and diisocyanoadociane **18**, were shown to prevent peroxidative and glutathione-mediated breakdown of heme. Molecular modeling techniques demonstrated a correlation between antiplasmodial activity and binding to hemin.

15.3 COMPOUNDS WITH OTHER MODES OF ACTION

15.3.1 QUASSINOIDS

In 1947, Spencer et al. screened extracts from some 600 species of higher plants for *in vivo* antimalarial activities. This work was done using birds infected with malaria parasites, including canaries infected with *Plasmodium gallinaceum*, and it is interesting to note that studies on the malaria parasite genome have shown that the genome of the latter species shows a close relationship with *P. falciparum*, while the mouse malaria *P. berghei*, which is usually used nowadays, is more distantly related (Su and Wellems, 1998). Of the 126 families represented in the above screen, two in particular, the Simaroubaceae and the Amaryllidaceae, contained species with antimalarial activity.

Species of the Simaroubaceae have been used in traditional medicines for the treatment of protozoal diseases (malaria, amoebic dysentery) in Asia and Africa, as well as in South and Central America. Their antimalarial activity is mainly due to the presence of quassinoids, a group of oxygenated terpenoids related to quassin **19**, the first member of the group to be characterised, although quassin itself has no antimalarial activity. Alkaloids are also present in the Simaroubaceae, but they are less active against malaria parasites than the quassinoids. Species investigated included *Brucea javanica*, known as Yadanzi in China, *Eurycoma longifolia* (Malaysia), *Ailanthus altissima*

(India), *Simarouba amara* (Central America), and *Picramnia antidesma* (Central America), and this led to the isolation of 40 individual quassinoids. Ten of these were found to be 10-fold more active against multi-drug-resistant *P. falciparum* (strain K1) than chloroquine diphosphate (Phillipson et al., 1992).

Structure–activity relationship studies revealed that in the most active quassinoids (e.g., brusatol **20**, glaucarubinone **25**), a methylene–oxygen bridge is found between C-8 and C-11 or C-13. The oxidation state and substitution of ring A influences activity, the most potent compounds being hydroxylated at C-1 or C-3 with an adjacent carbonyl conjugated to a double bond. In many of the active compounds an ester is present at C-15 or occasionally at C-6, and loss of the esterifying acid leads to a marked reduction in antiplasmodial activity; however, an ester function is not necessarily a prerequisite for antimalarial activity as several quassinoids with 14,15-diol functions (e.g., bruceine D **22**, from *B. javanica*), but without esters, are highly active.

A few quassinoid glycosides have been assessed for *in vitro* antiplasmodial activities, but they were much less potent compared to their corresponding aglycones. Selected quassinoids were assessed for *in vivo* antimalarial activities against *P. berghei* in mice using Peters' 4-day suppressive test, and although several compounds were found to have potent activities (e.g., the effective dose 90% (ED_{90}) of glaucarubinone from *Simarouba glauca* was 3.43 mg kg^{-1} day^{-1}), they were also toxic to the mice at slightly higher doses (glaucarubinone was toxic to mice at 9 mg kg^{-1} day^{-1}), indicating that selectivity was poor. Quassinoids have potent cytotoxic activities, and a comparison of antiprotozoal activities of a series of quassinoids with their cytotoxic activities against KB cells (human mouth cancer) revealed that while all of the compounds were more toxic to malaria parasites than to the cells, the cytotoxicity/antiplasmodial activity ratios were low, the best being 285 for glaucarubinone (Wright et al., 1993). Glaucarubinone also showed the best selective activity against *Entamoeba histolytica* and against intracellular *Toxoplasma gondii*, but the toxicity/activity ratios were only 7 and 6, respectively. All of the compounds tested were more toxic to KB cells than to *Giardia intestinalis*.

A comparison of the antiplasmodial and cytotoxicity data for the quassinoids shows that cytotoxicity does not necessarily parallel antiplasmodial activity, thus opening up the possibility that the structural modification of the quassinoids may result in enhanced selectivity against malaria parasites. In an attempt to achieve this, C-3 ester derivatives of brusatol **20** (isolated from *B. javanica*) were prepared. The C-3 *t*-butoxycarbonyldecanoate ester of brusatol (chosen since lipidic amino acid conjugates have been shown to modify drug action and may act as pro-drugs) and the C-3 acetyl derivatives were both found to have similar antiplasmodial activities to that of the parent, but cytotoxicity was reduced by two- and threefold, respectively. However, mild alcoholic ammoniolysis of brusatol resulted in the C-13 amide that was markedly less active against malaria parasites and cells, suggesting that this part of the molecule plays a key role in its biological activity (Allen et al., 1993).

Quassinoids have been shown to inhibit protein synthesis in mammalian cells (Liao et al., 1976) and in malaria parasites (Kirby et al., 1989). Quassinoids were shown to have little effect on glycolysis in malaria parasites (Kirby et al., 1992), while a delayed effect on nucleic acid synthesis consistent with inhibition secondary to protein synthesis was observed (Kirby et al., 1989). Two quassinoids, chapparin **23** (a constituent of *Castella nicholsoni*) and glaucarubol **24** (from *Simarouba glauca*) that have weak antiplasmodial activities were found to have little effect on protein synthesis. Like anisomycin, bruceantin **21**, a quassinoid from *B. javanica* that has potent antiplasmodial and cytotoxic activities, appears to act by binding at a site within the eukaryotic 60S ribosomal subunit and causes inhibition of ribosomal peptidyl transferase activity (Beran et al., 1980). As this group of compounds has a similar mode of action against mammalian cells and against malaria parasites, and since protein synthesis is very similar in both, it is likely to be difficult, if not impossible, to develop quassinoid analogues that are highly selective against malaria parasites.

15.3.1.1 Traditional Medicines Containing Quassinoids

When used in traditional medicines, herbs are usually prepared in the form of aqueous extracts or teas, but most of the quassinoids that have potent antimalarial activities are relatively nonpolar compounds poorly soluble in water, while the few glycosides tested to date have been found to be inactive against *P. berghei* in mice. Chromatographic analysis of an aqueous tea prepared from *B. javanica* showed the presence of only small quantities of nonpolar quassinoids but much larger amounts of the polar quassinoid glycosides. On acid hydrolysis, lipophilic quassinoids were released and these were extracted and found to have *in vitro* antiplasmodial activity 39 times greater than the freeze-dried aqueous tea. The freeze-dried hydrolysate was found to be 4.5 times more potent than the aqueous tea against *P. berghei* in mice, and it is possible that this increase in activity is due to *in vivo* hydrolysis of glycosides to more potent aglycones.

Another factor that may be important is the presence in herbal teas of other compounds that may act additively or synergistically with quassinoids. At least some of the Simaroubaceae species used traditionally to treat malaria contain β-carboline alkaloids that have some activity against *P. falciparum*; although this is less than that of chloroquine, it is possible that some synergism could occur. In contrast, *in vitro* studies have shown that the action of chloroquine may be antagonised by the aqueous tea prepared from *B. javanica* (Allen et al., 1994), which, if this occurs *in vivo*, could have implications for patients taking both of the above for malaria treatment.

15.3.2 CEPHAELIS AND STRYCHNOS ALKALOIDS

The genus *Cephaelis*, like *Cinchona*, belongs to the family Rubiaceae and the alkaloids of the two genera are related biosynthetically. Emetine **27**, from *Cephaelis ipecacuanha*, has potent *in vitro* activity against *P. falciparum* and has been claimed to be effective clinically in the treatment of malaria (James, 1985), but the cardiotoxicity and poor therapeutic/toxic ratio of this compound is well established. Cephaeline **26**, and tubulosine **28**, isolated from *Pogonopus tubulosus* (Rubiaceae) are also highly active *in vitro*, and tubulosine was active against *Plasmodium vinckei petteri* and *P. berghei* in mice (Sauvain et al., 1996). The genus *Strychnos* (Loganiaceae) contains many alkaloids, some of which are closely related to emetine and can be considered to be indole analogues of emetine. Several of 20 *Strychnos* alkaloids tested were found to have potent antiplasmodial activities ($IC_{50} < 1$ μ*M*), the most active being 3',4'-dihydrousambarensine **29**, an alkaloid isolated from *Strychnos usambarensis*, a species used by the Banyambo tribe of the Rwanda–Tanzania border for the preparation of an arrow poison (Wright et al., 1994). The former compound was markedly less toxic to KB cells than to malaria parasites (cytotoxic/antiplasmodial ratio = 1474), and when tested in mice infected with *P. berghei*, no toxicity was seen at 30 mg kg^{-1} day^{-1} given orally and by subcutaneous injection, but there was no effect on parasitaemia. Another alkaloid from *S. usambarensis*, strychnopentamine **30**, was similarly active *in vitro* but inactive against *P. berghei*; the reasons for the lack of the *in vivo* activities of the above alkaloids are unknown. Molecular modeling studies have shown that emetine and 3',4'-dihydrousambarensine are able to adopt a common conformation, and this suggests that, like emetine, the latter may act as protein synthesis inhibitors (Quetin-Le Clerq et al., 1991). Consistent with this hypothesis is the observation that all of the *Strychnos* alkaloids with potent activities against *P. falciparum in vitro* are dimeric indole alkaloids, while a number of monomeric indole alkaloids, including retuline **31** and isoretuline **32** (constituents of *Strychnos variabilis*), as well as holstiine **33** and holstiline **34** (constituents of *Strychnos henningsii*), were considerably less active (Wright et al., 1994). The major alkaloids of *Strychnos myrtoides*, strychnobrasiline **35** and malagashanine **36**, are also monomeric indole alkaloids that are structurally related to holstiline and, in common with the latter, have no significant activity against *P. falciparum in vitro*. However, the two major *S. myrtoides* alkaloids as well as the crude alkaloid extract of this species potentiate the antimalarial action of chloroquine both *in vitro* and *in vivo*, findings that support the use of *S. myrtoides* as an adjunct to chloroquine in the treatment of malaria (Rasoanaivo et al., 1994). A full account of *S. myrtoides* has been given in Chapter 9.

15.3.3 LAPACHOL AND DERIVATIVES

Lapachol **37** is a naphthoquinone found in the heartwood of South American species of the family Bignoniaceae that have been used traditionally in Brazil for the treatment of malaria and fevers (Carvalho et al., 1988). While lapachol itself was only weakly active *in vitro* against *P. falciparum* (Hudson et al., 1985), the related derivative lapinone **38** was shown to cure *Plasmodium vivax* malaria in man (Fawaz and Haddad, 1951). The improved activity of lapinone compared to some other naphthoquinones has been attributed to higher potency and increased resistance to metabolic degradation, as it has been shown that terminal oxidation of the naphthoquinone hydrocarbon side chain leads to loss of activity (Thompson and Werbel, 1972). Unfortunately, although lapinone was highly potent when given by parenteral administration, it had poor oral activity. The synthesis of naphthoquinone derivatives as potential antiamalarial agents began in the 1940s and more recent work in the 1980s led to the development of atovaquone **39**, now used clinically in combination with proguanil to prevent the high rate of recrudescence seen when atovaquone is used alone (reviewed by Vaidya, 1998).

Initial studies showed that hydroxynaphthoquinones inhibited mitochondial respiration, and more recent studies using isolated mitochondria revealed that these compounds inhibit cytochrome *c*, thus disrupting electron transport at the bc_1 complex (reviewed in Vaidya, 1998). Malaria parasite cytochrome *c* was found to be some 500-fold more sensitive to atovaquone than the enzyme from rat liver mitochondria, providing an explanation for the selectivity of hydroxynaphthoquinones against malaria parasites. The mitochondrion plays a central role in the malaria parasite, and one of the consequences of electron transport inhibition is the collapse of the mitochondrial membrane potential. Interestingly, in contrast to the mammalian mitochondrion, it appears that the parasite organelle cannot maintain its membrane potential in the absence of electron transport. The consequences of membrane potential collapse in the malaria parasite mitochondrion are likely to be diverse, but the initiation of programmed cell death via the release of cytochrome *c* into the cytosol followed by activation of a caspase is a possible scenario (reviewed in Vaidya, 1998).

The differences between the malaria parasite mitochondrion and the mammalian equivalent make it a viable biochemical target for antimalarial drugs, and the continued investigation of naturally occurring naphthoquinones, especially those from plants used traditionally in malaria treatment, would be worthwhile.

15.3.4 FEBRIFUGINE

The powdered roots of *Dichroa febrifuga* (Saxifragaceae), known in China as Ch'ang Shan, have been used traditionally for malaria treatment for many centuries (see Chapter 4). *D. febrifuga* contains quinazolone alkaloids, and the compound febrifugine **40** (β-dichroine) has been shown to have potent antimalarial activity against avian species of malaria in chicks and ducks and also has activity against murine (*P. berghei*) and simian (*P. cynomolgi*) malarias (Thompson and Werbel, 1972). In man, oral or parenteral treatment with the plant extract has been reported to be effective for malaria treatment, but nausea and vomiting occurred in some patients (Chang and But, 1986). However, the alkaloid is also reported to be toxic to the liver (Steck, 1972). Febrifugine and isofebrifugine have potent *in vitro* activities against both chloroquine-sensitive and chloroquine-resistant malaria parasites (Takaya et al., 1999), and febrifugine has been shown to enhance the production of nitric oxide in activated mouse peritoneal macrophages; two other alkaloidal constituents of *D. febrifuga* had similar but less potent effects (Murata et al., 1998). As nitric oxide production by activated macrophages may have cytotoxic effects against pathogenic organisms, enhancement of nitric oxide production may contribute to the *in vivo* antimalarial activity of

D. febrifuga alkaloids. Structure–activity relationship studies on a number of synthetic febrifugine analogues have shown that the hydroxy group on the piperidine ring of the molecule is essential for potent antimalarial activity (Thompson and Werbel, 1972). Some attempts to improve the antimalarial activity of febrifugine while decreasing its cytotoxicity by structural modifications have been attempted but have not been very successful (Takaya et al., 1999; Kobayashi et al., 1999).

15.4 CONCLUSION

With the exceptions of quinine and artemisinin, few detailed pharmacological investigations have been carried out on plant-derived antimalarial compounds. Such studies are important not only because they contribute to our knowledge of traditional antimalarial plants — which hopefully will lead to the more effective use of these species in malaria treatment — but they may also contribute to the development of new antimalarial agents. Plant species contain a wide diversity of chemical types and it is likely that some of these will have novel modes of action against malaria parasites, so that the elucidation of their mechanisms of action could lead to the identification of new biochemical targets against which new compounds may be developed.

REFERENCES

Aceti, A., Bonincontro, A., Cametti, C., Celestino, D., and Leri, O. (1990). Electrical conductivity of human erythrocytes infected with *Plasmodium falciparum* and its modification following quinine therapy. *Trans. R. Soc. Trop. Med. Hyg.*, 84, 671–672.

Alin, M.H. (1997). *In vitro* susceptibility of Tanzanian wild isolates of *Plasmodium falciparum* to artemisinin, chloroquine, sulfadoxine/pyrimethamine and mefloquine. *Parasitology*, 114, 503–506.

Alin, M.H., Ashton, M., Kihamia, C.M., Mtey, G.J.B., and Bjorkman, A. (1996). Multiple-dose pharmacokinetics of oral artemisinin and comparison of its efficacy with that of oral artesunate in falciparum-malaria patients. *Trans. R. Soc. Trop. Med. Hyg.*, 90, 61–65.

Allen, D., Toth, I., Wright, C.W., Kirby, G.C., Warhurst, D.C., and Phillipson J.D. (1993). *In vitro* antimalarial and cytotoxic activities of semisynthetic derivatives of brusatol. *Eur. J. Med. Chem.*, 28, 265–269.

Allen, D., Wright, C.W., Phillipson, J.D., Kirby, G.C., and Warhurst, D.C. (1994). The effectiveness of chloroquine as an antimalarial drug may be compromised by the co-administration of traditional medicines. *J. Pharm. Pharmacol.*, 46 (Suppl.), Abstract 16.

Arzel, E., Rocca, P., Grellier, P., Labaeid, M., Frappier, F., and Gueritte, F. (2001). New synthesis of benzo-δ-carbolines, cryptolepines, and their salts: *in vitro* cytotoxic, antiplasmodial and antitrypanosomal activities of δ-carbolines, benzo-δ-carbolines and cryptolepines. *J. Med. Chem.*, 44, 949–960.

Asawamahasakda, W., Ittarat, I., Chang, C.C., McElroy, P., and Meshnick, S.R. (1994a). Effects of antimalarials and protease inhibitors on plasmodial hemozoin production. *Mol. Biochem. Parasitol.*, 67, 183–191.

Asawamahasakda, W., Ittarat, I., Pu, Y.M., Ziffer, H., and Meshnick, S.R. (1994b). Reaction with antimalarial endoperoxides with specific parasite proteins. *Antimicrob. Agents Chemother.*, 38, 1854–1858.

Atamna, H. and Ginsburg, H. (1995). Heme degradation in the presence of glutathione. *J. Biol. Chem.*, 270, 24876–24883.

Beran, Y.M., Benzie, C.R., and Kay, J.E. (1980). Bruceantin, an inhibitor of the initiation of protein synthesis in eukaryotes. *Biochem. Soc. Trans.*, 8, 357–359.

Berman, P.A. and Adams, P.A. (1997). Artemisinin enhances heme-catalysed oxidation of lipid membranes. *Free Radic. Biol. Med.*, 22, 1283–1288.

Bhisutthibhan, J., Pan, X.-Q., Hossler, P.A., Walker, D.J., Yowell, C.A., Carlton, J., et al. (1998). The *Plasmodium falciparum* translationally controlled tumor protein homolog and its reaction with the antimalarial drug artemisinin. *J. Biol. Chem.*, 273, 16192–16198.

Bilia, A.R., Lazari, D., Messori, L., Taglioli, V., Temperini, C., and Vincieri, F.F. (2002). Simple and rapid physico-chemical methods to examine action of antimalarial drugs with hemin. Its application to *Artemisia annua* constituents. *Life Sci.*, 70, 769–778.

Bohle, D.S., Dinnebier, R.E., Madsen, S.K., and Stephens, P.W. (1997). Characterisation of the products of the heme detoxification pathway in malarial late trophozoites by x-ray diffraction. *J. Biol. Chem.*, 272, 713–716.

Bonjean, K., De Pauw-Gillet, M.C., Defresne, M.P., Colson, P., Houssier, C., Dassonneville, L. et al. (1998). The DNA intercalating alkaloid cryptolepine interferes with topoisomerase II and inhibits primarily DNA synthesis in 1316 melanoma cells. *Biochemistry*, 37, 5136–5146.

Brewer, T.G., Peggins, J.G., Grate, S.J., Petras, J.M., Levine, B.S., Weina, P.J., et al. (1994). Neurotoxicity in animals due to arteether and artemether. *Trans. R. Soc. Trop. Med. Hyg.*, 88 (Suppl. 1), 33–36.

Camacho Corona, M.del R., Croft, S.L., and Phillipson, J.D. (2000). Natural products as sources of antiprotozoal drugs. *Curr. Opin. Anti-Infect. Invest. Drugs*, 2, 47–62.

Carvalho, L.H., Rocha, E.M.M., Raslan, D.S., Oliveira, A.B., and Krettli, A.U. (1988). *In vitro* activity of natural and synthetic naphthoquinones against erythrocytic stages of *Plasmodium falciparum*. *Braz. J. Med. Biol. Res.*, 21, 485–487.

Chang, H.-M. and But, P.P. (1986). *Pharmacology and Applications of Chinese Materia Medica*, Vol. 1. World Scientific, Singapore, pp. 1–773.

Chawira, A.N., Warhurst, D.C., and Peters, W. (1986). Qinghaosu resistance in rodent malaria. *Trans. R. Soc. Trop. Med. Hyg.*, 80, 477–480.

Chawira, A.N., Warhurst, D.C., Robinson, B.L., and Peters, W. (1987). The effect of combinations of qinghaosu (artemisinin) with standard antimalarial drugs in the suppressive treatment of malaria in mice. *Trans. R. Soc. Trop. Med. Hyg.*, 81, 554–558.

Chen, P.Q., Li, G.Q., Guo, X.B., He, K.R., Fu, Y.X., and Fu, L.C. (1994). The infectivity of gametocytes of *Plasmodium falciparum* from patients treated with artemisinin. *Chin. Med. J.*, 107, 709–711.

Chen, Y., Zhu, S.-M., Chen, H.-Y., and Li, Y. (1998). Artesunate interaction with hemin. *Bioelectrochem. Bioenerg.*, 44, 295–300.

Chou, A., Chevli, R., and Fitch, C.D. (1980). Ferriprotoporphyrin IX fulfils the criteria for identification as the chloroquine receptor of malaria parasites. *Biochemistry*, 19, 1543–1549.

Dassoneville, L., Lansiaux, A., Wattelet, A., Wattez, N., Mahieu, C., Van Miert, S., et al. (2000). Cytotoxicity and cell cycle effects of the plant alkaloids cryptolepine and neocryptolepine: relation to drug-induced apoptosis. *Eur. J. Pharmacol.*, 409, 9–18.

Dorn, A., Stoffel, R., Matile, H., Bubendorf, A., and Ridley, R.G. (1995). Malarial haemozoin/β-haematin supports haem polymerisation in the absence of protein. *Nature*, 374, 269–271.

Durasingh, M.T., Roper, C., Walliker, D., and Warhurst, D.C. (2000). Increased sensitivity to the antimalarials mefloquine and artemisinin is conferred by mutations in the pfmdr1 gene of *Plasmodium falciparum*. *Mol. Microbiol.*, 36, 955–961.

Egan, T.J., Ross, D.C., and Adams, P.A. (1994). Quinoline anti-malarial drugs inhibit spontaneous formation of β-haematin (malaria pigment). *FEBS Lett.*, 352, 54–57.

Elford, B.C., Roberts, M.F., Phillipson, J.D., and Wilson, R.J.M. (1987). Potentiation of the antimalarial activity of qinghaosu by methoxylated flavones. *Trans. R. Soc. Trop. Med. Hyg.*, 81, 434–436.

Fawaz, G. and Haddad, F.S. (1951). The effect of lapinone (M-2350) on *P. vivax* infection in man. *Am. J. Trop. Med. Hyg.*, 31, 569–571.

Fishwick, J., Edwards, G., Ward, S.A., and McLean, W.G. (1998a). Binding of dihydroartemisinin to differentiating neuroblastoma cells and rat cortical homogenate. *NeuroToxicology*, 19, 405–412.

Fishwick, J., Edwards, G., Ward, S.A., and McLean, W.G. (1998b). Morphological and immunocytochemical effects of dihydroartemisinin on differentiating NB2a neuroblastoma cells. *NeuroToxicology*, 19, 393–403.

Fishwick, J., McLean, W.G., Edwards, G., and Ward, S.A. (1995). The toxicity of artemisinin and related compounds on neuronal and glial cells in culture. *Chem. Biol. Interact.*, 96, 263–271.

Fort, D.M., Litvak, J., Chen, J.L., Lu, Q., Phuan, P.-W., Cooper, R., and Bierer, D.E. (1998). Isolation and unambiguous synthesis of cryptolepinone: an oxidation artefact of cryptolepine. *J. Nat. Prod.*, 61, 1528–1530.

Gantner, F., Uhlig, S., and Wendel, A. (1995). Quinine inhibits release of tumour necrosis factor, apoptosis, necrosis and mortality in a murine model of septic liver failure. *Eur. J. Pharmacol.*, 294, 353–355.

Gay, F., Ciceron, L., Litaudon, M., Bustos, M.D., Astagneau, P., Diquet, B., et al. (1994). *In vitro* resistance of *Plasmodium falciparum* to qinghaosu derivatives in West Africa. *Lancet*, 343, 850–851.

Giao, P.T., Binh, T.Q., Kager, P.A., Long, H.P., Thang, N.V., Nam, N.V., and De Vries, P.J. (2001). Artemisinin for treatment of uncomplicated falciparum malaria: is there a place for monotherapy? *Am. J. Trop. Med. Hyg.*, 65, 690–695.

Grace, J.M., Aguilar, A.J., Trotman, K.M., and Brewer, T.G. (1998). Metabolism of β-arteether to dihydroqinghaosu by human liver microsomes and recombinant cytochrome P450. *Drug Metab. Dispos.*, 26, 313–317.

Gupta, S., Svensson, U.S.H., and Ashton, M. (2001). *In vitro* evidence for auto-induction of artemisinin metabolism in the rat. *Eur. J. Drug Metab. Pharmacokinet.*, 26, 173–176.

Halliwell, B. and Gutteridge, J.M.C. (1989). *Free Radicals in Biology and Medicine*, 2nd ed. Clarendon Press, Oxford.

Hindley, S., Ward, S.A., Storr, R.C., Searle, N.L., Bray, P.G., Park, B.K., et al. (2002). Mechanism-based design of parasite-targetted artemisinin derivatives: synthesis and antimalarial activity of new diamine containing analogues. *J. Med. Chem.*, 45, 1052–1063.

Hong, Y.L., Yang, Y.Z., and Meshnick, S.R. (1994). The interaction of artemisinin with malarial haemozoin. *Mol. Biochem. Parasitol.*, 63, 121–128.

Hudson, A.T., Randall, A.W., Fry, M., Ginger, C.D., Hill, B., Latter, V.S., et al. (1985). Novel antimalarial hydroxynaphthoquinones with potent broad spectrum antiprotozoal activity. *Parasitology*, 90, 45–55.

Inselburg, J. (1985). Induction and isolation of artemisinin resistant mutants of *Plasmodium falciparum. Am. J. Trop. Med. Hyg.*, 34, 417–418.

James, R.F. (1985). Malaria treated with emetine or metronidazole. *Lancet*, August 31, 8453, 498 (letter).

Kamchonwongpaisan, S., McKeever, P., Houssier, P., Ziffer, H., and Meshnick, S.R. (1997). Artemisinin neurotoxicity: neuropathology in rats and mechanistic studies *in vitro. Am. J. Trop. Med. Hyg.*, 56, 7–12.

Kirby, G.C., O'Neill, M.J., Phillipson, J.D., and Warhurst, D.C. (1989). *In vitro* studies on the mode of action of quassinoids with activity against chloroquine-reisistant *Plasmodium falciparum. Biochem. Pharmacol.*, 38, 4367–4374.

Kirby, G.C., Warhurst, D.C., and Phillipson, J.D. (1992). Mode of action of antimalarial quassinoids: limited effects *in vitro* upon glycolysis in *Plasmodium falciparum. Trans. R. Soc. Trop. Med. Hyg.*, 86, 343.

Klayman, D.L. (1985). Qinghaosu (artemisinin): an antimalarial drug from China. *Science*, 228, 1049–1055.

Kobayashi, S., Ueno, M., Suzuki, R., Ishitani, H., Kim, H.S., and Wataya, Y. (1999). Catalytic asymmetric synthesis of antimalarial alkaloids febrifugine and isofebrifugine and their biological activity. *J. Org. Chem.*, 64, 6833–6841.

Krogstad, D.J. and De, D. (1998). Chloroquine: modes of action and resistance and the activity of chloroquine analogs. In *Malaria, Parasite Biology, Pathogenesis and Protection*, Sherman, I.W., Ed. American Society for Microbiology Press, Washington, DC, pp. 331–340.

Kumar, N. and Zheng, H. (1990). Stage-specific gametocytocidal effect *in vitro* of the antimalarial drug qinghaosu on *Plasmodium falciparum. Parasitol. Res.*, 76, 214–218.

Lee, L.S. and Hufford, C.D. (1990). Metabolism of antimalarial sesquiterpene lactones. *Pharmacol. Ther.*, 48, 345–355.

Li, L.P., Liu, R.C., Xu, Y.S., Zhou, Z.Y., and Lu, L. (1981). Effect of sodium artesunate on the ultrastructure of erythrocytic form of *Plasmodium knowlesi. Zong Cao Yao*, 12, 175 (cited in Chen et al., 1998).

Liao, L.L., Kupchan, S.M., and Horowitz, S.B. (1976). Mode of action of the antitumour compound bruceantin, an inhibitor of protein synthesis. *Mol. Pharmacol.*, 12, 167–176.

Lisgarten, J.N., Coll, M., Portugal, J., Wright, C.W., and Aymami, J. (2002). The antimalarial and cytotoxic drug cryptolepine intercalates into DNA at cytosine-cytosine sites. *Nat. Struct. Biol.*, 9, 57–60.

McGready, R., Cho, T., Keo, N.K., Thwai, K.L., Villegas, L., Looareesuwan, S., et al. (2001). Artemisinin antimalarials in pregnancy: a prospective treatment study of 539 episodes of multidrug-resistant *Plasmodium falciparum. Clin. Infect. Dis.*, 33, 2009–2016.

Meshnick, S.R. (1994). The mode of action of antimalarial peroxides. *Trans. R. Soc. Trop. Med. Hyg.*, 88 (Suppl. 11), 31–32.

Meshnick, S.R. (1998). From quinine to qinghaosu: historical perspectives. In *Malaria, Parasite Biology, Pathogenesis and Protection*, Sherman, I.W., Ed. American Society for Microbiology Press, Washington, DC, pp. 341–353.

Meshnick, S.R., Thomas, A., Ranz, A., Xu, C.M., and Pan, H.Z. (1991.) Artemisinin (qinghaosu): the role of intracellular hemin in its mechanism of antimalarial action. *Mol. Biochem. Parasitol.*, 49, 181–190.

Meshnick, S.R., Yang, Y.Z., Lima, V., Kuypers, F., Kamchongwongpaison, S., and Yuthavong, Y. (1993). Iron dependent free radical generation from the antimalarial agent artemisinin (qinghaosu). *Antimicrob. Agents Chemother.*, 37, 1108–1114.

Miller, L.G. and Panosian, C.B. (1997). Ataxia and slurred speech after artesunate treatment for falciparum malaria. *N. Engl. J. Med.*, 336, 1328.

Muhia, D.H., Thomas, C.G., Ward, S.A., Edwards, G., Mberu, E.K., and Watkins, W.M. (1994). Ferriprotoporphyrin catalysed decomposition of artemether: anlaytical and pharmacological implications. *Biochem. Pharmacol.*, 48, 889–895.

Mukanganyama, S., Naik, Y.S., Widersten, M., Mannervi, B., and Hasler, J.A. (2001). Proposed reductive metabolism of artemisinin by glutathione transferases *in vitro*. *Free Radical Res.*, 35, 427–434.

Murata, K., Takano, F., Fushiya, S., and Oshima, Y. (1998). Enhancement of NO production in activated macrophages *in vivo* by an antimalarial crude drug, *Dichroa febrifuga*. *J. Nat. Prod.*, 61, 729–733.

Murphy, J., Clyde, D., Herringon, D., Bagar, S., Davis, J., Palmer, K., et al. (1990). Continuation of chloroquine-susceptible *Plasmodium falciparum* parasitaemia in volunteers receiving chloroquine. *Antimicrob. Agents Chemother.*, 34, 676–679.

Newman, D.J., Cragg, G.M., and Snader, K.M. (2000). The influence of natural products upon drug discovery. *Nat. Prod. Rep.*, 17, 215–234.

Noamesi, B.K., Larrson, B.S., Laryea, D.L., and Ullberg, S. (1991). Whole-body autoradiographic study on the distribution of [^{3}H]-cryptolepine in mice. *Arch. Int. Pharmacodyn.*, 313, 5–14.

Oduola, A.M.J., Sowunmi, A., Milhous, W.K., Kyle, D.E., Martin, R.K., Walker, O., et al. (1992). Innate resistance to new antimalarial drugs in *Plasmodium falciparum* from Nigeria. *Am. J. Trop. Med. Hyg.*, 86, 123–126.

Orjih, A.U. (1996). Haemolysis of *Plasmodium falciparum* trophozoite-infected erythrocytes after artemisinin exposure. *Br. J. Haematol.*, 92, 324–328.

Orjih, A.U., Banyal, H.S., Chevli, R., and Fitch, C.D. (1981). Hemin lyses malaria parasites. *Science*, 214, 667–669.

Pagola, S., Stephens, P.W., Bohle, D.S., Kosar, A.D., and Madsen, S.K. (2000). The structure of malaria pigment β-haematin. *Nature*, 404, 307–310.

Parapini, S., Basilico, N., Pasinin, E., Egan, T., Olliaro, P., Taramelli, D., and Monti, D. (2000). Standardisation of the physicochemical parameters to assess *in vitro* the β-haematin inhibitory activity of antimalarial drugs. *Exp. Parasitol.*, 96, 249–256.

Peters, W. (1987). *Chemotherapy and Drug Resistance in Malaria*, 2nd ed. Academic Press, London.

Peters, W., Lin., L., Robinson, B.L., and Warhurst, D.C. (1986). The chemotherapy of rodent malaria XL. The action of artemisinin and related sesquiterpenes. *Ann. Trop. Med. Parasitol.*, 80, 483–489.

Peters, W. and Robinson, B.L. (2000). The chemotherapy of rodent malaria. LVIII. Drug combinations to impede the selection of drug resistance. Part 2. The new generation: artemisinin or artesunate with long-acting blood schizontocides. *Ann. Trop. Med. Parasitol.*, 94, 23–35.

Phillipson, J.D., Wright, C.W., Kirby, G.C., and Warhurst, D.C. (1992). Tropical plants as sources of antiprotozoal agents. In *Phytochemical Potential of Tropical Plants, Recent Advances in Phytochemistry*, Vol. 27, Downum, K.R., Romeo, J.T., and Stafford, H.A., Eds. Plenum Press, New York, pp. 1–40.

Posner, G.H. (1998). Antimalarial peroxides in the qinghaosu (artemisinin) and yingzhaosu families. *Exp. Opin. Ther. Pat.*, 8, 1487–1493.

Price, R.N., Nosten, F., Luxemberger, C., ter Kuile, F.O., Paiphun, L., Chongsuphajaisiddhi, T., et al. (1996). Effects of artemisinin derivatives on malaria transmissibility. *Lancet*, 347, 1654–1658.

Qinghaosu Antimalarial Coordinating Research Group. (1979). Antimalarial studies on qinghaosu. *Chin. Med. J.*, 92, 811–816.

Quetin-Le Clerq, J., Dupont, J., Wright, C.W., Phillipson, J.D., Warhurst, D.C., and Angenot, L. (1991). Modelisation moleculaire de usambarensine, de la tubulosine et de l'emetine, alcaloides cytotoxiques et antiambiens. *J. Pharm. Belg.*, 46, 85–92.

Radloff, P.D., Phillips, J., Nkeyi, M., Stuchler, D., Mittelholzer, M.L., and Kremsner, P.G. (1996). Arteflene compared with mefloquine for treating *Plasmodium falciparum* malaria. *Am. J. Trop. Med. Hyg.*, 55, 259–262.

Ramu, K. and Baker, J.K. (1995). Synthesis, characterisation, and antimalarial activity of the glucuronides of the hydroxylated metabolites of arteether. *J. Med. Chem.*, 38, 1911–1921.

Rasoanaivo, P., Ratsimamanga-Urverg, S., Milijaona, R., Rafatro, H., Rakoto-Ratsimamanga, A., Galeffi, C., and Nicoletti, M. (1994). *In vitro* and *in vivo* chloroquine-potentiating action of *Strychnos myrtoides* alkaloids against chloroquine-resisitant strains of *Plasmodium* malaria. *Planta Med.*, 60, 13–16.

Reed, M.B., Saliba, K.J., Caruana, S.R., Kirk, K., and Cowman, A.F. (2000). PghI modulates sensitivity and resisitance to multiple antimalarials in *Plasmodium falciparum*. *Nature*, 403, 906–909.

Robert, A., Coppel, Y., and Meunier, B. (2002). Alkylation of heme by the antimalarial drug artemisinin. *Chem. Commun.*, 414–415.

Rosenthal, P.J. and Meshnick, S.R. (1998). Hemoglobin processing and the metabolism of amino acids, heme and iron. In *Malaria*, Sherman, I.W., Ed. American Society for Microbiology, Washington, DC, pp. 145–158.

Sauvain, M., Moretti, C., Bravo J.-A., Callapa, J., Munoz, V., Ruiz, E., et al. (1996). Antimalarial activity of alkaloids from *Pogonopus tubulosus*. *Phytother. Res.*, 10, 198–201.

Schmuck, G., Roehrdanz, E., Haynes, R.K., and Kahl, R. (2002). Neurotoxic mode of action of artemisinin. *Antimicrob. Agents Chemother.*, 46, 821–827.

Shen, J.X., Ed. (1989). *Antimalarial Drug Development in China*. National Institute of Pharmaceutical Research and Development, Beijing, pp. 31–95.

Shukla, K.L., Gundi, T.M., and Meshnick, S.R. (1995). Molecular modeling studies of the artemisinin (qinghaosu)-hemin interaction. Docking between the antimalarial agent and its putative receptor. *J. Mol. Graph.*, 13, 215–222.

Slater, A.F.G. and Cerami, A. (1992). Inhibition by chloroquine of a novel haem polymerase enzyme activity in malaria trophozoites. *Nature*, 355, 167–169.

Smith, S.L., Fishwick, J., McLean, W.G., Edwards, G., and Ward, S.A. (1997). Enhanced *in vitro* neurotoxicity of artemisinin derivatives in the presence of haemin. *Biochem. Pharmacol.*, 53, 5–10.

Spencer, C.F., Koniuszy, F.R., Rogers, E.F., Shavel, J., Jr., Easton, N.R., Kaczka, E.A., et al. (1947). Survey of plants for antimalarial activity. *Lloydia*, 10, 145–174.

Sponer, U., Prajakwong, S., Wiedermann, G., Kollaritsch, H., Wernsdorfer, G., and Wernsdorfer, W.H. (2002). Pharmacodynamic interaction of doxycycline and artemisinin in *Plasmodium falciparum*. *Antimicrob. Agents Chemother.*, 46, 262–264.

Steck, E.A. (1972). *The Chemotherapy of Protozoan Diseases*. Walter Reed Army Institute of Research, Washington, DC.

Su, X.-Z. and Wellems, T.E. (1998). Genome discovery and malaria research: current status and promise. In *Malaria*, Sherman, I.W., Ed. American Society for Microbiology Press, Washington, DC, pp. 253–266.

Sugioka, Y. and Suzuki, M. (1991). The chemical basis for the ferriprotoporphyrin IX-chloroquine complex induced lipid peroxidation. *Biochim. Biophys. Acta*, 1074, 19–24.

Takaya, Y., Tasaka, H., Chiba, T., Uwai, K., Tanitsu, M., Kim, H.-S., et al. (1999). New type of febrifugine analogues, bearing a quinolizidine moiety, show potent antimalarial activity against *Plasmodium* malaria parasite. *J. Med. Chem.*, 42, 3163–3166.

Thompson, P.E. and Werbel, L.M. (1972). *Antimalarial Agents, Chemistry and Pharmacology*. Academic Press, London, pp. 317–324.

Titulaer, H.A.C., Zuidema, J., Kager, P.F., Wetsteyn, J.C.F.M., Lugt, C.H.B., and Merkus, F.W.H.M. (1990). The pharmacokinetics of artemisinin after oral, intramuscular, and rectal administration to human volunteers. *J. Pharm. Pharmacol.*, 42, 810–813.

Tripathi, A.K., Gupta, A., Garg, S.K., and Tekwani, B.L. (2001). *In vitro* β-haematin formation assays with plasma of mice infected with *Plasmodium yoelii* and other parasite preparations. Comparative inhibition with quinoline and endoperoxide antimalarials. *Life Sci.*, 69, 2725–2733.

Tripathi, R., Dutta, G.P., and Vishwakarma, R.A. (1996). Gametocytocidal activity of alpha/beta arteeether by the oral route of administration. *Am. J. Trop. Med. Hyg.*, 54, 652–654.

Vaidya, A.B. (1998). Mitochondrial physiology as a target for atovaquone and other antimalarials. In *Malaria, Parasite Biology, Pathogenesis and Protection*, Sherman, I.W., Ed. American Society for Microbiology Press, Washington DC, pp. 365–368.

van Agtmael, M.A., Gupta, V., van der Wosten, T.H., Rutten, J.P.B., and van Boxtel, C.J. (1999). Grapefruit juice increases the bioavailability of artemether. *Eur. J. Clin. Pharmacol.*, 55, 405–410.

van Vugt, M., Price, R.N., Mann, C., Poletto, C., et al. (2000). A case-control auditory evaluation of patients treated with artemisinin derivatives for multidrug-resistant *Plasmodium falciparum* malaria. *Am. J. Trop. Med. Hyg.*, 62, 65–69.

Vyas, N., Avery, B.A., Avery, M.A., and Wyandt, C.M. (2002). Carrier-mediated partitioning of artemisinin into *Plasmodium falciparum* infected erythrocytes. *Antimicrob. Agents Chemother.*, 46, 105–109.

Walker, D.J., Pitsch, J.L., Peng, M.M., Robinson, B.L., Peters, W., Bhisutthiban, J., and Meshnick, S.R. (2000). Mechanisms of artemisinin resistance in the rodent malaria pathogen *Plasmodium yoelii. Antimicrob. Agents Chemother.*, 44, 344–347.

Wang, D.Y. and Wu, Y.L. (2000). A possible antimalarial action mode of qinghaosu (artemisinin) series compounds. Alkylation of reduced glutathione by C-centered primary radicals produced from anti-malarial compound qinghaosu and 12-(2,4-dimethoxyphenyl)-12-deoxoqinghaosu. *Chem. Commun.*, 22, 2193–2194.

White, N.J. (1994). Clinical pharmacokinetics and pharmacodynamics of artemisinin and derivatives. *Trans. R. Soc. Trop. Med. Hyg.*, 88 (Suppl. 1), 41–43.

White, N.J. (1998). Why is that antimalarial drug treatments do not always work? *Ann. Trop. Med. Parasitol.*, 92, 449–458.

Wilairatana, P. and Looareesuwan, S. (2002). The clinical use of artemisinin and its derivatives in the treatment of malaria. In *Artemisia, Medicinal and Aromatic Plants: Industrial Profiles*, Vol. 18, Wright, C.W., Ed. Taylor & Francis, London, pp. 289–307.

Woerdenbag, H.J. and Pras, N. (2002). Analysis and quality control of commercial *Artemisia* species. In *Artemisia, Medicinal and Aromatic Plants: Industrial Profiles*, Vol. 18, Wright, C.W., Ed. Taylor & Francis, London, pp. 51–77.

Wright, A.D., Wang, H., Gurrath, M., Konig, G.M., Kocak, G., Neumann, G., et al. (2001). Inhibition of heme detoxification processes underlies the antimalarial activity of terpene isonitrile compounds from marine sponges. *J. Med. Chem.*, 44, 873–885.

Wright, C.W. (2002). Antiprotozoal natural products. In *Pharmacognosy*, 15th ed., Evans, W.C., Ed. Harcourt, London, pp. 407–413.

Wright, C.W., Addae-Kyereme, J., Breen, A.G., Brown, J.E., Cox, M.F., Croft, S.L., et al. (2001). Synthesis and evaluation of cryptolepine analogues for their potential as new antimalarial agents. *J. Med. Chem.*, 44, 3187–3194.

Wright, C.W., Allen, D., Cai, Y., Chen, Z., Phillipson, J.D., Kirby, G.C., et al. (1994). Selective antiprotozoal activity of some *Strychnos* alkaloids. *Phytother. Res.*, 8, 149–152.

Wright, C.W., Anderson, M.M., Allen, D., Phillipson, J.D., Kirby, G.C., Warhurst, D.C., and Chang, H.R. (1993). Quassinoids exhibit greater selectivity against *Plasmodium falciparum* than against *Entamoeba histolytica, Giardia intestinalis* or *Toxoplasma gondii in vitro. J. Euk. Microbiol.*, 40, 244–246.

Wright, C.W., Phillipson, J.D., Awe, S.O., Kirby, G.C., Warhurst, D.C., Quetin-Leclercq, J., and Angenot, L. (1996). Antimalarial activity of cryptolepine and some other anhydronium bases. *Phytother. Res.*, 10, 361–363.

Wright, C.W. and Warhurst, D.C. (2002). The mode of action of artemisinin and its derivatives. In *Artemisia, Medicinal and Aromatic Plants: Industrial Profiles*, Vol. 18, Wright, C.W., Ed. Taylor & Francis, London, pp. 249–288.

Wu, W.-M., Wu, Y., Wu, Y.L., Yao, Z.-J., Zhou, C.-M., Li, Y., et al. (1998). United mechanistic framework for the Fe(II)-induced cleavage of qinghaosu derivatives/analogues. The first spin-trapping evidence for the previously postulated secondary C-4 radical. *J. Am. Chem. Soc.*, 120, 3136–3325.

Wu, W.-M., Yao, Z.-L., Wu, Y.-L., Jiang, K., Wang, Y.-F., Chen, H.-B., et al. (1996). Ferrous iron induced cleavage of the peroxy bond in qinghaosu and its derivatives and the DNA damage associated with this process. *Chem. Commun.*, 18, 2213–2214.

Yang, S.D., Ma, J.M., Sub, J.H., Chen, D.X., and Song, Z.Y. (1986). Clinical pharmacokinetics of a new effective antimalarial artesunate, a qinghaosu derivative. *Chin. J. Clin. Pharmacol.*, 1, 106–109.

Yao-De, W., Qi-Zhong, Z., and Jie-Sheng, J. (1992). Studies on the antimalarial action of gelatin capsule of *Artemisia annua. Chung Kuo Chi Sheng Ching Hsueh Yu Chi Sheng Chung Ping Tsa Chih*, 10, 290–294.

Zhang, L., Zhou, W.S., and Xu, X.X. (1998). A new sesquiterepene peroxide (yinghzaosu C) and sesquiterpenol (yinghaosu D) from *Artabotrys unciatus* (L.) Meer. *J. Chem. Soc. Chem. Commun.*, 523.

Zhao, Y., Hanton, W.K., and Lee, K.-H. (1986). Antimalarial agents. 2. Artesunate, an inhibitor of cytochrome oxidase activity in *Plasmodium berghei. J. Nat. Prod.*, 49, 139–142.

16 Guidelines for the Nonclinical Evaluation of the Efficacy of Traditional Antimalarials

Philippe Rasoanaivo, Eric Deharo,
Suzanne Ratsimamanga-Urverg, and François Frappier

CONTENTS

16.1 INTRODUCTION

For the foreseeable future, chemotherapy and impregnated bed nets will remain the two most useful tools for the control of the deadly disease malaria, which kills 2 million people each year. Paradoxically, only approximately 10 antimalarial drugs are available on the market for the prevention or the treatment of malaria, and the development of new ones is costly and time-consuming. The use of chemotherapy for controlling the pathogenic organism is further restricted by the development of drug resistance. As there is no longer a single drug that can prevent or cure all cases of malaria, researchers have reconsidered the entire therapeutic approach for the control of this old and most devastating tropical disease. A more flexible attitude should be adopted to this end. Whereas the urgent need for the discovery or design of new antimalarial drugs with different mechanisms of action is recognised, plant-based antimalarials form the basis of medicines used by the majority of people in most regions afflicted with malaria. Many have been shown in experimental studies to have antiplasmodial effects, and as such, they may offer viable alternatives to prescription drugs in the treatment of this life-threatening disease.

Before a traditional antimalarial plant can be used in primary health care, however, it is essential to adequately assess its efficacy. Indeed, insufficient evidence of the efficacy of an herbal product is not acceptable, particularly when that product may entail serious health risks. Evaluating the efficacy of antimalarial plants may be viewed at two levels. The first is to demonstrate that, in the form in which they are used in traditional medicine to treat malaria, they have beneficial effects. The second, assuming that chemically defined constituents are responsible for the observed activity, is to isolate these constituents for further investigation. Based on our own experience and our exchange of views with colleagues involved in a malaria research program, we will propose in this chapter guidelines for the preparation of extracts of plants, bioassay with crude extracts, and bioassay-guided fractionation procedures.

16.2 AN INTEGRATED APPROACH TO ANTIMALARIAL PLANTS

When dealing with traditional medicine, it is important to bear in mind that healers basically treat the symptoms of a disease, especially those that are apparent to them. As malaria can produce a wide variety of symptoms, over 1200 plant species are used to treat this disease (see Chapter 11). Medicinal plants considered effective in the treatment of malaria are therefore those observed by healers to alleviate or prevent one or more recognised symptoms of malaria. As malaria can occur concurrently with other infectious diseases, accurate diagnosis can be difficult to achieve, and this makes malaria symptoms somewhat complex. Ethnomedical beliefs of populations also play a role in the choice of plants for the treatment of malaria. Based on these considerations, antimalarial plants can be roughly divided into three categories:

1. Plants with a direct effect on the parasite, either at the erythrocytic stage (antiplasmodial drugs) or at the hepatic stage (preventive drugs)
2. Plants with effects on host–parasite relationships (immunostimulants, antipyretics, etc.)
3. Plants with no clear effects on malaria, but with probable psychosomatic action, the use of which originates from ethnomedical beliefs

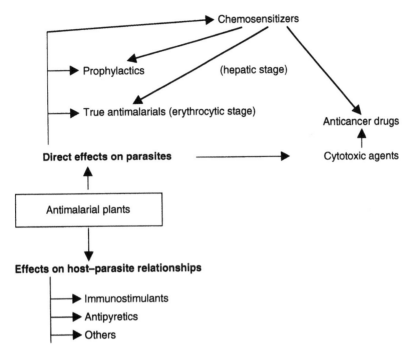

FIGURE 16.1 Components of the integrated approach based on known or probable biological activities of antimalarial plants.

This has led us to propose an integrated approach for investigating antimalarial plants (Rasoanaivo, 2002). The computerised compilation of all ethnomedical uses of Madagascan medicinal plants for the period 1500–2002 (Rasoanaivo, 2000) provides a good illustration of this approach. For example, medicinal plants used to treat malaria in one region are reported to be used as tonics in another, and even claimed to be toxic in yet other parts of the island. Similar variations in the use of medicinal plants may also be encountered in other countries. This integrated approach is summarised in Figure 16.1.

Which of these therapeutic lines should be prioritised when evaluating the efficacy of antimalarial plants? In our opinion, effort must of necessity be selective, focusing on one or two relevant biological activities. Progress in our understanding of the biology and biochemistry of malaria parasites during the last two decades clearly shows that the erythrocytic stage of *Plasmodium* is by far the most important target in malaria chemotherapy (Olliaro and Goldberg, 1995). We therefore suggest that antimalarial plants should first be evaluated for their ability to kill the parasite in the blood stage. Another important target is the hepatic stage of *Plasmodium*, in which there is a serious lack of appropriate drugs.

For efficacy evaluation, three key points must be considered: the extraction procedure, the choice of appropriate malaria-oriented bioassays, and the fractionation procedure.

16.3 PREPARATION OF EXTRACTS FOR ANTIMALARIAL SCREENING

A wide range of methods for extraction and preliminary fractionation are described in the literature dealing with plant chemistry. They are generally designed for a specific purpose such as pure phytochemical characterisation, biological screening, bioassay-guided fractionation, and selective extraction of one or more compounds. In demonstrating biological activities of medicinal plants, the choice of appropriate extraction and fractionation procedures is of paramount importance. We think that some projects have failed to detect useful compounds in medicinal plants in part because of shortcomings in the procedures employed for preparing the plant material for analysis.

16.3.1 PLANT MATERIAL

First of all, the botanical identity of the antimalarial plants collected must be authenticated by a taxonomist. So many mistakes have occurred in the past in this respect that proper identification of the plant material should be regarded as an essential prerequisite. It is also common practice to keep voucher specimens of plants examined at the institute where the work is carried out. Although generally omitted, the details of plant harvesting conditions, such as time of day, season, and stage of plant development, should be recorded in order to work out the optimum conditions for collecting the plant material.

Plant parts collected should be those used in traditional preparations. Ideally, fresh plants should be used to prevent enzymatic degradation or hydrolysis from occurring. Alternatively, plants may be dried before extraction, especially if they are not used fresh. If this has to be done, it is essential that the drying operation be carried out under controlled conditions so as to minimise postharvest chemical changes. To this end, plants should be dried as quickly as possible without using high temperature or sunlight, and preferably air-dried. Once thoroughly dried, plant material can be stored for long periods of time before analysis.

The plant material collected must be free from contamination with other plants. Mosses often grow in close association with higher plants, and it is essential to remove them. In the case of higher plants, mixtures of plants may sometimes be gathered in error. Additionally, two closely similar plants growing side by side in the field may be incorrectly assumed to be the same, or a plant may be collected without realising that it has a parasite intertwined with it. It is important to collect plants that are not affected by viral, bacterial, or fungal infections. Not only may compounds derived from microbial synthesis be detected in such plants, but also infection may seriously alter plant metabolism and unexpected products could be formed, possibly in large amounts.

16.3.2 EXTRACTION PROCEDURE

16.3.2.1 Ethnopreparation-Based Extraction

Traditional healers have their own methods for preparing antimalarial recipes, some of which have proved to be efficient and safe for use in humans over the thousands of years they have been in use. However, the mode of preparation of the formulation currently used by healers is commonly ignored when evaluating the efficacy of medicinal plants. Traditional antimalarial remedies should be prepared exactly as they are used in traditional medicine, preferably in close collaboration with healers, and used as such while evaluating them using *in vitro* and *in vivo* experimental models.

We wish to report here an interesting case that deserves attention. Ten years ago, we investigated a medicinal plant of Madagascar, *Strychnopsis thouarsii* baill. (Menispermaceae), which has been traditionally used in decoction form as an antimalarial for a long time, and as a chloroquine enhancer more recently. At that time, we were focusing on drugs that reverse chloroquine resistance in malaria. We were able to isolate alkaloids with chloroquine-potentiating effects (Ratsimamanga-Urverg et al., 1992). As these alkaloids did not show any promising results for drug development, we discontinued this work. At the inaugural meeting of Research Initiative on Traditional Antimalarial Methods (RITAM) in Moshi in 1999, we became aware of the availability of *in vitro* and *in vivo* tests at the hepatic stage of the malaria parasite. Our colleagues in Paris have since tested the decoction of the plant prepared in the traditional way and have found that this decoction killed hepatic schizonts (see Chapter 17). Bioassay-guided fractionation has shown that the active constituents would be located in the polar fraction. These results, as yet unpublished, clearly show that we would have missed interesting compound(s) if we restricted our work to easily isolated alkaloids without taking into consideration the traditional preparation as used by the local population.

If the extracts display significant antimalarial activity either in the preerythrocytic stage or in the erythrocytic stage of the malaria parasite, isolation of the active principles is considered a logical next step. On the other hand, lack of antimalarial activity in these two experimental models does not necessarily imply that the traditional remedies concerned must be discarded. A strategy for handling this possibility is proposed in the last paragraph of this chapter. It is suggested that observational clinical studies, discussed in detail elsewhere in this book, should be carried out instead, to assess the reported beneficial effects of the traditional preparations.

16.3.2.2 Solvent Extraction

In some cases, the plants used as antimalarials are known from the literature, but the exact recipe for preparing the traditional remedy is not clearly described. Scientists generally avoid using water extraction, which is the method used by healers in most cases, because of the complexity and difficulty involved in developing a suitable workup procedure with aqueous extracts. Organic solvent extractions are therefore used as a good alternative in evaluating the antimalarial activities of plants. To this end, alcohol or aqueous alcohol, in any case, is a good all-purpose solvent for preliminary extraction in a screening program. Particularly, methyl or ethyl alcohol has the ability to extract a broad spectrum of chemical substances. It has been used in screening programs searching for antimalarial compounds from plants (Leaman et al., 1995; Muñoz et al., 2000; Simonsen et al., 2001). In this single-solvent extraction procedure, the plant material is subjected to extraction exhaustively, by repeated maceration with alcohol or aqueous alcohol at room temperature. The alcohol fraction in the combined extracts is evaporated off under reduced pressure at a temperature not exceeding 45°C, and the residual water extract is freeze-dried or evaporated to dryness by azeotropic methods by repeatedly adding 95% ethanol to the residual water until this water is completely removed. As a general rule in our laboratory, approximately 25 g of dried plant material is used for extraction in the primary screening. Extraction with alcohol in a Soxhlet apparatus has been reported for various parts of an antimalarial plant (Sharma and Sharma, 1999), but in our own work we avoid the use of this technique because extracts are continuously boiled with the solvent for several hours, which may alter labile constituents.

Successive extractions with solvents in increasing order of polarity are also a useful practice followed in several laboratories (Gessler et al., 1994). In this procedure, plant material is defatted with petroleum ether, cyclohexane, or heptane, the use of hexane being avoided because of its toxicity and flammability. The residual powdered plant is then extracted, preferably with ethyl acetate because of its lower toxicity compared to chlorinated hydrocarbon solvents, or alternatively with dichloromethane or chloroform. Thereafter, the residue is extracted with methanol or ethanol, and finally with water. This procedure is based on the old Roman principle of solubility: *similia similibus solvuntur* (the similar dissolves the similar). Scientifically speaking, nonpolar solvents dissolve selectively nonpolar compounds, and polar solvents dissolve preferably polar compounds. A reasonable alternative is to shorten the procedure by using only one nonpolar solvent (ethyl acetate) and one polar solvent (methanol or water).

When using antimalarial screening tests involving inhibition of enzymes, all tannins and polyphenols should be removed from the extracts in order to avoid false positives due to interference with the enzymes. This is achieved preferably by passage through a polyamide chromatography column, or alternatively by precipitation with polyvinylpyrrolidone.

16.4 BIOASSAY OF PLANT EXTRACTS FOR ANTIMALARIAL ACTIVITY

In malaria chemotherapy, most research has been aimed at searching for drugs that could kill the malaria parasite in the erythrocytic stage of its development. Methods used for evaluating the efficacy of drugs, i.e., basic assessment of antiplasmodial activity or mechanism-based assays, have therefore been designed to meet this purpose. There are basic tools used in a primary screening to assess the ability of herbal antimalarials to kill the malaria parasite in the asexual stage. Regarding the hepatic stage, tests are also available for use in screening programs (see Chapter 17). The following paragraphs will be dedicated to experimental models for detecting antiplasmodial activity of plant extracts in the blood stage of the malaria parasite.

16.4.1 EXPERIMENTAL MODELS FOR DETECTING ANTIPLASMODIAL ACTIVITY OF PLANT EXTRACTS IN THE ERYTHROCYTIC STAGE OF MALARIA PARASITES

16.4.1.1 *In Vitro* Antiplasmodial Tests

The radioactive microdilution technique originally developed by Desjardins et al. (1979) and modified by Le Bras and Deloron (1983) and O'Neill et al. (1985), and later on by several scientists around the world to suit specific screening purposes, has proved to be very useful for the preliminary evaluation of *in vitro* antiplasmodial activity of plant extracts. It is based on the inhibition of tritiated hypoxanthine uptake by *Plasmodium falciparum* cultured in human blood. It is accurate, rapid, reproducible, easily automated, and only requires small amounts of extracts, which makes it well suited for bioassay-guided fractionation. Although it has been difficult to adapt this technique for high-throughput screening due to the limiting factor of culturing the parasite, one of its advantages is the possibility of discovering bioactive molecules with unexpected or novel mechanisms of action. The results are expressed as percentage inhibition with respect to controls for one single dose, generally 10 µg/ml, or as median inhibitory concentration (IC_{50}) obtained by linear regression methods (Huber and Koella, 1993). Its disadvantages include the need to use tritiated hypoxanthine (which is hazardous and expensive) and a liquid scintillation counter. Alternatively, viable parasites in each well can be stained with Giemsa or Diff-Quick® (a rapid staining kit) reagents and examined under the microscope. The parasites on each blood film are counted using a high-power microscope lens with oil immersion, and the percentage of growth inhibition with respect to the control is determined by a simple arithmetic calculation.

Colorimetric methods have also been developed and applied successfully to the assessment of *P. falciparum* drug susceptibility. These include the parasite lactate dehydrogenase (LDH) assay (Makler et al., 1993) and the microculture tetrazolium assay (Delhaes et al., 1999). These two enzyme assays are nonradioactive, rapid, reliable, and inexpensive to perform, suggesting their suitability for application in the screening of antimalarial drugs. However, to the best of our knowledge, there has hitherto been no report on the use of these two methods for the screening of plant extracts for antiplasmodial activity. At this point, further investigation is needed before they can be applied as routine methods.

In vitro tests are also useful for the detection of drug interaction, i.e., antagonism, synergism, and simple additive effects, by the isobologram method (Rasoanaivo et al., 1994). However, they have some limitations because they are not necessarily predictive of *in vivo* activity.

16.4.1.2 Mechanism-Based Assays

The progress in our understanding of the biology and biochemistry of malaria parasites during the last two decades has led to the identification of drug targets that are both parasite specific and essential for parasite growth and survival. There are several recent reports on chemotherapeutic targets for antimalarial drug discovery and development (Olliaro and Yuthavong, 1999; Jomaa et al.,

1999; Macreadie et al., 2000). Particularly, the heme polymerisation process has attracted much attention as a valid drug target (Monti et al., 1999; Kurosawa et al., 2000). However, one limiting factor is the use of the expensive radiolabeled ^{14}C-hematin in the experimental procedure. Recently, it has been shown that the detoxification mechanism of hematin in malaria parasite involves not a polymerisation, as was previously believed, but a dimerisation of hematin, which is called biocrystallisation (Hempelmann and Egan, 2002). This has led to the development of simple, nonradiolabeled techniques to assess extracts/compounds for their ability to inhibit the biocrystallisation process (Deharo et al., 2002; Sahal et al., 2003; Steele et al., 2002).

16.4.1.3 Cytotoxicity Tests

Cytotoxicity tests are not designed to screen extracts for antimalarial activity. The malaria parasite, like all living organisms, undergoes cell division from merozoites to schizonts. Extracts and compounds that inhibit cell division may also kill the parasite. Cytotoxic natural products may therefore give false positive results for antiplasmodial activity in screening programs using either the *in vitro* radioactive or the colorimetric methods. Any extract that inhibits the growth of *P. falciparum in vitro* should therefore be tested systematically for cytotoxicity. There are several techniques currently available (Husoy et al., 1993). To estimate the potential of a given extract to inhibit parasite growth without host toxicity, the selectivity index (SI) was introduced and defined as the ratio of IC_{50} in cytotoxicity to the IC_{50} in *P. falciparum*. The higher the SI, the higher is the selective antiplasmodial activity of a given extract or compound.

16.4.1.4 *In Vivo* Antiplasmodial Activity

The reference test for the blood schizontocidal activity of plant extracts is the 4-day suppressive test of Peters et al. (1975) using the rodent malaria model (see also Peters and Robinson, 1999, for more details). It evaluates the reduction in parasitaemia of mice infected by rodent malaria parasite following administration of a daily dose of plant extract. It is also useful to evaluate drug interactions. Results are expressed as percent of parasitaemia inhibition with respect to untreated controls. One hundred percent parasite inhibition is called parasite clearance.

Whenever possible, each *in vitro* or *in vivo* test involving the erythrocytic stages of *Plasmodium* malaria should include a positive control with a *Cinchona* sp. (*C. ledgeriana*, *C. succirubra*, *C. calisaya*). These plants offer the possibility of calculating the activity index (AI) defined as IC_{50} of a plant extract/IC_{50} of *Cinchona* sp. extract. If AI = 1, the plant is as active as the reference; if AI < 1, the plant is more active than the reference; if AI > 1, the plant is less active than the reference. This index would make results comparable from different laboratories.

16.4.2 RANKING THE EFFICACY RESULTS OF ANTIMALARIAL EXTRACTS

Assuming that antimalarial plants are effective in treating the disease, those acting on the erythrocytic stage of the malaria parasite are expected to have significant *in vitro* antiplasmodial activity. In a previous paper, we ranked the level of efficacy of extracts according to their IC_{50} values (Rasoanaivo et al., 2002). The proposed thresholds for *in vitro* antiplasmodial activity are summarised with some modifications in Table 16.1.

For single-dose testing, extracts that inhibit 80% or more of the parasite growth with a concentration of 10 µg/ml are considered worthy of further bioassay-guided fractionation.

Regarding *in vivo* activity in rodent malaria models, parasite clearance without recrudescence for a period of 60 days would be the ideal criterion. The suggested thresholds for *in vivo* antiplasmodial activity for a dose level of 250 mg/kg/day are listed in Table 16.2.

With respect to the selectivity index, SI > 300 should be attained on the basis of mean values reported in relevant papers (O'Neill et al., 1985; Schrével et al., 1994; Angerhofer et al., 1999).

TABLE 16.1

Proposed Thresholds for *In Vitro* Antiplasmodial Activity of Antimalarial Extracts

IC_{50} (µg/ml)	Level of Activity
<0.1	Very good
0.1–1.0	Good
	This is the concentration range that is generally considered active in screening programs for antimalarial activity, warranting bioassay-guided fractionation
1.1–10	Good to moderate
	This range may reasonably be considered for bioassay-guided fractionation
11–25	Weak
26–50	Very weak
>100	Inactive

TABLE 16.2

Proposed Thresholds for *In Vivo* Activity of Antimalarial Extracts at the Dose of 250 mg/kg/day

% Inhibition	Level of Activity
100–90	Very good to good activity
90–50	Good to moderate
50–10	Moderate to weak
0	Inactive

Ideally, extracts effective at the blood stage of the malaria parasite should have strong *in vitro* and *in vivo* antimalarial activities and should be devoid of cytotoxicity at concentrations of up to 100 µg/ml. Priority should be given to the safety evaluation of such extracts with the objective of conducting clinical trials of the traditional medicine or a derived phytomedicine. Extracts that meet these criteria also deserve further investigation aimed at isolating the bioactive compound(s), which may serve as candidate(s) for drug development in the Western pharmaceutical context.

16.5 BIOASSAY-GUIDED FRACTIONATION

It has been widely recognised in screening programs involving plant extracts, and our experience in screening for antimalarial activity confirms this (Rasoanaivo et al., 2002), that the percentage of plants with confirmed activity could be significantly increased if the plants are subjected to preliminary fractionation before screening (Statz and Coon, 1976). It is therefore advisable that extracts with moderate activity be subjected to bioassay-guided fractionation. If reasonably good antimalarial activity is not obtained at this stage, further work is discontinued since activity rarely improves significantly as additional fractionation progresses. During the phytochemical study, continuous monitoring for antimalarial activity is conducted with all the fractions obtained, avoiding exhaustive isolation and structure elucidation for those compounds not related to the antimalarial activity detected initially.

Fractionation should be initiated using simple methods, which entail minimum risk of destroying or altering active principles, thus leading to useless artifacts. As far as possible, no chemical reaction of any kind should be utilised during the fractionation procedure. The one exception is the conventional acid–base treatment when the active constituents are alkaloids that are presumably unaffected by this treatment.

When the single-solvent extraction procedure is used, there are several alternative methods of fractionation to choose from depending on the equipment available in the laboratory. Four different methods with which we have some experience in our screening programs are described in the next paragraphs.

16.5.1 Solvent Partition

Solvent partition is the simplest and least expensive method for a preliminary fractionation. Generally performed in a separating funnel, this method is based on the differential solubility of compounds between two immiscible solvent phases. It is used as early as possible in the processing of plant material in order to remove the maximum quantity of inactive constituents. An important feature of this method is its flexibility/adaptability, offering a good range of options for a preliminary fractionation. Modification or adaptation can be done gradually, at any time during the screening program, with the aim of developing a procedure that would give the highest percentage of active compounds for a given sample of antimalarial plants. Figure 16.2 shows a general procedure we presently use in our institute for extraction and solvent partition in screening programs.

Initially we used chloroform as a solvent of medium polarity (Rasoanaivo et al., 1999), but the reported toxicity and the difficulty in obtaining this hazardous chemical led us to discontinue its use in favor of ethyl acetate (Rasoanaivo et al., 2002). Our procedure is flexible, and in most cases ethanol extracts and ethyl acetate fractions are systematically tested. The other fractions are considered when ambiguity occurs in the test results.

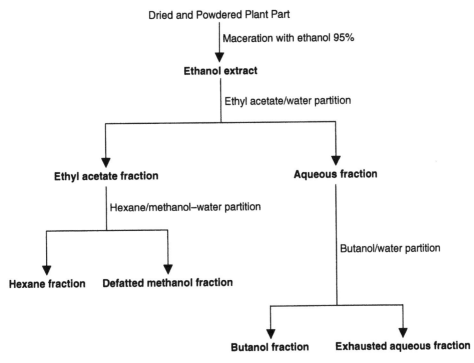

FIGURE 16.2 A general procedure used by Institut Malgache de Recherches Appliquées (IMRA) for extracting plant material and fractionating into different solubility classes.

Successive solid–liquid extractions using solvents in increasing order of polarity represent an acceptable alternative to liquid–liquid partition.

16.5.2 COUNTERCURRENT DISTRIBUTION

Although some of the inactive constituents are probably removed by solvent partition, fractions are still a mixture of several compounds. It is known that the biological activity is not always due to the main constituents but sometimes to minor ones, or even to the synergism of active and inactive constituents. One of the best methods for further fractionation of active extracts employs the countercurrent distribution (CCD) technique performed using an automated Craig apparatus. The principle of fractionation involves the partition of each of the substances in a mixture between two immiscible solvent phases. Basically, CCD fractionation is similar to solvent partition in a separating funnel, but the operation is automatically repeated several times in terms of transfers in the Craig apparatus. The choice of a suitable two-phase solvent system is crucial for a successful fractionation with CCD (Galeffi, 1980). After a judicious mixture of organic solvents and water is made, 100 to 200 transfers are run and fractions are gathered according to similarity pattern in thin-layer chromatography (TLC) and systematically tested. Further CCD fractionation can be performed on the active fraction(s) to isolate the active constituent(s). This technique presents several advantages over the solid–liquid fractionation methods: (1) no irreversible adsorption, (2) total recovery of introduced extract, (3) possibility of isolating very minor constituents with close polarity, and (4) minimum risk of sample degradation. Our Italian colleagues at the Istituto Superiore di Sanità in Rome have used this technique extensively for the bioassay-directed fractionation/separation of plant extracts. We have benefited from their expertise in the isolation of the minor cytotoxic constituents of *Kalanchoe tomentosa* (Rasoanaivo et al., 1993) and *Millettia pervilleana* (Galeffi et al., 1997) and in other work.

16.5.3 SOLID–LIQUID COLUMN CHROMATOGRAPHY

Silica gel column chromatography is a good method for fractionating active extracts if the compounds have a low to medium polarity. We have preferentially used this technique in most of our bioassay-guided fractionation. To this end, a coarse fractionation is first done using a system of solvents with increasing polarity, and the collected fractions are examined by TLC. Fractions with similar TLC patterns are combined and monitored by antimalarial tests. Active semipurified fraction(s) are further subjected to a more refined column chromatography.

In some cases, a combination of two or several analytical techniques is necessary to adequately locate and isolate the antimalarial substances. Samuelson and co-workers (1985, 1987) described a fractionation procedure for crude aqueous extracts using a combination of solvent partition, ion exchangers, and gel filtration, all fractions being subjected to bioassay. This method gives useful information about the physical properties of the active constituents, which might enable the development of suitable isolation methods.

16.5.4 PREPARATIVE HPLC FRACTIONATION

Preparative high-performance liquid chromatography (HPLC) could be used on a routine basis to fractionate extracts before subjecting them to bioassay (Sevenet et al., personal communication). However, the equipment required and the cost of columns that must be replaced every 40 passages would make this technique beyond the reach of many laboratories.

16.6 PROBLEMS ENCOUNTERED IN THE ANTIMALARIAL TESTING OF PLANT EXTRACTS

Plant extracts are generally mixtures of several compounds. Their composition as well as the concentration of individual components may vary depending on ecological conditions. Most pharmacological screening problems arise from this.

16.6.1 FAILURE TO ISOLATE ACTIVE CONSTITUENT(S) FROM AN ACTIVE EXTRACT

In a bioassay-guided fractionation, failure to isolate active constituents from an active extract is often a serious problem. There are some plausible explanations for this.

- The antimalarial compound is labile under certain conditions (of temperature, acidity, basicity, light, solvent used, etc.) and progressive degradation occurs during the fractionation procedure.
- Some compounds are inactive on their own but may act synergistically with other constituents. When they become separated, antimalarial activity generally decreases or disappears. In this particular case, the crude extracts are far more active than the individual compounds.
- The fractionation procedure devised to isolate the bioactive constituents is inadequate.

16.6.2 FAILURE TO OBTAIN POSITIVE RESULTS WITH EXTRACTS CONTAINING ANTIMALARIAL CONSTITUENTS

Although some antimalarial plants do not display significant activities on either the erythrocytic or the hepatic stages, they may unexpectedly contain useful active compounds. Failure to identify these compounds could be explained as follows:

- The active compounds are present in insufficient quantity in the crude extract to display activity at the dose levels employed.
- If the bioactive compound is present in sufficient quantity, another explanation would be that other constituents may exert antagonistic effects during testing procedure, or some substances may stimulate some biological parameters that counteract the action of the bioactive compound, and therefore negate the effect of inhibitory constituents.
- The pharmacological model used to demonstrate the biological activity would be inappropriate. Some extracts are active *in vitro* but lack *in vivo* activity because bioactive constituents would be metabolised *in vivo* into inactive compounds, or they do not enter the parasite particularly because of their hydrophilic property. On the other hand, some extracts are active only *in vivo* because some components would be metabolised *in vivo* into active compounds.

16.6.3 FALSE POSITIVE TESTS

One current false positive test is due to the erythrotoxicity of some constituents. This is the case of extracts containing saponins. Their strong *in vitro* activity is generally due to the known haemolytic action of some saponins, which affect the red blood cells.

16.6.4 Failure to Duplicate Antimalarial Results between Different Samples of the Same Plant

One explanation would be a variation in the concentration of active principles between samples as a result of environmental and genetic variability (period of collection, areas, developmental stages such as plant height, flowering time, size of vegetative features).

Another cause is the lack of ability to collect the same specimen on separate occasions (Rasoanaivo et al., 2002).

16.6.5 Discrepancies between Test Results for the Same Extract between Different Laboratories

Discrepancies have occurred in the results of *in vitro* antimalarial tests for the same extract between independent laboratories. Some discrepancies fortunately fall within the range of active extracts, which permits the bioassay-guided fractionation of the extracts. In other cases, however, extracts claimed to be active in one laboratory are found to be inactive in another and vice versa. This problem is still the subject of debate and controversies. Several parameters such as solvents used to dissolve the extracts (either dimethyl sulfoxide [DMSO] or methanol), parasitaemia, haematocrit, CO_2 concentration, stability of extracts, incubation time, and strains used may play an important role in the tests. Since it is practically impossible to harmonise all experimental procedures in various laboratories, discrepancies are unavoidable. The activity scaling proposed here for *in vitro* antiplasmodial activity should therefore be regarded as a flexible guide for selecting extracts for further investigation rather than a rigid rule to be followed in all cases.

16.7 SCENARIOS IN THE OUTCOME OF THE EFFICACY EVALUATION

There are several scenarios depending on the chemical structure of the bioactive individual compounds and their biological activity. The isolated antiplasmodial constituent may have:

- Both known structure and known antimalarial activity, i.e., isolating quinine in another plant species
- Known structure with antimalarial activity acting by a known validated mechanism
- Known structure with antimalarial activity acting by a novel mechanism
- New structure with antimalarial activity acting by a known validated mechanism
- Novel structure with antimalarial activity acting by a new mechanism

16.8 FURTHER INVESTIGATION OF ANTIMALARIAL PLANTS

Considering the international trend in malaria chemotherapy research, we have proposed in this paper two antimalarial screening targets, the erythrocytic and the hepatic stages of malaria parasites. However, unlike the tight selection of drug candidates for development in the pharmaceutical industry, the results should be handled with flexibility, taking into account the holistic approach of traditional medicine. There cannot be a rigid framework with defined decision points. The stages in the process of decision making toward a final product (efficacy evaluation, safety evaluation, standardisation, galenical formulation, clinical observation, and clinical trials) will be different for each plant that is investigated, and it is necessary to spend more time on the study of individual plants that have strong ethnobotanical evidence of usefulness in the treatment or prevention of malaria. Particularly, the high frequency of indications of a certain plant must encourage further insight into the study of antimalarial activity of a plant, with more appropriate methods. Several scenarios must be considered carefully before rejecting any extracts.

16.8.1 EXTRACTS WITH GOOD *IN VITRO* POTENCY BUT LACKING *IN VIVO* ACTIVITY

This is sometimes encountered with herbal antimalarials. One explanation is that the active constituent(s) is metabolised *in vivo* into inactive compound(s). Another plausible explanation is the unsuitability of the *in vivo* rodent malaria models to demonstrate the expected activity. Additional *in vivo* models may be needed to adequately evaluate these antimalarial plants (Dow et al., 1999).

16.8.2 EXTRACTS WITH GOOD *IN VIVO* ACTIVITY BUT LACKING *IN VITRO* ACTIVITY

To the best of our knowledge, this case is rare. One logical explanation is that constituents may act as pro-drugs, and *in vivo* metabolisation is required to yield active compounds. The cases of proguanil (Carrington et al., 1951; Crowther and Levi, 1953) and primaquine (Russell et al., 2003) illustrate this possibility. It is also possible that extracts act by an unknown or unexpected mechanism. The investigation of these extracts may open up new lines of research, possibly leading to the development of new tools for the treatment of malaria.

16.8.3 EXTRACTS WITH NEITHER *IN VITRO* NOR *IN VIVO* ACTIVITY

Assuming that the plants are claimed to be efficient in treating malaria in a broad sense, there are many alternatives to deal with such a situation, and this has been explained in the integrated approach to antimalarial plants (Rasoanaivo et al., 2002). Particularly, the possible stimulation of the immune system of the infected host by antimalarial extracts is a relevant area to explore further (Foldes and Matyi, 1994; Murata et al., 1999). In some cases, successive treatments with various plants are frequently done until the patient is cured; thus, possible additive or synergistic activities should also be considered.

16.8.4 PHYTOMEDICINES WITH SEVERAL PLANT INGREDIENTS

Some antimalarial phytomedicines containing ingredients from several plants are used in primary health care (Gasquet et al., 1993). Each individual plant may have biological activities that fall into the categories discussed in this paper. It is also possible that some herbs have synergistic effects, and other materials may counteract the potential toxicity of other ingredients in the formulation. In our opinion clinical investigation should be tailored to adequately evaluate these traditional medicines and phytomedicines, and phytochemical studies from plant mixtures should not be discarded, but one must keep in mind that to isolate one or more active compounds from one plant is already difficult work; thus, to determinate interactions between a mixture of plants is even harder.

16.9 CONCLUSIONS

Because of the complex nature of biological systems, no kind of test can be expected to function perfectly. In other words, it is impossible to devise a test or series of tests that will identify all active substances with no false positives or false negatives. Therefore, scientists performing screening programs must show vigilance, ingenuity, imagination, and common sense in order to minimise the shortcomings of fractionation and tests so as not to allow true biological activity to go undetected. There is no one definitive or rigid scheme for efficacy evaluation of antimalarial plants; this is a dynamic process, and the rule adopted is learning by doing.

During the past 5 years, several screening programs of plant extracts for antiplasmodial activities have been published by or in collaboration with Third World scientists (Rasoanaivo et al., 2002; Antoun et al., 2001; Simonsen et al., 2001; Muñoz et al., 2000; Traore-Keita et al., 2000; Omulokoli et al., 1997). Furthermore, many active compounds have been discovered in laboratories based in developing countries. We believe that in the next decades to come, many useful antimalarial compounds will be isolated from tropical plants. The majority of those discoveries will be made by or in collaboration with competent and highly motivated scientists in developing countries.

REFERENCES

Angerhofer, C.K., Guinaudeau, H., Wongpanich, V., Pezzuto, J.M., and Cordell, G.A. (1999). Antiplasmodial and cytotoxic activity of natural bisbenzylisoquinoline alkaloids. *J. Nat. Prod.*, 62, 59–66.

Antoun, M.D., Ramos, Z., Vazques, J., Oquendo, I., Proctor, G.R., Gerena, L., and Franzblau, S.G. (2001). Evaluation of the flora of Puerto Rico for *in vitro* antiplasmodial and antimycobacterial activities. *Phytother. Res.*, 15, 638–642.

Carrington, H.C., Crowther, A.F., Davey, D.G., Levi, A.A., and Rose, F.L. (1951). A metabolite of "Paludrine" with high antimalarial activity. *Nature*, 168, 1080.

Crowther, A.F. and Levi, A.A. (1953). Proguanil, the isolation of a metabolite with high antimalarial activity. *Br. J. Pharmacol.*, 8(1), 93–97.

Deharo, E., Garcia, N.R., Oporto, P., Gimenez, A., Sauvain, M., Jullian, V., and Ginsburg, H. (2002). A non-radiolabelled ferriprotoporphyrin IX biomineralisation inhibition test for the high throughput screening of antimalarial compounds. *Exp. Parasitol.*, 100, 252–256.

Delhaes, I., Lazaro, J.E., Gay, F., Thellier, M., and Danis, M. (1999). The microculture tetrazolium assay (MTA): another colorimetric method of testing *Plasmodium falciparum* chemosensitivity. *Ann. Trop. Med. Parasitol.*, 93, 31–40.

Desjardins, R.E., Canfield, C.J., Haynes, J.D., and Chulay, J.D. (1979). Quantitative assessment of antimalarial activity *in vitro* by an automated dilution technique. *Antimicrob. Agents Chemother.*, 16, 710–718.

Dow, G.S., Reynoldson, J.A., and Thompson, R.C. (1999). *Plasmodium berghei*: a new rat model for assessment of blood schizonticidal activity. *Exp. Parasitol.*, 93, 92–94.

Foldes, J. and Matyi, A. (1994). The immunomodulating effect of a new polyamine (the MAP-1987) administered with chloroquine in plasmodia infected mice. *Acta Microbiol. Immunol. Hung.*, 41, 73–82.

Galeffi, C. (1980). New trends in the separation of active principles from plants. *J. Ethnopharmacol.*, 2, 127–134.

Galeffi, C., Rasoanaivo, P., Federici, E.G., Pallazino, G., Nicoletti, M., and Rasolondratovo, B. (1997). Two prenylated isoflavanones from *Millettia pervilleana*. *Phytochemistry*, 45, 189–192.

Gasquet, M., Delmas, F., Timon-David, P., Keita, A., Guindo, M., Koita N., Diallo, D., and Doumbo, O. (1993). Evaluation *in vitro* and *in vivo* of a traditional antimalarial "Malarial 5." *Fitoterapia*, 65, 423–426.

Gessler, M.C., Nkunya, M.H., Mwasumbi, L.B., Heinrich, M., and Tanner, M. (1994). Screening Tanzanian medicinal plants for antimalarial activity. *Acta Trop.*, 56, 65–77.

Hempelmann, E. and Egan, T.J. (2002). Pigment biocrystallisation in *Plasmodium falciparum*. *Trends Parasitol.*, 18, 11.

Huber, W. and Koella, J.C. (1993). A comparison of three methods of estimating EC_{50} in studies of drug resistance of malaria parasites. *Acta Trop.*, 55, 257–261.

Husoy, T., Syversen, T., and Jenssen, J. (1993). Comparison of four *in vitro* cytotoxicity tests: the MTT assay, NR assay, uridine incorporation and protein measurements. *Toxicity In Vitro*, 7, 149–154.

Jomaa, H., Wiesner, J., Sanderbrand, S., Altincicek, B., Weidemeyer, C., Hintz, M., Turbachova, I., Eberl, M., Zeidler, J., Lichtenthaler, H.K., Soldati, D., and Beck, E. (1999). Inhibitors of the nonmevalonate pathway of isoprenoid biosynthesis as antimalarial drugs. *Science*, 285, 1573–1576.

Kurosawa, Y., Dorn, A., Kitsuji-Shirane, M., Shimada, H., Satoh, T., Matile, H., Hofheinz, W., Masciadri, R., Kansy, M., and Ridley, RG. (2000). Hematin polymerisation assay as a high-throughtput screen for identification of new antimalarial pharmacophores. *Antimicrob. Agents Chemother.*, 44, 2638–2644.

Leaman, D.J., Arnason, J.T., Yusuf, R., Sangat-Roemantyo, H., Soedjito, H., Angerhofer, C.K., and Pezzuto, J.M. (1995). Malaria remedies of the Kenyah of the Apo Kayan, East Kalimantan, Indonesian Borneo: a quantitative assessment of local consensus as an indicator of biological efficacy. *J. Ethnopharmacol.*, 49, 1–16.

Le Bras, J. and Deloron, P. (1983). In vitro study of drug sensitivity of *Plasmodium falciparum*: evaluation of a new semi-micro test. *Am. J. Trop. Med. Hyg.*, 32, 447–451.

Macreadie, I., Ginsburg, H., Sirawaraporn, W., and Tilley, L. (2000). Antimalarial drug development and new targets. *Parasitol. Today*, 16, 438–443.

Makler, M.T., Ries, J.M., Williams, J.A., Bancroft, J.E., Piper, R.C., Gibbins, B.L., and Hinrichs, D.J. (1993). Parasite lactate dehydrogenase as an assay for *Plasmodium falciparum* drug sensitivity. *Am. J. Trop. Med. Hyg.*, 48, 739–741.

Monti, D., Vodopivee, B., Basilico, N., Olliaro, P., and Taramelli, D. (1999). A novel endogenous antimalarial: Fe (II)-protoporphyrin IXα (heme) inhibits hematin polymerization to β-hematin (malaria pigment) and kills malaria parasites. *Biochemistry*, 38, 8858–8863.

Muñoz, V., Sauvain, M., Bourdy, G., Callapa, J., Bergeron, S., Rojas, I., Bravo, J.A., Balderrama, L., Ortiz, B., Gimenez, A., and Deharo, E. (2000). A search for natural bioactive compounds in Bolivia through a multidisciplinary approach. Part I. Evaluation of the antimalarial activity of plants used by the Chacobo Indians. *J. Ethnopharmacol.*, 69, 127–137.

Murata, K., Takano, F., Fushiya, S., and Oshima, Y. (1999). Potentiation by febrifugine of host defense in mice against *Plasmodium berghei* NK65. *Biochem. Pharmacol.*, 58, 1593–1601.

Olliaro, P.L. and Goldberg, D.E. (1995). The *Plasmodium* digestive vacuole: metabolic headquarters and choice drug target. *Parasitol. Today*, 11, 294–297.

Olliaro, P.L. and Yuthavong, Y. (1999). An overview of chemotherapeutic targets of antimalarials drug discovery. *Pharmacol. Ther.*, 81, 91–110.

Omulokoli, E., Khan, B., and Chhabra, S.C. (1997). Antiplasmodial activity of four Kenyan medicinal plants. *J. Ethnopharmacol.*, 56, 133–137.

O'Neill, M.J., Bray, D.H., Boardman, P., Phillipson, J.D., and Warhust, D.C. (1985). Plants as source of antimalarial drugs. Part 1. *In vitro* test method for the evaluation of crude extracts from plants. *Planta Med.*, 5, 394–398.

Peters, W., Portus, J.H., and Robinson, B.L. (1975). The chemotherapy of rodent malaria, XXII. The value of drug resistant strains of *P. berghei* in screening for blood schizontocidal activity. *Ann. Trop. Med. Parasitol.*, 69, 155–171.

Peters, W. and Robinson, B.L. (1999). Malaria. In *Handbook of Animal Models of Infection*, Zak, O. and Sande, M.A., Eds. Academic Press, London, pp. 757–773.

Rasoanaivo, P. (2000). Une banque de données sur les plantes médicinales de Madagascar, *Info-Essences*, 15, 5–6.

Rasoanaivo, P., Galeffi, C., Multari, G., Nicoletti, M., and Capolongo, L. (1993). Kalanchoside, a cytotoxic bufadienolidic glycoside from *Kalanchoe tomentosa. Gaz. Chim. Ital.*, 123, 539–541.

Rasoanaivo, P., Oketch-Rabah H., Willcox, M., Hasrat, J., and Bodeker, G. (2003). Preclinical considerations on antimalarial phytomedicines. Part I. Efficacy evaluation. *Fitoterapia*, in press.

Rasoanaivo, P. (2002). Pre-clinical evaluation of traditional antimalarials: guidelines and recent results. Abstracts of the third MIM Pan-African Malaria Conference, November 17–22, Arusha, Tanzania, p. 88.

Rasoanaivo, P., Ramanitrahasimbola, D., Rafatro, H., Rakotondramanana, D., Robijaona, B., Rakotozafy, A., Ratsimamanga-Urverg, S., Labaïed, M., Greller, P., Allorge, L., Mambu, L., and Frappier, F. (2004). Screening plant extracts of Madagascar for the search of antiplasmodial compounds. *Phytotherapy Research* (in press).

Rasoanaivo, P., Ratsimamanga-Urverg, S., Milijaona, R., Rafatro, H., Galeffi, C., and Nicoletti, M. (1994). *In vitro* and *in vivo* chloroquine potentiating action of *Strychnos myrtoides* alkaloids against chloroquine-resistant strain of *Plasmodium* malaria. *Planta Med.*, 60, 13–16.

Rasoanaivo, P., Ratsimamanga-Urverg, S., Ramanitrahasimbola, D., Rafatro, H., and Rakoto-Ratsimamanga, A. (1999). Criblage d'extraits de plantes de Madagascar pour recherche d'activité antipaludique et d'effet potentialisateur de la chloroquine, *J. Ethnopharmacol.*, 64, 117–127.

Ratsimamanga-Urverg, S., Rasoanaivo, P., Ramiaramanana, L. Milijaona, R., Rafatro, H., Verdier, F., Rakoto-Ratsimamanga, A., and Le Bras, J. (1992). *In vitro* antimalarial activity and chloroquine-potentiating action of BBIQ enantiomers from *Strychnopsis thouarsii* and *Spirospermum penduliflorum. Planta Med.*, 58, 540–543.

Russell, B., Kaneko, O., Jenwithisuk, R., et al. (2003). Antimalarial Activity on Liver and Blood Stage *Plasmodium vivax*, as Measured by Real Time PCR and IFA Methods. Paper presented at the International Conference on Malaria: Current Status and Future Trends, Chulabhorn Research Institute, Bangkok, Thailand, February 16–19.

Sahal, D., Kannan, R., and Chauhan, V.S. (2003). Applying malaria parasite's heme detoxification system for screening potential antimalarial drugs. *Anal. Biochem.*, 312, 258–260.

Samuelson, G. (1987). Plants used in traditional medicine as sources of drugs. *Bull. Chem. Soc. Ethiopia*, 1, 47–54.

Samuelson, G., Kyerematen, G., and Farah, M. (1985). Preliminary chemical characterisation of pharmacologically active compounds in aqueous plant extracts. *J. Ethnopharmacol.*, 14, 193–201.

Schrével, J., Sinou, V., Grellier, P., Frappier, F., Guénard, D., and Potier, P. (1994). Interactions between docetaxel (taxotere) and *Plasmodium falciparum*-infected erythrocytes. *Proc. Natl. Acad. Sci. U.S.A.*, 91, 8472–8476.

Sharma, P. and Sharma, J.D. (1999). Evaluation of *in vitro* schizontocidal activity of plant parts of *Calotropis procera*: an ethnobotanical approach. *J. Ethnopharmacol.*, 68, 83–95.

Simonsen, H.T., Nordskjold, J.B., Nyman, U., Palpu, P., Joshi, P., and Varughese, G. (2001). *In vitro* screening of Indian medicinal plants for antiplasmodial activity. *J. Ethnopharmacol.*, 74, 195–204.

Statz, D. and Coon, F.B. (1976). Preparation of plant extracts for antitumor screening. *Cancer Treat. Rep.*, 60, 999–1005.

Steele, J.C.P., Phelps, R.J., Simmonds, M.S.J., Warhurst, D.C., and Meyer, D.J. (2002). Two novel assays for the detection of haemin-binding properties of antimalarials evaluated with compounds isolated from medicinal plants. *J. Antimicrob. Chemother.*, 50, 25–31.

Traore-Keita, F., Gasquet, M., Di Giorgio, C., Ollivier, E., Delmas, F., Keita, A., Doumbo, O., Balansard, G., and Timon-David, P. (2000). Antimalarial activity of four plants used in traditional medicine in Mali. *Phytother. Res.*, 14, 45–47.

17 Models for Studying the Effects of Herbal Antimalarials at Different Stages of the *Plasmodium* Life Cycle

Dominique Mazier, J.F. Franetich, M. Carraz, O. Silvie, and P. Pino

CONTENTS

17.1 INTRODUCTION

Malaria remains a major threat to public health with 40% of the world's population currently at risk, 300 to 500 million cases of clinical malaria, and 1 to 2 million deaths each year. The dramatic extension of plasmodial resistance to most antimalarial drugs is directly responsible for the widespread persistence of this high level of malaria endemicity and related morbidity in tropical areas (World Health Organization, 2001). Such a situation, together with the difficulty in developing vaccines (Silvie et al., 2002), highlights the urgent need for novel antimalarial drugs (Ridley, 2002). In this regard, natural products of vegetal origin remain a poorly exploited source of potentially active molecules, even though the two most effective drugs for malaria originate from plants: quinine from the bark of the Peruvian cinchona tree and artemisinin from the Chinese antipyretic *Artemisia annua*. It is likely that other plants contain as yet undiscovered antimalarial substances.

There is controversy over the most reliable way to validate the clinical effectiveness of herbal antimalarials as claimed by traditional healers. There are two approaches that can be followed in a complementary manner: 1) clinical trials and 2) biological *in vitro* and *in vivo* assays (whose interpretation is more difficult than usually thought).

Discrepancies often exist when comparing field and laboratory results. One of the most important reasons is the difficulty in selecting appropriate laboratory models. This difficulty is linked to the complexity of the *Plasmodium* cycle and the variety of pathophysiological manifestations of the disease. Although the parasite has four stages that could be targeted by drugs (see Figure 17.1), most of the assays performed are restricted to the asexual erythrocytic stage of the parasite. This is mainly due to the large-scale *in vitro* culture of this stage in several laboratories, in contrast to the other developmental stages of the parasite, which are almost always ignored. Yet the antimalarial activity of a product may be restricted to an effect interfering with another part of the cycle. Schematically, a plant unable to kill the erythrocytic parasite may still prevent clinical manifestations (1) by impeding blood stage infection (activity at the hepatic stage), or (2) by interfering with any consequences of blood stage multiplication (i.e., secretion of cytokines, adhesion on endothelial cells). In this chapter we will concentrate on the assays to be performed even when the plant or its fractions have no effect on the erythrocytic multiplication of the parasite.

17.2 ASEXUAL ERYTHROCYTIC STAGE SCREENING

The first screening usually performed aims to identify products able to kill the parasite during its erythrocytic multiplication. This has already been discussed in more detail in Chapter 16. In summary, different assays are available, all of them rendered possible because *Plasmodium falciparum*-infected red blood cells can be maintained in culture, in contrast to other species (Trager, 1994). One of the most used assays has been described by Desjardins et al. (1979) and consists in the quantitative assessment of antimalarial activity *in vitro* by a semiautomated microdilution technique. Assays are usually performed using *P. falciparum* laboratory strains known for their sensitivity or resistance to antimalarials such as chloroquine and artemisinin used as reference. When a plant of interest is identified, it may be investigated *in vivo*. Rodent models include *Plasmodium vinckei petteri* in mice (Fleck et al., 1997) and the more sophisticated *P. falciparum* in immunocompromised BXN (bg/bg xid/xid nu/nu) mice (Moreno et al., 2001). Primate models have also been used (Wengelnik et al., 2002). When no effect is observed on the erythrocytic multiplication, it is still crucial to test the plant in other models that we will describe in more detail.

17.3 HEPATIC STAGE SCREENING

Before multiplying in the blood (the process responsible for pathology), the parasite undergoes multiplication in the liver where it can be a target for drug intervention. As causal prophylactics and antirelapse agents, only two related drugs are available that target the parasite during the hepatic phase of development, primaquine and tafenoquine. However, they have recognised haematological toxicity, and their use is restricted, particularly in Africa because of the frequency of G6PD deficiency, not to mention their unaffordable cost. Atovaquone activity on the first step of the parasite development in the liver is now clear, but this drug is not an antirelapse agent (Shanks et al., 1998).

The development of pharmaceutical products capable of inhibiting parasite maturation inside hepatocytes could be relevant for three main reasons.

First, they could act as causal prophylactic agents by preventing the parasite from reaching the blood. Such drugs would be particularly needed for (1) protection in endemic areas, of people at risk (pregnant women), especially in areas where parasites display high levels of drug resistance; and (2) people moving into an endemic region for malaria (refugees, European travelers).

Second, it is generally accepted that a very small number of sporozoites are injected by *Anopheles* mosquitoes. Therefore, the probability of any given mutation being present in liver stage parasites is many orders of magnitude smaller than that of the same mutation being found in blood stages (the number of mitotic divisions is much larger during erythrocytic schizogony than during liver schizogony). It follows that the chance of mutants emerging under pressure from a drug aimed at the liver stage is very small, making them an attractive target (Mazier and Doerig, 2000). Resistance to drugs that are strictly specific for exo-erythrocytic stages is therefore unlikely to emerge.

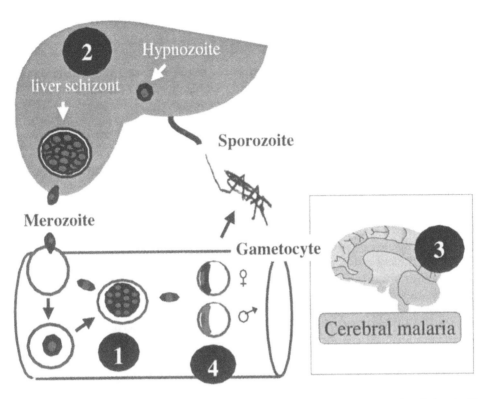

FIGURE 17.1 The parasite cycle starts in man with the injection of *Plasmodium* sporozoites during the blood meal of an infected female anopheline mosquito. Sporozoites rapidly join the liver and invade hepatocytes, where they differentiate into liver schizonts. In *P. vivax* and *Plasmodium ovale*, a proportion of sporozoites transform into hypnozoites; these latent forms can persist for months or even years in hepatocytes, before starting to divide and give rise to the late relapses characteristic of infection with these species. After approximately 7 days of parasite multiplication in the liver, mature schizonts rupture and release thousands of merozoites in the circulation. Each merozoite can invade an erythrocyte, where it differentiates into a schizont. After 48 to 72 hours, the mature erythrocytic schizonts rupture and release new merozoites. These merozoites invade erythrocytes and initiate other rounds of erythocytic multiplication, or differentiate into sexual stages, the gametocytes. When ingested by a blood-feeding mosquito, the gametocytes turn to male and female gametes, which combine to form a diploid zygote in which, 10 to 15 days later, sporozoites will differentiate and then migrate to the salivary glands where they can be inoculated into a new host. Four stages in this cycle could be inhibited by drugs: The *asexual erythrocytic stage* is responsible for the pathology. Different possibilities exist for interfering with the disease: one is to develop drugs able to kill the intraerythrocytic parasite (1); another is to prevent clinical manifestations, mainly cerebral malaria, which is, with anemia, one of the most tragic complications of falciparum malaria (3). The *hepatic stage* (2) is asymptomatic but is the starting point of parasite proliferation. This truly prophylactic approach, while preventing any erythrocytic growth and thus malaria pathology, could also limit the appearance of drug resistance. This approach appears to be possible when referring to primaquine, the only available molecule that is active against both the multiplication of the liver parasites (schizonts) and the latent form (hypnozoite) responsible for relapsing infections observed with *P. vivax* and *P. ovale*. The *sexual erythrocytic stage* (4) is asymptomatic but responsible for parasite transmission to mosquitoes.

Third, they could act as antirelapse agents that could constitute a radical cure against the recrudescence of *Plasmodium vivax*. *P. vivax* is the most widely distributed human malarial parasite, causing an estimated 70 to 80 million clinical cases annually (Mendis et al., 2001). In some parts of the world, including Latin America, this is the most prevalent species of human malaria. Moreover, *P. vivax* has now developed resistance to chloroquine in Southeast Asia and in South America. Unlike most *Plasmodium* species, *P. vivax* produces hypnozoites, dormant liver stages that upon reactivation cause new clinical attacks or relapses years after the original infection (Krotoski, 1985). An important point is the absence of hypnozoite sensitivity to molecules active against the multiplying hepatic forms.

The ability to culture plasmodia in hepatocytes has opened new perspectives in research programs directed to the discovery of new bioactive natural products. Different *in vitro* and *in vivo* models using rodent and human plasmodia have been developed (Mazier et al., 1984, 1985) and have turned out to be useful for the investigation of antiplasmodial molecules (Basco et al., 1999; Marussig et al., 1993).

Preliminary screening of chosen plants or fractions is usually performed in *in vitro* models against rodent parasites (*Plasmodium yoelii* or *Plasmodium berghei*) in primary rodent hepatocytes or hepatoma cell lines, which are easier to use than cultures of human parasites in human hepatocytes. When interesting results are obtained, they have to be confirmed with *P. falciparum*: primary cultures of human hepatocytes infected with *P. falciparum* sporozoites obtained from *Anopheles stephensi* mosquitoes after membrane feeding on cultured gametocytes (Ponnudurai et al., 1982). If molecules of interest are identified, they have to be investigated *in vivo* in rodents (Gregory and Peters, 1970). If these results are also promising, the drug can then be tested in monkey models such as *Aotus* or *Saimiri* monkeys infected with one of the previously monkey-adapted *P. vivax* and *P. falciparum* strains (Collins et al., 1988, 1994, 1996).

According to this approach, in a collaborative project with the Institut Malgache de Recherches Appliquées, Madagascar (P. Rasoanaivo), and the Laboratoire de Chimie des Substances Naturelles of the Muséum National d'Histoire Naturelle (F. Frappier, M. Carraz), *Strychnopsis thouarsii* has been chosen and studied for the discrepancy between its efficacy when administered to malaria patients and its weak activity when tested on the erythrocytic multiplication of the parasite (Ratsimamanga-Urverg et al., 1992). Interestingly, a decoction of this plant has been found to inhibit the growth of the parasite in hepatocyte cultures. A preliminary bioassay-guided fractionation led to a fraction with enhanced activity.

17.4 ANTIDISEASE SCREENING

The severity of a cerebral *P. falciparum* infection depends largely upon the ability of infected red blood cells (RBCs) to adhere to the endothelium of microvessels. This phenomenon can lead to the occlusion of the affected microvessels. Moreover, activation of endothelial cells resulting from the adhesion of infected erythrocytes leads to an overexpression of different mediators such as cytokines and contributes by itself to the pathology. Thus, plant products able to interfere either with adherence of RBCs to endothelial cells or with damage resulting from adherence phenomena might be of interest.

Figure 17.2 schematically depicts the interior of a microvessel, showing the players potentially implicated in cerebral pathology and thus proposing a panel of molecules, the induction of which is possibly inhibited by plants. Among the receptors expressed on the surface of endothelial cells, some are known to interact with parasitised red blood cells (PRBCs), mainly via PfEMP1, a highly polymorphic parasite ligand. A number of the involved receptors have been identified: thrombospondin (TSP), CD36, ICAM-1, E-selectin, VCAM-1, PECAM-1/CD31, P-selectin, $\alpha_v\beta_3$, and thrombomodulin through its chondroitin sulfate moiety (CSA). This has been reviewed by Craig and Scherf (2001). These adhesion molecules are constitutive or inducible, their expression modulated by various cytokines, notably tumor necrosis factor-alpha (TNF-α), interferon-gamma (IFN-γ), interleukin 1 beta (IL-1β), and interleukin 4 (IL-4), and whose increased levels have been demonstrated in cerebral malaria. In contrast, nitric oxide (NO) reduces the expression of ICAM-1 and VCAM-1 (Takahashi et al., 1996). The parasite itself, and more precisely glycosylphosphatidylinositol (GPI), causes an increased expression of adhesion molecules via a signal transduction pathway mediated by a tyrosine kinase and the activation of NF kappa B/c-rel. This increase is achieved either directly or through the intermediates TNF-α and IL-1β, secreted by monocytes and

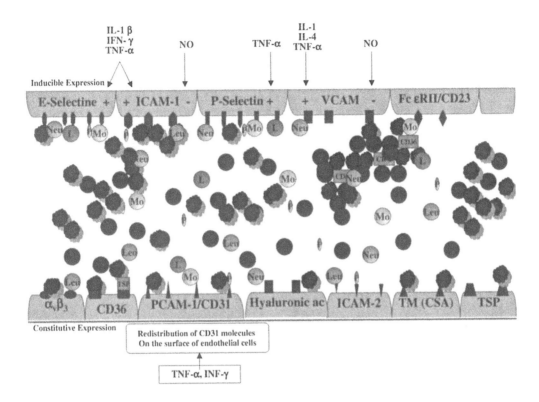

FIGURE 17.2 Schematic section of a microvessel showing possible interactions between endothelial receptors, parasitised or unparasitised red blood cells, leukocytes, and platelets. Receptors whose expression has been stimulated (+) or inhibited (–) by cytokines or other chemical mediators are symbolised at the surface of the endothelial cell. The potentially adherent cells and molecules are abbreviated as follows: CR-1, complement receptor-1; CSA, chondroitin sulfate; RBC, red blood cell; PRBC, parasitised red blood cell; L, lymphocytes; Leu, leukocytes; Mo, monocytes; Neu, neutrophils; P, platelets; TM, thrombomodulin; TSP, thrombospondin; STSP, soluble thrombospondin.

macrophages (Schofield et al., 1996). Very recently, we have shown that *P. falciparum* induces the expression of receptors with a weak affinity for members of the immunoglobulin E (IgE) family (CD23), on the surface of human endothelial cells; this stimulation provokes the induction of the inducible NO synthase (iNOS) and the synthesis of NO (Pino et al., 2003). In theory, all these molecules, implicated either in adhesion, in modulation of this adhesion, or in activation of the expression of various endothelial mediators, should be considered as targets possibly inhibited by phytochemicals.

Based on a cytoadherence model established in our laboratory and using a primary culture of human lung microvessel endothelial cells (Muanza et al., 1996) (see Figure 17.3), two enzyme-linked immunosorbent assays (ELISAs) assays are now available: The first one quantifies parasitic lactate dehydrogenase (LDH) activity, thus allowing the identification of products interfering with the cytoadherence of *Plasmodium* to endothelial cells. A second assay, measuring the production of mono- and oligonucleosomes in the cytoplasm of apoptotic cells (Salgame et al., 1997), quantifies the damage resulting from adhesion mechanisms. It has recently confirmed its interest as a good indicator of cellular death by apoptotic phenomena in our model of cytoadherence of *P. falciparum*-infected RBCs to endothelial cells (Pino et al., 2003).

Moreover, beside an effect on the adhesion phenomenon, cytokines such as TNF-α (McGuire et al., 1998) and IL-1β (Harpaz et al. 1992) are known to induce fever. Interfering with their secretion by monocytes or macrophages could be one explanation of the clinical effect observed

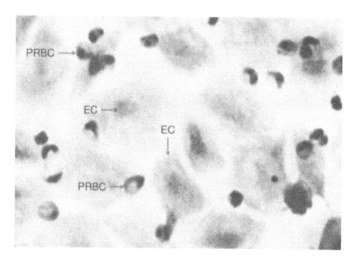

FIGURE 17.3 Co-culture of endothelial cells and *P. falciparum*-infected red blood cells. Endothelial cells (EC) are seeded and cultured in 96-well plates. At confluence, *P. falciparum* parasitised red blood cells (PRBCs) are added.

in populations treated with plants. Co-culture assays associating *P. falciparum*-infected RBCs and monocytes/macrophages (Pichyangkul et al., 1994) and appreciating the effect of plant products on cytokine secretion are a realistic approach.

17.5 THE SEXUAL ERYTHROCYTIC STAGE

Like the hepatic stage, the sexual erythrocytic stage is asymptomatic. However, preventing the formation of gametocytes or destroying them could break off parasite transmission to mosquitoes. This would be particularly valuable in the case of parasite genomes carrying drug resistance alleles. Additionally, resistance to antigametocyte compounds could not be selectively amplified, due to the nondividing nature of sexual forms. Assays can be performed in rodent models (Ramkaran and Peters, 1970) or *in vitro*, using cultures of sexual stages of *P. falciparum* (Hogh et al., 1998). In both cases, the potential effect on gametocytes is appreciated after feeding mosquitoes with blood containing sexual stages.

17.6 CONCLUSION

Besides fundamental research on the biology of *Plasmodium* as a means to identify novel drug targets and their putative specific inhibitors, it is important to validate in laboratory models the efficacy of herbal antimalarials observed in human trials. In this view, we emphasise several points in conclusion. The development of novel drugs targeting multiplication of the parasite in erythrocytes is an obvious necessity, and most resources available to malaria research are devoted to this aim. We have seen, however, the interest of drugs targeting nonpathogenic stages of the malaria parasite (Mazier and Doerig, 2000), hepatic and sexual stages. A second point concerns the way to proceed. A possible algorithm could be first an assay on erythrocytic multiplication. When negative, assays then have to be planned to search an effect at the hepatic level and on cytoadherence. As a third point we highlight the need to assess the efficacy of herbal antimalarials as they are used in real life, for example, a decoction as traditionally prepared. Bioassay-guided fractionation has to be planned only as a second step, with the serious threat that following isolation and purification, activity, which often results from the synergy of different compounds, will be lost. This additional difficulty must not impede research to identify valuable molecules that may be used as original leads for the development of novel antimalarials.

ACKNOWLEDGMENTS

We thank Merlin Willcox for critical reading of the manuscript. Part of the research was supported by VIH-PAL and INSERM PROGRES. P. Pino received a scholarship from VIH-PAL.

REFERENCES

Basco, L.K., Ringwald, P., Franetich, J.F., and Mazier, D. (1999). Assessment of pyronaridine activity *in vivo* and *in vitro* against the hepatic stages of malaria in laboratory mice. *Trans. R. Soc. Trop. Med. Hyg.*, 93, 651–652.

Collins, W.E., Galland, G.G., Sullivan, J.S., Morris, C.L., and Richardson, B.B. (1996). The Santa Lucia strain of *Plasmodium falciparum* as a model for vaccine studies. II. Development of *Aotus vociferans* as a model for testing transmission-blocking vaccines. *Am. J. Trop. Med. Hyg.*, 54, 380–385.

Collins, W.E., Morris, C.L., Richardson, B.B., Sullivan, J.S., and Galland, G.G. (1994). Further studies on the sporozoite transmission of the Salvador I strain of *Plasmodium vivax*. *J. Parasitol.*, 80, 512–517.

Collins, W.E., Skinner, J.C., Pappaioanou, M., Broderson, J.R., Filipski, V.K., McClure, H.M., et al. (1988). Sporozoite-induced infections of the Salvador I strain of *Plasmodium vivax* in *Saimiri sciureus boliviensis* monkeys. *J. Parasitol.*,74, 582–585.

Craig, A. and Scherf, A. (2001). Molecules on the surface of the *Plasmodium falciparum* infected erythrocyte and their role in malaria pathogenesis and immune evasion. *Mol. Biochem. Parasitol.*, 115, 129–143.

Desjardins, R.E., Canfield, C.J., Haynes, J.D., and Chulay, J.D. (1979). Quantitative assessment of antimalarial activity *in vitro* by a semiautomated microdilution technique. *Antimicrob. Agents Chemother.*, 16, 710–718.

Fleck, S.L., Robinson, B.L., Peters, W., Thevin, F., Boulard, Y., Glenat, C., et al. (1997). The chemotherapy of rodent malaria. LIII. 'Fenozan B07' (Fenozan-50F), a difluorinated 3,3'-spirocyclopentane 1,2,4-trioxane: comparison with some compounds of the artemisinin series. *Ann. Trop. Med. Parasitol.*, 91, 25–32.

Gregory, K.G. and Peters, W. (1970). The chemotherapy of rodent malaria. IX. Causal prophylaxis. I. A method for demonstrating drug action on exo-erythrocytic stages. *Ann. Trop. Med. Parasitol.*, 64, 15–24.

Harpaz, R., Edelman, R., Wasserman, S.S., Levine, M.M., Davis, J.R., and Sztein, M.B. (1992). Serum cytokine profiles in experimental human malaria. Relationship to protection and disease course after challenge. *J. Clin. Invest.*, 90, 515–523.

Hogh, B., Gamage-Mendis, A., Butcher, G.A., Thompson, R., Begtrup, K., Mendis, C. et al. (1998). The differing impact of chloroquine and pyrimethamine/sulfadoxine upon the infectivity of malaria species to the mosquito vector. *Am. J. Trop. Med. Hyg.*, 58, 176–182.

Krotoski, W.A. (1985). Discovery of the hypnozoite and a new theory of malarial relapse. *Trans. R. Soc. Trop. Med. Hyg.*, 79, 1–11.

Marussig, M., Motard, A., Renia, L., Baccam, D., Lebras, J., Charmot, G., et al. (1993). Activity of doxycycline against preerythrocytic malaria. *J. Infect. Dis.*, 168, 1603–1604.

Mazier, D., Beaudoin, R.L., Mellouk, S., Druilhe, P., Texier, B., Trosper, J., et al. (1985). Complete development of hepatic stages of *Plasmodium falciparum in vitro*. *Science*, 227, 440–442.

Mazier, D. and Doerig, C. (2000). Drugs targeting non-pathogenic stages of the malaria parasite. *J. Altern. Complementary Med.*, 2, 195–207.

Mazier, D., Landau, I., Druilhe, P., Miltgen, F., Guguen-Guillouzo, C., Baccam, D., et al. (1984). Cultivation of the liver forms of *Plasmodium vivax* in human hepatocytes. *Nature*, 307, 367–369.

McGuire, W., D'Alessandro, U., Stephens, S., Olaleye, B.O., Langerock, P., Greenwood, B.M., et al. (1998). Levels of tumour necrosis factor and soluble TNF receptors during malaria fever episodes in the community. *Trans. R. Soc. Trop. Med. Hyg.*, 92, 50–53.

Mendis, K., Sina, B.J., Marchesini, P., and Carter, R. (2001). The neglected burden of *Plasmodium vivax* malaria. *Am. J. Trop. Med. Hyg.*, 64(Suppl.), 97–106.

Moreno, A., Badell, E., Van Rooijen, N., and Druilhe, P. (2001). Human malaria in immunocompromised mice: new *in vivo* model for chemotherapy studies. *Antimicrob. Agents Chemother.*, 45, 1847–1853.

Muanza, K., Gay, F., Behr, C., and Scherf, A. (1996). Primary culture of human lung microvessel endothelial cells: a useful *in vitro* model for studying *Plasmodium falciparum*-infected erythrocyte cytoadherence. *Res. Immunol.*, 147, 149–163.

Pichyangkul, S., Saengkrai, P., and Webster, H.K. (1994). *Plasmodium falciparum* pigment induces monocytes to release high levels of tumour necrosis factor-alpha and interleukin-1 beta. *Am. J. Trop. Med. Hyg.*, 51, 430–435.

Pino, P., Vouldoukis, I., Dugas, N., Conti, M., Nitcheu, J., Traore, B., Danis, M., Dugas, B., and Mazier, D. (2003). Nitric oxide production by CD23-bearing human endothelial cells down-regulates the expression of ICAM-1 and decreases cytoadhesion of *Plasmodium falciparum*-infected erythrocytes. *Cell. Microbiol.*, in press.

Pino, P., Vouldoukis, I., Kolb, J.P., Mahmoudi, N., Desportes-Livage, I., Bricaire, F., Danis, M., Dugas, B., and Mazier, D. (2003). *Plasmodium falciparum*-infected erythrocyte adhesion induces caspase activation and apoptosis in human endothelial cells. *J. Infect. Dis.*, 187, 1283–1290.

Ponnudurai, T., Meuwissen, J.H., Leeuwenberg, A.D., Verhave, J.P., and Lensen, A.H. (1982). The production of mature gametocytes of *Plasmodium falciparum* in continuous cultures of different isolates infective to mosquitoes. *Trans. R. Soc. Trop. Med. Hyg.*, 76, 242–250.

Ramkaran, A.E. and Peters, W. (1970). Action of chloroquine on infectivity of gametocytes of rodent malarias. *Trans. R. Soc. Trop. Med. Hyg.*, 64, 8.

Ratsimamanga-Urverg, S., Rasoanaivo, P., Ramiaramanana, L., Milijaona, R., Rafatro, H., Verdier, F., et al. (1992). *In vitro* antimalarial activity and chloroquine potentiating action of two bisbenzylisoquinoline enantiomer alkaloids isolated from *Strychnopsis thouarsii* and *Spirospermum penduliflorum*. *Planta Med.*, 58, 540–543.

Ridley, R.G. (2002). Medical need, scientific opportunity and the drive for antimalarial drugs. *Nature*, 415, 686–693.

Salgame, P., Varadhachary, A.S., Primiano, L.L., Fincke, J.E., Muller, S., and Monestier, M. (1997). An ELISA for detection of apoptosis. *Nucleic Acids Res.*, 25, 680–681.

Schofield, L., Novakovic, S., Gerold, P., Schwarz, R.T., McConville, M.J., and Tachado, S.D. (1996). Glycosylphosphatidylinositol toxin of *Plasmodium* up-regulates intercellular adhesion molecule-1, vascular cell adhesion molecule-1, and E-selectin expression in vascular endothelial cells and increases leukocyte and parasite cytoadherence via tyrosine kinase-dependent signal transduction. *J. Immunol.*, 156, 1886–1896.

Shanks, G.D., Gordon, D.M., Klotz, F.W., Aleman, G.M., Oloo, A.J., Sadie, D., et al. (1998). Efficacy and safety of atovaquone/proguanil as suppressive prophylaxis for *Plasmodium falciparum* malaria. *Clin. Infect. Dis.*, 27, 494–499.

Silvie, O., Danis, M., and Mazier, D. (2002). Malaria: the disease and vaccine development. *Vaccines Child. Pract.*, 5, 9–13.

Takahashi, M., Ikeda, U., Masuyama, J., Funayama, H., Kano, S. and Shimada, K. (1996). Nitric oxide attenuates adhesion molecule expression in human endothelial cells. *Cytokine*, 8, 817–821.

Trager, W. (1994). Cultivation of malaria parasites. *Methods Cell Biol.*, 45, 7–26.

Wengelnik, K., Vidal, V., Ancelin, M.L., Cathiard, A.M., Morgat, J.L., Kocken, C.H., et al. (2002). A class of potent antimalarials and their specific accumulation in infected erythrocytes. *Science*, 295, 1311–1314.

World Health Organization. (2001). *The Use of Antimalarial Drugs*. Report of an Informal Consultation, WHO/CDS/RBM. WHO, Geneva, 33, 1–143.

18 Guidelines for the Preclinical Evaluation of the Safety of Traditional Herbal Antimalarials

*Merlin Willcox, Shingu K. Gamaniel,
Motlalepula G. Matsabisa, Jean René Randriasamimanana,
Charles O.N. Wambebe, and Philippe Rasoanaivo*

CONTENTS

18.1 INTRODUCTION

Herbal remedies for malaria are already in popular use in developing countries, and many have been shown to have antiplasmodial activities in experimental studies. However, few attempts appear to have been made to determine the safety of medicinal plants used traditionally to treat malaria, with the aim of bringing them into clinical trials and general use. Most work has been directed by academic interests, emphasising the evaluation of *in vitro* and *in vivo* activity in experimental models. Presently, the scanty data available indicate that preclinical safety testing in Africa, and probably in other developing countries, follows a pattern of "What can we do?" instead of "What ought to be done?" and "How should it be done?"

The lack of access to effective, cheap, safe, and user-friendly medicines has resulted in an ever-increasing number of people using herbal medicines. There is a belief that these medicines are safe because they are natural. This situation has hampered the process of phytomedicine development, especially in disease-endemic countries (DECs). Some of the major factors responsible for this situation are lack of access to information, lack of knowledge, lack of appropriate resources, and inexperience in this area on the part of scientists working in DECs. The current scarcity of such information is also a reflection of the extent to which this field has been neglected by research funders and policy makers.

Preclinical safety evaluation is the first major step in the development of antimalarial phytomedicines. Before a phytomedicine is brought into a clinical trial program, its safety profile must be assured by conducting a series of nonclinical pharmacological and toxicological investigations. Naturally, the expected benefits of a herbal remedy should outweigh its potential risks. Indeed, insufficient evidence of safety of an herbal product is enough to stop further development, as it may carry serious health risks. The goal of preclinical safety evaluation is to characterise the possible toxic effects with respect to target organs (toxicity studies), and to determine the influence of treatment on physiological functions, especially those involving vital organs and systems, such as the cardiovascular system, the respiratory system, the central nervous system (CNS), and hepatic and renal function (safety pharmacology). This chapter will provide an outline of considerations to be taken into account when designing a preclinical safety-testing program.

18.2 CURRENT REGULATORY PROVISIONS FOR PRECLINICAL
SAFETY EVALUATION

There are a number of documents, publications, and guidelines that may provide information on preclinical safety testing, but in many instances, access to them has been limited. Based on the need for increased awareness and access to adequate information on preclinical safety evaluation for scientists working in the DECs, it has become necessary to assemble and briefly summarise the relevant regulatory provisions. Generally, the regulatory guidelines may be divided into two broad categories:

1. Guidelines for preclinical safety investigations of pharmaceuticals and chemicals in Western countries
2. World Health Organization (WHO) guidelines for evaluating the safety of herbal medicines

18.2.1 GUIDELINES FOR PRECLINICAL SAFETY INVESTIGATIONS OF PHARMACEUTICALS AND CHEMICALS IN WESTERN COUNTRIES

All Western countries have their own regulatory guidelines for the conduct of toxicity studies, but unfortunately they are not easily accessible. More general considerations on how to design and develop a toxicological program appear to be useful at a first stage. Three relevant sources of information should be consulted for basic knowledge on the subject. They all are available on the Internet, and this will facilitate access to the information:

1. The European Union (EU) has developed guidelines called *EU/CPMP's Notice to Applicants*, which can be found at http://pharmacos.eudra.org/F2/eudralex/vol-2/home.htm#ntactd.
2. The U.S. Food and Drug Administration has developed the draft *Redbook II*, available in hard copy for the public. The agency is now making the *Redbook 2000* on toxicological principles for the safety of food ingredients, available at http://www.cfsan.fda.gov/~redbook/red-toca.html.
3. The International Conference on Harmonisation of Technical Requirements for Registration of Pharmaceuticals for Human Use, Common Technical Document (ICH-CTD) is a project that brings together tripartite regulatory authorities and experts (EU, Japan, and U.S.) to discuss scientific and technical aspects of product registration and to make recommendations on ways to achieve greater harmonisation in the interpretation and application of technical guidelines and requirements for product registration in order to reduce or obviate the need to duplicate the testing carried out during the research and development of new drugs. Box 18.1 reports the organisation of nonclinical study reports in the ICH guidelines and ICH-CTD, which can be found at http://www. ich.org.

Box 18.1 provides guidelines for conducting preclinical safety investigations (ICH, 2000). They homogenise the data requirements for particular types of toxicity studies in the form of generic standards or standard packages of studies for different countries. It is worth pointing out that these toxicity studies are not designed to demonstrate that a chemical is safe, but rather to characterise what toxic effects it can produce.

Not all the study types may be necessary, and scientific judgment has to be exercised to adapt these guidelines on "What ought to be done" in the design of the preclinical testing program in a specific case. Furthermore, legislation does not really describe what preclinical toxicity studies should be conducted for a particular product, but it expects a proper choice of studies based on state-of-the-art science, which can evolve faster than laws can be changed.

18.2.2 WHO GUIDELINES FOR EVALUATING THE SAFETY OF TRADITIONAL HERBAL MEDICINES

WHO has published several guidelines on the safety evaluation of traditional medicines (WHO, 1993, 1998, 2000). These set out the methods to be followed for different toxicological tests. However, the preamble states, "not all tests are necessarily required for each herbal medicine intended for human study." These guidelines aim to help researchers decide which, if any, toxicological tests are necessary before embarking on observational clinical studies (phase I), as described in Chapter 21.

WHO (1996) also states that "if the product has been traditionally used without demonstrated harm, no specific restrictive regulatory action should be undertaken unless new evidence demands a revised risk-benefit assessment." Further recommendations were issued by WHO in 1998 and are reproduced in Box 18.2. WHO maintains the position that there is no requirement for preclinical

Box 18.1: ICH-CTD Recommended Structure for Reporting Nonclinical Safety Studies on a Drug

1. Pharmacology
   1.1. Primary pharmacodynamics
   1.2. Secondary pharmacodynamics
   1.3. Safety pharmacology
   1.4. Pharmacodynamic drug interactions
2. Pharmacokinetics
   2.1. Analytical methods and validation reports
   2.2. Absorption
   2.3. Distribution
   2.4. Metabolism
   2.5. Excretion
   2.6. Pharmacokinetic drug interactions (nonclinical)
   2.7. Other pharmacokinetic studies
3. Toxicology
   3.1. Single-dose toxicity (in order by species, by route)
   3.2. Repeat-dose toxicity (in order by species, by route, by duration, including supportive toxicokinetics evaluations)
   3.3. Genotoxicity
       3.3.1. *In vitro*
       3.3.2. *In vivo* (including supportive toxicokinetics evaluations)
   3.4. Carcinogenicity (including supportive toxicokinetics evaluations)
       3.4.1. Long-term studies (in order by species, including range-finding studies that cannot appropriately be included under repeat-dose toxicity or pharmacokinetics)
       3.4.2. Short- or medium-term studies (including range-finding studies that cannot appropriately be included under repeat-dose toxicity or pharmacokinetics)
       3.4.3. Other studies
   3.5. Reproductive and developmental toxicity (including range-finding studies and supportive toxicokinetics evaluations) (if modified study designs are used, the following subheadings should be modified accordingly)
       3.5.1. Fertility and early embryonic development
       3.5.2. Embryo–fetal development
       3.5.3. Prenatal and postnatal development, including maternal function
       3.5.4. Studies in which the offspring (juvenile animals) are dosed and further evaluated
   3.6. Local tolerance
   3.7. Other toxicity studies (if available)
       3.7.1. Antigenicity
       3.7.2. Immunotoxicity
       3.7.3. Mechanistic studies (if not included elsewhere)
       3.7.4. Dependence
       3.7.5. Metabolites
       3.7.6. Impurities
       3.7.7. Other
4. Key literature references

Box 18.2: Requirements for Registration of Traditionally Used Medicinal Herbal Products (WHO, 1998)

a. Name of product
b. List of ingredients (active and inactive) of the product with scientific names, part of plant used, and quantity, and with reference to the source text for the prescription, if available
c. The list of plant ingredients of the product with taxonomic classification, including species, genus, and family
d. Methods and technology used in manufacture
e. Physical and chemical identification tests
f. Quality standards for the ingredients when necessary (which may include the limit of residue of heavy metals and pesticides, insecticides, and herbicides)
g. Quality standards for the products
h. Stability tests
i. Therapeutic uses and dosage
j. Evidence of traditional use or recent clinical experience with the product in the form proposed, to support the safety and efficacy of the product
k. Package and packaging materials
l. Content on label or package insert

---

toxicity testing, but rather that evidence of traditional use or recent clinical experience is sufficient. Preclinical toxicity testing is only required for new medicinal herbal products that contain herbs with no traditional history of use. It is therefore a matter of rational judgment to fill the gaps in the selection of tests for the preclinical safety evaluation of herbal medicines.

18.3 OBSERVED TOXICITY OF ANTIMALARIAL MEDICINAL PLANTS

In spite of the widespread use of plants for the treatment of malaria, there have been few published reports of toxicity from these plants. Before filling the gaps, it is useful to have a background of toxic effects associated with the use of antimalarial plants. A thorough literature search was conducted to find any published material, using the databases MEDLINE, EMBASE, AMED, and TOXBASE, and experts were contacted to pass on any unpublished findings. The major findings are summarised here.

18.3.1 Severe Toxicity

Ransome-Kuti (1986) describes the use of a "poisonous" mixture containing cows' urine, tobacco, and herbs that is administered to convulsing children. Although he states that the child improves in most cases, he also makes the unsubstantiated claim that the poison is a major contributor in 60% of malaria deaths. Serious cases of poisoning with margosa oil (*Azadirachta indica*), producing a Reye's-like syndrome, have been observed in infants. These are discussed in more detail in the Chapter 6.

18.3.2 Cardiovascular Effects

Decoction of the leaves of *Pauridiantha* (= *Urophyllum*) *lyallii* is largely known for its use in the treatment of malaria in Madagascar (Boiteau, 1999; Rahantamalala, 2000), but for unclear reasons probably related to charlatanism, it has recently appeared to be a herbal tea used to prevent or cure nearly 50 different ills. Rabezanahary (1990) reports 14 cases of patients admitted to the hospital who happened to have been taking this herbal medicine, and focuses on the electrocardiogram (ECG) and electrolyte levels of these patients. As a result, ECG and electrolyte abnormalities were observed for all patients. Almost all of the cases concerned ingestion over a prolonged period for hypertension or other chronic diseases, rather than acute ingestion over a short period, as would be used for the treatment of malaria.

Aerial parts of *Vernonia trinervis* are reported to be an antimalarial in combination with *Vernonia polygalaefolia*, but are also used alone as a diuretic, as vermifuge in veterinary medicine, and for the treatment of blennorrhagia (Rasoanaivo, 2000). Rahantamalala (2000) reports the case of a patient who had taken *V. trinervis* for 4 months and presented with hypotension, fatigue, dizziness, blurred vision, polyuria, and hyponatraemia. These symptoms are not uncommon side effects of diuretics, usually when used over a prolonged period. It is unlikely that these symptoms would occur in a short course of treatment for malaria. There were inverted T waves in V2 to V6 that resolved after stopping ingestion of the herbal medicine. This may be related to a direct cardiac activity of the plant and requires further investigation.

All these cases underline the importance of performing regular electrocardiograms on patients before, during, and after treatment with herbal medicines in observational clinical trials.

18.3.3 Respiratory Effects

Jarikre and Bazuaye (1986) reported two cases of acute pulmonary edema associated with a traditional treatment for malaria. In both cases, the treatment consisted of inhalation, bathing, and drinking of a decoction of leaves of *Carica papaya*, *Citrus limon*, and *Cymbopogon citratus* twice daily for 3 days. The first case resolved on treatment with conventional antimalarials and antihistamines; the second improved after treatment with antibiotics, diuretics, and steroids, and aspiration of bilateral pleural effusions. The authors excluded most other causes of pulmonary edema, although this can be a complication of malaria itself. They hypothesised that in these cases the pulmonary edema was caused by a toxic alveolitis, with increased alveolar capillary permeability, due to thermal damage from the inhalation of steam and vapours whose temperature was probably close to 100°C. There may or may not have been additional damage from toxins contained in the vapours and firewood smoke in the enclosed room. Anah (1982) observed two similar cases, both of which were diagnosed as allergic alveolitis secondary to herbal inhalations, and responded to steroid therapy. Clearly, inhalation of vapours at high temperature and of smoke is likely to be harmful and of little benefit in the treatment of malaria.

Birbeck (1999) reports cases of oral lesions and aspiration pneumonia in children with febrile seizures who had been given boiling hot medicinal teas by mouth while unconscious. Here again, the major damage was probably caused by the physical temperature of the water and the fact that the patients were unable to protect their airway while unconscious, rather than any intrinsic toxicity of the tea.

18.3.4 Neurological Effects

Central nervous system depression has been observed after the accidental ingestion of 10 ml of eucalyptus oil by a 3-year-old child (Patel and Wiggins, 1980). He made a full recovery within 24 hours. The antimalarial preparation of eucalyptus is as a decoction of the leaves and is used in a much lower dose.

18.3.5 Hepatotoxicity

Neem seed oil administered to infants can cause acute hepatotoxicity, and a syndrome akin to Reye's syndrome (see Chapter 6, Section 6.2.5.2), although there is no evidence of similar toxicity for neem leaf teas.

Datta et al. (1978) report three cases of hepatic failure after ingestion of *Heliotropium eichwaldi* for 20 to 50 days, as a treatment for epilepsy or vitiligo. This plant is also used for the treatment of fevers in India.

18.3.6 HAEMATOLOGICAL EFFECTS

Razafimanantsoa (1991) reports three cases of childhood haematological problems attributed to the ingestion of medicinal herbs. The first is a case of leukemia that was associated with the ingestion of the decoction of stem barks of *Cedrelopsis grevei*. This species is reputed for its multifarious uses in traditional medicine in Madagascar, among these as a febrifuge and for the treatment of malaria anaemia (Rasoanaivo, 2000). Another case is reported for a child who had taken *Psiadia altissima* and *Physalis peruviana* for the treatment of malaria and was admitted for hemorrhages and anaemia, but there is no evidence that these were caused by the plant rather than by the disease for which they were given. The other case of hemorrhage and fever was associated with the administration of aspirin as well as the long-term ingestion of an herbal tea. It could be that this was an undiagnosed case of Reye's syndrome, related to the ingestion of aspirin rather than the herbal tea (Willcox, 2001). In any case, these reports underline the importance of monitoring full blood counts in clinical trials of herbal antimalarials to observe for any haematological effects.

18.3.7 MISCELLANEOUS PLANTS THAT ARE ALSO USED AS ANTIMALARIALS

Ramanantenasoa (1988) reports cases of acute toxicity observed in the clinical toxicology department of a university hospital in Madagascar. The two most important concerned plants were *Manihot utilissima* (cassava), and *Ricinus communis*. All the cases of intoxication from cassava concern the ingestion of the root, whereas it is a decoction of the leaf that is used as an antimalarial (Dalziel, 1937; Rasoanaivo et al., 1992). Some varieties of cassava plant are known to contain cyanogenic glycosides, which are destroyed by proper preparation.

All but one of the cases of intoxication from *R. communis* reported by Ramanantenasoa (1988) concern ingestion of the seed, either accidentally or in the belief that it was a food. One case of dizziness, blurred vision, and visual hallucinations is reported following the ingestion of a large quantity of a leaf infusion of *R. communis*, in order to provoke an abortion. The most common mode of use of this plant as an antipyretic is topical application of the leaves, sometimes applied with castor oil, to the forehead and other parts of the body (Anis and Iqbal, 1986; Comerford, 1996; Singh and Anwar Ali, 1994). Vongo (1999) reports that a decoction of roots can be used to treat malaria. However, the use of the fruits in combination with the leaves and seeds of other plants has been reported by Adjanohoun (1993) as a treatment for fever, and the use of seeds is mentioned by Coe and Anderson (1996), although their route of administration is not clear. Evidently the oral ingestion of *R. communis* seeds is not safe.

The other cases of intoxication from food plants with claimed antimalarial properties (*Mangifera indica, Capsicum annuum, Psidium guajava*) concern the ingestion of excessive quantities of the fruits as foods, rather than the use of other plant parts as medicines.

Rahantamalala (2000) reports the case of a patient who had taken *Mystroxylon aethiopicum* for several years (as a matter of habit) and was admitted with edema and hypertension. This plant is claimed to have antimalarial activity. It is not clear whether the congestive cardiac failure in this hypertensive patient was related to the herbal tea that she had been drinking for several years. This also merits further research.

It could be postulated that human beings have evolved to avoid toxic plants. This is supported by the observation that deaths from plant poisoning are rare, except when people move to an unknown area (Reynolds and Cousins, 1993). In general, traditional herbalists inherit from their elders the knowledge of which plants are safe and which are unsafe to use.

Although there are few reports of serious adverse effects of traditional antimalarials, many occurrences are undoubtedly unreported. Nevertheless, most of the reports that do exist are concerned with alternative uses of the plants (for food, for other medical problems, over a long period of time, or accidental ingestion). Importantly, as an acute disease, malaria does not require long-term ingestion of medicines (unless taken prophylactically), thus reducing the risk of chronic toxicity.

18.4 DECISION CRITERIA FOR THE SELECTION OF ANTIMALARIAL CANDIDATES FOR PRECLINICAL SAFETY EVALUATION

Prior to preclinical evaluation of traditional herbal medicines for malaria, one must decide on which material to evaluate and on what basis. Initial careful selection is important. Priority should be given to antimalarial plants or products that already have evidence of safety and efficacy based on local use or published data. Endemic plants and threatened species or plants reported to contain novel compounds may also be considered to be of high priority.

In the Western style of drug discovery process, the selection of a candidate for preclinical studies is achieved through very tight criteria. In phytomedicine development, candidates are selected more on a case-by-case basis rather than in a definitive or rigid scheme. Ideally, priority should be given to antimalarial phytomedicines that:

1. Have strong evidence of efficacy in traditional uses
2. Display strong evidence of *in vitro* and *in vivo* efficacy in the experimental models, as discussed in Chapter 15
3. Are devoid of cytotoxic effects, which is predictive of a more antimalarial selectivity

These criteria should be taken as a guideline to maintain an element of flexibility. The emphasis on a flexible approach should be seen as creating a situation in which the selection of the candidate is conducted as a scientific exercise rather than a set of stereotyped criteria to be followed in a routine manner.

18.5 PRELIMINARY CONSIDERATIONS ON DESIGNING PRECLINICAL TOXICITY STUDIES ON HERBAL ANTIMALARIALS

18.5.1 Evidence of Traditional Use

Estimative pharmacovigilance has been proposed by Koumare (1995) as a useful approach in determining the safety of traditional medicines. This is simply a retrospective survey of cases treated with a particular medicine by traditional healers, and the healers' observations of possible side effects or contraindications to the medicine. This information can be used to guide both preclinical studies and inclusion and exclusion criteria for clinical observational studies (see Chapter 21). Koumare proposes that it should be used in conjunction with "real pharmacovigilance;" that is, prospective surveillance of cases treated, in the same way that postmarketing surveillance is conducted for any drug.

There must be documented experience of long-term use without evidence of safety problems, specifying the period of use, the health disorders treated, the number of users, and the countries or regions with such experience (WHO, 1996). Most crucial of all (although not mentioned by WHO) is to record the methods of preparation and the doses used. Most useful medicinal plants, like most useful pharmaceuticals, may be toxic if prepared incorrectly or given in an incorrect dose, because they contain pharmacologically active compounds. The old adage *dosis sola facit venenum* (only the dose makes a poison) holds true for herbal antimalarials. Indeed, the Greek word φάρμακον has the dual meaning of "remedy" and "poison."

Such evidence will be easier to demonstrate in cultures with a written tradition (such as traditional Chinese medicine and Ayurveda) than in those with an oral tradition (such as African traditional medicine). Nevertheless, substantial ethnobotanical research already exists (see Chapter 11) to support such evidence-based studies. The more widespread the use of a given preparation, the more likely it is to be effective and the less likely it is to be toxic at the prescribed dose. True traditional healers are an integral part of their community, and their usual aim is to provide the best possible treatments for their patients, who are most often their neighbors and relatives. By the same token, they would soon know if any of their treatments caused acute toxic effects.

18.5.2 LITERATURE REVIEW

Most common medicinal plants have been the subject of scientific research. There is a large body of literature on the chemical constituents of plants and their pharmacological activities. This literature should be consulted before embarking on any study that is aimed at answering questions such as those in Box 18.3. Even when a particular species has not been studied, it is likely to contain similar principles to other plants in the same genus and family. Thus, the principle of chemotaxonomic classification can be used to predict which chemical compounds a plant will contain. However, the quantity of toxic compounds in a plant may vary considerably according to the time of year and time of day (Huxtable, 1990). Susceptibility to toxic effects also varies according to age, sex, nutritional status, and genetic predisposition.

BOX 18.3: THE AIMS OF A LITERATURE SEARCH ON SAFETY OF A PLANT

1. Are there any reports of human toxicity associated with ingestion of the plant? If so, which part of the plant, in what preparation, and at what dose?
2. Have any laboratory studies of toxicity been carried out on the relevant preparation of the plant? If so, what did the results show?
3. What pharmacologically active compounds does this plant species (or genus) contain? In which parts of the plant are they found? What are their principal pharmacological effects, and at what doses?

None of the existing databases can cover all published information in a given topic, and it is therefore advised to consult as many databases as possible. The first step in a literature search is to consult data freely accessible to the Internet, for example, PubMed (www.ncbi. nlm.nih.gov/PubMed) and the more general internet search engine Google (www.google.com). Helpful information on a poisonous plant bibliography may be found at the U.S. Food and Drug Administration's website (http://vm.cfsan.fda.gov/~djw/A-ANN.html). References are given in author alphabetical order, starting from A–ANN, and the other author references can be downloaded by selecting the appropriate initials at the end of the listing. The WHO and six of the biggest medical journal publishers launched in July 2001 a unique initiative to make full-text access free of charge (or at low cost depending on gross national product (GNP) per capita) to nearly 1500 top international medical journals to medical schools and research institutes in low-income countries. This initiative is known as the Health InterNetwork Access to Research Initiative (HINARI), and it is said to be the biggest step ever taken toward reducing the health information gap between rich and poor countries. More information can be obtained at hinari@who.int.

Second, a reference text on plant toxicology (Frohne and Pfände, 1983; Newall et al., 1996) and a pharmacopoeia or similar monographs (WHO, 1999) can be consulted. Then databases such as EMBASE, NAPRALERT, or Biological Abstracts may be consulted. The *Chapman & Hall Chemical Database Natural Products Selection*, available on CD, also gives relevant information on plants. A new database on the safety of traditional medicines is currently being developed by the Commonwealth Working Group on Complementary Medicine, but is not yet available.

18.6 SUGGESTED PRELIMINARY TOXICITY TESTING MODELS FOR HERBAL ANTIMALARIALS

Toxicity studies have several parts, and the duration and other details are usually determined by the regulatory authorities. Overall, guidelines give the whole array of investigations to be conducted, but do not provide rational and concrete guidance about the tests and assays that should be regarded as necessary for a specific drug in a given situation. The planning of a scientifically justifiable testing program should first identify those assays and tests that are absolutely necessary for the target drug product and disease, without including tests that would not add value or necessary information. We therefore suggest that minimal toxicity testing requirements for herbal antimalarials should be based on three main factors: the observed toxicity of the herbal antimalarial product or its components, preliminary information on traditional use and evidence from the literature, and the duration of treatment required for the disease in question.

18.6.1 SINGLE-DOSE TOXICITY STUDIES (ACUTE TOXICITY)

A striking change has occurred in the acute lethal toxicity studies in the last decade (Lorke, 1983), and attempts have been made to reform the determination of lethal dose 50% (LD_{50}) values. As a result of progress in animal ethics, some alternative toxicological tests have been developed to minimise the number of animals used (Reader et al., 1989; Bracher et al., 1987). Some acute toxicity tests are designed to determine the LD_{50}. WHO (2000) describes the acute toxicity test, using at least five rodents of each sex and at least two nonrodents, at several dose levels, in order to determine the approximate lethal dose of a herbal medicine. According to the FDA *Redbook II*, information on LD_{50} can be used to:

- Deal with cases of accidental ingestion of a large amount of the material
- Determine possible target organs that should be scrutinised or special tests that should be conducted in repeated-dose toxicity tests
- Select doses for short-term toxicity and subchronic toxicity tests

WHO (1998) states that acute toxicity testing is only necessary in the case of new medicinal herbal products that contain herbs with no traditional history of use. Furthermore, it must be emphasised that this is a crude test; many important observations of toxicity, for example, morbidity and pathogenesis, which may have more significance than mortality, are not represented by LD_{50} values. The main focus of the acute toxicity test should be on observing the symptoms and recovery of the test animals, rather than on determining the LD_{50} of the drug.

The limit test suggested in the FDA *Redbook II* would be appropriate for herbal antimalarials that are not expected to be particularly toxic. In this test, 5 g/kg of the test antimalarial is administered orally to five to seven animals that are observed closely for up to 14 days. Symptoms of toxicity and recovery are noted. Gross and histopathological examination of the test animals at the end of the study may help identify toxic effects on target organs. If no animals die as a result of this dose, there is no need to test higher dosages because this dose is the practical upper limit that can be administered orally to a rodent. The acute toxicity of the herbal antimalarial can be expressed as being greater than 5 g/kg. If there are deaths following administration of this dose, another alternative would be the maximal tolerated dose (MTD), which could be determined with lower dosages. The MTD is defined as the dose that, when given for the duration of the chronic study as the highest dose, will not shorten the treated animals' longevity from any toxic effects other than the induction of neoplasms (see *Redbook II*, p. 113).

18.6.2 28-DAY REPEATED-DOSE TOXICITY STUDIES

Although acute toxicity will rarely be missed by traditional healers, chronic toxicological risks may pass unrecognised. While this does not appear to be very relevant to the treatment of malaria, where most effective treatments will be short term (7 days or less), some unexpected adverse effects may occur. Children are particularly vulnerable to malaria, and this would justify their inclusion in the clinical trials of drugs in uncomplicated malaria (WHO, 1996). Potential organ damage should therefore be properly assessed in long-term toxicity studies. The duration and other details of long-term toxicity studies are usually determined by national regulatory authorities/agencies. The WHO (2000) general guidelines recommend repeated-dose long-term toxicity tests in at least 10 rodents of each sex and at least 3 nonrodents of each sex, in at least 3 dose levels, for 2 to 4 weeks if the period of administration of the drug is to be less than 1 week (see Table 18.1).

Based on the observed toxic effects of herbal antimalarials, special attention should be paid to the influence of the test antimalarial on physiological functions (neurological, cardiovascular, respiratory, renal, hepatic) in the 28-day repeated-dose toxicity studies. Three doses should be used by oral administration with respect to the effective dose (ED). Generally, apart from the effective (nontoxic) dose of the test substance, another dose level is chosen that is capable of producing obvious overt toxic effects such that between this dose and the nontoxic dose, the addition of at least one more dose will enhance the possibility of observing a dose–response relationship for toxic manifestations. In some cases, the formulae ED $\times$ 5 and ED $\times$ 10 may be appropriate, although this may not be appropriate for drugs with a low therapeutic index.

TABLE 18.1
Commonly Used Ranges of Administration Periods for Repeated-Dose Studies

Expected Period of Clinical Use	Administration Period for the Toxicity Study
Single or repeated administration for:	
Less than 1 week	2–4 weeks (14–28 days)
1–4 weeks, repeated	4–12 weeks (28–84 days)
1–6 months, repeated	3–6 months (90–180 days)
Over 6 months, long-term repeated	9–12 months (270–365 days)

Note: The test substance should be administered 7 days a week and the administration periods for toxicity must be recorded in each result.

The study should include:

- General examination:
  - General appearance
  - Activity
  - Gait
  - Behaviour
  - Body fur coat
  - Mucous membranes
  - Body orifices
  - Body temperature
  - Sign of morbidity
  - Body weights
  - Food consumption
  - Water consumption
- Laboratory parameters:
  - Urinalysis
  - Haematology
  - Clinical chemistry
  - Sacrifice and autopsy
  - Histopathology

18.6.3 REPRODUCTIVE AND DEVELOPMENTAL TOXICITY (TERATOGENICITY) STUDIES

Pregnant women are also vulnerable to malaria. In practice, the initial clinical trials of any anti-malarial will exclude pregnant women, both because of the potential severity of malaria in pregnancy and because of the fear of adverse effects from the treatment. Only when a remedy has been found to be highly safe and effective in uncomplicated malaria would its use in pregnancy be contemplated.

Usually, reproductive and teratogenicity studies are required to be undertaken for drugs that are to be used in women of childbearing age. As for other toxicological tests, WHO (1998) only recommends reproductive toxicology tests for new herbal medicines, not for traditional medicines. In practice, traditional healers may know if a particular herb is to be avoided during pregnancy. Some traditional antimalarials are widely regarded to have abortifacient properties (Nicholas, 1999). In the case of one particular Ugandan herb, some women believed it to be abortifacient (Ndyomugyenyi et al., 1998) whereas a midwife believed it to be safe (Willcox, 1999). In these cases, depending on the extent of experience of traditional healers with the remedy in pregnant women, the view of regulatory authorities, and ethics committees, it may or may not be deemed necessary that reproductive and developmental toxicology studies be conducted. Methods for such studies have been described in detail elsewhere (WHO, 1993; ICH, 1993; *Redbook II*; *Redbook 2000*).

18.6.4 CARCINOGENICITY AND GENOTOXICITY STUDIES

ICH (1995) states that carcinogenicity tests are necessary for pharmaceuticals whose expected clinical use is continuous for at least 6 months, or when used frequently in the treatment of chronic or recurrent conditions such as allergic rhinitis or depression. This is unlikely to apply to antimalarials, except those taken regularly as a prophylactic. Carcinogenicity studies are also recommended if there is any cause for concern about carcinogenic potential (Box 18.4). The procedures for evaluation of carcinogenicity have been extensively described elsewhere (ICH, 1997a; WHO, 1993; *Redbook II*, *Redbook 2000*) and do not need to be repeated here.

BOX 18.4: CAUSES FOR CONCERN ABOUT CARCINOGENIC POTENTIAL (ICH, 1995)

1. Previous demonstration of carcinogenic potential in the product class that is considered relevant to humans
2. Structure–activity relationship suggesting carcinogenic risk
3. Evidence of preneoplastic lesions in repeated-dose toxicity studies
4. Long-term tissue retention of parent compound or metabolite(s) resulting in local tissue reactions or other pathophysiological responses

---

Genotoxicity studies are mainly used to evaluate the potential for carcinogenesis, although if a compound is genotoxic, there is also a risk of heritable germline mutations. ICH (1997b) suggests the following three tests as a standard screen for genotoxicity:

1. A test for gene mutation in bacteria
2. An *in vitro* test with cytogenetic evaluation of chromosomal damage with mammalian cells or an *in vitro* mouse lymphoma tk assay
3. An *in vivo* test for chromosomal damage using rodent haematopoietic cells

18.6.5 DRUG INTERACTION STUDIES

Malaria often co-occurs with bacterial infections, necessitating simultaneous treatment with antibiotics. Malarious patients may also be taking long-term medications such as contraceptives, anticonvulsants, oral hypoglycemics, antihypertensives, anticoagulants, or indeed other traditional herbal medicines. At this point, the danger of drug interactions in herbal medicines has been raised by Izzo et al. (2002). It is therefore useful to assess interactions of antimalarial phytomedicines with other drugs, but bearing in mind that it would be impossible to investigate every possible drug interaction. Drug interaction studies should be done when the value of the antimalarials has been finally established and their form of presentation ascertained.

Drug interactions may be pharmacokinetic, when the absorption, distribution, or elimination of the object drug is altered by the precipitant drug, or pharmacodynamic, when the precipitant drug alters the effect of the object drug at its site of action (Grahame-Smith and Aronson, 1992).

18.6.6 PHARMACOKINETIC INTERACTIONS

18.6.6.1 Absorption Interactions

Mahmoud et al. (1994) showed that traditional Sudanese beverages, including *Tamarindus indica*, which is used for the treatment of malaria, reduced the oral bioavailability of chloroquine ingested at the same time. Tsakala et al. (1990) showed that a traditional antidiarrhoeal treatment (made with earth from termite mounds), like kaolin, adsorbs chloroquine, and so would also reduce its oral bioavailability. Other traditional medicines (and foods) may interfere with the absorption of modern drugs. This can best be detected by measuring plasma levels of the target drug in patients having ingested the drug simultaneously with the traditional medicine or food in question (Mahmoud et al., 1994).

18.6.6.2 Protein-Binding Displacement Interactions

These interactions are rarely of clinical importance (Grahame-Smith and Aronson, 1992). Although the plasma concentration of unbound drug may be increased by displacement from plasma proteins, the rate of clearance increases in proportion to the degree of displacement, and soon an equilibrium is reached. A traditional medicine with components that bind to plasma proteins could potentially alter plasma levels of warfarin, phenytoin, and tolbutamide.

18.6.6.3 Metabolism Interactions

The most common target for metabolic interactions is phase I oxidation, the first stage of hepatic metabolism of many drugs, which is carried out by cytochrome P450 and NADPH cytochrome c reductase (Grahame-Smith and Aronson, 1992). This system may be induced or inhibited by other drugs. For example, alcohol induces the metabolism of anticoagulants and phenytoin, thus reducing their plasma concentration and their efficacy. Conversely, metronidazole inhibits the metabolism of alcohol and warfarin, thus increasing their plasma concentration and effects.

Iwu et al. (1986) describe a method for evaluating the effect of a medicinal plant on the activity of NADPH cytochrome c reductase in rats. An extract of *A. indica* was injected intraperitoneally once a day for 4 days to rats in the experimental group, while those in the control group were injected with distilled water. The animals were sacrificed 24 hours after the last dose, and the liver removed. Microsomes were isolated from liver homogenates by differential centrifugation and precipitation with calcium chloride. The microsomal protein was mixed with a buffer, cytochrome c, and NADPH. The rate of formation of ferrocytochrome c was monitored by the increase in absorbance at 55 nm for 10 minutes at 27°C. The activity of other enzymes in the microsomes can also be assessed. Using this method Iwu et al. (1986) found that *A. indica* leaf extracts inhibit the activity of NADPH cytochrome c reductase by 75%, but increase the activity of aniline hydroxylase by 23%.

Iwu et al. (1986) suggest that oxidative stress in infected cells caused by a lack of NADPH may explain the antimalarial activity of *A. indica*. Theoretically this could lead to haemolytic anaemia in patients with G6PD deficiency. Through their effect on NADPH cytochrome c reductase, *A. indica* leaf extracts may also reduce the metabolism of other drugs. However, the clinical significance of this interaction will not be clear until human pharmacokinetic studies have been undertaken.

18.6.6.4 Excretion Interactions

Some drugs may compete with each other for secretion by carrier systems in the renal tubules, and also for secretion into bile. For example, quinidine and quinine reduce the biliary secretion of digoxin, and quinidine (but not quinine) reduces the renal tubular secretion of digoxin (Hedman et al., 1990). Thus quinidine causes a twofold rise in plasma concentration of digoxin, which may provoke toxicity. To prevent this, the dose of digoxin should be halved for patients started on quinidine or *Cinchona* bark preparations.

These interactions are best studied in pharmacokinetic studies in humans. Models exist to study the rate of uptake of drugs into isolated hepatocytes, the rate of biliary excretion from isolated rat liver, and excretion by guinea pigs *in vivo*, but there may be differences between pharmacokinetics in animals and humans (Hedman et al., 1990).

18.6.7 PHARMACODYNAMIC INTERACTIONS

These interactions alter the effect of the object drug at its site of action. Direct pharmacodynamic interactions are those where the two drugs act antagonistically or synergistically at the same site, or on different sites with a similar end result (Grahame-Smith and Aronson, 1992). For example, Allen et al. (1994) found that a decoction of *Brucea javanica* fruits, which is traditionally used to treat malaria in Thailand and China, antagonises the effect of chloroquine *in vitro*. Conversely, Rasoanaivo et al. (1994) have found that the bark of *Strychnos myrtoides* potentiates the effect of chloroquine on resistant parasites *in vitro* and *in vivo* (see Chapter 9). These effects can be investigated *in vitro* by comparing the inhibitory concentration 50% (IC_{50}) of the drugs separately and in combination.

Indirect pharmacodynamic interactions result from an effect of the precipitant drug altering the effect of the object drug, although the two effects are not related. For example, hypokalaemia secondary to diuretics (or possibly certain herbal medicines) increases the risk of arrhythmias with quinidine and other antiarrhythmic drugs (Mehta, 2002). Therefore, it is important to assess the effect of new drugs on serum electrolytes.

18.7 CONCLUSION

There have been few reports of toxicity from the use of traditional herbal antimalarials. Minimum sets of toxicity tests are proposed here as guidelines, bearing in mind that although there are similarities in function between human and animal species, some limitations of animal testing are obvious. Obviously, adequate infrastructure, qualified resource persons, and budget are prerequisite to carry on these toxicity tests, and all this may be out of reach of scientists in some developing countries. One good initiative of the WHO/AFRO (Africa Regional Office) is the proposal for the identification of centers of excellence in Africa that may use their expertise to help other centers.

Since it is practically impossible to harmonise all cases of antimalarial phytomedicines, these guidelines should be regarded as flexible rather than definitive rules to be followed routinely. If the medicine is found to be safe and efficacious, it may later be decided to investigate its use as a prophylactic or as a treatment for malaria in pregnancy. At this stage it may or may not be deemed necessary to embark on tests of chronic and reproductive toxicity.

The first and most important step is to establish good evidence of traditional use. A literature search is also necessary to see whether cases of toxicity have been reported for the preparation in question, or whether there is any other justification for conducting laboratory tests of toxicity. Preclinical safety data should not be viewed in isolation, but should be complemented by careful clinical observations in treated patients.

REFERENCES

Adjanohoun, J.E., Ahyi, M.R.A., Aké Assi, L., et al. (1993). *Traditional Medicine and Pharmacopoeia: Contribution to Ethnobotanical and Floristic Studies in Uganda*. OAU/STRC, Lagos.

Allen, D., Wright, C.W., Phillipson, J.D., Kirby, G.C., and Warhurst, D.C. (1994). The effectiveness of chloroquine as an antimalarial may be compromised by the co-administration of traditional medicines. *J. Pharm. Pharmacol.*, 42 (Suppl. 2), 1044.

Anah, C.O. (1982). Extrinsic allergic alveolitis associated with concoction for treating malaria traditionally. *Nig. J. Microbiol.*, 2, 121–124.

Anis, M. and Iqbal, M. (1986). Antipyretic activity of some Indian plants used in traditional medicine. *Fitoterapia*, 57, 52–54.

Birbeck, G.L. (1999). Traditional African medicines complicate the management of febrile seizures. *Eur. Neurol.*, 42(3), 184.

Boiteau, P. (1999). *Dictionnaire des Noms Malgaches des Végétaux*, Vol. I. Editions Alzieu, Grenoble, France.

Bracher, M., Faller, C., Spengler, J., and Reinhardt, C.A. (1987). Comparison of *in vitro* cell toxicity with the *in vivo* eye irritation. *Mol. Toxicol.*, 1, 561–570.

Coe, F.G. and Anderson, G.J. (1996). Ethnobotany of the Garífuna of eastern Nicaragua. *Econ. Bot.*, 50, 71–107.

Comerford, S.C. (1996). Medicinal plants of two Mayan healers from San Andrés, Petén, Guatemala. *Econ. Bot.*, 50, 327–336.

Dalziel, J.M. (1937). *The Useful Plants of West Tropical Africa*. Crown Agents for the Colonies, London.

Datta, D.V., Khuroo, M.S., Mattocks, A.R., Aikat, B.K., and Chhuttani, P.N. (1978). Herbal medicines and veno-occlusive disease in India. *Postgrad. Med. J.*, 54, 511–515.

Frohne, D. and Pfände, H.J. (1983). *A Colour Atlas of Poisonous Plants*. Wolfe, London.

Grahame-Smith, D.G. and Aronson, J.K. (1992). *Oxford Textbook of Clinical Pharmacology and Drug Therapy*. OUP, Oxford.

Hedman, A., Angelin, B., Arvidsson, A., Dahlqvist, R., and Nilsson, B. (1990). Interactions in the renal and biliary elimination of digoxin: stereoselective difference between quinine and quinidine. *Clin. Pharmacol. Ther.*, 47, 20–26.

Huxtable, R.J. (1990). The harmful potential of herbal and other plant products. *Drug Saf.*, 5 (Suppl. 1), 126–136.

ICH. (1993). ICH Harmonised Tripartite Guideline. Detection of Toxicity to Reproduction for Medicinal Products. Available at www.ich.org.

ICH. (1995). ICH Harmonised Tripartite Guideline. Guideline on the Need for Carcinogenicity Studies of Pharmaceuticals. Available at www.ich.org.

ICH.(1997a). ICH Harmonised Tripartite Guideline. Non-Clinical Safety Studies for the Conduct of Human Clinical Trials for Pharmaceuticals. Available at www.ich.org.

ICH. (1997b). ICH Harmonised Tripartite Guideline. Testing for Carcinogenicity of Pharmaceuticals. Available at www.ich.org.

ICH. (1997c). ICH Harmonised Tripartite Guideline. Genotoxicity: A Standard Battery for Genotoxicity Testing of Pharmaceuticals. Available at www.ich.org.

Iwu, M.M., Obeoda, O., and Anazodo, M. (1986). Biochemical mechanism of the antimalarial activity of *Azadirachta indica* leaf extract. *Pharmacol. Res. Commun.*, 18, 81–91.

Izzo, A.A., Borrelli, F., and Capasso, R. (2002). Herbal medicine: the dangers of drug interaction. *Trends Pharmacol. Sci.*, 23, 358–359.

Jarikre, L.N. and Bazuaye, E.A. (1986). Acute pulmonary oedema (toxic alveolitis) associated with traditional fever therapy. *East Afr. Med. J.*, 63, 656–659.

Koumare, M. (1995). Contrôle de qualité et mise sur le marché des médicaments traditionnels Africains. Réunion du Groupe d'Experts sur la Promotion et le Développement de l'Utilisation Industrielle des Plantes Médicinales en Afrique. WHO/AFRO, Brazzaville, Congo.

Lorke, D.A. (1983). New approach to practical acute toxicity testing. *Arch. Toxicol.*, 54, 275–287.

Mahmoud, B.M., Ali, H.M., Homeida, M.M.A., and Bennett, J.L. (1994). Significant reduction in chloroquine bioavailability following coadministration with the Sudanese beverages Aradaib, Karkadi and Lemon. *J. Antimicrob. Chemother.*, 33, 1005–1009.

Mehta, D.K., Ed. (2002). *British National Formulary.* British Medical Association, London.

Ndomugyenyi, R., Neema, S., Magnussen, P. (1998). The use of formal and informal services for antenatal care and malaria treatment in rural Uganda. *Health Policy Planning*, 13(1), 94–102.

Newall, C.A., Anderson, L.A., and Phillipson, J.D. (1996). *Herbal Medicines: A Guide for Health Care Professionals.* The Pharmaceutical Press, London.

Nicolas, J.P. (1999). *Plantes Médicinales des Mayas K'iché de Guatemala.* Ibis Press, Paris.

Patel, S. and Wiggins, J. (1980). Eucalyptus oil poisoning. *Arch. Dis. Child.*, 55, 405–406.

Rabezanahary, M. (1990). Problèmes posés par la prise de décoction d'une plante médicinale malgache: le *Pauridiantha lyalli* (Tamirova). Thèse de Doctorat Médecine, Antananarivo, Madagascar, No. 1995.

Rahantamalala, C. (2000). Contribution à l'inventaire et à l'évaluation en laboratoire de quelques plantes médicinales utilisées dans le traitement du paludisme. Thèse pour obtenir le grade de Docteur en Médecine, Faculté de Médecine, Université d'Antananarivo, Madagascar.

Ramanantenasoa, C. (1988). Contribution à l'étude des intoxications aigues par les plantes médicinales malgaches. Thèse pour l'obtention du doctorat en médecine, Faculté de Médecine, Université d'Antananarivo, Madagascar.

Ransome-Kuti, O. (1986). Child health in Nigeria: past, present and future. *Arch. Dis. Child.*, 61, 198–204.

Rasoanaivo, P. (2000). Une banque de donnés sur les plantes médicinales de Madagascar, *Info-Essences*, 15, 5–6.

Rasoanaivo, P. (2002). Médecine Traditionnelle et Plantes Médicinales de Madagascar, Compilation sur cédérom des croyances et pratiques ethnomédicales des Malgaches.

Rasoanaivo, P., Petitjean, A., Ratsimamanga-Urverg, S., and Rakoto-Ratsimamanga, A. (1992). Medicinal plants used to treat malaria in Madagascar. *J. Ethnopharmacol.*, 37, 117–127.

Rasoanaivo, P., Ratsimamanga-Urverg, S., Milijaona, R., et al. (1994). *In vitro* and *in vivo* chloroquine-potentiating action of *Strychnos myrtoides* alkaloids against chloroquine-resistant strains of *Plasmodium* malaria. *Planta Med.*, 60, 13–16.

Razafimanantsoa, M.O. (1991). Les Tambavy Administrés aux enfants: Résultats d'enquête préliminaire. Thèse pour l'obtention du doctorat en médecine, Faculté de Médecine, Université d'Antananarivo, Madagascar.

Reader, S.J., Blackwell, V., O'hhara, R., Clothier, R.H., Griffin, G., and Balls, M.A. (1989). Vital dye release method for assessing the short-term cytotoxic effects of chemicals and formulations. *ALTA*, 17, 28–37.

Reynolds, P. and Cousins, C.C. (1993). *Lwaano Lwanyika: The Tonga Book of the Earth.* London, Panos Publications.

Singh, V.K. and Anwar Ali, Z. (1994). Folk medicines in primary health care: common plants used for the treatment of fevers in India. *Fitoterapia*, 65, 68–74.

Tsakala, M., Tona, L., Tamba, V., et al. (1990). Etude *in vitro* de l'adsorption de la chloroquine par un remede antidiarréique employé traditionellement en Afrique. *J. Pharm. Belg.*, 45, 268–273.

Vongo, R. (1999). The Role of Traditional Medicine on Antimalarials in Zambia. Paper presented at the First International Meeting of the Research Initiative on Traditional Antimalarials. Available at http://mim.nih.gov/english/partnerships/ritam_program.pdf.

WHO. (1993). *Research Guidelines for Evaluating the Safety and Efficacy of Herbal Medicines.* WHO Regional Office for the Western Pacific, Manila.

WHO. (1996). WHO Expert Committee on Specifications for Pharmaceutical Preparations, 34th Report, WHO Technical Report Series 863. WHO, Geneva.

WHO. (1998). *Guidelines for the Appropriate Use of Herbal Medicines*, Western Pacific Series 23. WHO Regional Publications, WHO Regional Office for the Western Pacific, Manila.

WHO. (1999). *WHO Monographs on Selected Medicinal Plants*, Vol. 1. WHO, Geneva.

WHO. (2000). General Guidelines for Methodologies on Research and Evaluation of Traditional Medicine, WHO/EDM/TRM/2000.1. WHO, Geneva.

Willcox, M.L. (1999). A clinical trial of 'AM', a Ugandan herbal remedy for malaria. *J. Public Health Med.*, 21, 318–324.

Willcox, M.L. (2001). Salicylates, nitric oxide, malaria, and Reye's syndrome. *Lancet*, 357, 1881–1182 (letter).

Part 5

Clinical Research

19 An Overview of Clinical Studies on Traditional Herbal Antimalarials

Merlin Willcox and Gerard Bodeker

CONTENT

19.1 INTRODUCTION

The use of traditional herbal antimalarials is widespread, and there is some laboratory data for their efficacy, as demonstrated in the preceding chapters. However, the most important question to answer is whether traditional preparations are safe and effective against malaria in human patients. Some clinical data have already been mentioned in the case studies at the beginning of this book, but it would be useful to compare the efficacy of these preparations with each other and with any others that have been tested clinically. It would also be useful to review what methods have been used to conduct clinical trials on traditional herbal antimalarials and to reflect on how these could be improved.

    This review aims systematically and critically to appraise clinical research on the safety and efficacy of traditional plant-based treatments for malaria. It aims to identify important findings, highlight methodological challenges, and suggest priorities for future research in this field.

19.2 METHODS

A literature search was performed using MEDLINE, CAB, SOCIOFILE, and EMBASE to find all articles published up to and including 2003 referring to traditional herbal remedies for malaria (key words: malaria; traditional medicine; malaria, therapy). References of relevant articles were searched, and some journals were searched by hand, to try to identify as many relevant articles as possible. The Research Initiative on Traditional Antimalarial Methods (RITAM) network (consisting of over 200 researchers in the field of traditional antimalarials) was consulted and conferences attended in an attempt to identify all published and unpublished case reports and clinical studies on herbal antimalarials.

Studies were categorised according to methodology (case reports, cohort studies, controlled trials) and according to species of malaria. Each study was scrutinised for measures of efficacy (most commonly parasite clearance and clearance of fever or symptoms) and safety (reports of side effects or changes in haematological, biochemical, or electrocardiographic parameters). This information is summarised in Table 19.1 to Table 19.5.

19.3 RESULTS

Eighteen case reports (Table 19.1) were identified of herbal antimalarials for the treatment of malaria (falciparum in 14 of these cases). There were 17 cohort studies of herbal remedies for the treatment of falciparum malaria (Table 19.2), 12 of remedies for vivax malaria (Table 19.3), and 5 of remedies for malaria of undefined species (Table 19.4). Eight controlled trials were also identified, two for falciparum malaria and six for vivax (Table 19.5).

There was often limited information about the method of preparation of the remedies, which would make them very difficult to replicate. In some cases, this is a deliberate intention of the authors, in order to protect intellectual property rights (Ajaiyeoba et al., 2000; Araoye, personal communication; Bitahwa et al., 1997; Willcox, 1999).

19.3.1 SAFETY AND TOLERABILITY

Specific case reports of toxicity from herbal antimalarials have been reviewed in Chapter 18. Of the studies reviewed here, few reported data on side effects of the herbal medicines: 3 (17%) of the case studies, 13 (38%) of the cohort studies, and 4 (50%) of the controlled trials did so. In the other studies, it seems that patients were not questioned about adverse effects or new symptoms since starting treatment.

Only three cohort studies (9%) and three of the controlled trials (38%) reported effects on blood parameters (most commonly liver function tests), and two studies monitored electrocardiograms (ECGs). No cases of toxicity were reported.

Minor side effects can be important. For example, Willcox (1999) found that almost half of the patients taking the herbal remedy AM experienced one or more side effects. In about 8% of cases, these were unpleasant enough to cause patients to stop taking the treatment. Many herbal antimalarials have a very bitter taste, which can make them difficult to administer to children. Doses often need to be taken more often, and the volume to ingest is often larger than with conventional drugs (Valecha et al., 2000; Guindo, 1988).

19.3.2 EFFICACY

19.3.2.1 Case Reports

The 18 case reports identified are summarised in Table 19.1, presented in order of efficacy. Unfortunately, the reports are generally of inadequate quality. They state that patients improved or that parasites cleared, but often without quantitative data to show this, and with no information on

TABLE 19.1
Clinical Case Reports of Malaria Patients Taking Traditional Herbal Remedies

Parasite Clearance by Day 7[a]	Symptoms[b]	Treatment	No. of Subjects	Malaria Species[c]	Side Effects[d]	Country	Reference
100%	100% improved	AM + rema	4	P.f.	NA	Uganda	Bitahwa et al., 1997
100%	100% cured	B. cathartica root + S. africana root decoction	2	P.f.	NA	Mozambique	Jurg et al., 1991
82.6%	Improved	WPE 1	5	P.f.	NR	Nigeria	Ajaiyeoba et al., 2000
60% by 48 hours	100% cured; Rapid relief (within 30 minutes)	Ajadilopea		NA	Pruritus in 1 case	Nigeria	Araoye (personal communication)
50%	NA	Crotalaria monteiro root + Clematis viridiflora root and leaf	2	P.f.	NA	Mozambique	Jurg et al., 1991
46.1%	Improved	WPE 5		P.f.	NR	Nigeria	Ajaiyeoba et al., 2000
38%	38% cured; 50% improved; 12% no effect	Azadirachta indica bark tincture	8	P.f.	Sweating, laxative	India	Deare et al., 1916
25%	17% cured; 50% improved; 33% no effect	A. indica bark tincture	12	P.v.	Sweating, laxative	India	Deare et al., 1916
23.1%	Improved	WPE 7		P.f.	NR	Nigeria	Ajaiyeoba et al., 2000
17.4%	Improved	WPE 2		P.f.	NR	Nigeria	Ajaiyeoba et al., 2000
17%	NA	Strychnos heningsii root bark	6	P.f.	NA	Mozambique	Jurg et al., 1991
7.6%	Improved	WPE 6		P.f.	NR	Nigeria	Ajaiyeoba et al., 2000
7.6%	Improved	WPE 4		P.f.	NR	Nigeria	Ajaiyeoba et al., 2000
7.6%	Improved	WPE 8		P.f.	NR	Nigeria	Ajaiyeoba et al., 2000
0%	Improved	WPE 3		P.f.	NR	Nigeria	Ajaiyeoba et al. 2000
0%	NA	Momordica balsamina leaf	1	P.f.	NA	Mozambique	Jurg et al., 1991
NA	80% improved	Tinospora cordifolia stem decoction	5	NA	NA	India	Vedavathy & Rao, 1995
NA	100% improved	Calotropis procera flowers + black pepper	10	NA	NA	India	Vedavathy & Rao, 1995

a Percent of patients clear of parasites by day 7.
b Cured = afebrile, symptoms completely resolved; improved = partial resolution of symptoms.
c P.f. = Plasmodium falciparum; P.v. = Plasmodium vivax; NA = not reported.
d NA = Not available; NR = none reported.

TABLE 19.2
Clinical Cohort Studies of Patients Taking Herbal Remedies for Falciparum Malaria

Parasite Clearance[a]	Symptoms[b]	ACR[c] (%)	Treatment[d]	N	Age (years)	Side Effects	Country	Reference
100% (d 4)	100% FC (d 4)	NA	Totaquina I and II	73	≥16	10–15% vomiting 2–9% vertigo	India	Hicks and Diwan Chand, 1935
100% (d 4)	100% cure (d 4)	NA	A. annua infusion	5	>10	NA	Democratic Republic of Congo	Mueller et al., 2000
100% (d 4)	NA	NA	Parinari exelsa Sabin stem bark decoction	203	1–72	NA	Tanzania	Lugakingira, unpublished b
100% (d 3–9)	100% cure (d 3–9)	NA	Ayush-64	4	3–65	None reported	India	Sharma et al., 1981
100% (d 3) in adults Decline of parasitaemia in children	NA	NA	Morinda lucida leaf extract	NA	Adults Children	NA	Nigeria	Makinde et al., unpublished
70% (d 4) 100% (d 7)	NA	NA	Alocacia macrorhiza root decoction	20	15–50	Few (not detailed)	Laos	Phetsouvanh, 1991
95% (d 5)	98.5% FC (d 5)	NA	Totaquina I	155	NA	Vomiting	Europe, North Africa, Asia	Pampana, 1934
92% (d 4)	77% cure (d 4)	NA	A. annua decoction	48	>10	25% nausea 2% vomiting	Democratic Republic of Congo	Mueller et al., 2000
92% (d 4)	92% cure (d 4)	NA	Cassia spectabilis twig decoction	60	NA	NA	Congo	Mabiala, 1991
86% (d 4)	89% FC (d 4)	NA	Totaquina II	261	NA	Vomiting	Europe, North Africa, Asia	Pampana, 1934

Percent of patients clear of parasite count to insignificant levels (d 6)[a]	[b]	[c]	Treatment[d]	N	Age	Side effects	Country	Reference
39% (d 7)	90% FC (d 7)	9%	Butyrospermum parkii seed oil (external application)	13	0.5–16	NA	Burkina Faso	Bugmann, 2000
29% (d 7)	60% FC (d 7)	12%	Ficus gnaphalocarpus + Glinus lotoides +/− B. parkii	17	0.5–16	NA	Burkina Faso	Bugmann, 2000
14% (d 7)	80% FC (d 7)	5%	Combinations of Combretum micranthum, Combretum glutinosum, Guiera senegalensis, Ziziphus mauritiana	86	0.5–16	NA	Burkina Faso	Bugmann, 2000
8% (d 7)	92% FC (d 7)	55%	AM + rema	19	1–50	52% — none / 48% — one or more minor side effects	Uganda	Willcox, 1999
100%: decline of parasite count to insignificant levels (d 6)	100% cured (d 3) / 100% FC (d 3) (antipyretics in 26)	NA	Terraplis interretis decoction	53	3 months–40 years	NA	Burkina Faso	Thiombiano, 1991
NA	79% cure (d 3)	NA	D. febrifuga root extract	14	NA	Abdominal pain, vomiting	China	Tsu, 1947
NA	63% cure	NA	D. febrifuga root extract	253	NA	Abdominal pain, vomiting	China	Tsu, 1947

[a] Percent of patients clear of parasites by specified day.

[b] Cured = afebrile, symptoms completely resolved; FC = fever clearance.

[c] ACR = Adequate clinical response at day 14 (afebrile regardless of parasitaemia, or aparasitaemic, regardless of temperature — see Chapter 21 for full definition); NA = data not available.

[d] See Chapter 2 for full description of totaquina preparations and Chapter 5 for full description of Ayush-64.

TABLE 19.3
Clinical Cohort Studies of Patients Taking Herbal Remedies for Vivax Malaria

Parasite Clearance[a]	Symptoms[b]	Treatment[c]	No. of Subjects	Age (years)	Side Effects[d]	Country	Reference
100% (d 4)	100% FC (d 4)	Totaquina I and II	84	≥16	10–15% vomiting 2–9% vertigo	India	Hicks and Diwan Chand, 1935
100% (d 7) (9% relapse rate)	88% improved (d 6)	D. febrifuga root extract	67	NA	Abdominal pain, vomiting	China	Tsu, 1947
99.5% (d 5)	99.5% FC (d 5)	Totaquina I	175	NA	Vomiting	Europe, North Africa, Asia	Pampana, 1934
94.5% (d 4)	96% FC (d 4)	Totaquina II	290	NA	Vomiting	Europe, North Africa, Asia	Pampana, 1934
86% (d 7)	NA	Ayush-64	57	>12	NA	India	CCRAS, 1987, pp. 102–112
82% (d 9)	86% cured (d 9)	Ayush 64	28	3–65	None reported	India	Sharma et al., 1981
65% (d 10)	65% cured (d 10) 15% improved	Caesalpinia crista	20	8–55	NA	India	Panda, 1998
55% (d 4)	55% cured (d 3)	Sudarsana churna + seethamsurasa	20	15–50	No effect on LFTs	India	Venkataraghavan et al., 1982
53% (2 months)	NA	Totaquina I	110	≥16	NA	India	Acton, 1920
27% (2 months)	NA	Totaquina I	110	NA	Nausea, vomiting	India	Sinton and Bird, 1929
NA	89.1% response	Ayush-64	1442	NA	NA	India	CCRAS, 1987, pp. 88–89
NA	67% cured	D. febrifuga root extract	322	NA	Abdominal pain, vomiting	China	Tsu, 1947

[a] Percent of patients clear of parasites by specified day.
[b] Cured = afebrile, symptoms completely resolved; improved = partial resolution of symptoms; FC = fever clearance.
[c] See Chapter 2 for full description of Totaquina preparations and Chapter 5 for full description of Ayush-64. Sudarsana churna = Ayurvedic preparation containing 54 ingredients, principally Swertia chirata (33%). Seethamsurasa = Ayurvedic preparation containing Azadirachta indica bark and arsenic.
[d] LFTs = liver function tests.

TABLE 19.4
Clinical Cohort Studies of Patients Taking Herbal Remedies for Malaria (Indeterminate Species)

Parasite Clearance[a]	Symptoms[b]	Treatment[c]	N	Malaria Species[d]	Age (years)	Side Effects[e]	Country	Reference
100% (d 2)	NA	*Maytenus senegalensis* stem bark and root bark decoction	120	NA	<5 to >30	NA	Tanzania	Lugakingira, unpublished a
100%, takes 1 more day than quinine	FC equivalent to quinine	*D. febrifuga* root extract	13	P.f./P.v.	NA	Nausea and vomiting	China	Jang et al., 1946
69% (d 7) (19% recrudescence by d 14)	88% FC (d 7) (15% recrudescence by day 14)	*Cissampelos mucronata* *Cassia abbreviata* *Mwinga mbunda*	26	NA	NA	No effect on AST	Tanzania	Matthies, 1998
3% (but decline in parasite counts)	Antipyretic effect	*Cochlospermum tinctorium* root decoction	88	NA	6–14	NA	Burkina Faso	Ouedraogo et al., 1992
NA	100% FC (d 3)	*S. myrtoides* + CQ SM + Q SM + SP	14 4 2	NA	16–47	No effect on FBC, U+Es, LFTs, ECG	Madagascar	Randriamahary-Ramialiharisoa et al., 1994

[a] Percent of patients clear of parasites by specified day.
[b] Cured = afebrile, symptoms completely resolved; improved = partial resolution of symptoms; FC = fever clearance.
[c] CQ = chloroquine; Q = quinine; SP = sulfadoxine-pyrimethamine.
[d] P.f. = *Plasmodium falciparum*; P.v. = *Plasmodium vivax*.
[e] AST = Aspartate aminotransferase; FBC = full blood count; U+Es = urea and electrolytes; LFTs = liver function tests; ECG = electrocardiogram.

TABLE 19.5
Controlled Trials of Herbal Antimalarials

Treatments[a]	Parasite Clearance[b]	Symptoms[c]	No. of Subjects	Species[d]	Age	Side Effects[e]	Country	Reference
A. annua capsules with oil	PCT = 33 hours	FCT = 18 hours	103	P.v.	18–30	NA	China	Yao-De et al., 1992
CQ	PCT = 50 hours	FCT = 24 hours	20					
C. sanguinolenta aqueous root extract	PCT = 3.3 days	FCT = 36 hours	12	P.f.	NA	Fewer than CQ	Ghana	Boye, 1989
CQ	PCT = 2.2 days	FCT = 48 hours	10					
D. febrifuga root extract	100% (d 7)	92% FC (d 2)	12	P.f./P.v.	NA	Abdominal pain, vomiting	China	Tsu, 1947
Placebo	13% (d 7)	13% FC (d 2)	8					
Ayush-64	NA	No significant difference between treatments	30	P.v.	NA	NA	India	CCRAS, 1987, p. 87
CQ + PQ			28					
Ayush-64	95% (d 6)	95% improved d 6	58	P.v.	>12	NA	India	CCRAS, 1987, pp. 90–102 (outpatients)
CQ + PQ	100% (d 6)	100% improved d 6	60					
Ayush-64	72% (d 6)	72% improved d 6	30	P.v.	>12	No effect on FBC, LFTs, ECG	India	CCRAS, 1987, p. 90–102 (inpatients)
CQ + PQ	100% (d 6)	100% improved d 6	30					
Ayush-64	49% (d 28)	Slow recovery	54	P.v.	18–60	No effect on FBC, biochemistry 3 GI effects (A64 group)	India	Valecha et al., 2000
CQ	100% (d 28)	Fast recovery	50					
Malarial	58% < 100 on d 7	59% FC on d 7	36	P.f.	5–45	1 constipation 3 allergy	Mali	Koita, 1991
CQ	92% < 100 on d 7	50% FC on d 7	17					Guindo, 1988

a CQ = chloroquine; PQ = primaquine.
b Percent of patients clear of parasites by specified day. PCT = parasite clearance time.
c Cured = afebrile, symptoms completely resolved; improved = partial resolution of symptoms; FC = fever clearance.
d P.f. = Plasmodium falciparum; P.v. = Plasmodium vivax.
e FBC = full blood count; LFTs = liver function tests; ECG = electrocardiogram.

laboratory methods and quality control. Jurg et al. (1991) report temperature and parasitaemia for patients taking one remedy (*Bridelia cathartica* root and *Spirostachys africana* root decoction), and Ajaiyeoba et al. (2000) report parasite clearance rates for the eight remedies they studied. They also reported mean parasite counts for all patients, which declined.

The danger of case reports is that the authors can select "successful" cases, which do not necessarily represent the overall efficacy of a remedy. The AM remedy from Uganda was reported to clear parasites in all four cases, including three with drug-resistant malaria (Bitahwa et al., 1997). However, a subsequent cohort study on the same remedy with the members of the same rural population (Willcox, 1999) found that although parasite counts declined, clearance only occurred in 8% of a random sample of 19 patients.

However, case reports can help researchers to select remedies for more detailed study. The *B. cathartica* and *S. africana* remedy reported by Jurg et al. (1991) was found to completely cure the two patients in the report, and *in vitro* studies subsequently showed that extracts from these plants inhibited the growth of *Plasmodium falciparum* in a dose-dependent manner. Remedy WPE 1 produced parasite clearance in 83% of patients (Ajaiyeoba et al., 2000). "Ajadilopea," from Nigeria, was reported to provide symptomatic relief within 30 minutes of administration and rapid parasite clearance (Araoye, unpublished report). Unfortunately, no further research on these remedies was found in the literature.

19.3.2.2 Cohort Studies

Cohort studies have been grouped according to the malaria species being treated and are summarised in Table 19.2 to Table 19.4. They are presented in order of efficacy. However, outcome measures were not standardised, and few patients were followed up beyond the end of their treatment. The most common measure of parasitological efficacy was parasite clearance, but even this was measured at different times after the start of treatment. Few studies reported an outcome in terms of adequate clinical response (ACR) at day 14 (see Chapter 21).

19.3.2.2.1 Cohort Studies on Falciparum Malaria (Table 19.2)

Of the 17 studies on falciparum malaria, 6 reported 100% parasite clearance on days 4 to 7 and a further 3 reported parasite clearance rates above 90%. Follow-up beyond day 7 is only available for two of these nine studies. Patients treated with *Alocacia macrorhiza* root decoction showed no recrudescence over 21 days of follow-up (Phetsouvanh, 1991). However, this study has not been published or replicated. Recrudescence rates for patients treated with *Artemisia annua* preparations are relatively high (see Chapter 3).

Gametocytemia was only measured in one of the trials. In 12 patients treated with Ayush-64, gametocytes were cleared in 83% by day 9 (Sharma et al., 1981). This has been discussed in more detail in Chapter 5.

One problem with cohort studies is that the population chosen may be semi-immune to malaria and may clear parasites and symptoms even without effective treatment. Therefore, high parasite clearance rates are not necessarily indicative of efficacy. For example, parasites were cleared in 100% of adults treated with a leaf extract of *Morinda lucida*, but not in any children, who have a less well-developed immune response (Makinde et al., unpublished). Yet this is the group at greatest risk of mortality and in whom efficacy is most important.

In highly endemic areas (as in much of sub-Saharan Africa) children are considered to have a good immune response to malaria above the age of 5. In areas with high transmission of malaria, achieving complete parasite clearance for any length of time may not be realistic. In these circumstances, the World Health Organization (WHO) recommends that adequate clinical response is a more useful measure of treatment efficacy. In order to qualify for ACR, a patient must either be aparasitaemic on day 14 or be afebrile (regardless of parasitaemia), without previously having met the criteria for an early treatment failure (see Chapter 21).

Some remedies may produce low rates of parasite clearance, but higher rates of ACR. For example, Thiombiano (1991) found that parasitaemia declined to insignificant levels and that patients were clinically cured after treatment with a decoction of *Terraplis interretis*. Willcox (1999) found that the Ugandan AM remedy cleared parasites in only 8% of patients, but that parasitaemia declined to lower levels, and that 55% of patients had an adequate clinical response.

19.3.2.2.2   Cohort Studies on Vivax Malaria (Table 19.3)

Of these twelve studies, five concern totaquina preparations (see Chapter 2) and three concern the ayurvedic drug Ayush-64 (see Chapter 5). Initial parasite clearance rates of up to 100% are reported for totaquina, which is significantly better than any of the other remedies tested. Two studies on totaquina followed patients for 2 months and found that 27 to 53% were still clear at this time. This is to be expected, as the active compounds in *Cinchona* are not effective against hypnozoites of *Plasmodium vivax*. None of the other studies followed patients beyond 10 days, so their effect on long-term relapses is unknown.

19.3.2.2.3   Cohort Studies on Malaria of Indeterminate Species (Table 19.4)

One cohort study (Jang et al., 1946) reported figures for tertian malaria — a term that presumably includes *P. vivax* (benign tertian) and *P. falciparum* (malignant tertian). Four did not specify the malaria species. These studies are of debatable value because the natural history, prognosis, and public health importance of malaria differs greatly between infections caused by *P. falciparum* and those caused by the other species. Nevertheless, *Maytenus senegalensis* stem and root bark decoction and *Dichroa febrifuga* root extract were both reported to clear parasites in all treated patients, and a Tanzanian mixture of plants cleared parasites in 69% of patients. These results would justify further investigation of these remedies.

Randriamahary-Ramialiharisoa et al. (1994) investigated the potential of *Strychnos myrtoides* to reverse resistance to chloroquine, quinine, and sulfadoxine-pyrimethamine. All patients became asymptomatic by day 3, but there was no report of parasite counts. A larger-scale randomised controlled trial of *S. myrtoides* as a resistance reverser for chloroquine was carried out in Madagascar in 2001, and preliminary results are published in Chapter 9 of this volume.

19.3.2.3   Controlled Trials

Eight controlled trials have been reported and are summarised in Table 19.5. Not all of these were randomised or double blind. The four trials of Ayush-64 were reported to be double blind, as both the herbal medicine and the drug were administered in identical capsules (CCRAS, 1987; Valecha et al., 2000). Koita (1991) randomised patients to receive either a traditional herbal decoction or chloroquine tablets; patients and doctors could not be blinded, but the laboratory technicians performing the parasite counts were blinded. Randomisation and blinding are not reported for the other trials. Only one trial was placebo controlled (Tsu, 1947).

Of the trials for falciparum malaria, the most promising is that of *Cryptolepis sanguinolenta*, in which parasite clearance took only 1 day longer than with chloroquine, and in which fever clearance was 12 hours faster (see Chapter 8). Also of interest was the trial of Malarial; although parasites were not cleared completely by the herbal medicine, there was a good clinical response (see Chapter 7).

Five of the trials were on vivax malaria, and four of these concerned the Ayurvedic remedy Ayush-64 (see Chapter 5). The initial parasite clearance observed is good, but many of the patients relapse by day 28. Last but not least, Yao-De et al. (1992) showed that oil-based *A. annua* capsules cleared parasites and fever more rapidly than chloroquine. There is a clear case for further investigation of this preparation (see Chapter 3).

19.4　DISCUSSION

It is clear that there is only very scarce clinical research on herbal remedies for malaria. This may be partly due to publication bias: more such studies may exist, but in local languages and journals, or presented at regional conferences, which do not reach the mainstream databases. There may also be a lack of funding for this type of research, because there are no profits to be made from promoting the cultivation and use of plants by ordinary people.

Reports of studies that have been found often contain incomplete information, particularly with regard to side effects and the evaluation of efficacy. Few remedies have gone through the steps of classical clinical testing. Therefore, an approach of comparison close to meta-analysis is not yet possible. Nevertheless, some important points emerge from these studies.

19.4.1　Ethics

In most of the reported observational studies (case reports and cohort studies), traditional remedies have been investigated clinically without preliminary extensive toxicity studies. As the patients are choosing to take the herbal remedy of their own free will, and the remedies have been in use for a long time, this is ethically acceptable (WHO, 1993). However, patients should be closely monitored, both clinically and with blood tests and ECGs, to screen for possible adverse effects. This would serve the purpose of a phase I-type study before proceeding to a phase II-style randomised controlled trial (see Chapter 21).

In randomised controlled trials, the control group should be treated with the local standard first-line regime for malaria. It would not be ethical to treat with a placebo, although this has been done in the past. It is important to gauge the efficacy of herbal remedies in the current situation and to compare them with locally used drugs, especially in the presence of drug resistance.

19.4.2　Safety and Tolerability

Too few clinical studies recorded or reported adverse effects, or monitored blood parameters or electrocardiograms. The little data that does exist suggest that severe adverse reactions are infrequent — none were recorded in the studies reviewed. The most common adverse effects were minor gastrointestinal symptoms (nausea, vomiting, abdominal pain, diarrhoea), but the fact that most patients tolerated them implies that they must perceive the treatment to be effective. The issue of palatability is also important, especially where young children are concerned. As might be expected for mixtures containing alkaloids, some were bitter, making them difficult to administer to children.

Future clinical studies on herbal antimalarials should include monitoring for side effects, both major and minor. Recording detailed data on possible adverse effects (symptoms; ECGs; liver, kidney, and bone marrow function tests) is very important, especially in cohort studies and randomised controlled trials (see Chapter 21).

19.4.3　Efficacy

Clinical observations in the form of case reports and cohort studies are useful for directing further research. However, methodological problems were very common.

First, dosages are rarely defined, and there is a lot of scope for variability, from differing recipes of compound preparations, different preparation methods, and different quantities administered. The method of preparation of the herbal medicine needs to be fully recorded and, in the case of cohort studies, standardised. In order to protect intellectual property rights, investigators may not wish to publish the formulation. However, it is important to keep a record of the method in a safe place as a reference for future research.

Secondly, cases should be selected on the basis of predefined inclusion and exclusion criteria. Some of the case reports and cohort studies did not define these, and it is possible that the authors simply selected the cases with the best responses to treatment.

Thirdly, outcome measures need to be clearly defined and reported. Parasite clearance was the most common measure used, but different trials reported this at different times after initiation of treatment, and rarely later than day 7. It is important to follow up patients for at least 14 days, and if possible, up to 28 days, in order to measure recrudescence rates. Polymerase chain reaction (PCR) could be used to differentiate between reinfections and recrudescences. In endemic areas, adequate clinical response should also be reported as an outcome measure, as this is now the standard for monitoring the efficacy of conventional antimalarial drugs (WHO, 1996).

Fourthly, natural immunity to malaria may account for the clinical improvement, especially in adults in endemic areas. In order to overcome this problem, randomised controlled trials are needed. Nevertheless, if a study is designed simply to validate the use of a traditional remedy in a local setting, good results from a cohort study may be sufficient to justify this. Good results from a randomised controlled trial would be needed, however, to justify the recommendation of the use of a herbal preparation in areas where it is not in traditional use.

The controlled trials reviewed also faced many of the methodological problems described above. It was not clear in all cases that patients were randomly allocated to treatment groups. It was not always possible to blind patients to the treatments they received (if one is a herbal decoction and the other is a tablet), but at least the clinicians and laboratory staff assessing the outcomes could have been blinded.

Although chloroquine was the most common treatment given to control groups, none of the trials was conducted recently in areas with significant drug resistance, and only two of the trials concerned *P. falciparum* exclusively. There is clearly a need for controlled trials of herbal remedies in areas where there is significant resistance to first-line drugs, but drug policy cannot be changed because of budgetary constraints. Some herbal remedies may prove a valuable and viable alternative in these circumstances. Furthermore, in remote areas where no orthodox health care is available, there would be value in a readily available remedy that prevents mortality while the patient goes to seek further assistance.

In an endemic area where there are good levels of immunity, and most people carry parasites asymptomatically, it could be argued that a herbal remedy that reduces parasitaemia to an asymptomatic level is sufficiently effective. This has been put into practice in India, where Ayush-64 has been used on a large scale in malaria epidemics (Bhatia, 1997), and in Mali, where Malarial is listed in the national formulary (Ministère de la Santé, 1998).

Thus herbal remedies could be evaluated using a protocol similar to that recommended by WHO for the assessment of therapeutic efficacy of antimalarial drugs (WHO, 1996), which takes a pragmatic clinical approach. Guidelines for case reports, cohort studies, and controlled trials have been prepared by the RITAM and are published in Chapter 21 of this book.

To ask whether traditional herbal remedies are safe and effective is to ask the same about chemotherapy. The answer, in all probability, will turn out to be similar: certain treatments, in certain places, and for certain patients are safe and effective; others are not. The difference is that in the case of chemotherapy, we already know which treatments are safe and effective, in which circumstances, and for which patients. For herbal treatments, we have almost no information — and there are many more herbal treatments used throughout the world than varieties of antimalarial drugs.

19.5 CONCLUSION

Clinical observations on traditional remedies are feasible and useful, and this chapter has presented clinical evidence that some herbal remedies are safe and effective against malaria. However, better evidence from randomised clinical trials is necessary before the use of an herbal remedy can be recommended on a large scale.

As randomised controlled trials are expensive and time-consuming, it is important to prioritise remedies for clinical investigation according to existing data from sociological, ethnobotanical, pharmacological, and preliminary clinical observational studies. In remote, resource-poor settings where even current antimalarials are not steadily available, evidence-based traditional medicine deserves recognition and expanded support by policy makers and funders. Feedback of research results may improve the quality of traditional medicine use and reduce mortality and morbidity from malaria.

ACKNOWLEDGMENTS

The authors thank Rosemary Byrd for providing some of the database searches and members of the RITAM network for providing some of the unpublished studies. They thank two anonymous reviewers for their constructive criticisms and comments.

REFERENCES

Acton, H.W. (1920). Researches on the treatment of benign tertian fever. *Lancet*, 198, 1257–1261.

Ajaiyeoba, E.O., Falade, C.O., Fawole, O.L., et al. (2000). A New Approach to the Development of Antimalarial Drugs from Nigerian Ethnomedicine. Paper presented at the WHO/TDR meeting on Natural Products for the Treatment of Tropical Diseases, Geneva, August 26–28.

Bhatia, D. (1997). Role of Ayush-64 in malaria epidemic. *J. Res. Ayurveda Siddha*, XVIII, 71–76.

Bitahwa, N., Tumwesigye, O., Kabariime, P., Tayebwa, A.K.M., Tumwesigye, S., and Ogwal-Okeng, J.W. (1997). Herbal treatment of malaria: four case reports from the Rukararwe Partnership Workshop for Rural Development (Uganda). *Trop. Doct.*, (Suppl. 1), 17–19.

Boye, G.L. (1989). Studies on the antimalarial action of *Cryptolepis sanguinolenta* extract. In *Proceedings of International Symposium on East-West Medicine, Seoul, Korea*, pp. 242–251.

Bugmann, N. (2000). Le concept du paludisme, l'usage et l'efficacité *in vivo* de trois traitements traditionnels antipalustres dans la région de Dori, Burkina Faso. Thesis, Medical Faculty, University of Basel, Switzerland.

CCRAS. (1987). *Ayush-64: A New Antimalarial Herbal Compound*. CCRAS, Delhi.

Deare, B.H., Mathew, C.M., Bahadur, R.C.L.B., Gage, A.T., and Carter, H.G. (1916). *The Third Report of the Indigenous Drugs Committee*. Superintendent Government Printing, Calcutta.

Guindo, M. (1988). Contribution à l'étude du traitement traditionnel du "Suma" (Paludisme). Thèse pour obtenir le grade de Docteur en Pharmacie, Ecole Nationale de Médecine et de Pharmacie du Mali.

Hicks, E.P. and Diwan Chand, S. (1935). The relative clinical efficacy of totaquina and quinine. *Rec. Malaria Surv. India*, 5, 39–50.

Jang, C.S., Fu, F.Y., Wang, C.Y., Hunag, K.C., Lu, G., and Chou, T.C. (1946). Ch'ang shan, a Chinese antimalarial herb. *Science*, 103, 59.

Jurg, A., Tomás, T., and Pividal, J. (1991). Antimalarial activity of some plant remedies in use in Marracuene, southern Mozambique. *J. Ethnopharmacol.*, 33, 79–83.

Koita, N. (1991). A comparative study of the traditional remedy "suma-kala" and chloroquine as treatment for malaria in the rural areas. In *Proceedings of an International Conference of Experts from Developing Countries on Traditional Medicinal Plants, Arusha, Tanzania, February 18–23, 1990*, Mshigeni, K.E., Nkunya, M.H.H., Fupi, V., Mahunnah, R.L.A., and Mshiu, E., Eds. Dar Es Salaam University Press, Tanzania.

Lugakingira, E.S. (Unpublished a). Clinical Experience Report: Treatment of Malaria Using *Maytenus senegalensis* Lam. Ministry of Health, Dar Es Salaam.

Lugakingira, E.S. (Unpublished b). Treatment of Malaria Using *Parinari exelsa* Sabin: A Clinical Experience Report. Ministry of Health, Dar Es Salaam.

Mabiala, J.B. (1991). Paludisme et Médecine traditionelle au Congo. Séminaire CREDES, Paris, November 4–9.

Makinde, J.M., Laoye, O.J., and Salako, L.A. (Unpublished). In Vivo Study of *Morinda lucida*.

Matthies, F. (1998). Traditional Herbal Antimalarials: Their Role and Their Effects in the Treatment of Malaria Patients in Rural Tanzania. Ph.D. dissertation, University of Basel, Switzerland.

Ministère de la Santé, des Personnes Agées et de la Solidarité. (1998). *Formulaire Thérapeutique National*. Editions Donnaya, Bamako, Mali.

Mueller, M.S., Karhagomba, I.B., Hirt, H.M., Wernakor, E., Li, S.M., and Heide, L. (2000). The potential of *Artemisia annua* L. as a locally produced remedy for malaria in the tropics: agricultural, chemical and clinical aspects. *J. Ethnopharmacol.*, 73, 487–493.

Ouedraogo, J.B., Guigemde, T.R., Trsore, M., Traore, S.A., Dakuyo, Z., and Sanou, A. (1992). Etude de l'efficacité parasitologique du N'dribala dans le traitement du paludisme. *Sci. Tech.*, 20, 45–53.

Pampana, E.J. (1934). Clinical tests carried out under the auspices of the Malaria Commission. *League of Nations: Q. Bull. Health Org.*, III, 328–343.

Panda, P.K. (1998). Clinical study of Caesalpinia crista L. (Lata Karanja) in malaria patients. *J. Res. Ayurveda Siddha*, 19, 122–127.

Phetsouvanh, T. (1991). Rapport succinct du traitement du paludisme en RDP Lao par le rhizome *d'Alocacia macrorhiza* L. Présentation au Séminaire CREDES, Paris, November 4–9.

Randriamahary-Ramialiharisoa, A., Ranaivoravo, J., Ratsimamanga-Urverg, S., Rasoanaivo, P., and Rakoto-Ratsimamanga, A. (1994). Evaluation en clinique humaine de l'action potentialisatrice d'une infusion de *Strychnos myrtoides* vis-à-vis d'antipaludéens. *Rev. Méd. Pharm. Afr.*, 8, 123–131.

Sharma, K.D., Kapoor, M.L., Vaidya, S.P., and Sharma, L.K. (1981). A Clinical Trial of "Ayush 64" (A coded antimalarial medicine) in cases of malaria. *J. Res. Ayur Siddha*, 2, 309–326.

Sinton, J.A. and Bird, W. (1929). Studies in malaria, with special reference to treatment. Part XI. The cinchona alkaloids in the treatment of benign tertian malaria. *Indian J. Med. Res.*, 16, 725–746.

Thiombiano, A. (1991). Etudes des propriétés antipaludéenes de *Terraplis interretis* chez l'homme. Présentation au Séminaire CREDES, Paris, November 4–9.

Tsu, C.F. (1947). Chang shan in the treatment of malaria. *Trop. Med. Hyg.*, 50, 75–77.

Valecha, N., Devi, C.U., Joshi, H., Shahi, V.K., Sharma, V.P., and Lal, S. (2000). Comparative efficacy of Ayush-64 vs. chloroquine in vivax malaria. *Curr. Sci.*, 78, 1120–1122.

Vedavathy, S. and Rao, D.N. (1995). Herbal folk medicine of Tirumala and Tirupati region of Chittoor district, Andhra Pradesh. *Fitoterapia*, 66, 167–171.

Venkataraghavan, S., Seshadri, C., Ramakrishna, B., Gowri, N., Revathi, R., and Chari, M.V. (1982). Effect of Sudarsana churna and Seetamsurasa in Plasmodium vivax malaria: a preliminary report. *Nagarjun*, 25, 99–100.

WHO. (1993). *Research Guidelines for Evaluating the Safety and Efficacy of Herbal Medicines*. WHO Regional Office for the Western Pacific, Manila.

WHO. (1996). Assessment of Therapeutic Efficacy of Antimalarial Drugs, WHO/MAL/96.1077.

Willcox, M.L. (1999). A clinical trial of 'AM', a Ugandan herbal remedy for malaria. *J. Public Health Med.*, 21, 318–324.

Yao-De, W., Qi-Zhong, Z., and Jie-Sheng, W. (1992). Studies on the antimalarial action of gelatin capsule of *Artemisia annua*. *Chung Kuo Chi Sheng Chung Hsueh Yu Chi Sheng Chung Ping Tsa Chih*, 10, 290–294.

20 Observational Methods for Assessing Traditional Antimalarials

Bertrand Graz

CONTENTS

20.1 INTRODUCTION

Observational (or nonexperimental) clinical studies are field research methods that can provide reliable information on the effects of traditional antimalarial treatments on human beings at relatively low cost. In this chapter, two widely held views over clinical studies will be challenged: the hierarchy of clinical studies with randomised controlled trials (RCTs) being the gold standard, and the concept that clinical studies are to be conducted only after preclinical studies. We will thus show that:

1. An observational study can, under some circumstances, yield scientific results that are as reliable as those of RCTs, and sometime more so. In addition, some important data can *only be obtained* through observational studies.
2. A lot of time and energy can be saved by also conducting observational studies *before preclinical studies*.

The last section of the chapter presents some practical recommendations for designing appropriate and reliable observational studies, and a couple of innovative approaches to traditional antimalarial research are suggested.

BOX 20.1: DEFINITIONS OF TERMS AND PRINCIPLES USED IN THIS CHAPTER

Observational methods: Quantitative, epidemiological methods in which data are collected through a rigorous and formalised observation. Common observational methods are cohort studies (prospective or retrospective), case-control studies, case series, and case reports.

Traditional antimalarials: Treatments used for a long time in a particular area for treating or preventing a set of symptoms or problems related to human *Plasmodium* infection.

Assessing, assessment: These terms refer to the research activity leading to a "judgement of proof beyond a reasonable doubt in the evaluation of interventions" (Hennekens and Buring, 1994). The main points of assessment are treatment effectiveness, safety, and access.

Clinical studies, a scientific perspective: Conducting scientific research requires a neutral view over the phenomenon under study. For clinical studies, this means that we have no preconception toward a treatment, whether developed by a laboratory or by generations of traditional healers. With such principles, it is even feasible to study, with full scientific rigor, the effectiveness and safety of a traditional treatment that varies in composition and whose constituents are not yet known, as will be shown in this chapter.

20.2 WHY WE NEED OBSERVATIONAL STUDIES

When evaluating the effect of a medical intervention on malaria patients, the dilemma is often: "observational study or RCT?" We still have difficulties in coming out of the exclusive confidence in RCTs, dating to the excessive enthusiasm of some authors in the period 1960–1980 and still held by some influential bodies (Concato et al., 2000). There are, however, serious limitations with RCTs that can be overcome by carefully designed observational studies. We may eventually come to a point where observational studies can be seen as complementary to RCTs, rather than as a poor alternative. Observational studies, besides their specific remits, are also useful guides to the design of new controlled trials and may test whether results of experimental studies translate into routine practice (Pocock and Elbourne, 2000).

20.3 FEASIBILITY

An observational study can be feasible where and when experimentation (such as an RCT) is not. An experimental design may be "unnecessary, inappropriate, impossible, or inadequate" (Black, 1996). For example, if we want to evaluate rapidly and with limited means a traditional antimalarial in the area where it is used, it might be feasible and appropriate to follow up a few dozen patients using the treatment as usual. World Health Organization (WHO)-issued guidelines proposed that positive results of such a small trial can then serve for local recommendations (see Chapter 21). Toxicology studies are generally not necessary before observational studies of traditional treatments that have long been in use (see Chapters 18 and 21).

20.4 SPECIFICITY

For some research questions, observational studies appear as the design of choice. For example, a follow-up study could answer the question "Does this traditional healer have a cure rate with malaria patients that is similar to local outcomes with modern drugs?" A case-control study could answer the following: "Is it true that those seen at this hospital with severe malaria did not use a particular traditional antimalarial, while those suffering only from a mild episode did?" (If true, this could be an index of protective activity against severe forms of the disease.) A large follow-up study would be needed to answer "Are there some rare but severe adverse events with a certain treatment?" Indeed, postmarketing surveillance through large cohort studies is indisputably more sensitive than medium-size randomised controlled trials for the detection of rare adverse effects.

20.5 EXTERNAL VALIDITY

As compared with RCTs, observational methods can be conducted in conditions closer to real practice. An RCT is not predictive of the treatment success in either a particular patient or in a population different from the one under study. This relates to a problem that has received little attention so far: the lack of external validity (or generalisability) of RCTs. Three main reasons for this weakness of RCTs have been shown: health care professionals conducting the experiment may be unrepresentative; patients included in the study (through an often low, and unreported, recruitment rate and restrictive criteria) may be very different from those treated in practice; and treatment may be atypical (with more care, etc.), compared to usual practice (Black, 1996).

This is a matter of difference between *efficacy* (referring to positive results within a definite research setting) and *effectiveness* (positive results in real-world practice). In short, RCTs usually offer an indication of efficacy, and observational methods offer an indication of effectiveness.

As a general rule, we should keep in mind that following work by David Hume and Karl Popper, predictions from a clinical study are at best tentative and can only be established within a framework of assumed regularity (of disease, health care, patients, environment, etc.). Misplaced belief in induction (such as issuing universal statements about the effectiveness of a given treatment) is responsible for many false notions regarding clinical trials (Senn, 1991).

20.6 UNDER WHAT CONDITIONS CAN WE TRUST OBSERVATIONAL STUDIES?

What is the reliability of observational studies? In a study comparing them with RCTs when applied to the same subjects, it appears that "well-designed observational studies (with either a cohort or a case-control design) do not systematically over-estimate the magnitude of the effects of treatment as compared with those in randomised, controlled trials on the same topic" (Concato et al., 2000). It appeared that carefully conducted observational studies had less variability in point estimate (i.e., less heterogeneity of results) than RCTs on the same topics. The authors regretted that data based on weak forms of observational studies were often mistakenly used to criticise all observational research.

They consequently developed the concept of the restricted cohort design (Concato et al., 2000), referring to follow-up studies conducted with strict principles, as follows:

- Identify a zero time for determining a patient's eligibility and baseline features.
- Choose inclusion and exclusion criteria similar to those in experimental trials.
- Adjust for differences in baseline susceptibility to the outcome (for example, time since arrival in endemic area is important for susceptibility to severe malaria).
- Use statistical methods similar to an RCT, including intention to treat analysis.

Today, detailed practical recommendations exist for conducting high-quality observational research of various types in the field of traditional antimalarials (see Chapter 21).

Numerous important pieces of knowledge on malaria have been gained through observational studies. Here are a few examples from recent studies:

- During a follow-up study after malaria treatment, symptoms were perceived and reported by only 19.5% of the individuals at the time of first recurrent parasitaemia (Owusu-Agyei et al., 2001).
- In another prospective study, it was observed that after radical antiparasitic treatment in endemic populations, exposure to high vs. low entomologic inoculation rates was not significantly correlated with time to reappearance of malaria parasitaemia (Sokhna et al., 2001).
- An example of a long-term longitudinal study is the one that has shown that combining mefloquine with artesunate could in some situations halt the progression of mefloquine resistance (Nosten et al., 2000).
- A retrospective study found that mefloquine was used significantly more often by those travelers suffering from neuropsychiatric symptoms, strengthening the suspicion that mefloquine was a causal factor (Potasman et al., 2000).

20.7 WHY OBSERVATIONAL STUDIES ALSO NEED TO BE CONDUCTED *BEFORE* PRECLINICAL STUDIES

At first this seems to be a paradox: preclinical studies by definition are those conducted before clinical studies. The word *preclinical* is regrettable because it is misleading. As Mamadou Koumaré, former director of traditional medicine for the World Health Organization (African region), said, "After one has spent millions on animal research it might still not work" (Movich, 1998). In past decades, thousands of plants have been randomly tested *in vitro* and on animals for their potential antimalarial properties, with quite few results. The standard process of preclinical studies before clinical studies has yielded disappointing results, which may at least partly be explained as follows.

First, the chances are very low that a plant, among the hundreds showing *in vitro* antimalarial properties, will lead to a widely recommendable antimalarial. The plant (or its extracted active constituents) might be too toxic, unstable, not absorbed, metabolised, and so on. Thus, there is an enormous amount of research with negative results.

Second, some plants may have no *in vitro* antimalarial properties, and even no effect on laboratory animals; they are therefore discarded from further research programs. The very same plant, however, might well be effective in human beings. A constituent may be activated through a human-specific metabolic pathway or antimalarial activity may be mediated by complex physiologic processes such as the immune system. Thus, inappropriate screens lead to false negatives: effective traditional treatments may be missed.

There is a third situation where an active traditional antimalarial can be missed or forgotten: the active constituent has been known (and patented) for a long time; as such it is uninteresting for commercial purposes. The same antimalarial, however, may be of public health interest if it provides a first-aid treatment available in remote areas where shortages of essential drugs are frequent (see Chapters 2 and 3).

Of course, this is not to say that preclinical studies are useless. They are of great interest and use. However, their yield could be enhanced by the appropriate use of clinical observational studies. For this matter, it does not make a difference whether the research goal is to discover new molecules or to test effectiveness and safety of traditional treatments currently consumed. In both cases, it is desirable to accelerate the research process in order to meet public health needs.

So far, traditional antimalarials for preclinical study have mainly been selected through ethnopharmacological studies or at random. Ethnologic clues already enhance the chances of finding a treatment that will prove effective (see Chapter 11). Then, before proceeding to preclinical studies, a selected traditional treatment could be rapidly evaluated through a small-scale, nonexperimental clinical study. This can save a lot of time, resources, and eventually lives.

If a cohort study of 20 malaria patients finds a 90 or 95% clinical cure within 2 days, this is enough to point the treatment used as having a high potential. In this case, if subsequent preclinical studies show a very low *in vitro* antimalarial activity, this will not lead to the abandonment of the studied product. On the contrary, a more complex human-specific therapeutic activity can be suspected and searched for.

In summary, observational studies should be conducted before preclinical studies because they help to identify promising traditional treatments, especially when time and resources are limited.

BOX 20.2: QUESTIONS TO BE ANSWERED THROUGH OBSERVATIONAL STUDIES

1. What is already known about the treatment under study? From the international literature in medicine, geography, ethnology, toxicology, botany, phytochemistry, parasitology, and related fields.
2. How do local names for the disease and malaria (as defined by microscopic examination or clinical syndrome) overlap?
3. What is the local effectiveness of a treatment? (Note: Traditional antimalarials may be effective in very different ways, as discussed in the last section of this chapter.)
4. What are the side effects of a treatment and the traditional precautions?
5. What are the difficulties encountered while performing clinical follow-up of patients at the traditional practice level?
6. What are the side effects of the study? Hypotheses: patient referral has improved; contact between primary health care professionals and traditional healers is more frequent; young local people have started to use or learn local treatment; traditional healers have raised their prices; plants used as antimalarials are becoming rare — or start to be cultivated; and so on.
7. What is learned from experience to improve recommendations for locally *and* internationally sound research?
8. What is the generalisability of results? This needs to be discussed in detail, as well as conditions for applying results elsewhere (foreseeable role of main factors of variability, recommended surveillance).

20.8 EXAMPLE OF OBSERVATIONAL STUDIES CONDUCTED ON TRADITIONAL ANTIMALARIALS

Thirty-four cohort studies on traditional antimalarials have been retrieved from a comprehensive literature search (see Chapter 19). Five of them examined Ayurvedic preparations used in India. Others investigated traditional remedies in Uganda, Tanzania, and Madagascar. The Chinese herb *Artemisia annua*, the source of artemisinin and its derivatives, was studied in the Democratic Republic of Congo. These studies provide indications not only on therapeutic and secondary effects but also on the potential to reverse resistance to chloroquine and the feasibility of plant cultivation and preparation.

The interpretation of results focused on the improvement of clinical symptoms (fever, headache, etc.) and parasite clearance. Lack of parasite clearance was interpreted as a failure; this should be questioned in view of local epidemiology: when rapid reinfection is the rule (especially in areas of stable transmission), parasite clearance is not necessarily a desirable outcome. This is so because in areas where complete eradication of malaria is not achievable in the near future, a low parasitaemia may contribute to maintaining immunity, prevent severe forms of the disease, and is often tolerated without any clinical symptoms (Bell, 1995).

20.9   RECOMMENDATIONS FOR CLINICAL PRACTICE

How can we use results from observational studies to draw recommendations for clinical practice? Are such results sufficient for an official health authority to recommend and support the use of a traditional antimalarial? These questions have been debated in several meetings, such as the one organised by the World Health Organization, African region, in a regional workshop in Madagascar on November 20–24, 2000.

When speaking of uncomplicated malaria, it is proposed that good results from a cohort study (demonstrating a treatment failure rate of less than 25%) are sufficient to recommend local use in an area where the plant has traditionally been used (see Chapter 21).

It was, however, proposed that good results from a randomised controlled trial are also necessary "before wider use of the herbal preparation can be recommended." This appears inconsistent with what was said about the lack of external validity of RCTs and might reflect the need of those studying traditional antimalarials to comply to commonly held (but, as we have tried to show, erroneous) views on the hierarchy of research designs. In the case of malaria, it could be argued that observational studies also lack external validity, in that resistance rates and the effect of immunity vary very widely from one community to another. Unfortunately, an RCT cannot be more valid externally because the randomised controlled part of the study does not eliminate variance due to these external factors. An RCT determines what part of the observed cure rate is due to the treatment itself (= treatment efficacy), but this efficacy may change from one site to another, precisely because of external factors such as local plasmodium resistance to the treatment under study. On the other hand, an observational study may show that all the patients get better (which means that the pragmatic goal is attained), but there is no way of knowing whether this is simply because of their acquired immunity, or whether the medicine (be it herbal or conventional) has an additional effect.

Ideally, when applying recommendations derived from RCT results, we should conduct a small observational study to verify that the theoretical treatment efficacy is applicable locally (= local effectiveness). Experimental and nonexperimental designs need to be viewed as complementary methods (Black, 1996; Concato et al., 2000).

The definitions of good results and of treatment failure are matters of concern: parasite count can vary greatly within days and hours, even among asymptomatic patients, and is poorly correlated with clinical status (Guiguemde et al., 1992). In fact, the very definition of malaria is problematic, especially in populations with a high transmission rate and immunity level, where a large proportion of the population has a positive parasitaemia at any given time (Rogier et al., 2001). For this reason it is particularly important to design studies with sufficient attention to the two areas of main concern: effects of maternal infection and events related to severe malaria (including rate of convulsions, coma, survival, sequelae, and suspected secondary effects).

20.10   HOW TO CHOOSE THE APPROPRIATE STUDY DESIGN

When designing a research protocol, it is highly advisable to contact specialists in epidemiology and statisticians, in order to discuss the planned study and ensure its scientific value. The few days spent in discussing a protocol beforehand can save months, if not years; it can make the difference between a major medical discovery and useless research. Protocol discussions and support may be possible with a local university department or through multidisciplinary groups such as the Research Initiative on Traditional Antimalarial Methods (Willcox and Bodeker, 2000).

The choice of an appropriate research protocol, then the design and conduct of the study itself, is made easier by practical recommendations issued by WHO (see next chapter). These guidelines are meant for study of natural products and herbal preparations, but other types of traditional treatment could be studied along the same principles.

To choose an appropriate research design, we need to make the research question and objectives very clear and explicit. We must be sure that all partners fully agree with the chosen objectives. This has to be accomplished before opting for an experimental or observational research design. Doing so will allow epidemiologists and statisticians to help to their full potential and design adequate proposals.

20.11 RESEARCH ON TRADITIONAL ANTIMALARIALS: OLD AND NEW QUESTIONS

We can observe that while the total amount of research is enormous, we badly lack some information that could be obtained through relatively simple studies. For example, we could study correlations between domestic practices and severe malaria: Do some of them appear dangerous, or protective? Such information could lead to research of utmost public health importance, because it may uncover factors (e.g., in case of severe malaria, a simple way of avoiding extreme hypoglycaemia and acidosis during referral) that could be applied on a large scale and make important changes in terms of mortality rates.

Most research on traditional antimalarials has so far had one or both of the following objectives:

- To test the hypothesis that a local treatment can be safely used for uncomplicated malaria
- To discover new molecules with antiplasmodial properties

The main burden of malaria is related to severe episodes and maternal infection. Since malaria eradication is in most places not feasible in the foreseeable future, we are left with trying to moderate its impact on individuals and populations. As long as we see traditional antimalarials only as first-line treatments for uncomplicated malaria (or as a potential source of patentable molecules), we do not approach a broad solution to the malaria problem. That is why it seems important also to look for other potential utilities of traditional antimalarials.

Could some locally used antimalarials prevent severe malaria? Severe malaria can in principle be avoided by the early and adequate treatment of uncomplicated malaria; studies are necessary to verify whether this is also true with traditional herbal antimalarials, even when they do not lead to parasite clearance. Could a traditional antimalarial have a protective effect during pregnancy, so as to treat anaemia, increase birth weight, or prevent abortion? It is always particularly difficult to prove the safety of a drug during pregnancy, because experiments are out of the question. There are nonexperimental study possibilities with products used traditionally by pregnant women: when a control group exists (nonusers of the product of interest) correlations between use/nonuse and adverse events can be studied. This type of study design led, for example, to the discovery of the importance of folic acid in nutrition for the prevention of spina bifida.

For ethical reasons, questions on malaria and pregnancy or severe malaria treatment can be approached often only with observational studies, such as retrospective or case-control studies. Retrospective studies would be acceptable, prospective would not. Several studies of that kind have already been conducted focusing on factors affecting the prognosis of severe malaria (Nacher et al., 2001; Snow et al., 1998). The proposal here is that such studies should include the use of traditional antimalarials as a potential protective factor.

Risk (or harm) reduction can be a valuable approach for very common diseases responsible for important mortality and morbidity. Some risk/protective factors may come up as a surprise: in malaria, it appeared (thanks to patient follow-up and a cross-sectional survey) that concomitant intestinal worm (helminth) infection is a protective factor (Murray et al., 1978; Nacher et al., 2001). It might well appear that the use of some traditional antimalarials is a protective factor in the case of severe malaria and pregnancy. It must be stressed here that this is *not* a suggestion that experimental clinical trials must be conducted on herbal medicines for severe malaria or malaria in pregnancy. It is unethical to conduct any human experiment unless there is strong evidence that a treatment is the best we have to avoid suffering or to increase survival. And some evidence on traditional medicines so far has tended to show the opposite: De Francisco et al. (1994) found that 62% of children who died of malaria had consulted a traditional healer (but most had also consulted a health care professional); traditional medicine was described as often inappropriately used for the treatment of cerebral malaria with convulsions, in Tanzania (Winch et al., 1996). It is nevertheless legitimate to keep hope in more research and conduct *nonexperimental* studies, which are ethically acceptable and may bring promising information.

From a public health perspective, the main question remains: Are some traditional antimalarials able to treat or prevent severe and deleterious forms of the disease? To raise ambitions when studying antimalarials increases the potential importance of research findings but decreases the chances of finding them. This seems necessary, however, if the desired outcome of research is to reach public health importance. Decreasing mortality is rarely expected as an outcome of traditional treatment but, as was proposed here, it could well be. We need to register overall mortality at all ages and not only child mortality. Studying under-5 mortality only can be a source of error, since children protected from infection during their first 5 years of life may suffer later from more severe malaria.

In summary, new research questions and hypotheses should be considered when designing studies on traditional antimalarials, because of their immense public health importance:

- The hypothesis of a potential reduction in risk of severe malaria and mortality correlated with the use of certain traditional antimalarials, *even if they do not clear parasitaemia*
- The hypothesis of possible correlations between pregnancy outcomes and local factors, including consumption of traditional antimalarials, nutritional specificities, and other local practices

When looking for preliminary clues of important effects of a traditional treatment, observational studies will often be the only feasible and appropriate research method, with a retrospective or case-control design.

Funding for observational studies has been scarce in recent years. Large parts of the means attributed to malaria research (Medicines for Malaria Venture's declared objectives are U.S. $30 million a year) are used for laboratory research and randomised controlled studies, rather than observational research (exact amounts not found, despite extensive bibliographic research and contacts with WHO). This can be in part a result of research policies of funding agencies; it can also be due to the rarity of submitted projects of case-control or follow-up studies.

It is urgent to make it possible for policy makers and funding agencies to understand that observational methods are powerful tools for approaching their objectives. For this, physicians and other scientists should plan and submit research projects that take advantage of the possibilities offered by observational designs. This is so because "new directions in clinical evaluation must be forged by researchers who are able to transcend limitations in research orthodoxy in the interest of providing sound information to the public of what constitutes good healthcare" (Bodeker, 2000).

ACKNOWLEDGMENTS

The author would like to thank Jacques Falquet and Merlin Willcox for their help in the conception and writing of this chapter.

REFERENCES

Bell, R.D. (1995). *Tropical Medicine*, 4th ed. Blackwell Science, Oxford.

Black, N. (1996). Why we need observational studies to evaluate the effectiveness of health care? *Br. Med. J.*, 312, 1215–1218.

Bodeker, G. (2000). Complementary medicine and evidence. *Ann. Acad. Med. Singapore*, 29, 3–6.

Concato, J., Shah, N., and Horwitz, R.I. (2000). Randomised controlled trials, observational studies, and the hierarchy of research designs. *N. Engl. J. Med.*, 342, 1887–1892.

De Francisco, A., Schellenberg, J.A., Hall, A.J., Greenwood, A.M., Cham, K., and Greenwood, B.M. (1994). Comparison of mortality between villages with and without primary health care workers in Upper River Division, the Gambia. *J. Trop. Med. Hyg.*, 97, 69–74.

Guiguemde, T.R., Toe, A.C., Sadeler, B.C., Gbary, A.R., Ouedraogo, J.B., and Louboutin-Croc, J.P. (1992). [Variation of the parasite density of *Plasmodium falciparum* in asymptomatic carriers: consequences for malaria chemoresistance studies]. *Med. Trop.*, 52, 313–315 (in French).

Hennekens, C.H. and Buring, J.E. (1994). Observational evidence: doing more good than harm. *Ann. N.Y. Acad. Sci.*, 703, 19–23.

Movich, R. (1998). Malian medicinals. In *Institute of Current World Affairs Letters*, January, ICWA, Hanover, NH, pp. 3–7.

Murray, J., Murray, A., Murray, M., and Murray, C. (1978). The biological suppression of malaria: an ecological and nutritional interrelationship of a host and two parasites. *Am. J. Clin. Nutr.*, 31, 1363–1366.

Nacher, M., Singhasivanon, P., Vannaphan, S., Treeprasertsuk, S., Phanumaphorn, M., Traore, B., Looareesuwan, S., and Gay, F. (2001). Socio-economic and environmental protective/risk factors for severe malaria in Thailand. *Acta Trop.*, 78, 139–146.

Nosten, F., van Vugt, M., Price, R., Luxemburger, C., Thway, K.L., Brockman, A., McGready, R., Kuile, F., Looareesuwan, S., and White, N.J. (2000). Effects of artesunate-mefloquine combination on incidence of *Plasmodium falciparum* malaria and mefloquine resistance in western Thailand: a prospective study. *Lancet*, 356, 297–302.

Owusu-Agyei, S., Koram, K.A., Baird, J.K., Utz, G.C., Binka, F.N., Nkrumah, F.K., Fryauff, D.J., and Hoffman, S.L. (2001). Incidence of symptomatic and asymptomatic *Plasmodium falciparum* infection following curative therapy in adult residents of northern Ghana. *Am. J. Trop. Med. Hyg.*, 65, 197–203.

Pocock, J.S. and Elbourne, D.R. (2000). Randomised trials or observational tribulations? *N. Engl. J. Med.*, 342, 1907–1909 (editorial).

Potasman, I., Beny, A., and Seligmann, H. (2000). Neuropsychiatric problems in 2500 long-term young travellers to the tropics. *J. Travel Med.*, 7, 5–9.

Rogier, C., Henry, M.C., and Spiegel, A. (2001). [Diagnosis of malaria outbreaks in endemic areas: theoretical aspects and practical implications]. *Méd. Trop.*, 61, 27–46 (in French).

Senn, S.J. (1991). Falsificationism and clinical trials. *Stat. Med.*, 10, 1679–1692.

Snow, R.W., Peshu, N., Forster, D., Bomu, G., Mitsanze, E., Ngumbao, E., Chisengwa, R., Schellenberg, J.R., Hayes, R.J., Newbold, C.I., and Marsh, K. (1998). Environmental and entomological risk factors for the development of clinical malaria among children on the Kenyan coast. *Trans. R. Soc. Trop. Med. Hyg.*, 92, 381–385.

Sokhna, C.S., Faye, F.B.K., Spiegel, A., Dieng, H., and Trape, J.F. (2001). Rapid assessment of *Plasmodium falciparum* after drug treatment among Senegalese adults exposed to moderate seasonal transmission. *Am. J. Trop. Med. Hyg.*, 65, 167–170.

WHO — African Region. (2000). Guidelines for Evaluating the Efficacy and Safety of Herbal Preparations Used for the Treatment of Uncomplicated Malaria. Paper presented at a regional workshop on the Methodology for Evaluation of Traditional Medicines, Antananarivo, Madagascar, November 20–24.

Willcox, M.L. and Bodeker, G. (2000). Plant-based malaria control: research initiative on traditional antimalarial methods. *Parasitol. Today*, 16, 220–221.

Winch, P.J., Makemba, A.M., Kamazima, S.R., et al. (1996). Local terminology for febrile illnesses in Bagamoyo district, Tanzania and its impact on the design of a community-based malaria control programme, *Soc. Sci. Med.*, 42, 1057–1067.

21 Guidelines for Clinical Studies on Herbal Antimalarials

Merlin Willcox and Idowu Olanrewaju

CONTENTS

21.1 BACKGROUND

Traditional herbal medicines are widely used to treat malaria (see Chapter 10), yet very few have been clinically evaluated. When clinical evaluation has taken place, it has often been of a standard unacceptable to doctors, policy makers, and those planning malaria control programs (see Chapter 19). These guidelines aim to motivate further research in this field and to help researchers plan studies that are both affordable and of adequate quality.

It is important to emphasise that there has been little funding for this type of research, making it almost impossible to conduct studies of the same scale as drug trials. Pharmaceutical companies will not be interested in paying for trials of traditional medicines that they cannot patent and that will compete with their own products; therefore, the main source of funding for this sort of study will need to be governmental and nongovernmental medical research organisations.

Case reports can be useful to guide future research and can be used as an initial screening mechanism to determine whether a remedy merits further laboratory or clinical investigation. Cohort studies are the next step in clinical research, to obtain a better idea of the efficacy, safety, and tolerability of a remedy. In order to eliminate sources of bias and to convincingly demonstrate efficacy, the decisive step is the randomised controlled trial.

There have been a number of case reports, cohort studies, and randomised controlled trials on herbal antimalarials, but many are of poor quality (see Chapter 19). It is important that each of these methods be clearly defined, to help primary health care workers and researchers improve the standard of work in this field. These guidelines are therefore proposed for the evaluation of traditional herbal antimalarials, to guide this type of research.

21.2 METHODS OF GUIDELINE DEVELOPMENT

Initial guidelines were drafted on the basis of the World Health Organization (WHO) protocol "Assessment of Therapeutic Efficacy of Antimalarial Drugs for Uncomplicated Falciparum Malaria in Areas with Intense Transmission" (WHO, 1996). The clinical development group of the Research Initiative on Traditional Antimalarial Methods (RITAM) discussed and developed these guidelines. They were then presented and debated at the WHO regional workshop on the Methodology for Evaluation of Traditional Medicines, held in Antananarivo, Madagascar, November 20–24, 2000 (Figure 21.1), and at the 10th International Workshop on Natural Products, held in Ile-Ife, Nigeria, in December 2000. WHO-AFRO (Africa Regional Office) has been using these guidelines as a basis to develop its own, which have not yet been published.

Certain aspects of the guidelines were piloted during a clinical trial of *Strychnos myrtoides* extract in Madagascar during 2001 (see Chapter 9). Further modifications have been made on the basis of this experience and of evaluations of WHO guidelines by Bloland et al. (1998), Plowe et al. (2001), and White (2002).

FIGURE 21.1 The WHO Regional Workshop on the Methodology for Evaluation of Traditional Medicines, November 20–24, 2000, Antananarivo, Madagascar, where a preliminary version of these guidelines was discussed and adopted. (Copyright 2000, Merlin Willcox.)

21.3 GENERAL POINTS

Investigators should consult the Standard Operating Procedures elaborated by TDR (Karbwang and Pattou, 1999). This sets out the procedures to be followed by investigators for all clinical trials conducted in partnership with TDR. In addition, WHO guidelines for good clinical practice in trials should be followed (WHO, 1995). These set out not only the responsibilities of the investigator, but also those of the sponsor and monitor, as well as considerations on ethics, protocol development, safety, record keeping, statistics, and quality assurance.

These general points do not need to be reiterated here. The purpose of these guidelines is to standardise methodology for the clinical evaluation of traditional herbal antimalarials, in order to improve the quality and comparability of trials in this field. The emphasis is on aspects specific to traditional medicines for the treatment of malaria, which have not been addressed elsewhere.

21.3.1 GUIDELINE FOR CASE REPORTS (ETHNOMEDICAL EVIDENCE)

21.3.1.1 Study Site

Any place where traditional herbal remedies are used for the treatment of malaria should be considered as a potential study site for initial observations. However, the level of immunity of the population varies widely from area to area, according to the level of malaria transmission. Therefore, it is important to record the level of endemicity of malaria in the area as background information. Observed improvements are more likely attributable to the treatment in areas of unstable transmission where populations have lower levels of immunity. It is also important to record whether the area is rural or urban, and approximately what percentage of the population use traditional herbal remedies as a first-line treatment for malaria.

21.3.1.2 Herbal Medicine to Be Tested

This should be guided by ethnobotanical information and an overview of the literature. If the herbal remedy has been traditionally used according to a credible informant, with anecdotal evidence of safety and efficacy, there should be no need for laboratory studies before initial observational studies (see Chapter 18). The traditional health practitioner's observations on safety in pregnant women and children should be taken into consideration when deciding whether to exclude these groups. This is in line with WHO guidelines for the appropriate use of herbal medicines (WHO, 1998). The observation of patients who choose to take these remedies can serve as an initial screening to determine whether further laboratory and clinical studies would be warranted.

The following information about the medicine must be recorded:

1. Name of product
2. Evidence of traditional use
3. List of ingredients, with scientific names, part(s) of the plant(s) used, and quantity (governed by a formal agreement of confidentiality between the investigator and informant if necessary)
4. Where and in which season the plants are collected
5. How the remedy is prepared, stored, and administered

In order to preserve intellectual property rights, the above information need not always be published; however, it should be recorded and retained in at least two copies by the principal investigator and the informant in safe places for future reference, should further research be necessary. The investigators should satisfy themselves that the herbal preparation given to all subjects has been prepared according to the recorded method.

21.3.1.3 Ethical Considerations

The case reports should be conducted under the supervision of qualified staff, whose priority at all times must be the welfare of subjects enrolled in the study. In order to facilitate this sort of research, primary health care staff such as nurses, pharmacists, and medical assistants, as well as traditional health practitioners, could be trained to carry out case reports. Subjects should be asked for their informed consent. The information sheet and consent form should follow the WHO guidelines for good clinical practice (WHO, 1995). A sample information sheet is in Appendix 21.1, and a sample consent form is in Appendix 21.2. Patients will be monitored closely as described below. Patients may if necessary be reimbursed for travel to and from their clinic for follow-up. Those who do not improve or develop signs of severe malaria will be treated appropriately.

It is important at this initial stage to define who owns the intellectual property rights to the remedy — whether this belongs to an individual health practitioner, a group of health practitioners, or a whole community or ethnic group. A formal agreement should be signed by the investigator(s) and informant(s) to protect their intellectual property rights, and to agree on benefit-sharing arrangements, should the research proceed further.

21.3.1.4 The Test Procedure

1. Clinical assessment before commencing treatment
   a. History
      - Background: age, sex, contact details (physical and postal address, telephone, fax, and e-mail where applicable)
      - Previous antimalarial medication (detail: when taken, what medication and dose, for how long)
         - Enumerate symptoms of malaria from checklist: fever, rigors, headache, joint pains, convulsions, fatigue, loss of appetite, dizziness, vomiting
         - Ask questions to exclude other causes of febrile illness:
           Respiratory: cough/dyspnea
           Gastrointestinal: abdominal pain/diarrhoea
           Genito-urinary: dysuria, urinary frequency/vaginal or penile discharge
           Otolaryngological: sore throat, rhinorrhoea, earache
           Any other symptoms
   b. Full physical examination system by system: to exclude other causes of febrile illness
      - Temperature (axillary)
      - Physical examination
         - General: Jaundice? Anemia?
         - Respiratory: Elevated respiratory rate? Signs of respiratory distress? Focal added sounds?
         - Heart: Murmurs?
         - Abdomen: Tenderness?
         - Ear, nose, and throat: Inflamed or opaque tympanic membranes? Rhinitis? Red throat?
2. Laboratory investigations
   a. Thick and thin blood films to determine parasitaemia (if any), species, and density — before and after initial treatment
      - Microscopy of each blood film should be performed independently by at least two blinded, experienced microscopists, and the geometric mean taken of the two readings, unless they differ by greater than a factor of 2 between the two observers. In this case the reading should be taken by a third, more experienced, blinded microscopist.
   b. Urine dipstick or microscopy: to exclude urinary tract infection

3. Inclusion criteria
   - Parasitaemia (>1000/µl of a defined species; each species should be assessed separately)
   - Axillary temperature ≥37.5°C or history of fever in the last 48 hours
   - Absence of signs of severe malaria or of other pyrexial illnesses
   - Ability to return for follow-up
   - Informed consent of patient or parent/guardian
   - If subjects do not fulfill the inclusion criteria, they should be treated appropriately with standard locally available medication.

4. Follow-up (see Appendixes 21.3 and 21.6)
   - Administration of herb as per traditional regime for the duration of the study, ensuring that the patient takes the right dose.
   - Repeat blood smears at days 3, 7, and 14, and any other day if patient develops danger signs or fever.*
   - Repeat axillary temperature at days 1, 2, 3, 7, and 14* and questioning about side effects. Questioning about other concomitant treatment and exclusion if taking drugs that might alter the response.
   - Screening for danger signs at any visit, and appropriate alternative treatment if necessary.
   - Patients are allowed to withdraw voluntarily from the study without compromising their continued care.

5. Outcome measures
   a. Primary outcome measure: **clinical response**, as defined below
      - **Early treatment failure** = Development of danger signs on day 1, 2, or 3 in the presence of parasitaemia; axillary temperature ≥37.5°C on day 2 with parasitaemia > day 0 count; axillary temperature ≥37.5°C on day 3 with parasitaemia. (Afebrile patients with parasitaemia on day 3 ≥25% of count on day 0 will *not* be counted as early treatment failures, but will be observed closely.)
      - **Late treatment failure** = Development of any danger signs or signs of severe malaria in the presence of parasitaemia on any day from day 4 to day 14, without previously meeting any of the criteria of early treatment failure; axillary temperature ≥37.5°C with parasitaemia on any day from day 4 to day 14, without previously meeting any of the criteria of early treatment failure.
      - **Adequate clinical response** = Absence of parasitaemia on day 14 irrespective of axillary temperature, without previously meeting any of the criteria of early or late treatment failure; axillary temperature <37.5°C irrespective of the presence of parasitaemia, without previously meeting any of the criteria of early (modified) or late treatment failure.
   b. Secondary outcome measures
      - **Parasitological response** = The WHO classification is useful (Bruce-Chwatt, 1986; Rieckmann, 1990):
         - **S** = Clearance of asexual parasites within 7 days of initiation of treatment, without subsequent recrudescence
         - **RI** = Clearance of asexual parasites within 7 days of initiation of treatment, followed by recrudescence
         - **RII** = Marked reduction of asexual parasitaemia, but no clearance
         - **RIII** = No marked reduction of asexual parasitaemia
      - Side effects
      - Withdrawal from study

---

* Follow-up at least to day 28 is required for infections with *Plasmodium vivax* and *Plasmodium ovale* (White, 2002).

6. Treatment of failures
   a. Patients with signs of uncomplicated malaria should be treated with the locally available first-line medicine.
   b. Patients developing signs of severe malaria should be treated with parenteral or intrarectal antimalarials according to local guidelines, and referred for admission to hospital.

21.3.2 Guideline for Cohort (Phase 1) Studies

21.3.2.1 Herbal Preparation to Be Tested

These studies should be carried out on the basis of good results from case studies or other convincing ethnomedical evidence. The herbal preparation will need to undergo quality control assessments, and the dose will need to be standardised according to previous observations. There is no need for formal laboratory evidence of safety or efficacy at this stage, providing there is sufficient ethnomedical evidence of its safety and efficacy.

21.3.2.2 Study Site

Any place where traditional herbal remedies are used for the treatment of malaria should be considered as a potential study site for initial observations.

The following characteristics are necessary:

1. An established population who live close enough to return for follow-up, with enough cases of malaria in the time allotted for the study
2. Basic clinical and diagnostic facilities, to enable diagnosis of malaria and treatment of patients

21.3.2.3 The Test System

This should be done as for case reports, with the following extra considerations:

1. Sample size: Assuming that a useful medicine would have a failure rate less than 25%, to detect this with 95% confidence and 80% power would require the following (WHO, 1996):
   a. Initial sample of 16:
      i. If 0 failures, treatment is promising and requires further investigation.
      ii. If >5 failures, there must be at least 25% failures.
   b. Full sample of 42 is necessary if there were 1 to 5 failures in the original group of 16:
      i. <6 failures = low failure rate → move on to randomised controlled trial
      ii. >6 failures = high failure rate → investigate alternative preparations or doses
2. Parasite species: Each patient in a cohort should have a pure infection with the same predefined species. In practice this will usually be *Plasmodium falciparum* or *Plasmodium vivax*, as the other species are rare. The natural history and prognosis of these infections are different. Follow-up beyond day 28 may be necessary for vivax infections, according to the relapse interval of the local strain (White, 2002).
3. Follow-up to 28 days. If possible, polymerase chain reaction (PCR) on blood samples to distinguish between reinfections and recrudescences.
4. Investigations to include, if possible (see Figure 21.2 and Appendixes 21.6 and 21.7):
   a. Full blood count (or haematocrit) on day 0, last day of treatment, and days 14 and 28
   b. Urea and electrolytes, and liver function tests on day 0, last day of treatment, and day 28, if possible
   c. ECG on day 0 and last day of treatment, to look for signs of cardiotoxicity

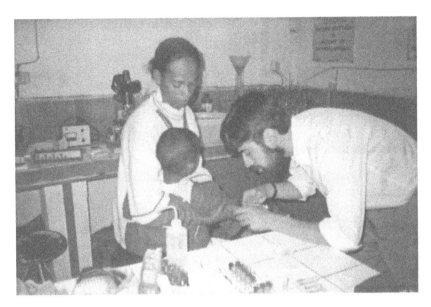

FIGURE 21.2 A venous blood sample needs to be taken from subjects in order to measure haematological and biochemical parameters. With appropriate equipment, these tests can be performed even in remote locations, as illustrated here in the course of a trial of a *Strychnos myrtoides* extract in Ankazobe, Madagascar. (Copyright © Philippe Rasoanaivo; see Chapter 9.)

5. Questioning about side effects, in some detail. The checklist in Appendix 21.4 may be useful (based on George, 1974, quoted in Fernex, 1984), although it may need to be abbreviated in order to save time. If using this, or an abbreviated form, patients should be asked about each of these symptoms on day 0 before starting the medication and on each subsequent follow-up visit. New symptoms that develop after day 0 must be assumed to be side effects of the treatment administered (although in some cases they may be due to the disease itself). Adverse events should be recorded on an adverse event record (Appendix 21.5). Symptoms that were present on day 0 but which subsequently disappear must be assumed to be symptoms of the illness that have resolved.

6. The proportion of patients in each category is the key outcome measure for cohort studies. The same outcome categories may be used as for the case reports (see 21.3.1.4.5), or the modified WHO outcome measures may be used (White, 2002). In areas of low to moderate transmission, follow-up should be to 28 days. Two extra categories may be included:

   a. **Adequate clinical and parasitological response** (ACPR): Absence of parasitaemia and fever at days 14 to 28

   b. **Late parasitological failure** (LPF): Presence of parasitaemia at days 14 to 28 and measured axillary temperature <37.5°C, without meeting the criteria of early or late treatment failure

7. The percentage of patients experiencing side effects must also be recorded, categorised by system. The following definitions of side effects can be used (Appendixes 21.4 and 21.5):

   a. Mild: tolerable, patient up and about

   b. Moderate: causes discomfort, but up and about

   c. Severe: interferes with the patient's activities

   d. Subjects will be withdrawn from the study if they have a moderate or severe adverse reaction. The reaction should be recorded, and the treatment should be stopped or changed.

8. Comparison with historical controls. If comparable local data are available about levels of efficacy of conventional antimalarials, it has been proposed that this may be used to compare with the results from the cohort study. However, these comparisons should be interpreted with extreme caution, as levels of immunity and drug resistance may vary widely at different times and in different patient populations.

21.3.3 Guideline for Randomised Controlled Trials (Phase III)

21.3.3.1 Herbal Preparation to Be Tested

Good evidence of safety and efficacy from ethnomedical information and a cohort study (see above) would be needed to justify a randomised controlled clinical trial. The hypotheses to be tested are that the herbal preparation will be safe and its efficacy will not be inferior to the local first-line treatment for uncomplicated malaria.

The following are necessary preconditions for a randomised clinical trial:

- A standard preparation administered orally according to a regimen defined by ethnomedical observations.
- Sustainable supply of the medicinal plant raw material, and conservation measures to protect this.
- Laboratory data to demonstrate lack of toxicity (see Chapter 18): cytotoxicity; teratogenicity; acute, subacute, and subchronic toxicity (2 to 4 weeks); LD_{50} — mouse and guinea pig; or maximum tolerated dose. This list is not compulsory. As many as possible of these tests should be done, according to resources available and country drug registration legislation. Good data from a cohort/phase I study may serve as an alternative to animal data.
- Clinical data to demonstrate safety and tolerability in humans.

The following characteristics are desirable but not essential:

- Laboratory evidence of efficacy as an antimalarial (according to guidelines in Chapter 16)
- No local use as a treatment (because if people are taking the plant anyway, this will interfere with the trial)

21.3.3.2 Study Site

The study should be conducted at a health facility. Two or more sites should be used if possible. The following characteristics are necessary:

- An established population who live close enough to return for follow-up, with enough cases of malaria in the time allotted for the study.
- Basic clinical and diagnostic facilities, to enable diagnosis of malaria and treatment of patients.
- A multidisciplinary team, including traditional health practitioners, a medical doctor, and laboratory staff (and nurses and pharmacists if necessary). The health facility should have experience of this type of study.

21.3.3.3 The Test System

This should be done as for cohort studies, with the following extra considerations.

21.3.3.3.1   Exclusion Criteria

- Signs of severe malaria (Note: Patients with hyperpyrexia alone do not need to be excluded)
- Pregnant women
- Children below the age of 5*
- Elderly patients (age >65)
- Haemoglobin less than 5 g/dl

Some investigators may also wish to exclude patients with medical conditions necessitating the use of other medications and patients who have ingested antimalarial drugs in the past seven days. In practice, this may be difficult to achieve, as many patients self-medicate with chloroquine before presenting to a health facility. Nsimba et al. (2002) found that over 97% of children without a prior history of chloroquine treatment had the drug in their blood. Excluding all of these would mean that study subjects would be difficult to recruit and unrepresentative of the majority. WHO (1996) recommends that previous ingestion of drugs is recorded, but should not be an exclusion criterion. Bloland et al. (1998) found that prior use of chloroquine did not affect treatment outcome. Thus, patients having taken chloroquine can safely be included, but there is less data about other drugs. Investigators can decide what is best in a particular site, depending on local levels of drug use and resistance to antimalarials.

21.3.3.3.2   Sample Size

This should be calculated to be able to detect a difference of 20% or more between the herbal preparation and the standard local first-line antimalarial preparation, with confidence of 95% and power of 90%. A statistician should be involved at an early stage.

21.3.3.3.3   Randomisation

Patients should be randomised to receive either the herbal preparation or the control drug. A simple randomisation procedure may be used. Alternatively, a pseudo-randomisation procedure may be adopted by using balanced block randomisation where a balance will be achieved for every sixth patient. This method may be adopted due to time trends that may develop over the 28 days of follow-up. The patients would be randomised to ten blocks of six patients each on arrival to the hospital or the recruiting center. A block will be selected by use of a table of random numbers (Hulley and Cummings, 1988; ICH, 1998). It will be a single-blind trial where the clinician and laboratory technicians will not know the drug the patient is taking.

21.3.3.3.4   Control Treatment

The control group should receive the standard locally used first-line treatment for uncomplicated malaria.

21.3.3.3.5   Data Management and Analysis

A database can be created in a suitable program with screen formats matching the data collection forms. Control checks that include referential integrity, minimum and maximum values, and jumps can be included in order to reduce data entry errors.

The database to be created will include the following:

Demography and socioeconomic characteristics
Treatment evaluation forms — initial period
Treatment evaluation forms — follow-up period

---

* ICH (2000) states that "The presence of a serious or life-threatening disease for which the product represents a potentially important advance in therapy suggests the need for relatively urgent and early initiation of pediatric studies." Thus, if results are good in the first randomised controlled trial, a subsequent trial should be done including younger children.

Validation of the database should be conducted periodically by checking the entries against the raw data. Errors detected should be corrected immediately. An overall validation exercise can also be carried out at the end of the data collection exercise, after which frequencies of all variables can be produced and an electronic backup made.

The z-test can be used to compare the proportion of patients in each group with different outcomes.

22.3.3.6   Equipment and Funding

Appendix 21.8 contains a list of equipment needed for a randomised controlled trial for one treatment centre. This may serve as a checklist when preparing a budget and applying for funding. It can be adapted according to the number of patients to be included in the trial and the particular characteristics of the study site.

21.4   SUMMARY AND DISCUSSION

The above standardised scheme is suggested for the clinical investigation of traditional herbal antimalarials. It is proposed that a case report, if containing adequate information, can provide the initial justification for further research. The next suggested step is a cohort study of 16 to 42 people recording more detailed information about clinical response and adverse effects. If results are promising, this could lead to a randomised controlled trial comparing the traditional herbal preparation to the locally used first-line medication for uncomplicated malaria.

It is proposed that good results from a cohort study (demonstrating a treatment failure rate of less than 25%) are sufficient to endorse local use, in an area where the remedy has traditionally been used. Conversely, results showing a treatment failure rate greater than 25%, or significant adverse effects, would lead to a recommendation against use of the remedy at the specified dose. If there is sufficient ethnomedical evidence, the study could be repeated using a different dosage or preparation of the plant(s). This scheme would facilitate the rapid evaluation of herbal preparations already in use. However, it is proposed that good results from a randomised controlled trial, showing noninferiority or superiority to standard first-line treatments (and a failure rate of less than 25%), together with a good safety profile, are necessary before wider use of the herbal preparation can be recommended.

REFERENCES

Bloland, P.B., Kazembe, P.N., Oloo, A.J., Himonga, B., Barat, L.M., and Ruebush, T.K. (1998). Chloroquine in Africa: critical assessment and recommendations for monitoring and evaluating chloroquine therapy efficacy in sub-Saharan Africa. *Trop. Med. Int. Health*, 3, 543–552.

Bruce-Chwatt, L.J. (1986). *Chemotherapy of Malaria*. WHO, Geneva.

Fernex, M. (1984). Clinical trials: phases I and II. In *Antimalarial Drugs I*, Peters, W. and Richards, W.H.G., Eds. Springer-Verlag, Berlin.

Hulley, S.B. and Cummings, S.R. (1988). *Designing Clinical Research*. Williams & Wilkins, Baltimore.

ICH. (1998). *Statistical Principles for Clinical Trials*, ICH, Geneva. www.ich.org.

ICH. (2000). *Clinical Investigation of Medicinal Products in the Pediatric Population*, ICH, Geneva. www.ich.org.

Karbwang, J. and Pattou, C. (1999). Standard Operating Procedures for Clinical Investigators, TDR/TDR/SOP/99.1.

Nsimba, S.E.D., Massele, Y., Eriksen, J., Gustafsson, L.L., Tomson, G., and Warsame, M. (2002). Case management of malaria in under-fives at primary health care facilities in a Tanzanian district. *Trop. Med. Int. Health*, 7, 201–209.

Plowe, C.V., Doumbo, O.K., Djimde, A., etal. (2001). Chloroquine treatment of uncomplicated *Plasmodium falciparum* malaria in Mali: parasitologic resistance versus therapeutic efficacy. *Am. J. Trop. Med. Hyg.*, 64, 242–246.

Rieckmann, K.H. (1990). Monitoring the response of malaria infections to treatment. *Bull. WHO*, 68, 759–760.

White, N.J. (2002). The assessment of antimalarial drug efficacy. *Trends Parasitol.*, 18, 458–464.

WHO. (1995). Guidelines for Good Clinical Practice (GCP) for Trials on Pharmaceutical Products, WHO Technical Report Series 850, Annex 3.

WHO. (1996). Assessment of Therapeutic Efficacy of Antimalarial Drugs for Uncomplicated Falciparum Malaria in Areas with Intense Transmission, WHO/MAL/96.1077.

WHO. (1998). Guidelines for the Appropriate Use of Herbal Medicines, Western Pacific Series 23. WHO Regional Publications, WHO Regional Office for the Western Pacific, Manila.

APPENDIX 21.1: PATIENT INFORMATION SHEET

(To be translated into appropriate local language)

A CLINICAL TRIAL OF A HERBAL PREPARATION FOR MALARIA

Background

Malaria is an important disease in this country, affecting many people and causing some deaths. The standard medicines are not always effective, so we need to look for other medicines that are equally effective and affordable. Many traditional medicines are used to treat malaria, but most have not been tested scientifically to see whether they work well.

Aims

We are performing a study to see if a certain traditional plant preparation can treat and cure malaria that is not severe in nature. The use of this medicine to treat a variety of ailments is not new, and all we are doing now is assessing it more carefully using scientific principles. Studies performed in the laboratories show that the preparation is effective against malaria and that it is safe and nontoxic. So we would like now to do a study in people infected with a mild form of malaria to see if the herbal preparation is able to cure them. Your participation in this study is voluntary, and you can withdraw at any time without affecting the care you receive.

Procedures to Be Followed

1. If you agree voluntarily to participate in this study, we shall perform a physical examination and laboratory investigations to determine whether you qualify to be in the study.
2. If you qualify to be in this study you may be given the herbal preparation or the standard treatment, tablets. Neither you nor the staff treating you will decide which treatment you will receive — this will be chosen beforehand at random. The dosage of the treatments is as follows: _ or _.
3. You will come to the clinic every day to receive your treatment (or if applicable, stay at our research ward for at least 7 days). After this we shall follow you up for another 3 weeks. During the initial treatment period, we shall take blood samples on the first day, the third day, and the seventh day. This should cause only small discomfort. This will help us to know whether the herbal preparation is effective against malaria and whether it is affecting the function of your kidneys, bone marrow, or liver. We will also do electrical tracings of your heart (a painless procedure) to see whether its function is affected by the herbal preparation. Similar tests will then be done on days 14 and 28. *It is very important that you return to the clinic for these examinations, because it is the only way that we can find out if you still have malaria and whether further treatment is needed.*

Benefits of Participating in the Study

The results of this study will help us to find better malaria treatments for yourself and others in the future. The treatment you receive during the study will be free of charge, and you will receive a small amount of money to compensate for your travel to the clinic for the follow-up visits. You will be followed more closely than usual, with the regular checks and blood tests as described above. If the treatment fails, or your condition worsens, you will be given an alternative treatment for free at the clinic.

Risks and Inconveniences of Participating in the Study

As explained above, you will have to return several times to be checked by the investigators and for repeat blood tests. If you are unable to return for these visits, we regret that we cannot include you in the study.

Although we have not detected any toxic effects of the herbal preparation in the laboratory studies and the preliminary human studies that we have done, it may still cause adverse effects. We will follow you closely, and if we detect any adverse effects through your reports, the blood tests, or heart tracing, we will stop the treatment immediately, and then continue to observe the functioning of your organs. We shall make sure that no harm is done to you purposely.

Your records will remain confidential. Only authorised people will be able to view them for the analysis of results and monitoring of the trial.

Alternatives

This is a research study and you are under no obligation to participate. If you choose not to participate, you will be treated as usual by the staff. If you have chosen to participate, you are still free to stop participating at any time and for any reason. You do not have to give a reason, but if you are stopping because of an adverse effect, we ask that you inform the research staff, because we need to find out if the medicine has any adverse effects. Your decision will not affect your treatment, your rights, or the care you receive.

Further Information

We hope that the above information has cleared any fears or doubts you might have. However, during the entire period of the study, do not hesitate to ask us any questions that might come up. For further information, advice, or alternative treatment, please contact the doctor treating you or the principal investigators (names).

APPENDIX 21.2: CONSENT FORM

(To be translated into appropriate local language)

I, Mr./Mrs./Miss _____, do hereby give consent/permission to Prof./Dr./Mr./Mrs./Miss _____
to include and carry out research on me/my child* in the intended research protocol as explained to me and understood by me. I have understood the patient information sheet after reading it/hearing it. I understand that the research involves me taking either a tablet or an herbal preparation for treating malaria from which I am suffering.

I have also been made to understand the implications, benefits, and risks of participating in the trial. I accept the tests and treatment to be carried out and the risks thereof. I have made this decision freely and understand that I am under no obligation to participate in the study.

I understand that I have the right to withdraw from the research at any time, for any reason, without penalty or harm. Although I do not have to give a reason for withdrawing, I agree to inform the investigators if I have experienced an adverse effect from the treatment. If I withdraw, I understand that I will be cared for by the doctors like any other patient.

All the above conditions have been explained to me in the _____ language, in which I am fluent. Data/biological samples will be coded and remain confidential (name not disclosed).

Signed: _____
Signature of Patient (or parent/guardian if patient is under 16 years of age)

Date _____

Signature of Witness _____
Name and address of Witness: _____
_____

    I certify that I have fully explained the trial to the above patient and that I have not put them under any pressure to participate.

Signature of Investigator _____

APPENDIX 21.3: CLINICAL FOLLOW-UP FORM

Day	0	1	2	3		7		14		28
Date										
Time										
Danger Signs										
Unable to drink										
Vomiting everything										
Convulsions										
Lethargic/unconscious										
Unable to sit/stand										
Oliguria/anuria										
Spontaneous bleeding										
Symptoms in Preceding 24 Hours										
History of fever										
Sweating										
Rigours										
Headache										
Joint pains										
Weakness										
Tiredness										
Dizziness										
Appetite ($\downarrow$, N, $\uparrow$)										
Nausea										
Vomiting										
Other (specify)										
Side Effects (specify)										
Examination										
Axillary temperature (°C)										
Respiratory rate										
Treatment Given										
Time										
Vomiting within 30 minutes										
Other treatment taken										
Reasons for Exclusion or Loss to Follow-Up										

Overall Assessment:
ETF/LTF/ACR/exclude/loss to follow-up

APPENDIX 21.4: CHECKLIST OF POSSIBLE SIDE EFFECTS

Day	0	1	2	3	4	5	6	7	14	28
Nervous System										
Drowsiness										
Anxiety										
Insomnia										
Nightmares										
Shakiness										
Numbness										
Tinnitus										
Blurred vision										
Unpleasant taste										
Thirst										
Cardiovascular										
Fast heartbeat										
Irregular heartbeat										
Respiratory										
Cough										
Chest pain										
Stuffy nose										
Gastrointestinal										
Heartburn										
Abdominal pain										
Diarrhoea										
Constipation										
Intestinal wind										
Black stools										
Genito-Urinary										
Dysuria										
Nocturia										
Dark urine										
Change in sexual ability/desire										
Muco-Cutaneous										
Skin rash										
Pruritus										
Easy bruising										
Dry mouth										
Others (Specify)										

APPENDIX 21.5: ADVERSE EVENT RECORD

Patient record number: _____

Date event started: _____ **Date event stopped:** _____

Severity of adverse event:

☐ Mild, tolerable, patient up and about
☐ Moderate, causes discomfort, but patient is up and about
☐ Severe, interferes with the patient's activities
☐ Fatal

Nature of adverse event: Describe symptoms, system by system, using previous checklist as a guide.

_____

_____

_____

_____

_____

_____

_____

_____

Causality:

☐ Definitely *not* caused by the herbal preparation
☐ Possibly caused by the herbal preparation
☐ Probably caused by the herbal preparation
☐ Definitely caused by the herbal preparation

Did this adverse event cause withdrawal from the study? Y/N

APPENDIX 21.6: LABORATORY INVESTIGATIONS RESULTS SHEET

Patient record number: _____

Case reports: Tests 1 to 3 only need be done.

X = Not to be done on this day

Day	0	1	2	3		7	14	28
Date								
Slide Code								
Species (A)		X	X					
Parasitaemia (A)		X	X					
Initials of A								
Species (B)		X	X					
Parasitaemia (B)		X	X					
Initials of B								
Parasitaemia (C)								
Parasitaemia — geometric mean		X	X					
Biochemistry								
Glucose		X	X	X	X	X	X	X
Bilirubin		X	X					
AST		X	X					
ALT		X	X					
Albumin		X	X					
Total protein		X	X					
Sodium		X	X					
Potassium		X	X					
Creatinine		X	X					
Haematology								
Hb (g/dl)		X	X					
Platelets $\times 10^9/l$		X	X					
White cell count $\times 10^9/l$		X	X					

APPENDIX 21.7: ECG RESULTS SHEET

Each ECG should be given a code number and read independently and blindly by two researchers. In case of disagreement, the trace should be read by a third, more experienced, blinded researcher.

If the ECG is normal on days 0, 3, and 7, there is no need to repeat it again on days 14 and 28.

Day	0	3	7	14	28
Date					
ECG code number					
Rate (per minute)					
Rhythm					
Axis					
PR interval (msec)					
QRS interval (msec)					
QT interval (msec)					
QTc (seconds) = QT/$\sqrt[3]{RR}$					
P waves					
QRS complexes					
ST segments					
T waves					
U waves					
Interpretation					

APPENDIX 21.8: PLANNING A BUDGET FOR A RANDOMISED CONTROLLED TRIAL

1. Calculate the number of patients needed (see sample size calculations, above).
2. Decide how many sites are needed (according to number of patients presenting with malaria per month at a site, length of time allotted for study, and number of patients required).

EQUIPMENT NEEDED PER STUDY SITE

Clinical:
  1 stethoscope
  2 thermometers (electronic if possible, with spare batteries)
  1 balance
  1 sphygmomanometer
  1 otoscope (+ spare batteries)
  1 ECG machine
  1 examination room (3 chairs, 2 tables, examination couch, redecoration if necessary)
Laboratory:
  1 suitable room (4 tables or working surfaces, 4 chairs)
  1 microscope
  1 centrifuge
  1 portable glucometer
  1 spectrophotometer for biochemical analyses
  2 tally counters
  2 laboratory timers
  2 slide trays
  3 staining jars
  2 measuring cylinders, 10 and 500 ml
  4 nets (to protect slides from flies while drying)

Record Keeping and Analysis

1 computer and printer
Data management program: for example, Excel
Statistics package: for example, Epi-info or SPSS

SUPPLIES NEEDED PER 100 PATIENTS

Clinical:
  75 courses of herbal preparation (include costs of collecting, processing, extraction, and
    quality control)
  75 courses of control drug from a certified source
  100 courses of next line alternative drug
  20 ampoules quinine for i.m. injection
  200 tablets of cotrimoxazole
  200 doses of amoxycillin
  200 tablets of paracetamol
  20×2 ml disposable syringes, 20×5 ml syringes (for injections)
  600×10 ml syringes or vacutainers, for blood samples (or 20-ml syringes if doing *in vitro*
    chloroquine sensitivity testing)
  500 sterile lancets
  800 alcohol swabs
  200 urine dipsticks (including strips for blood, protein, leukocytes, nitrites)
  75 urine pregnancy tests
  ECG paper sufficient for 500 readings
  500 glucose test strips for glucometer
  600 blood bottles for haematology assays
  600 blood bottles for biochemical assays
  600 blood bottles for *in vitro* chloroquine sensitivity tests, if applicable
  Cotton wool — 2 packets of 500 g
Laboratory:
  3000 microscope slides, frosted edge
  Giemsa stain stock solution, 500-ml bottle
  20 buffer tablets, pH 7.2
  20 l of distilled water
  Methanol, 500-ml bottle
  Immersion oil, 50-ml bottle
  Reagents for 600 tests for each of the following: bilirubin, AST, ALT, γ-GT, alkaline phos-
    phatase, albumin, total protein, Na, K, creatinine, haemoglobin
  Reagents for *in vitro* sensitivity testing, if applicable
  Reagents for 200 urine tests for chloroquine, if applicable
  300 pieces of filter paper for PCR, if applicable

Record Keeping

150 patient record booklets
4 notebooks
20 pens

PATIENT EXPENSES PER 100 PATIENTS

Compensation and transport costs for seven visits per patient
Cost for hospital admission for seven days, if applicable
Cost provision for inpatient management of cases developing severe malaria

STAFF COSTS PER STUDY SITE

Salary for one doctor (principal investigator)
Compensation for resident doctor (for collaboration)
Salary for one nurse, medical assistant, or medical student
Salaries for two laboratory technicians
Transport, accommodation, and food costs for staff

STAFF COSTS OF REFERENCE CENTER

Administration expenses (salary of administrator, stationery, telephone, fax and post costs, etc.)
Payment of traditional health practitioner(s) (as per benefit-sharing agreement)
Payment per hour of a statistician (to provide advice on study design and data analysis)
Payment per hour of a third, most experienced laboratory technician to check slides on which the first two readers disagree
Payment per hour of a cardiologist to read ECGs on which the first two readers disagree
Payment of trial monitor
Payment of pharmacologist (for preparation, standardisation, and quality control of herbal preparation)

Part 6

Repellence and Vector Control

22 An Overview of Plants Used as Insect Repellents

Sarah J. Moore and Annick D. Lenglet

CONTENTS

Mankind and mosquitoes must have lived in close association since our ancestors first evolved. The deadly diseases carried by these insects and the annoyance they cause is likely to have encouraged the discovery of methods of personal protection. The use of plant substances to protect us from insects probably developed early, but there is no archaeological evidence of this. However, Capuchin monkeys have been observed rubbing the millipede *Orthoporus dorsovittatus* onto their coat during the period of maximum mosquito activity (Valderrama et al., 2000). This species contains chemicals called benzoquinones that have known repellent activity.

The ancient Chinese were accomplished entomologists and their literature contains many references to repellent and insecticidal plants. Medical folklore of ancient Egypt mentions hanging plant stalks in doors and windows to deter insects from entering houses. Many other plants with reputed insecticidal and repellent properties were noted by Greek and Roman classical writers. They included absinthe (*Artemisia absinthium*), asafoetida (*Ferula asafoetida*), bay (*Laurus nobilis*), cassia (*Cassia* spp.), cedar (*Thuja occidentalis*), lemon (*Citrus medica*), cumin (*Cuminum cyminum*), elder (*Sambucus nigra*), fig (*Ficus carica*), garlic (*Allium sativum*), heliotrope (*Heliotropum arborescens*), hellebore (*Veratrum album*), ivy (*Hedera helix*), oak (*Quercus robur*), oregano (*Oreganum vulgare*), pomegranate (*Punica* spp.), and squill (*Urgenia maritima*). Many concoctions made from these plants are found in the *Geoponika* (1806), an anthology published in the sixth or seventh century including work by authors who lived between 200 B.C. and 200 A.D. The *Geoponika* also recommends burning animal feces, bones, horns, ivory, and various plants against a variety of insect pests. Many of the fumigants used produce smoke that is obnoxious to man, and thus was assumed to be repellent to insects. However, several of the plants used have insecticidal properties: hellebore contains several insecticidal alkaloids in its rhizome; hemlock (*Conium maculatum*) and squill contain compounds poisonous to invertebrates and vertebrates; and some species of lupin (*Lupinus* spp.) were used, which also contain insecticidal ingredients (Wyrostkiewicz et al., 1996).

Smoke is a common method of repelling biting insects that is used throughout the world. Fresh or dried plants are frequently added to fires to enhance the repellent properties of the smoke. Other methods are hanging the plants around the house or sprinkling leaves on the floor. Mosquito coils made from dried plants and combustible material such as sawdust are also a cheap and often an effective method of repelling mosquitoes. They are probably derived from the incense used in religious ceremonies by Hindus, Buddhists, and the followers of Confucius. In Java today, the same incense used in ceremonies to honor ancestors is also used on a daily basis to repel mosquitoes (Sangat-Roemantyo, 1990).

Our ancestors depended entirely on the use of plants to kill and repel insects and agricultural and medical arthropods. In the modern world, many societies still use plants for a wide range of medicinal and entomological uses, yet very little research has been put into finding out which of these plants are effective, and control now relies mainly on synthetic chemicals. Pyrethroids are synthetic analogues of pyrethrins contained in the flower heads of members of the chrysanthemum family. They have now replaced organochlorines and organophosphorus compounds as the most widely used group of insecticides, yet their cost is frequently prohibitive. Utilising homegrown repellents may reduce the need for foreign imports where exchange rate inequalities and transport costs inflate expenditure. As well as being cheap and locally available, natural repellents are normally culturally acceptable and locally known. Increasing hopes are being placed on repellents as a method of preventing malaria. They are particularly promising in regions where vector mosquitoes bite early in the evening and in regions where vector mosquitoes have exophagic (outdoor) feeding habits, so that bed net use is unlikely to be very effective. They are also important where drug-resistant parasites and insecticide-resistant vectors prevail. However, traditional repellents may have disadvantages, as they tend to last for a shorter period than synthetic preparations such as deet (N,N-diethyl-m-toluamide), therefore necessitating frequent reapplication. Other disadvantages include strong odors, skin irritation, and possible health effects since they have generally not been evaluated for toxicity.

22.1 NATURALLY OCCURRING REPELLENT AND INSECTICIDAL CHEMICALS

Plants contain many chemicals, which are important in their defense against insects. These fall into several categories, including repellents, feeding deterrents, toxins, and growth regulators. There are thousands of chemical compounds that act in one or more of these ways. Most can be grouped into five major chemical categories: nitrogen compounds (primarily alkaloids), terpenoids, phenolics, proteinase inhibitors, and growth regulators. Although these compounds arose as defenses against phytophagous insects, many are also effective against mosquitoes and other biting diptera.

22.1.1 ALKALOIDS

These compounds are insecticidal at low concentrations and frequently toxic to vertebrates. Their mode of action varies, but many affect acetylcholine receptors in the nervous system (e.g., nicotine) or membrane sodium channels of nerves (e.g., veratrin). Insecticidal examples include nicotine (*Nicotinia* spp.), anabasine (*Anabasis aphylla*), veratrin (*Schoenocaulon officinale*), and ryanodine (*Ryania speciosa*). Physostigmine, which served as the model compound for the development of the carbamate insecticides, is an alkaloid isolated from the calabar bean (*Physostigma venenosum*) (Stedman and Barger, 1925). Although these chemicals are not volatile, they may be used as repellents by burning plant material, either on a fire or in a mosquito coil to create an insecticidal smoke that repels the insects through direct toxicity. Alkaloids are found in large quantities in many members of the Berberidaceae, Fabaceae, Solanaceae, and Ranunculaceae families, all of which are used extensively as traditional insect repellents (Johnson, 1998; Secoy and Smith, 1983).

22.1.2 TERPENOIDS

These are among the most widespread and structurally diverse of the plant products. Monoterpenes and sesquiterpenes are major components of many essential oils. Common examples include myrcene (bay leaves), geraniol (citronella), eucalyptol, also known as cineole (*Eucalyptus* spp.), and linalol (found in many plants). Open-chain structures include menthol (mint family), camphor (sagebrush), α and β pinene, and limonene (common in many plants). One important group of monoterpenes is the insecticidal pyrethrins, which are harvested from the dried heads of flowers in the *Chrysanthemum* genus. These plants are still widely cultivated in Kenya, Tanzania, Ecuador, Brazil, the former USSR, Japan, and India for use in mosquito coils and sprays. Monoterpenes are present in many members of the Lamiaceae, Myrtaceae, and Poaceae. They are present in plants to deter herbivores, and some exhibit considerable toxicity to insects while having low mammalian toxicity (Golob et al., 1999). Many of these molecules are volatile oils and deter phytophagous insects by acting in the vapour form on olfactory receptors. The effect of these compounds on haematophagous insects is possibly an evolutionary hangover. Plants containing terpenes may be used as repellents without modification by rubbing fresh leaves onto the skin to release the oils, or they may be bruised to release the oils, then hung around the home. Other uses may be as fumigants when the fresh leaves are burned or the oils evaporated.

There are several important groups in the triterpene category: triterpenes, steroids, saponins, sterolins, and cardiac glycosides. The widely publicised compound Azadirachtin, derived from the neem tree (*Azadirachta indica*), is a triterpenoid (see Chapter 6). Azadirachtin and saponins (also found in the neem tree) are insect growth regulators (phytoecdysones).

Common triterpenes include ursolic and oleanic acid, limonins, and cucurbitacins. Many are bitter and toxic to fish. Triterpenes are the constituents of many folk remedies, particularly in Asia. Plants of the family Asteraceae have many members that contain these compounds and are widely used in mosquito control.

22.1.3 PHENOLICS

Phenolics are nonnitrogenous compounds attached to benzene rings. The functions of phenolics are diverse, contributing to cell wall structure, flower colour, and defense against both vertebrate and invertebrate herbivores. Important phenolics in terms of insecticidal and repellent function are the flavonoids, which are characteristic compounds of higher plants. There are three important insect repellent flavonoid groups. First, the flavones that are found in the Labatiae, Umbelliferae, and Compositae and are quite new in evolutionary terms. The second important group is the isoflavonoids, found mainly in the Leguminosae, an example of which is the insecticidal compound rotenone present in *Derris eliptica*. Rotenone is a potent mitochondrial poison (Haley, 1978). The other main group of phenolics important in repelling insects is the tannins. They are found throughout the plant kingdom and exhibit toxicity by binding to proteins (Schultz, 1989).

22.2 ESSENTIAL OILS

Essential oils are derived by steam distillation from plants in several families. The Lamiaceae family includes basil (*Ocimum basilicum*), mint (*Mentha* spp.), hyptis (*Hyptis suaveolens*), lavender (*Lavandula* spp.), sage (*Salvia* spp.), and thyme (*Thymus* spp.). The Myrtaceae family includes eucalyptus (*Eucalyptus* spp.) and tea tree (*Melaleuca* spp.), and the Poaceae includes citronella, lemongrass, and palmarosa (*Cymbopogon* spp.). Table 22.1 shows the average protection times against *Aedes* mosquitoes of the most common essential oils.

Prior to the discovery of effective synthetic repellents, aromatic oils were used as repellents by the military. The British Indian army was issued with a cream composed of citronella, camphor, and paraffin, but it was only effective for 2 hours (Covell, 1943).

Indications that essential oils prevent malaria are available, although few in number. Philip et al. (1945) reported lower spleen indices in women than men in Southern Madras. An *in vivo* study of the local malaria vector *Anopheles fluviatilis* showed that the mosquitoes were biting men preferentially. The women of the region smeared themselves with turmeric (*Curcuma longa*), galangal (*Kaempferia galanga*), and mustard oil (*Brassica juncea*) before bathing. Tawatsin et al. (2001) found that the steam distillate of turmeric plants provided 8 hours protection against *Anopheles dirus*, and a hexane fraction of galangal provided 3 hours protection from *Aedes aegypti* in cage experiments (Choochote et al., 1999). Mustard oil provided 2.1 hours protection in field tests against *Anopheles culicifaces* (Ansari and Razdan, 1985). It is possible that the lower spleen indices in these women was due to their use of plant oils, particularly as *An. fluviatilis* bites for only a few hours early in the evening (Nagpal and Sharma, 1995). However, it is unlikely that the burden of malaria in the region is reduced by the use of these oils since the mosquitoes were presumably diverted to biting the women's unfortunate husbands.

Karen women on the Thai–Myanmar border use thanaka, a cosmetic preparation made from the pulp of the wood apple tree, *Limonia acidissima* (see Figure 22.1). This preparation is slightly repellent at high concentrations and enhances the repellency of deet when the two are mixed together

TABLE 22.1
The Repellency of Essential Oils (100% Concentration) to *Aedes* Mosquitoes from USDA (1943–1967)

Compound	Duration of Protection (hours)
Terpenene	0
Citronellal	<1
Limonene	≤1
Myrcene	≤1
α Pinene	≤1
Citronellol	1–2
Eugenol	1–2
Linalool	1–2
β Terpeneol	1–2
Geraniol	2–3
Citral	2–3

FIGURE 22.1 Burmese woman wearing thanaka. (Copyright 1996–2000, Naomi Suzuki.)

(Lindsay et al., 1998). In a follow-up clinical trial, pregnant women using a mixture of thanaka and deet experienced a 28% greater reduction in incidence of falciparum malaria than women using thanaka alone (sample size 897), although this did not reach statistical significance with a log rank test (McGready et al., 2001). The authors suggest that the combination of thanaka and deet could be useful in this region in areas of low malaria transmission. Local vector mosquitoes bite early in the evening and the prevalence of multi-drug-resistant *Plasmodium falciparum* is extremely high. Thus, the use of repellents for pregnant women is strongly recommended. In addition, significantly more women expressed a preference to the thanaka and repellent mixture, compared to repellents alone.

22.2.1 LEMON EUCALYPTUS

The lemon eucalyptus extract comes from the plant *Corymbia citriodora* (synonyms include *Eucalyptus citriodora* and *Eucalyptus maculata* var. *citriodora*) originating from China. The essential oil extract was determined to have mosquito-repelling properties. Several other *Eucalyptus* species are also repellent to mosquitoes, although this is one of the most powerful (Golob et al., 2002). Chemical analysis of the extract showed that it contained citronella, citronellol, geraniol, isopulegol, delta pinene, and sesquiterpene (Curtis et al., 1991). When Li et al. (1974) tested this full extract on *Ae. aegypti* mosquitoes, the protection only lasted 1 hour. However, *p*-menthane-3,8-diol (PMD) was discovered in the waste distillate of the extract of the lemon eucalyptus plant, which was determined to be the active ingredient for the repellent activity of mosquitoes. It was given the Chinese name *Quwenling*, which means "effective repeller of mosquitoes."

PMD has undergone several trials in different parts of the world. Laboratory studies by Trigg and Hill (1996) showed that 30% PMD was almost as effective as deet, the most widely available synthetic repellent, against *Anopheles gambiae* Giles, which is the main malaria vector in sub-Saharan Africa. It was determined that PMD-impregnated towelettes (0.575 g) applied to the arms of human volunteers provided 90 to 100% protection against mosquitoes from laboratory-raised colonies of *Anopheles arabiensis* (Govere et al., 2000a).

Field studies in China showed that the protection time from *Aedes vexans* and *Aedes albopictus* was 2 and 5.5 hours, respectively, when PMD was used in a 20 to 30% glycerol or alcohol formulation (Li et al., 1974). In Tanzania, 50% PMD in isopropanol provided over 6 hours of protection from local malaria vectors *An. gambiae* and *Anopheles funestus* (Trigg, 1996). In the Bolivian Amazon, Moore et al. (2002) showed that 30% PMD in an alcohol base provided 96.9% protection for up to 4 hours postapplication from all mosquito species, compared to 84.8% protection by 15% deet. It is worth noting that 81.3% of the mosquito catch in the study area was comprised of *Anopheles darlingi*, which is the principal malaria vector in the whole of the Amazon region.

PMD is a promising new "natural" product that appears to be extremely effective in the laboratory as well as in the field. In addition, acute toxicity studies show limited toxicity, with an oral LD_{50} (lethal dose for 50%) of 2408 mg/kg and a dermal LD_{50} of >2000 mg/kg in rats (Trigg, 1996). For these reasons, the potential for commercial exploitation is high. Currently, Quwenling is available commercially in several countries in Europe and the U.S.

Interestingly, more recent research has looked at a low technology application of *C. citriodora*, by thermally expulsing volatiles from the fresh leaves using a metal plate over a traditional cooking fire in western Kenya (Seyoum et al., 2002b). *An. gambiae* showed a 74.5% reduction in biting, which is comparable to insecticidal mosquito coils (Charlwood and Jolley, 1984).

22.2.2 CYMBOPOGON SPP. (CITRONELLA GROUP) FAMILY: POACEAE

This genus contains several plants that are used throughout the world as insect repellents. They are rapidly growing grasses with distinctive aromatic foliage. Originating in India, the group is widely cultivated throughout the Tropics. These plants contain varying amounts of several insect repellent chemicals, although environmental conditions cause the content of volatile oils in plants to vary greatly. Repellent compounds contained in this group include alpha pinene, camphene, camphor, geraniol, and terpenen-4-ol. The most abundant repellent molecules found in the group are citronellal, citronellol, and geraniol (Duke, 2000).

The best known is citronella (*Cymbopogon nardus*), which is used in many commercial repellent preparations. These are particularly marketed for use on children, as natural repellents are perceived to be safer for use on children than deet. It is recognised as one of the best natural repellents. Curtis et al. (1987) calculated the ED_{50} (effective dose for 50%) of citronella in laboratory experiments to be 11.8 nl/cm² for *Anopheles stephensi*, 20.2 nl/cm² for *Anopheles albimanus*, and 42.1 nl/cm² for *An. gambiae*, which are similar to the effective doses of freshly applied deet. The U.S. Department of Agriculture (USDA, 1947–1964) reports longevity at 2 hours for 100% essential oil concentration. As initial effectiveness was similar, this indicates that the rapid loss of repellency is due to faster evaporation of citronella than deet.

Cymbopogon martinii martini (palmarosa) is a perennial grass, widely distributed throughout the Tropics. It contains between 750 and 4750 ppm geraniol (Duke, 2000), which gives it a sweet scent, and the oil is used in traditional Indian mosquito repellent preparations (Parrotta, 2001). From field-testing palmarosa against *Anopheles* mosquitoes in India, it was reported that the pure oil provided absolute protection for 12 hours (Ansari and Razdan, 1994). However, the tests utilised pairs of volunteers: one acting as bait and the other as collector — who wore no repellent. Large numbers of mosquitoes are diverted to the unprotected member of a pair, which reduces the number of mosquitoes attacking the protected individual, giving an inflated measure of repellency (Moore, in preparation). *Cymbopogon citratus* (lemongrass) is also traditionally used as a mosquito repellent in India (Parrotta, 2001). It has been evaluated as a repellent by Leal and Uchida (1998) and elicited similar responses in electroantennogram experiments as deet. Field tests in Bolivia showed that 25% *C. citratus* in ethanol provided 77.93 and 90.67% protection for 2 hours against *An. darlingi* and *Mansonia* spp., respectively (Moore, in preparation). It is rich in citral (70%), and many other repellent terpenes are present, including alpha pinene, citronellal, citronellol, and geraniol (Duke, 2000). The steam-extracted essential oil of *Cymbopogon winterianus* has been evaluated, mixed with 5% vanillin, against *Ae. aegypti*, *Culex quinquefasciatus*, and *Anopheles dirus*. It compared favorably with 25% deet, giving greater than 6 hours protection against all three mosquito species in laboratory experiments (Tawatsin et al., 2001). Another related plant is *Cymbopogon flexuosus*, which contains between 875 and 2500 ppm geraniol (Duke, 2000), although it does not seem to have been evaluated as a repellent. However, *Cymbopogon excavatus* is used in South Africa as a mosquito repellent. When evaluated in the laboratory against *An. arabiensis* by Govere et al.

(2000b), it gave good protection for 2 hours, but declined to 59.3% after 4 hours. This compares favorably with *C. nardus*.

The use of the *Cymbopogon* genus as an insect repellent is widespread throughout the world. It has many advantages for use in the prevention of malaria in areas of low transmission, where vector species are active early in the evening. The repellent oils are effective provided that they are reapplied frequently. Four hours' protection may be sufficient if people use the repellents before retiring to a bed net. The grasses grow readily and rapidly throughout much of the Tropics, and a simple steam distillation is sufficient to extract the repellent fractions. The plants in this family are pleasant smelling and are widely used in traditional medicine, making them acceptable for use. In addition, their high citronellal content makes the plants of this genus potential candidates for PMD production since citronellal is a precursor of this molecule.

22.2.3 PELARGONIUM SPP.

In Mpumalanga, South Africa, Govere et al. (2000b) determined through interviews with local people that *Pelargonium reniforme* (rose geranium) is seen as an effective plant at repelling mosquitoes. The leaves of this plant release a highly pungent odor. When tested, an alcohol formulation (200 mg/ml) made from the fresh leaves provided 63.3 and 59.3% protection after 3 and 4 hours, respectively, against laboratory raised *An. arabiensis*.

In Europe and Northern America the plant *Pelargonium citrosum* is being marketed as a mosquito-repelling plant, since the leaves release a citronella-like odor. It is said that if planted, it will repel mosquitoes within a 0.93-m² area (Matsuda et al., 1996). In field experiments with human subjects in Illinois, it proved ineffective at repelling *Ae. vexans* and *Aedes triseratus* mosquitoes (Jensen et al., 2000). Similarly, it did not protect human subjects against *Ae. albopictus* and *Culex quinquefasciatus* in Florida, nor *Aedes* spp. in Michigan and Canada (Cilek and Schreiber, 1994; Cummings and Craig, 1995; Matsuda et al., 1996). The essential oil constituents were analysed and compared to essential oils of the *Cymbopogon* species. It was determined that *P. citrosum* contains trace amounts of citronellal and large amounts of linalool, whereas this ratio is reversed in *C. winterianus* and *C. nardus* oils. However, linalool is repellent, and the plant contains citronellol (20.82%) and geraniol (22.57%) (Matsuda et al., 1996). It is possible that applying the essential oil to the skin or evaporating it into the air would provide protection. As essential oils are only repellent in the vapour phase, it is not surprising that the unbruised leaves of the plant provided no protection from mosquitoes. Seyoum et al. (2002b) bruised the leaves of live potted plants before testing them and showed a significant repellent effect with several plants against *An. gambiae*.

22.3 THE USE OF SMOKE AS A MOSQUITO REPELLENT

Smoke is the most widely used means of repelling mosquitoes utilised in the rural Tropics. Waste plant materials are frequently burned in Sri Lanka as a mosquito repellent, even though indoor residual spraying has been carried out by the government for many years (Silva, 1991). In rural Guinea-Bissau, 86% of residents used an unimpregnated bed net in conjunction with mosquito coils or plant-based smoke (Pålsson and Jaenson, 1999a). In the Solomon Islands, a recent survey revealed that fire with coconut husks and papaya leaves was the most prevalent form of personal protection from mosquitoes, being used by 52% of residents (Dulhunty et al., 2000). Tests in Papua New Guinea found that the traditional practice of burning mango wood (*Magifera* spp.), coconut husks (*Cocos nutifera*), wild ginger leaves (*Alpinia* spp.), and betelnut leaves (*Areca catechu*) repelled mosquitoes (Vernede et al., 1994).

Ongore et al. (1989) surveyed a population in Kenya and discovered that 16% burned *Lantana rhodesciense* leaves and 16% burned various waste plant materials, including sisal leaves and rice husks. In Sierra Leone and Ghana people burn orange peels to drive out mosquitoes, while in Ghana and the Gambia the leaves of neem (*Azadirachta indica*) and the baobab tree (*Adansonia digitata*)

FIGURE 22.2 Dai villagers in southern China adding plant material to their cooking fires to drive away mosquitoes. (Copyright 2000, Sarah Moore.)

are burned (Aikins et al., 1994). Members of the Lamiaceae (Section 22.3.1) and *Lippia* spp. (Section 22.3.4) are frequently burned as protection from mosquitoes throughout Africa.

There have been few studies to measure whether the use of smoke actually prevents malaria. Snow et al. (1987) showed that churai (*Daniellia oliveri*) smoke used as a repellent in the Gambia did not significantly reduce the incidence of malaria in children (see Section 22.3.2). However, a study in Sri Lanka found that the use of traditional fumigants did protect against malaria (van der Hoek et al., 1998). The difference in the results of the two studies may be related to erratic use of churai in the Gambia. It is also very likely that the massive entomological inoculation rate (EIR) in the Gambia requires more effective means of personal protection from infected bites than offered by the churai. In Sri Lanka the lower EIR and less anthropophilic vector mosquito *An. culicifaces* may mean that burning local plant repellents is a viable means of malaria prevention. In Sri Lanka, Holy Basil (*Ocimum sanctum*), *tulsi* (Sanskrit name), is commonly burned, as are the seed husks from oil extraction from neem (*A. indica*) and the butternut tree (*Madhuca longifolia*) (Silva, 1991). These seeds contain insecticidal and repellent agents.

Burning wood and adding repellent plants to it probably works in several ways. First, the smoke may disguise human kairomones and disrupt convention currents essential in mosquito host location. Second, burning may release repellent or irritant molecules; the molecules released by the plants also may be insecticidal, e.g., pyrethrum found in *Chrysanthemum* spp. The dried flowers are commonly made into joss sticks for use against mosquitoes and are extremely effective, providing up to 86% reduction in biting when used over an 8-hour period (Charlwood and Jolley, 1984).

22.3.1 LAMIACEAE FAMILY

Plants from the Basil family are used commonly in East and West Africa as mosquito repellents (Dalziel, 1937; Kokwaro, 1976). Many species of the Lamiaceae are strongly aromatic and toxic to insects, such as *Ocimum* spp. and *Hyptis* spp. (cited in Pålsson and Jaenson, 1999a).

FIGURE 22.3 *Ocimum basilicum.* (Copyright 2003, Merlin Willcox.)

22.3.1.1 *Ocimum* spp.

The essential oils from the species of this genus contain linalool, linalol, linoleic acid, *p*-cymene, estragole, eucalyptol, eugenol, citral, thujone, ocimene, camphor, methyl chavicol, oleic acid, and many other terpenes, all of which are effective repellents. It grows rapidly under a range of climatic conditions, although it is best adapted to a drier climate. The essential oil of *Ocimum basilicum* (see Figure 22.3) is larvicidal, producing 100% mortality of *Culex pipiens fatigans* at 0.12% concentration (Chavan and Nikam, 1982).

In Tanzania, traditionally fresh *Ocimum* spp., kivumbasi (local name), are burned, and freshly cut twigs of *O. suave* and *O. canum* are placed in the corners of rooms to prevent mosquitoes from entering (Stephens et al., 1995; White, 1973). The latter method was field-tested in Guinea-Bissau, West Africa, by Pålsson and Jaenson (1999b), who showed that fresh *O. canum* (also known as *O. americanum*) provided 63.6% protection from mosquito biting for 2 hours.

In Zimbabwe, *Ocimum* spp. leaves are rubbed on the skin as a method of repelling mosquitoes (Lukwa, 1994). When the juices from the leaves of *O. suave* and *O. canum* were spread on the legs of human volunteers, there was approximately 50% reduction in the proportion of female *An. gambiae* mosquitoes that were engorged with blood (White, 1973). A 250 mg/ml concentration of dried *O. canum* leaves in ethanol provided 70% repellency against *Ae. aegypti* for 2 hours (Lukwa et al., 1996). In Thailand, a 25% concentration of *O. canum* essential oil in ethanol was tested on three mosquito species. This formulation provided 3, 4, and 8 hours protection from the bites of *Ae. aegypti*, *An. dirus*, and *Cx. quinquefasciatus*, respectively (Tawatsin et al., 2001). Interestingly, when mixed with 5% vanillin, the protection times increased greatly for each mosquito species since it reduces the evaporation rates of repellents (Spencer, 1974).

22.3.1.2 *Hyptis* spp.

In the Brazilian Amazon *Hyptis* sp., locally called Hortelã-do-campo, is traditionally burnt and leaves are rubbed on the skin in order to keep mosquitoes away (Sears, 1996). Here, its repellent activity is associated with its strong smell. In West Africa the fresh plant is sometimes used or the aerial parts of the *H. suveolens* are placed on charcoal and the resulting smoke repels the mosquitoes (Pålsson and Jaenson, 1999a), although thermal expulsion of the plant volatiles actually attracted mosquitoes (Seyoum et al., 2002b). In Tanzania, freshly picked and bruised sprigs of *H. suaveolens*, *hangazimu*, are hung in the house to try to prevent mosquitoes from entering (Curtis et al., 1991). However, fresh *H. suaveolens* did not cause reduction in biting when hung in an experimental hut (Curtis and Lines, 1986, unpublished). In comparison, when tested in Guinea-Bissau, the fresh plant was able to provide approximately 70% protection from biting for 2 hours (Pålsson and Jaenson, 1999b). The smoldering plant provides the most effective protection. Nicholson and Lines (1987, unpublished) showed that there was a 10-fold reduction in biting in the presence of *hangazimu* smoke. Similarly, Pålsson and Jaenson (1999b) showed that smoldering *H. suaveolens* provided approximately 84% protection for 2 hours against *An. gambiae*, whereas Seyoum et al. (2002b) found only a 20.8% reduction in biting.

22.3.1.3 *Mentha*

There are few published accounts of these species of plants being used traditionally as personal protection against mosquitoes. In the Brazilian Amazon, the leaves are either rubbed on the skin or burnt to produce smoke (Sears, 1996). Barnard (1999) tested several concentrations of the *Mentha piperita* essential oil against *Ae. aegypti* and determined that with 100% concentration the protection time was 45 minutes; this was reduced to 30 minutes when the concentration was 25%. Field tests of *M. piperata* in India against *An. culicifaces*, *Anopheles annularis*, and *Anopheles subpictus* provided 85% protection over 11 hours (Ansari et al., 2000). However, these results are inflated as insect collectors not wearing repellent collected the mosquitoes from baits wearing repellent. It therefore appears that this is not a very effective repellent. Evaporating the essential oil of *M. piperita* at room temperature caused knockdown of several mosquito species (ICMR, 2000). A related plant, *Mentha arvensis* (Japanese mint), has extremely high vapour toxicity to insects (see Figure 22.4). It is widely grown throughout the Tropics for its essential oil. The leaves yield up to 80% menthol, which is the insecticidal ingredient contained in this species. Mosquito repellent chemicals are also contained in the plant, including menthone, limonene, beta pinene, alpha pinene, and linalool (Lee et al., 2001). Fumigants evaporated using heat or mosquito coils containing the plant may repel and kill mosquitoes. Evaporating the pure essential oil with a kerosine lamp caused a 51.94% reduction in mosquito biting in field trials (Moore, in preparation). The essential oil also has larvicidal effects on mosquitoes, its LD_{50} being 83.8 ppm against *An. stephensi* (Kumar and Dutta, 1987). The plant is able to reproduce asexually via a rhizome, making propagation easy and rapid.

22.3.1.4 *Thymus*

Thyme oil at 100% is repellent against *Aedes quadrimaculatus*, *An. albimanus*, and *Ae. aegypti* for at least 30 minutes, when applied to cloth (USDA, 1947–1964). Most recently, varying concentrations of the essential oil of red thyme were tested in the laboratory against *Ae. aegypti* and *An.*

FIGURE 22.4 *Mentha arvensis*. (Copyright 1999, Food and Agriculture Organization of the United Nations, Agricultural Services Bulletin 137.)

albimanus (Barnard, 1999). At 100% concentration it provides 135 and 105 minutes protection against *Ae. aegypti* and *An. albimanus*, respectively, and at 25% concentration, the protection time was 45 minutes for both species of mosquitoes. It was also determined that mixtures of essential oils were, in fact, no more effective than the essential oils alone. In the former USSR, a local method against biting insects was tying thyme stick, *Thymus serpyllum*, with thick cotton, drying these, and then burning them. Rubtzov tested this method and reported 85 to 90% protection for 60 to 90 minutes in the open air (Rubtzov, 1946, cited in Curtis et al., 1991).

22.3.2 *Daniellia oliveri* (Fabaceae)

The local names *churai*, *santango*, and *santão* refer to resins and wood commonly burnt indoors in western Africa to prevent mosquitoes from entering at night (Pålsson and Jaenson, 1999a; Lindsay and Janneh, 1989; Bockarie et al., 1994). In several field trials, it was determined to be an effective, accepted, and cheap form of personal protection. In Guinea-Bissau, smoke from the burning bark of *D. oliveri* reduced biting from mosquitoes by 74.7 and 77.9% in comparison to the control in two separate field experiments (Pålsson and Jaenson, 1999b). In Banjul, Gambia, *santango* reduced biting on human subjects by 77%, which was more than a permethrin mosquito coil, but less than deet soap (Lindsay and Janneh, 1989).

It is essential, however, to determine whether the product can reduce malaria as well as mosquito biting. Bockarie et al. (1994) performed entomological studies comparing households where wood had been burnt and where wood was not burnt. There was a significantly higher number of female *An. gambiae* in exit traps of houses where wood had been burnt. This showed that the environment created indoors while burning wood was not favorable for resting mosquitoes. However, the proportion of blood-fed mosquitoes was not greatly different between the two types of households, which indicated that they were still successful in blood feeding. Laboratory experiments by Curtis and Hill (1986, unpublished) showed that burning *churai* does not inhibit feeding success of mosquitoes. *Churai* is therefore reducing the apparent nuisance of mosquitoes, but not necessarily reducing the biting index sufficiently. This finding is substantiated by evidence from the Gambia, where it was determined that the burning of *D. oliveri* did not have any impact on the incidence of malaria in children (Snow et al., 1987).

22.3.3 *TAGETES* SPP.

Tagetes species contain monoterpenoid esters (Hogstad et al., 1984) and their larvicidal and insecticidal activity is well established (Green et al., 1991; Perich et al., 1994; Sharma and Saxena, 1994; Macedo et al., 1997). The essential oil was determined not to be a mosquito repellent (McCullough and Waterhouse, 1947, cited Green et al., 1991). However, studies in Zimbabwean communities (Lukwa et al., 1999) showed that people do use fresh plant material of *T. minuta* as a form of personal protection. It is used by crushing it and applying it to the skin, burning the plant material, or as whole-plant material. Okoth (1973) tested the effectiveness of whole-plant material of *T. minuta* against mosquitoes in Uganda. The field site had large numbers of *Mansonia uniformis* Theobald and *Anopheles marshalli* Theobald. Human landing catches were performed in a tent in which 4 kg of fresh *T. minuta* whole-plant material had been placed 1 hour previously, and in a control tent with no plants. Fewer mosquitoes were recorded biting and resting in the tent where the *T. minuta* plant material had been than in the untreated tent. Preliminary laboratory tests also showed that the plants had repellency in a choice test and significant toxicity when mosquitoes and plant parts were put in containers together.

More recently, Tyagi et al. (1997) carried out cage tests using the essential oil of *Tagetes minuta*. After 6 hours 86.4% protection was provided against *An. stephensi*; 84.2% against *Cx. quinquefasciatus*, and 75% against *Ae. aegypti*. Steam distillate of *T. minuta* evaporated at room temperature caused rapid knockdown of mosquitoes, including *An. culicifacies* and *An. stephensi* (ICMR, 2000). These results suggest that this plant has excellent potential as a mosquito repellent; further testing is required.

22.3.4 *LIPPIA* SPP.

In the Gambia *L. cheraliera* leaves are traditionally used as mosquito repellents. *L. javanica* is commonly found in southern Africa and is frequently used as a repellent (Lukwa, 1994). Hot leaf infusions of the leaves of these plants are traditionally used as remedies for a variety of ailments, including malaria (see Chapter 7). The leaves have a strong lemon smell (van Wyk et al., 1997), to which belief in its healing abilities can probably be attributed. *L. cheraliera* is also burned in the Gambia as a mosquito repellent smoke (Aikins et al., 1994). A thorough study carried out in Zimbabwe by Lukwa et al. (1999) revealed that 29% of the population used plants to protect themselves from mosquitoes, mainly by burning the leaves of *L. javanica*. The main constituents of the essential oils of this plant are monoterpenoids, such as myrcene, caryophyllene, linalool, *p*-cymene, and ipsdienone.

Govere et al. (2000b) tested an alcohol extract of dried *L. javanica* leaves on human subjects against *An. arabiensis* mosquitoes in the laboratory. The protection was 76.7% after 4 hours and 59.3% after 5 hours. Alcohol extracts of the dried leaves applied to the skin were also shown to provide 100% protection for 2 hours against *Ae. aegypti* (Lukwa, 1994). Work using the related *L. uckambensis* by Seyoum et al. (2002a) has shown that the release of volatiles from the leaves through thermal expulsion reduces *An. gambiae* biting by 49.5%.

22.3.5 *ARTEMISIA* SPP.

Members of this genus are found all over the world from tropical India to Siberia. They are low-growing perennial herbs in the family Asteraceae. The plants are aromatic, can tolerate poor conditions, and provide good cattle fodder, and these plants have been used against insects for centuries. In China, bundles of dried *A. vulgaris* are burned to repel biting insects. This observation led to an investigation by Hwang et al. (1985), which revealed that *A. vulgaris* contains insect repellents that can be released from the plant by combustion. The compounds isolated and found repellent to *Ae. aegypti* were camphor, linalool, terpenen-4-ol, α and β thujone, β pinene, myrcene, limonene, and cineol. The plants are also burned in Central Asia, Bolivia, and India to repel mosquitoes. They were burned to drive away mosquitoes by many Native Americans, including the Shuswap, Thompson, and Blackfoot tribes (Moerman, 1998). Extracts of *A. vulgaris* are also highly toxic to mosquito larvae (Deshmukh and Renapurkar, 1987).

Artemisia absinthium (absinthe) is a native of Europe, Central Asia, and Africa, yet it was only used as an insecticide in Europe (Smith and Secoy, 1981) and India (Chopra et al., 1965). It is insecticidal (Facknath and Kawol, 1993) and contains many repellent chemicals, including thujone, terpinen-4-ol, linalool, nerol, geraniol, pinene, and 1,8-cineole (Nin et al., 1995). Although it is reported as a mosquito repellent (Morton, 1981), this plant does not appear to have been evaluated against mosquitoes. Interestingly, the antimalarial compound artemisinin is found in *Artemisia annua* (see Chapter 3).

22.4 NEEM

The neem tree (*Azadirachta indica*) has become a focus of attention with regard to the control of agricultural pests, and more recently against medically important insects. It originates from India where it has been used to control and repel insects for thousands of years (see Chapter 6), and it has now been introduced to drier parts of Central and South America, Africa, Australia, and Southeast Asia, notably southern China, where extensive plantations may now be found. Neem is widely used in its raw form as an agricultural pesticide (Forster and Moser, 2000) and burned to repel mosquitoes. The trees can grow in depleted and saline soil, making them an excellent method of regenerating desertified or marginal land. They are fast growing and can be used for a multitude of purposes besides insect control, for instance, firewood, fodder for livestock, and as a shade tree.

Extensive research has been carried out into the effect of botanical derivatives of the neem tree and its relatives (see for reviews Mulla and Su, 1999; Schmutterer, 1990; Mordue and Blackwell, 1993). *A. indica* contains at least 35 biologically active principles, of which azadirachtin is the predominant active ingredient, found in the seed, leaves, and bark. The azadirachtin content of neem oil is positively correlated with its effect against insects (Isman et al., 1990). The seeds may contain up to 10 g/kg of seed kernels (Schmutterer, 1990). The active components exhibit actions against insects that can be grouped into six categories: antifeedency, growth regulation, fecundity suppression, sterilisation, oviposition repellency or attraction, and changes in biological fitness.

The repellency of neem oil to hematophagous insects has been tested, although the results have been variable. Burning and thermal expulsion of the leaves produce only a modest reduction (<25%) in biting (Seyoum et al., 2002b). However, experiments using neem oil derived from the seeds have shown better protection. Sharma et al. (1993a) tested neem oil in coconut oil against *An. culicifacies* at concentrations of 0.5, 1.0, and 2%. Two percent oil was reported to provide 100% protection for 12 hours. However, biting densities of mosquitoes in the area were extremely low. Only 144 anophelines were captured on control subjects in 288 man-hours. Pandian and Devi (1998) tested neem oil in coconut oil against *Armigeres subalbatus* and found that it provided only 124 minutes protection. Caraballo (2000) tested Neemos® gel in Venezuela and found that it offered 98.2% protection against *Anopheles darlingi* for 8 hours. Numbers of mosquitoes were high (217 anophelines per man-hour); however, each volunteer lay in a cot and had landing mosquitoes removed by another volunteer who was not wearing any repellent. Mosquitoes may have been diverted to the unprotected individual, which reduces the reliability of these results. Kant and Bhatt (1994) carried out another field test of neem oil in India. Two percent neem oil in coconut oil provided 98.0% protection against *An. culicifacies*. Unfortunately, numbers of mosquitoes were low: Only 1065 mosquitoes were captured on the control in 12 nights. Prakash et al. (2000) recorded 66.7% protection after 9 hours using 2% neem oil diluted in mustard oil. Again, numbers of mosquitoes were low, and mosquitoes were caught using an unprotected collector and a bait wearing repellent. In a test in the Bolivian Amazon with high densities of *An. darlingi* (mean = 71 mosquitoes/man-hour), 2% neem oil in ethanol provided only 56.7% protection 3 and 4 hours after application (Moore et al., 2002).

The most effective result was obtained by vapourising neem oil from mats. Five percent neem oil was more effective at reducing both biting and numbers of resting mosquitoes than 4% allethrin on mats (Sharma et al., 1993b). Sharma and Ansari (1994) proposed repelling mosquitoes using neem in kerosene. The mixture is used in kerosene lamps popular as a light source in rural areas throughout the developing world. Work was carried out in India between March and September 1992 and found that 1% neem oil in kerosene provided up to 84.6% protection from bites. Unfortunately, the paper is not clear as to whether treatments and control collections were carried out on the same day, neither is any mention made of baseline mosquito numbers. This method was used by Pates et al. (2002) to release 1% transfluthrin and produced only 43.8% reduction in biting. If neem oil in kerosene is effective in repelling mosquitoes, this has important implications for malaria control due to the ease of application of this method. Neem oil is cheap and freely available throughout India and many other regions of the world. Perhaps a better way of releasing the volatile repellent might be to place the repellent and oil mixture above the flame and not in the kerosene itself (Jo Lines and Helen Pates, personal communication). In this way the optimal temperature for release of the volatile repellent substance can be better regulated. Kerosene lamps are widely used, and the method is extremely simple and convenient.

22.5 GARLIC

It is a common misconception that eating garlic, *Allium sativum*, will make the skin unpalatable to mosquitoes. Stjernberg and Berglund (2000) claimed that the consumption of 1200 mg of garlic per day provides significant protection from tick bites. However, the accuracy of the study has been contested since the findings are exaggerated statistically due to the incorrect use of the collected data (Ranstam, 2001). Garlic does in fact have insecticidal properties, containing allicin, as well as the repellent compounds geraniol, linalol, caffeic acid, and ferulic acid. Commercial insecticide/repellent preparations based on garlic are available, certified for use against mites, nematodes, and mosquito larvae (Garlic Barrier AG, EPA 66352-2 from Allium Associates). Garlic was reported to be effective at repelling mosquitoes (Greenstock and Larrea, 1972) when rubbed on the skin, although the smell of the substance may limit its practicality. This is because alliin needs to mix with aliinase to form sulfenic acid and ammonia, which combine to form allicin. In Brazil whole cloves of garlic are hung around the neck as a repellent (Sears, 1996).

22.6 *ANNONA SQUAMOSA* (CUSTARD APPLE)

This plant is widely cultivated and naturalised throughout the Tropics, where it is grown for its fruit. The dried, powdered unripe fruit are traditionally used as an insecticide, particularly against lice (Parrotta, 2001). The leaves are used as an insecticide in the Gambia (Oliver-Bever, 1986). Steam distillation of the fruit peel yields an essential oil containing insect repellent chemicals, including α pinene, β pinene, limonene, ocimene, borneol, and camphor (Oliveros-Belardo, 1977). As the peel is a waste product, the oil may prove economical as well as useful as a repellent, although this has not been evaluated. In addition, seed extracts are potent against both agricultural pests (Sharma et al., 1999) and mosquito larvae (Saxena et al., 1993).

22.7 *LANTANA CAMARA*

L. camara is a hardy weed plant found throughout the Tropics and cultivated as an ornamental. It contains α pinene, caryophyllene, cineole, citral, eugenol, geraniol, linalool, *p*-cymene, terpeneol, and up to 80% caryophyllene, which is an insectifuge, although the relative compositions of these oils vary according to geographic location (Duke, 2000). The closely related *Lantana indica* is native to India and contains 46 different terpenes, giving the plant a pleasant smell of black currants. *L. camara* is regarded as a pest in many regions as it is prolific and toxic to cattle, although they will only feed on it if forage is sparse. Furthermore, *L. camara* is a refuge for tsetse flies and is implicated in epidemics of sleeping sickness (Syed and Guerin, 2004). This plant is useful as it can grow on depleted soils, covering the ground with fine leaf mulch that serves to retain water and improve the soil, and it is also efficient at checking soil erosion. Dua et al. (1996) tested a methanolic extract of the flowers in coconut oil (1.8% w/v) against *Aedes albopictus*, finding that >50% protection was provided for 4 hours. For the first hour 94.5% protection was provided with very high mosquito numbers — 1132 mosquitoes caught on the two control individuals. It would be interesting to repeat these experiments with malaria vectors, in comparison with deet, as this plant appears to be a good candidate natural repellent. The flowers are available in Central India between September and January. In addition, the leaves are repellent, giving 42.4% protection against *An. gambiae* when used as an ambient repellent (Seyoum et al., 2002a). This is comparable with some brands of mosquito coil containing synthetic insecticides.

22.8 *VITEX NEGUNDO* (INDIAN PRIVET)

This plant is widely naturalised and cultivated throughout the Old World, including areas of Africa, the Middle East, India, and Southeast Asia. It is a deciduous aromatic shrub, common in sandy, moist soils. The leaves are used as a mosquito repellent in India (Parrotta, 2001). In Maharashtra State, tribal people use smoke from the leaves of this plant at night to protect them from mosquito bites. Hebbalkar et al. (1992) extracted fractions from shade-dried leaves that contained terpinines, terpeneol, and sesquiterpene alcohols. The fraction containing monoterpenes and sesquiterpene alcohols protected for up to 3 hours in cage tests with blood-starved *Ae. aegypti*. There were 1000 to 1500 mosquitoes in each cage representing 10 to 20 bites per minute on control subjects, since only a 2-cm^2 area was offered for the mosquitoes to feed upon.

22.9 CONCLUSION

The World Health Organization (WHO) recommends the use of personal protection (Barnard, 2000). Although the short duration of natural repellents is noted, the WHO also states that "the use of any of those that can be shown to be effective should be encouraged." Repellents may be especially useful for residents of malaria-endemic countries where mosquito vectors bite early in the evening before people retire to the protection of their bed nets or where the vectors display exophagic feeding habits. This includes large areas of the Amazon, Southeast Asia, and Central Asia. However, there is some concern that use of repellents will drive anthropophilic mosquitoes from treated individuals to untreated individuals and, as such, not effect disease incidence. In contrast, in areas where the disease vectors are less anthropophilic, like *An. culicifaces* in rural India and neighboring countries, the use of repellents could divert biting from humans to animals and thereby reduce the incidence of human disease.

There are thousands of plants used in traditional mosquito repellent preparations and many more that could be used for this purpose. The majority of the plants are mentioned here because they are commonly used and, when tested, were shown to be effective in the short term. Plants used as repellents are particularly beneficial for rural areas where access to commercial preparations is minimal and expense is a limiting factor.

The only plant-based repellent that has been reported to be as effective and long lasting as deet is PMD (Quwenling). However, it is essential that further research in this field is stimulated in order to reveal other effective plant-based repellents. In the meantime, the plant-based repellents that are currently in use or being tested provide a cheap and accessible form of disease prevention. Several different types of repellent may be used at once, for instance, a fumigant concurrent with a skin repellent to provide extra protection, although the use of a partially effective method is better than no protection at all. The use of smoke as a repellent is not desirable due to the respiratory problems it can cause. However, the protection from malaria this practice may provide (van der Hoek et al., 1998) goes some way to balance out the health risks associated with this method, particularly as the primary use of the fires is for cooking. A possible new, low-technology method to release plant volatiles is thermal expulsion where the plant material is placed on a metal plate over a fire (Seyoum et al., 2002b). It appears to provide greater repellency than direct burning of plant material, possibly since it releases different repellent compounds, and may carry a lower risk of causing respiratory disease since less smoke is produced this way.

There are many characteristics required of repellent plants that have a potential role in malaria prevention. First, they obviously have to be effective. For skin repellents, the guideline is that they repel more than 60% of mosquitoes under natural conditions for about 4 hours. The greater the repellency and the longer the duration, the better, as people are notoriously bad at remembering to reapply repellents. Extended protection exceeding 90%, such as that exhibited by PMD, is desirable but very rare, as the essential oils that provide the protection are so volatile and rapidly evaporate. A similar level of protection is desirable for fumigants and coils, although this is more difficult to measure experimentally. Second, they must not irritate the skin. They must be safe and pleasant to use in order to guarantee compliance. Although plants with a disagreeable odor may be used under conditions of severe mosquito nuisance, those with a pleasant smell will be used more often. Plants suitable for use against mosquitoes need to be naturally abundant or easy to cultivate. Ideally, the plant is hardy and grows rapidly. The source of repellent must be sustainable, so replaceable parts are preferable, e.g., leaves or seeds rather than parts that when removed kill or damage the plant, such as roots or shoots. The parts utilised must be available all year round — or at least during the malaria transmission season — or alternatively easy to harvest and store. Finally, plant-based repellents should be easy to use, either by rubbing on the skin directly, by throwing them on the fire, or through simple procedures such as steam distillation or petroleum ether extraction. These factors would allow individuals and communities to grow or harvest plants for their own use, and allow easy, localised commercialisation.

At present there is limited evidence to show that the use of repellents can reduce the prevalence of malaria. The use of *churai* in the Gambia did not significantly reduce the incidence of malaria in children (Snow et al., 1987). However, a more recent controlled clinical trial by Rowland et al. (in preparation) showed that the use of a synthetic, deet-and-permethrin-based soap could reduce the incidence of malaria. This emphasises the need for further research into the use of cheap, locally available methods of reducing man–vector contact.

ACKNOWLEDGMENT

The authors thank Dr. Nigel Hill and Prof. Chris Curtis for their comments and suggestions on the manuscript.

REFERENCES

Aikins, M.K., Pickering, H., and Greenwood, B.M. (1994). Attitudes to malaria, traditional practices and bednets (mosquito nets) as vector control measures: a comparative study in five West African countries. *J. Trop. Med. Hyg.*, 97, 81–86.

Ansari, M.A. and Razdan, R.K. (1985). Relative efficacy of various oils in repelling mosquitoes. *Indian J. Malariol.*, 32, 104–111.

Ansari, M.A. and Razdan, R.K. (1994). Repellent action of *Cymbopogon martinii* Stapt var. Sofia oil against mosquitoes. *Indian J. Malariol.*, 31, 95–102.

Ansari, M.A., Vasudevan, P., Tandon, M., and Razdan, R.K. (2000). Larvicidal and mosquito repellent action of peppermint (*Mentha piperata*) oil. *Bioresour. Technol.*, 71, 267–271.

Barnard, D. (1999). Repellency of essential oils to mosquitoes (Diptera: Culicidae). *J. Med. Entomol.*, 36, 625–629.

Barnard, D.R. (2000). Global Collaboration for Development of Pesticides for Public Health (GCDPP). Repellents and Toxicants for Personal Protection, WHO/CDS/WHOPES/GCDPP/2000.5.

Bockarie, M.J., Service, M.W., Barnish, G., Momoh, W., and Salia, F. (1994). The effect of woodsmoke on the feeding and resting behaviour of *Anopheles gambiae s.s. Acta Trop.*, 57, 337–340.

Caraballo, A.J. (2000). Mosquito Repellent of Neemos®. *J. Am. Mosq. Control Assoc.*, 16, 45–46.

Charlwood, J.D. and Jolley, D. (1984). The coil works (against mosquitoes in Papua New Guinea). *Trans. R. Soc. Trop. Med. Hyg.*, 78, 678.

Chavan, S.R. and Nikam, S.T. (1982). Mosquito larvicidal activity of *Ocimum basilicum* Linn. *Indian J. Med. Res.*, 75, 220–222.

Choochote, W., Kanjanapothi, D., Panthong, A., Taesotikul, T., Jitpakdi, A., Chaithong, U., and Pitasawat, B. (1999). Larvicidal, adulticidal and repellent effects of *Kaempferia galanga. Southeast Asian J. Trop. Med. Public Health*, 30, 470–476.

Chopra, R.N., Badhwar, R.L., and Gosh, S. (1965). *Poisonous Plants of India*. National Printing Works, Delhi.

Cilek, J.E. and Schreiber, E.T. (1994). Failure of the "mosquito plant" *Pelargonium citrosum* (van Leenii) to repel adult *Aedes albopictus* and *Culex quinquefasciatus* in Florida. *J. Am. Mosq. Control Assoc.*, 10, 473–476.

Covell, G. (1943). *Anti-Mosquito Measures with Special Reference to India*, Health Bulletin 11, Malaria Bureau 3, 6th ed. Government of India Press, Calcutta.

Cummings, R.J. and Craig, G.B., Jr. (1995). The citrosa plant as a mosquito repellent? Failure in field trials in Upper Michigan. *Vector Control Bull. North Cent. States*, 4, 16–28.

Curtis, C.F., Lines J.D., Ijumba, J., Callaghan, A., Hill, N., and Karimzad, M.A. (1987). The relative efficacy of repellents against mosquito vectors of disease. *Med. Vet. Entomol.*, 1, 109–119.

Curtis, C.F., Lines, J.D., Lu, B., and Renz, A. (1991). Natural and synthetic repellents. In *Control of Disease Vectors in the Community*, Curtis, C.F., Ed. Wolfe, London, pp. 75–92.

Dalziel, J.M. (1937). *The Useful Plants of West Tropical Africa*. Crown Agents, London.

Deshmukh, P.B. and Renapurkar, D.M. (1987). Insect growth regulatory activity of some indigenous plant extracts. *Insect Sci. Appl.*, 8, 81–83.

Dixit, R.S., Perti, S.L., and Agarwal, P.N. (1965). New repellents. *Lab. Dev. J. Sci. Technol. B*, 3, 273–274.

Dua, V.K., Gupta, N.C., Pandey, A.C., and Sharma, V.P. (1996). Repellency of *Lantana camara* (Verbenaceae) flowers against *Aedes* mosquitoes. *J. Am. Mosq. Control Assoc.*, 12, 406–408.

Duke, J.A. (2000). Dr. Duke's Phytochemical and Ethnobotanical Databases. Available at http://www.ars-grin.gov/cgi-bin/duke/.

Dulhunty, J.M., Yohannes, K., Kourleoutov, C., Manuopanagai, V.T., Polyn, M.K., Parks, W.J., and Bryan, J.H. (2000). Malaria control in Central Malaita, Solomon Islands 2. Local perceptions of the disease and practices for its treatment and prevention. *Acta Trop.*, 75, 185–196.

Facknath, S. and Kawol, D. (1993). Antifeedant and insecticidal effects of some plant extracts on the cabbage webworm, *Crocidolomia binotalis. Insect Sci. Appl.*, 14, 571–574.

Forster, P. and Moser, G. (2000). *Status Report on Global Neem Usage*. Universum Verlagsansalt, Wiesbaden, Germany.

Geoponika [Agricultural and Domestic Pursuits], Owen, T., Trans. (1806). London(translated from Greek).

Golob, P., Moss, C., Dales, M., Fidgen, A., and Evans, J. (1999). The Uses of Spices and Medicinals as Bioactive Protectants for Grains, FAO Agricultural Services Bulletin 137. FAO, Rome.

Golob, P., Nishinura, H., and Satoh, A. (2002). Eucalyptus in insect and plant pest control. In *Eucalyptus*, Coppen, J.J.W., Ed. Taylor & Francis, London.

Govere, J., Durrheim, D.N., Baker, L., Hunt, R., and Coetzee, M. (2000a). Efficacy of three insect repellents against the malaria vector *Anopheles arabiensis*. *Med. Vet. Entomol.*, 14, 441–444.

Govere, J., Durrheim, D.N., Du Toit, N., Hunt, R.H., and Coetzee, M. (2000b). Local plants as repellents against *Anopheles arabiensis*, in Mpumalanga Province, South Africa. *Cent. Afr. J. Med.*, 46, 213–216.

Green, M.M., Singer, J.M., Sutherland, D.J., and Hibben, C.R. (1991). Larvicidal activity of *Tagetes minuta* (marigold) toward *Aedes aegypti*. *J. Am. Mosq. Control Assoc.*, 7, 282–286.

Greenstock, D.L. and Larrea, Q. (1972). *Garlic as an Insecticide*. Doubleday Research Association, Braintree, England, 12 pp.

Haley, T.J. (1978). A review of the literature of rotenone 1,2,12,12a-tetrahydro-8,9-dimethoxy-(2-(1-methyl-ethenyl)-1-benzopyrano[3,5-*B*]fluoro[2,3,*H*][1]-benzopyran-6(6*h*)-one. *J. Environ. Pathol. Toxicol.*, 1, 315–337.

Hebbalkar, D.S., Hebbalkar, G.D., Sharma, R.N., Joshi, V.S., and Bhat, V.S. (1992). Mosquito repellent activity of oils from *Vitex negundo* Linn. leaves. *Indian J. Med. Res.*, 95, 200–203.

Hogstad, S., Johansen, G.L., and Anthonsen, T. (1984). Possible confusion of pyrethrins with thiopenes in *Tagetes* species. *Acta Chem. Scand.*, 38, 902–904.

Hwang, Y.S., Wu, K.H., Kumamoto, J., Axelrod, H., and Mulla, M.S. (1985). Isolation and identification of mosquito repellents in *Artemisia vulgaris*. *J. Chem. Ecol.*, 11, 1297–1306.

ICMR. (2000). Air-Borne Toxicity of Plant Extracts against Mosquitoes, Indian Council of Medical Research Annual Report 2000. Available at www.icmr.in/annual/mrc.pdf.

Isman, M.B.O., Koul, A., Luczynski, A., and Kaminski, J. (1990). Insecticidal and antifeedant activities of neem oils and their relationship to azadirachtin content. *J. Agric. Food Chem.*, 38, 1406–1411.

Jensen, T., Lampman, R., Slamecka, M.C., and Novak, R.J. (2000). Field efficacy of commercial antimosquito products in Illinois. *J. Am. Mosq. Control Assoc.*, 16, 148–152.

Johnson, T. (1998). *CRC Ethnobotany Desk Reference*. CRC Press, Boca Raton, FL.

Kant, R. and Bhatt, R.M. (1994). Field evaluation of mosquito repellent action of neem oil. *Indian J. Malariol.*, 31, 122–125.

Kokwaro, J.O. (1976). *Medicinal Plants of East Africa*. East African Literature Bureau, Nairobi.

Kumar, A. and Dutta, G.P. (1987). Indigenous plant oils as larvicidal agents against *Anopheles stephensi* mosquitoes. *Curr. Sci.*, 56, 959–960.

Leal, W.S. and Uchida, K. (1998). Application of GC-EAD to the determination of mosquito repellents derived from a plant *Cymbopogon citratus*. *J. Asia-Pac. Entomol.*, 1, 217–221.

Lee, S.E., Lee, B.H., Choi, W.S., Park, B.S., Kim, J.G., and Campell, B.C. (2001). Fumigant toxicity of volatile natural products from Korean spices and medicinal plants towards the rice weevil, *Sitophilus oryzae*. *Pest Manage. Sci.*, 57, 548–553.

Li, Z., Yang, J., Zhuang, X., and Zhang, Z. (1974). Studies on the repellent *Quwenling, Malaria Res.*, Vol 1, p. 6 (in Chinese).

Lindsay, S.W., Ewald, J.A., Samung, Y., Apiwathnasorn, C., and Nosten, F. (1998). Thanaka (*Limonia acidissima*) and deet (di-methyl benzamide) mixture as a mosquito repellent for use by Karen women. *Med. Vet. Entomol.*, 12, 295–301.

Lindsay, S.W. and Janneh, L.M. (1989). Preliminary field trials of personal protection against mosquitoes in the Gambia using deet or permethrin in soap, compared with other methods. *Med. Vet. Entomol.*, 3, 97–100.

Lukwa, N. (1994). Do traditional mosquito repellent plants work as mosquito larvicides. *Cent. Afr. J. Med.*, 40, 306–309.

Lukwa, N., Masedza, C., Nyazema, N.Z., Curtis, C.F., and Mwaiko, G.L. (1996). Chemistry, biological and pharmacological properties of African medicinal plants. In *Proceedings of the First International IOCD-Symposium, Victoria Falls, Zimbabwe, February 25–28*, Hostettmann, K., Chinyanganya, F., Maillard, M., and Wolfender, J.L., Eds. pp. 321–325.

Lukwa, N., Nyazema, N.Z., Curtis, C.F., Mwaiko, G.L., and Chandiwana, S.K. (1999). People's perceptions about malaria transmission and control using mosquito repellent plants in a locality in Zimbabwe. *Cent. Afr. J. Med.*, 45, 64–68.

Macedo, M.E., Consoli, R.A., Grandi, T.S., dos Anjos, A.M., de Oliveira, A.B., Mendes, N.M., Queiroz, R.O., and Zani, C.L. (1997). Screening of Asteraceae (Compositae) plant extracts for larvicidal activity against *Aedes fluviatilis* (Diptera: Culicidae). *Mem. Inst. Oswaldo Cruz*, 92, 565–570.

Matsuda, B.M., Surgeoner, G.A., Heal, J.D., Tucker, A.O., and Maciarello, M.J. (1996). Essential oil analysis and field evaluation of the citrosa plant *"Pelargonium citrosum"* as a repellent against populations of *Aedes* mosquitoes. *J. Am. Mosq. Control Assoc.*, 12, 69–74.

McGready, R., Simpson, J.A., Htway, M., White, N.J., Nosten, F., and Lindsay, S.W. (2001). A double-blind randomised therapeutic trial of insect repellents for the prevention of malaria in pregnancy. *Trans. R. Soc. Trop. Med. Hyg.*, 95, 137–138.

Moerman, D.E. (1998). *Native American Ethnobotany*. Timber Press, Portland, OR.

Moore, S.J., Lenglet, A., and Hill, N. (2002). Field evaluation of three plant based insect repellents against malaria vectors in Vaca Diez Province, the Bolivian Amazon. *J. Am. Mosq. Control Assoc.*, 18, 107–110.

Mordue (Luntz), A.J. and Blackwell, A. (1993). Azadirachtin: an update. *J. Insect Physiol.*, 39, 903–924.

Morton, J.F. (1981). *Atlas of Medicinal Plants of Middle America*. Charles C Thomas, Springfield, IL.

Mulla, M.S. and Su, T. (1999). Activity and biological effects of neem products against arthropods of medical and veterinary importance. *J. Am. Mosq. Control Assoc.*, 15, 133–152.

Nagpal, B.N. and Sharma, V.P. (1995). *Indian Anophelines*. Science Publishers, Inc., Lebanon, NH.

Nin, S., Arfaioli, P., and Bosetto, M. (1995). Quantitative determination of some essential oil components of selected *Artemisia absinthium* plants. *J. Essential Oil Res.*, 7, 271–277.

Okoth, J. (1973). *Tagetes minuta* L. as a repellent and insecticide against adult mosquitoes. *East Afr. Med. J.*, 50, 317–322.

Oliver-Bever, B. (1986). *Medicinal Plants in Tropical West Africa*. Cambridge University Press, Cambridge, U.K.

Olivers-Belardo, L. (1977). The predominant terpenes in the fruit peel oil of *Annona squamosa*. *Philipp. J. Sci.*, 106, 37–47.

Ongore, D., Kamunvi, F., Knight, R., and Minawa, A. (1989). A study of knowledge, attitudes and practices (KAP) of a rural community on malaria and the mosquito vector. *East Afr. Med. J.*, 66, 79–90.

Pålsson, K. and Jaenson, T.G.T. (1999a). Comparison of plant products and pyrethroid treated bed nets for protection against mosquitoes (Diptera: Culicidae) in Guinea-Bissau, West Africa. *J. Med. Entomol.*, 36, 144–148.

Pålsson, K. and Jaenson, T.G.T. (1999b). Plant products used as mosquito repellents in Guinea-Bissau, West Africa. *Acta Trop.*, 72, 39–52.

Pandian, R.S. and Devi, T.S. (1998). Repellent action of plant oils on mosquito. *Insect Environ.*, 4, 58.

Parrotta, J.A. (2001). *Healing Plants of Peninsular India*. CABI Publishing, Oxon, U.K.

Pates, H.V., Lines, J.D., Keto, A.J., and Miller, J.E. (2002). Personal protection against mosquitoes in Dar es Salaam, Tanzania, by using a kerosene oil lamp to vapourise transfluthrin. *Med. Vet. Entomol.*, 16, 227–284.

Perich, M.J., Wells, C., Bertsch, W., and Tredway, K.E. (1994). Toxicity of extracts from three *Tagetes* against adults and larvae of yellow fever mosquito and *Anopheles stephensi* (Diptera: Culicidae). *J. Med. Entomol.*, 31, 833–837.

Philip, M.I., Ramakrishna, V., and Venkat Rao, V. (1945). Turmeric and vegetable oils as repellents against Anopheline mosquitoes. *Indian Med. Gaz.*, 80, 343–344.

Prakash, A., Bhattacharya, D.R., Mahapatra, P.K., and Mahanta, J. (2000). A preliminary field study on repellency of neem oil against *Anopheles dirus* (Diptera: Culicidae). *Assam J. Commun. Dis.*, 32, 145–147.

Ranstam, J. (2001). Garlic as a tick repellent: letter to the editor. *JAMA*, 285, 42.

Sangat-Roemantyo, H. (1990). Ethnobotany of Javanese incense. *Econ. Bot.*, 44, 413–416.

Saxena, R.C., Harshan, V., Saxena, A., Sukumaran, P., Sharma, M.C., and Kumar, M.L. (1993). Larvicidal and chemosterilant activity of *Annona squamosa* alkaloids against *Anopheles stephensi*. *J. Am. Mosq. Control Assoc.*, 9, 84–87.

Schmutterer, H. (1990). Properties and potential of natural pesticides from the neem tree, *Azadirachta indica*. *Annu. Rev. Entomol.*, 35, 271–297.

Schultz, J.C. (1989). Tannin-insect interactions. In *The Chemistry and Significance of Condensed Tannins*, Hemingway, R.W. and Karchesy, J., Eds. Plenum Press, New York, pp. 417–433.

Sears, R. (1996). An ethnobotanical survey of insect repellents in Brazil. *TRI News*, Spring 1996, Yale School of Forestry and Environmental Studies, New Haven, CT.

Secoy, D.M. and Smith, A.E. (1983). Use of plants in control of agricultural and domestic pests. *Econ. Bot.*, 37, 28–57.

Seyoum A., Kabiru, E.W., Lwande, W., Killeen, G.F., Hassanali, A., and Knols, B.G.J. (2002a). Repellency of live potted plants against *Anopheles gambiae* from human baits in semi-field experimental huts. *Am. J. Trop. Med. Hyg.*, 67, 191–195.

Seyoum A., Palsson, K., Kung'a, S., Kabiru, E.W., Lwande, W., Killeen, G.F., Hassanali, A., and Knols, B.G.J. (2002b). Traditional use of mosquito-repellent plants in western Kenya and their evaluation in semi-field experimental huts against *Anopheles gambiae*: ethnobotanical studies and application by thermal expulsion and direct burning. *Trans. R. Soc. Trop. Med. Hyg.*, 96, 225–231.

Sharma, H.C., Sankaram, A.V.B., and Nwanze, K.F. (1999). Utilisation of natural pesticides derived from neem and custard apple in integrated pest management. In *Azadirachta indica A. Juss.*, Singh, R.P. and Saxena, R.C., Eds. Science Publishers Inc., Enfield, pp. 199–213.

Sharma, M. and Saxena, R.C. (1994). Phytotoxicological evaluation of *Tagetes erectes* on aquatic stages of *Anopheles stephensi*. *Indian J. Malariol.*, 31, 21–26.

Sharma, V.P. and Ansari, M.A. (1994). Personal protection from mosquitoes (Diptera: Culicidae) by burning neem oil in kerosene. *J. Med. Entomol.*, 31, 505–507.

Sharma, V.P., Ansari, M.A., and Razdan, R.K. (1993a). Mosquito repellent action of neem (*Azadirachta indica*) oil. *J. Am. Mosq. Control Assoc.*, 9, 359–360.

Sharma, V.P., Nagpal, B.N., and Srivastava, A. (1993b). Effectiveness of neem oil mats in repelling mosquitoes. *Trans. R. Soc. Trop. Med. Hyg.*, 87, 626.

Silva, K.T. (1991). Ayurveda, malaria and the indigenous herbal tradition in Sri Lanka. *Soc. Sci. Med.*, 33, 153–160.

Smith, A.E. and Secoy, D.M. (1981). Plants used for agricultural pest control in western Europe before 1850. *Chem. Ind.*, 1981, 12–17.

Snow, R.W., Bradley, A.K., Hayes, R., Byass, P., and Greenwood, B.M. (1987). Does woodsmoke protect against malaria? *Ann. Trop. Med. Parasitol.*, 81, 449–451.

Spencer, T. (1974). In Khan, A.A., Maibach, H.I., and Skidmore, D.L. (1975). Addition of vanillin to mosquito repellents to increase protection time. *Mosq. News*, 35, 223–225.

Stedman, E. and Barger, G. (1925). Physostigmine (eserine). Part III. *J. Chem. Soc.*, 127, 247–258.

Stephens, C., Masamu, E.T., Kiama, A.J., Kinenekejo, M., Ichimori, K., and Lines, J. (1995). Knowledge of mosquitoes in relation to the public and domestic control activities in the cities of Dar es Salaam and Tanga. *Bull. WHO*, 73, 97–104.

Stjernberg, L. and Berglund, J. (2000). Garlic as an insect repellent. Research letter. *J. Am. Mosq. Control Assoc.*, 284, 831.

Syed, Z. and Guerin, P.M. (2004). Tsetse flies are attracted to the invasive plant *Lantana camara*. *J. Insect. Physiol.*, 50(1), 43–50.

Tawatsin, A., Wratten, S.D., Scott, R.R., Thavara, U., and Techadamrongsin, Y. (2001). Repellency of volatile oils from plants against three mosquito vectors. *J. Vector Ecol.*, 26, 76–82.

Trigg, J.K. (1996). Evaluation of a eucalyptus-based repellent against *Anopheles* spp. in Tanzania. *J. Am. Mosq. Control Assoc.*, 12, 243–246.

Trigg, J.K. and Hill, N. (1996). Laboratory evaluation of a eucalyptus-based repellent against four biting arthropods. *Phytother. Res.*, 10, 313–316.

Tyagi, B.K., Ramnath, T., and Shahi, A.K. (1997). Evaluation of repellency of *Tagetes minuta* (Family: Compositae) against the vector mosquitoes *Anopeles stephensi* Liston, *Culex quinquefasciatus* Say and *Aedes aegypti* (L.). *Int. Pest Control*, 39, 184–185.

U.S. Department of Agriculture (USDA). (1947–64). Results of Screening Tests with Materials Evaluated as Insecticides, Miticides and Repellents.

Valderrama, X., Robinson, J.G., Attygale, A.B., and Eisner, T. (2000). Seasonal anointment with millipedes in a wild primate: a chemical defense against insects? *J. Chem. Ecol.*, 26, 2781–2790.

van der Hoek, W., Konradsen, F., Dijkstra, D.S., Amerasinghe, P.H., and Amerasinghe, F.P. (1998). Risk factors for malaria: a microepidemiological study in a village in Sri Lanka. *Trans. R. Soc. Trop. Med. Hyg.*, 92, 265–269.

van Wyk, B.E., van Oudtshoorn, B., and Gericke, N. (1997). *Medicinal Plants of South Africa*. Briza Publications, Pretoria, pp. 168–169.

Vernede, R., van Meer, M.M.M., and Alpers, M. (1994). Smoke as a form of personal protection against mosquitoes, a field study in Papua New Guinea. *Southeast Asian J. Trop. Med. Public Health*, 25, 771–775.

White, G.B. (1973). The insect repellent value of *Ocimum* spp. (Labiatae): traditional anti-mosquito plants. *East Afr. Med. J.*, 50, 248–252.

Wyrostkiewicz, K., Wawrzyniak, M., et al. (1996). An evidence for insecticide activity of some preparations from alkaloid-rich lupin seeds on Colorado potato beetle (*Leptinotarsa decemlineata* Say), larvae of the large white butterfly (*Pieris brassica* L.), black bean aphid (*Aphis fabae* Scop.) and on their parasitoids (Hymenoptera: Parasitica) populations. *Bull. Pol. Acad. Sci. Biol. Sci.*, 44, 29–39.

23 Guidelines for Studies on Plant-Based Insect Repellents

Sarah J. Moore and RITAM Vector Control Group

CONTENTS

23.1 INTRODUCTION

It has now been established through a cluster controlled clinical trial in Pakistan (Rowland et al., 2004) that the use of an insect repellent in the form of Mosbar soap containing 20% deet (N,N-diethyl-3-methyl-benzamide) and 5% permethrin significantly reduced malaria incidence. It may also be able to prevent other vector-borne diseases such as leishmaniasis and dengue, although further research is required to establish this. In order to produce a community-wide reduction in disease incidence as seen in Pakistan, the majority of individuals must use repellent regularly, as mosquitoes will be diverted from repellent users to nonusers. In contrast, in areas where the disease vectors are less anthropophilic, like *Anopheles culicifaces* and *Anopheles stephensi* in rural India and Pakistan (Reisen and Boreham, 1982), the use of repellents could divert biting from humans to animals.

Insect repellents have a unique role in regions where mosquito vectors bite in the early evening since people are often outdoors at this time. Even if people sit indoors, their housing may still allow mosquitoes entry if it is poorly constructed or improperly screened. Physical barriers such as long clothing are unpopular in the Tropics since the evenings are so warm, and people do not want to sit beneath their bed nets early in the evening. Repellents may therefore provide a valuable supplement to bed net use. Indeed, it may be that in areas where the local vectors feed in the early evening, this is the only means of securing a reduction in the level of malaria transmission. Repellents are also useful for people in situations where they have to be outdoors at night or in the early morning.

The *Anopheles minimus* and *Anopheles dirus* species complexes of South and Southeast Asia commonly transmit drug-resistant *Plasmodium falciparum*, as does *Anopheles darlingi* in the Amazon region. They are efficient vectors, and in some regions they have adapted their feeding habits, with peaks in biting activity occurring in the early evening and early morning (Baimai et al., 1988; Ismail et al., 1978; Tadei et al., 1998). *An. minimus* and *An. darlingi* have become solely exophilic (outdoor feeders) in response to indoor residual spraying in some regions (Meek, 1995; Tadei et al., 1998), since this method reduced the numbers of endophilic (indoor feeding) species types only (Suthas et al., 1986). *An. dirus s.l.* is mainly exophilic (Zhou et al., 1998; Oo et al., 2003), and this species breeds in forested areas, limiting the impact of conventional control strategies (indoor residual spraying and larviciding). A similar problem is encountered in the Southwest Pacific Islands, where the use of insecticide-treated bed nets imparted a significantly greater impact on vector infectivity than DDT spraying (Hii et al., 1993). However, this study also showed that this measure was insufficient to reduce the number of malaria infective bites received by each individual to a level where malaria could be controlled without additional countermeasures. The primary vector mosquito of this region, *Anopheles farauti s.l.*, has a biting density of more than 60 times greater at dusk than midnight (Beebe et al., 2000), which may explain the failure of bed nets alone to eliminate malaria. In addition, the introduction of bed nets in Papua New Guinea resulted in a shift toward earlier biting of *An. farauti* (Charlwood and Graves, 1987). This evidence all suggests that in this ecological context, the protection afforded by bed nets is lower than in regions where vector mosquitoes bite late at night, and the population is at risk of contracting malaria in the early evening, unless an additional form of personal protection, such as a repellent, is utilised.

Synthetic repellents containing deet or other chemicals are effective and widely available, yet their retail price may be too high for daily use. Plant-based repellents have the advantage that they can be produced locally, reducing their cost, and may help to boost the local economy. They may also be more culturally acceptable in communities with a tradition of plant use, where chemicals may be perceived as unhealthy or unpleasant smelling.

The reduction in biting nuisance that repellents achieve is welcomed by everybody. If indeed widespread use of repellents reduces the prevalence of disease, and if effective, affordable repellents are available, their use may be motivated mainly by relief from insect-biting nuisance (Gambel et al., 1998). However, it may actually achieve a valuable impact on disease incidence.

23.2 MECHANISM OF ACTION OF INSECT REPELLENTS

A repellent is "a chemical that, acting in the vapour phase, prevents an insect from reaching a target to which it would otherwise be attracted" (Browne, 1977). Repellents take two forms: those that are applied directly to the skin or clothing, and those that are released into the air through evaporation or burning.

Skin repellents interfere with the way mosquitoes perceive host stimuli. They are effective in the vapour phase and work only on the short-range approach, alighting and exploration. If a mosquito probes the host, the repellent has failed. No one knows exactly how they work, and different chemicals are likely to act in different ways. The five main hypotheses (Davis, 1985), based on work with deet, postulate that they interfere with sensory neurones that respond to host cues, and possibly stimulate other behaviour receptors simultaneously. A repellent generally turns a mosquito aside just before it lands. The mosquito's feeding response depends on automatic responses occurring in a definite sequence, so that anything that interrupts the sequence will prevent feeding. Coils, smoke, and fumigants are frequently insecticidal or irritant. In addition, smoke may mask human kairomones (chemical emanations attractive to mosquitoes) and convection currents that mosquitoes need for short-range host location. Mosquitoes are very susceptible to desiccation and smoke production lowers humidity. These three methods may make the environment less suitable for mosquitoes.

23.3 METHODS FOR STUDYING INSECT REPELLENTS

23.3.1 *IN VITRO*

Some studies have utilised membrane blood feeders, commonly used for feeding mosquitoes in insectaries, to measure repellency. This should be avoided, as mosquitoes do not respond as enthusiastically to a feeder as they do to a living host, and there is much interspecific variation in readiness to feed from membrane feeders (Novak et al., 1991). Another testing method employs disks of paper impregnated with a test repellent, and the number of landings is counted. This is an excellent method for testing irritancy of a chemical, but is not a measure of repellency (Rutledge et al., 1999). Olfactometry may also be utilised, although again, the interplay between attractive host cues and repellent chemicals is missing, and it may be a more efficient method for testing either the irritancy or attractiveness of a substance. These methods are cheap and yield many results rapidly with no risk to human subjects, but they do not accurately mimic the conditions of repellent usage. Thus, different methodologies cannot be compared, and their results cannot be directly extrapolated to the end user. Therefore, these methods are not recommended for any repellent testing other than preliminary screening.

23.3.2 *IN VIVO*

Cork and Park (1996) chemically fractionated human sweat samples into acid and nonacid components. They measured the electrical response of sensillae in the antennae of mosquitoes and found that short-chained aliphatic acids (C_2 to C_8) elicited significantly greater responses than the longer-chained acids. Healy et al. (2002) found that these acids elicit a landing response, and Costantini et al. (2001) showed that they have a significant effect on mosquito host-seeking behaviour. Nicolaides et al. (1968) compared the skin of humans and domestic animals. They concluded that humans excrete mainly triglycerides and are therefore unique in having fatty acids as breakdown products on the skin surface. This means that short-chained aliphatic acids are reliable host cues for anthropophilic mosquitoes. Testing repellents on animals will not give representative data of how the repellent will perform when applied to human skin. In addition, the most efficient malaria vectors are extremely anthropophilic and will be less attracted to nonhuman hosts. Thus, this method may give a distorted measure of repellency.

23.4 LABORATORY STUDIES

Deet is the active ingredient of most commercially available skin repellents and is the most effective insect repellent available at this time (Fradin and Day, 2002). It should be the standard against which the effectiveness of alternative repellents is judged (WHO, 1996). It is always preferable to conduct tests on human volunteers for greatest accuracy, provided that laboratory-reared mosquitoes are used to eliminate the risk of disease transmission, and the volunteers selected show mild or no allergic reaction to mosquito bites. It is conventional to use *Aedes aegypti* mosquitoes for repellent testing, but people generally show milder reactions to *Anopheles* bites. *Ae. aegypti* are commonly used, as they are easy to rear under laboratory conditions and are avid biters. However, several other species also fulfill these criteria, including *An. stephensi*, *Anopheles gambiae s.s.*, *Anopheles arabiensis*, and *Anopheles albitarsis*. The U.S. EPA (1999) recommends using *Ae. aegypti* along with a representative human-biting species from both the *Anopheles* and *Culex* genera for laboratory studies of repellent efficacy. It is preferential to perform bioassays on the malaria vectors in the region for which the repellent is to be used since the sensitivity of different mosquito species to repellents varies (Curtis et al., 1987).

Tests are generally conducted with mosquitoes held in large laboratory cages (approximately $40 \times 40 \times 40$ cm) into which the forearm(s) of the volunteer is introduced with the hand protected by a glove to make it unattractive to mosquitoes since bites received on the hands may be uncomfortable for the volunteer. The whole forearm may be exposed, or a 25-cm² area of skin, the remainder being covered with a rubber sleeve. In some tests, the repellent is applied to a cotton stocking, as repellents are much more persistent on fabric than on skin, since loss of repellent through abrasion, skin absorption, evaporation, and sweating is reduced (Rozendaal, 1997). The stocking is drawn over another stocking, which has been drawn over the arm to prevent skin contact with a repellent compound. Different laboratories that carry out extensive comparative testing of repellents give different emphasis to either:

1. Percentage protection in relation to dose (percentage protection compared to the control, or effective dose required to prevent either 50 or 90% of bites — ED_{50} or ED_{90})
2. Protection time after treatment, with tests of only a single large dose (time to first bite or complete protection time)
3. Both of these parameters

The protocols used to measure these, recommended by WHO (1996), are as follows:

1. Choice or free-choice cage test on human arms. One arm of a volunteer is treated with a measured quantity of test material dissolved in isopropanol and the other with solvent only. Only 25 cm² of skin on each arm is exposed, and both arms are introduced simultaneously into the cage. The number of bites is counted in 5 minutes, during which mosquitoes that are not repelled feed to repletion. The percentage protection is calculated by comparing the counts on the two arms.
2. Time-to-first-bite cage test on human arms. One arm is treated with 1 ml of a 25% solution of the test compound in ethanol. The arm is exposed for 3 minutes, every 30 minutes, and the first time after treatment noted at which a bite occurs, provided that it is followed by a confirmatory bite in the same or the following exposure period.
3. Cotton fabric screening model. Treatment of 280 cm² of stocking with 1 g of test material dissolved in enough acetone to saturate the stocking, then exposure of an arm covered by the stocking to 1500 hungry mosquitoes for 1 minute at daily or longer intervals until at least five bites are obtained.
4. No-choice cage test on human arms. This requires sequential exposure of an arm with zero, and then progressively higher, doses of repellent for 30 seconds to cages each containing approximately 50 hungry *An. gambiae* (or 45 seconds with *An. stephensi*). The number biting at the end of the short exposure is quickly counted (preferably with the help of an assistant), and the mosquitoes are then shaken off before they can imbibe any blood. Hence, the same mosquitoes can be used for testing each dose, and their continued hunger can be checked by exposing the other untreated arm. Probit analysis is used to calculate the ED_{50} or ED_{90}. However, the ED_{90} gives a more practical reflection of protection time. After reaching a dose that gives 100% repellency, the arm is reexposed hourly until repellency declines to 50% compared with contemporary counts on the untreated arm to measure the duration of this protection.
5. No-choice room test on human legs. The feet and lower legs are both treated with a repellent, or the volunteer sits close to a source of repellent in a small mosquito-proof room. Twenty-five hungry mosquitoes are released into the room and the number of bites received on the lower leg in a 10-minute period is counted. This more closely stimulates a field test, but with the advantage that the number of mosquitoes is controlled, as are edaphic factors. Coils, vapourisers, smoke, and fumigants may also be tested in this way, although ventilation is essential, and careful recording of the number, position, and mass of each source of ambient repellent is necessary to be able to compare different experiments, since standardisation of test procedure for novel plant-based space repellents is difficult. This process is repeated every 30 minutes until protection wanes to 50% or 90%.

Not mentioned in the WHO (1996) document, but frequently encountered in the literature, is the use of a $4 \times 5 \times 18$ cm plastic cage containing 15 mosquitoes with five openings through which mosquitoes can choose to feed on one of five repellent-treated areas of skin when placed on the limb of a volunteer (ASTM E951-94). The number of probing mosquitoes in each area of skin is recorded after 25 minutes, and these data are used to construct a dose–response curve and subsequent ED since each of the five areas of skin have had a slightly different concentration of repellent applied.

23.5 FIELD STUDIES

Field evaluation of repellents (including skin and clothing treatment) and other means of personal protection (including coils, plant parts burned on charcoal, electrically heated mats, and liquid vapourisers) are conducted in and around houses depending on where and how they would normally be used. Assessment is by catches on human volunteers with both legs bared from the knee to ankle (man landing catches). The catchers should be working in their home villages so that their exposure to risks of infection is less than that for an individual new to the area. Available chemo-prophylaxis or vaccination against local insect-borne diseases must be provided and detailed in the ethical clearance and volunteer consent forms for the trial. Where possible, tests should be carried out at sites where there is no disease transmission, although this may not always be practical, since disease-free sites may have low densities of the relevant vector species. The timing of the tests depends on whether the target mosquitoes are day or night biters. Untreated (control) human subjects are placed at least 10 m from those treated with repellent or seated close to a spatial repellent source, since 10 m is the limit of short-range attraction (Gillies and Wilkes, 1970). For tests of protective devices (such as coils), pretreatment data and complementary assessment in untreated houses are used for comparison.

Each item for test, as well as the blank control, is rotated between different catchers and houses to compensate for variation in their attractiveness for mosquitoes. Catchers should be questioned about perceived adverse or beneficial side effects of the repellents or devices. Appropriate criteria for synthetic repellents are at least 80% reduction in biting for 6 to 8 hours after application without perceived adverse side effects (WHO, 1996). However, the protection provided by plant-based repellents is considered good if the protection provided is around 60% after 4 hours.

23.6 SOURCES OF VARIATION AND BIAS

Repellent dose, application method, and exposure time must be standardised and always reported. The use of scented cosmetic products 12 hours prior to and during testing should be avoided, as some of these may be slightly repellent, such as Avon Skin-So-Soft (Schreck and McGovern, 1989). Smoking must not be allowed, and there is evidence that alcohol consumption increases individual attraction to mosquitoes (Shirai et al., 2002). Ideally, the repellents should be tested on an experienced team of collectors to reduce variation in individual ability to collect mosquitoes. Each repellent must be tested by several individuals over a number of occasions since the attractiveness to mosquitoes of specific individuals varies (Lindsay et al., 1993) due to variations in skin emanations (review in Braks et al., 1999). Rutledge and Gupta (1999) have analysed data from 19 repellent tests in the literature. They show that the minimum number of subjects needed to determine protection levels of repellents increases with longer test duration. A minimum of 11 individuals is required to evaluate a statistically significant mean value for repellent protection over an 8-hour period. Statistical analysis should also include each individual as a variable in both laboratory and field tests. Bias may also be introduced in field tests if the position of catchers is not rotated daily, as proximity to mosquito breeding or resting sites may vary. Numbers of mosquitoes may fluctuate daily, so statistical analysis must also account for this potential source of variation. It is because of these numerous sources of variation that field tests need more replication than laboratory tests and require factorial analysis of variance after log transformation of individual scores. Field sites should have large numbers of local mosquito species in order to ensure that the repellent is effective against both vector species and nuisance species, since nuisance biting is the major motivation for repellent use.

Other sources of variation in laboratory tests are caused by differences in mosquito avidity related to their physiological state. The team from the USDA-Mosquito and Fly Research Unit have published a series of papers showing that mosquito attack rates and, consequently, repellent protection time are significantly influenced by mosquito body size (hence, larval nutrition), the age and parity of the mosquitoes, and the time of day (Xue and Barnard, 1996; Xue et al., 1995; Barnard et al., 1998). Allowing the mosquitoes access to sugar solution or blood will also decrease their avidity and, subsequently, the repellency of a chemical, since they will be partially, if not fully, engorged (Xu and Barnard, 1996). Mosquitoes used for repellent testing should therefore be nulliparous, aged between 3 and 10 days, and denied access to sugar prior to testing repellents.

Mosquitoes will always feed on the easiest option. In laboratory tests, choice boxes where mosquitoes are offered protected and unprotected areas of skin, or experiments where a protected and a control arm are offered simultaneously, should be avoided. A free-choice test calculated the ED_{50} of deet as 0.024 to 0.042 mg/cm^2 (Rutledge et al., 1978). A similar test with no choice calculated it as 0.35 mg/cm^2 (Cockcroft et al., 1998). In field tests, repellent should be applied to both legs. If it is applied to one limb, and the other limb is used as a control, both the longevity and percentage protection afforded by the repellent will be misleadingly inflated. When the protection afforded by the repellent wanes, mosquitoes will start to feed through the repellent. However, if there is an unprotected alternative, they will be diverted and feed upon it. This also applies in field tests if individuals are less than 10 m apart, since this is the limit of short-range attraction (Gillies and Wilkes, 1970). If pairs of collectors are utilised — one as bait, testing the repellent, and the other collecting the mosquitoes — then mosquitoes will be diverted to the unprotected collector. A field test of this found that 28.13% of mosquitoes were diverted from individuals wearing 15% deet to the unprotected individuals in pairs sitting 1 m apart ($p = .049$) (Moore, in preparation).

23.6.1 PROBLEMS WITH PROTOCOLS 1 TO 3 AND ASTM E951-94

Protocol 1 is a free-choice test that Rutledge et al. (1985) argued simulates the natural conditions of repellent use more closely than a no-choice design, since the mosquitoes are free to seek an untreated part of the host or an alternative untreated host. This has the effect of lowering the point of reference for the ED_{90} (Rutledge et al., 1976), which means that the repellent protection period is overestimated. This has serious implications, since under actual-use conditions the failure of repellent may result in the transmission of disease. The (ASTM E951-94) test is a free-choice test and as a result should be avoided, even though it has the advantage of allowing rapid gathering of data.

The time-to-first-bite method (protocol 2) also has problems, since failure of the repellent is measured using one or two individual mosquitoes that are in the upper limit of repellent tolerance distribution. In a natural population many individuals will have a far lower tolerance. This measurement looks at the response of the upper quartile of a population, rather than the more representative mean.

The cotton-stocking model (protocol 3) has the advantage that it prevents contact between the skin and the test compound. Therefore, this technique may be used to test compounds that may not have been thoroughly evaluated toxicologically, and it also allows higher doses to be used than may be tolerable if they were in direct contact with the skin. However, this method does not correlate well with results from tests where repellent is applied directly to the skin (Rutledge et al., 1989), and further studies need to be performed when the substance is deemed safe for use on the skin after toxicological evaluation.

23.7 TOXICOLOGICAL EVALUATION

This is a highly technical, as well as a highly contentious, area. There is a common belief that remedies of natural origin are harmless and carry no risk to the consumer, whereas man-made chemicals carry an inherent risk and require stringent testing. Nothing could be farther from the truth, since many phytochemicals are potently pharmacologically active. The WHO (1998) has published guidelines on the appropriate use of herbal medicines that have the potential to be applied to the use of plant-based repellents. The report highlights "the need for mechanisms to ensure that these products are safe and effective, yet remain broadly accessible." This is vital, since plant-based repellents may provide the only economically viable form of personal protection for the poorest rural communities in areas such as Southeast Asia, the Amazon Basin, and southwestern Pacific, where there is a real risk of malaria infection from early-biting vector mosquitoes. The WHO (2000) has also published guidelines for developing methodologies for research and evaluation of traditional medicine. This highlights the need for research of a similar nature to be carried out in the area of traditional and plant-based insect repellents since the information on this topic in the literature is scant. Toxicological evaluation of plant products is complicated by the fact that they are usually mixtures of numerous compounds with a composition that may vary considerably, depending on origin, harvest, and storage conditions. Contamination with other natural substances may also occur; thus, toxicological evaluation of plant-based repellents requires strict control of the identity and quality of plant material tested.

It can be argued that the testing and development of a traditional repellent does not require toxicological evaluation since the method has been used by many people for many years. It may be possible then to test a traditional method of repelling mosquitoes without a prior toxicological assessment, provided permission is granted by an ethical board. This must be accompanied by a full explanation of the nature of the material and potential risks to the volunteers taking part in the test, and this information should be included on consent forms. Every effort must be made to evaluate potential toxicity prior to testing, through literature search or chemical evaluation, for instance, and the tests must be conducted over a short period of time.

If the traditional repellent is deemed effective, then toxicology tests should be performed to assess risks that could be associated with long-term exposure.

In developed countries, repellents are registered as pesticides and go through stringent testing prior to obtaining registration. Even oil of citronella, which had been registered in the U.S. since 1948, was reregistered in 1994 (U.S. EPA, 1994) using more stringent test criteria. In order to obtain registration, the use profile of the chemical is considered. This is important since it will dictate the mode by which toxic principles may come into contact with humans. Formulation, method, and rate of application are reviewed and are followed by relevant mammalian and environmental toxicity studies. For skin repellents, it is most important to obtain acute oral toxicity, as well as acute dermal absorption and toxicity. Other common tests are for eye and dermal irritation and dermal sensitisation. The final tests look at subchronic toxicity, immunotoxicity, teratogenicity, developmental toxicity, reproduction toxicity, and genotoxicity. Once these tests are completed, a no-observed-adverse-effect level (NOAEL) and acceptable daily intake (ADI) value are established, and if these are sufficiently low, then the product is deemed safe for general use, in conjunction with correct labeling, which may include warnings.

23.8 SUMMARY AND RECOMMENDATIONS

The most representative measure of the effectiveness of a repellent will be obtained through biological assay (bioassay) on human subjects using protocol 4 for skin repellents or protocol 5 for space repellents, followed by a field test. Both of these tests employ a no-choice methodology, precluding the potential danger associated with overestimation of repellent efficacy. It is useful to perform an initial laboratory study to establish the effective dose (ED_{90}) and duration of the repellent since conditions are standardised and there is no disease risk. A follow-up field study is needed to reflect the performance of the chemical under representative user conditions.

The introduction of an untreated arm prior to the treated limb(s) measures the avidity of the mosquitoes. Poor rearing conditions will reduce the willingness of mosquitoes to bite, and factors such as time of day will also have an effect on mosquito response to host cues. Checking mosquito avidity prior to testing a repellent removes such problems since unresponsive mosquitoes may be removed and replaced by those that are ready to bite. It is desirable for a minimum of 80% of test mosquitoes to be avid prior to testing a repellent. In addition, the mosquito density is low in protocol 4. This is beneficial because it allows more accurate counting of landing and probing mosquitoes. Additionally, at high mosquito density, the willingness of the insects to bite is reduced in cage experiments (Barnard et al., 1998). It is therefore useful to state the size of cages used in experiments and to limit the number of mosquitoes used in each test. Using a low number of mosquitoes (between 10 and 50) in a cage with dimensions of $40 \times 40 \times 40$ cm will ensure that the mosquitoes behave in a more natural way.

A field study reflects the true performance of the repellent under the conditions in which it is required to perform, and accounts for many factors that affect the longevity of a repellent. Temperature, humidity, and wind affect evaporation of the repellent (Gabel et al., 1976; Khan et al., 1972; Wood, 1968), and perspiration and abrasion will also reduce the longevity of the repellent (Rutledge et al., 1985; Rueda et al., 1998). Also, a field study will test a repellent against representative mosquito species, which is important due to the differential interspecific responsiveness to repellents.

Seyoum et al. (2002) have tested mosquito repellent plants in a semifield experimental setup. At the Mbita Point Research and Training Center in Western Kenya, part of the International Centre of Insect Physiology and Ecology, tests are carried out in two screen-walled greenhouses, in which laboratory-reared mosquitoes are released. These are then caught by two individuals: one in an experimental hut with a charcoal-burning stove (control), and the other in an experimental hut where the stove had plant material added (treatment). The number of mosquitoes caught on each individual was compared. This methodology allows more natural simulation of field-testing conditions in the local area, since the individuals were in experimental huts modeled on local housing and larger numbers of mosquitoes were used than is possible in no-choice room tests. The only disadvantage is that the collections took place with a distance of less than 10 m between collectors. Since 10 m is the limit of short-range attraction, the tests were free choice and may have overestimated the effectiveness of the repellent plants, although the authors suggest it as a method for screening repellents before field-testing the most promising candidates. This method has the massive advantage that the individuals involved in collecting the mosquitoes are not exposed to disease, and should be considered as a methodology for use by other research institutes based in areas of high vector-borne disease incidence.

ACKNOWLEDGMENTS

Many thanks to all the members of the RITAM group who helped shape this chapter with many thoughtful and useful comments and ideas: Dr. Bodeker, Dr. Canyon, Professor Curtis, Dr. Gbolade, Dr. Gichini, Dr. Hill, Professor Fernando Dias de Avila Pires, Dr. Rabarison, Dr. Seyoum, and Dr. Willcox.

REFERENCES

Baimai, V., Kijchalao, U., Sawadwongporn, P., and Green, C.A. (1988). Geographic distribution and biting behaviour of four species of the *Anopheles dirus* complex (Diptera: Culicidae) in Thailand. *Southeast Asian J. Trop. Med. Public Health*, 19, 151–161.

Barnard, D.R., Posey, K.H., Smith, D., and Schreck, C.E. (1998). Mosquito density, biting rate and cage size effects on repellent tests. *Med. Vet. Entomol.*, 12, 39–45.

Beebe, N.W., Bakote'e, B., Ellis, J.T., and Cooper, R.D. (2000). Differential ecology of *Anopheles punctatus* and three members of the *Anopheles farauti* complex of mosquitoes on Guadalcanal, Solomon Islands, identified by PCR-RPLP analysis. *Med. Vet. Entomol.*, 14, 308–312.

Braks, M.A.H., Anderson, R.A., and Knols, B.G.J. (1999). Infochemicals in mosquito host selection: human skin microflora and *Plasmodium* parasites. *Parasitol. Today*, 15, 409–413.

Browne, L.B. (1977). Host related responses and their suppression: some behavioural considerations. In *Chemical Control of Insect Behaviour*. John Wiley & Sons, New York, pp. 117–127.

Charlwood, J.D. and Graves, P.M. (1987). The effect of permethrin-impregnated bednets on a population of *Anopheles farauti* in coastal Papua New Guinea. *Med. Vet. Entomol.*, 1, 319–327.

Cockcroft, A., Cosgrove, J.B., and Wood, R.J. (1998). Comparative repellency of commercial formulations of deet, permethrin and citronellal against the mosquito *Aedes aegypti*, using a collagen membrane technique compared with human arm tests. *Med. Vet. Entomol.*, 12, 289–294.

Costantini, C., Birkett, M.A., Gibson, G., Ziesmann, J., Sagnon, N.F., Mohammed, H.A., Coluzzi, M., and Pickett, J.A. (2001). Electroantennogram and behavioural responses of the malaria vector *Anopheles gambiae* to human-specific sweat components. *Med. Vet. Entomol.*, 15, 259–266.

Cork, A. and Park, K.C. (1996). Identification of electrophysiologically-active compounds for the malaria mosquito, *Anopheles gambiae*, in human sweat extracts. *Med. Vet. Entomol.*, 10, 269–276.

Curtis, C.F., Lines, J.D., Ijumba, J., Callaghan, A., and Hill, N. (1987). The relative efficacy of repellents against mosquito vectors of disease. *Med. Vet. Entomol.*, 1, 109–119.

Davis, E.E. (1985). Insect repellents: concepts of their mode of action relative to potential sensory mechanisms in mosquitoes (Diptera: Culicidae). *J. Med. Entomol.*, 22, 237–243.

Fradin, M.S. and Day, J.F. (2002). Comparative efficacy of insect repellents against mosquito bites. *N. Engl. J. Med.*, 347, 13–18.

Gabel, M.L., Spencer, I.S., and Akers, W.A. (1976). Evaporation rates and protection times of mosquito repellents. *Mosq. News*, 36, 141–146.

Gambel, J.M., Brundage, J.F., Burge, R.J., DeFraites, R.F., Smoak, B.L., and Wirtz, R.A. (1998). Survey of U.S. Army soldiers' knowledge, attitudes, and practices regarding personal protection measures to prevent arthropod-related diseases and nuisance bites. *Mil. Med.*, 163, 695–701.

Gillies, M.T. and Wilkes, T.J. (1970). The range of attraction of single baits for some West African mosquitoes. *Bull. Entomol. Res.*, 60, 225–235.

Healy, T.P., Copland, M.J.W., Cork, A., Przyborowska, A., and Halket, J.M. (2002). Landing responses of *Anopheles gambiae* elicited by oxocarboxylic acids. *Med. Vet. Entomol.*, 16, 126–132.

Hii, J.L.K., Kanai, L., Foligela, A., Kan, S.K.P., Burkot, T.R., and Wirtz, R.A. (1993). Impact of permethrin-impregnated mosquito nets compared with DDT house-spraying against malaria transmission by *Anopheles farauti* and *An. punctulatus* in the Solomon Islands. *Med. Vet. Entomol.*, 7, 333–338.

Ismail, I.A.H., Pinichpongse, S., and Boonrasri, P. (1978). Responses of *Anopheles minimus* to DDT residual spraying in a cleared forest foothill setting in central Thailand. *Acta Trop.*, 35, 69–82.

Khan, A.A., Maibach, H.I., and Skidmore, D.L. (1972). A study of insect repellents. 2. Effect of temperature on protection time. *J. Econ. Entomol.*, 66(2), 437–438.

Lindsay, S.W., Adiamah, J.H., Miller, J.E., Pleass, R.J., and Armstrong, J.R.M. (1993). Variation in attractiveness of human subjects to malaria mosquitoes (Diptera: Culicidae) in the Gambia. *J. Med. Entomol.*, 30, 368–373.

Meek, S.R. (1995). Vector control in some countries of Southeast Asia: comparing the vectors and the strategies. *Ann. Trop. Med. Parasitol.*, 89, 135–147.

Nicolaides, N., Levan, N.E., and Fu, H.C. (1968). The skin surface lipids of man compared with those of eighteen species of mammals. *J. Invest. Dermatol.*, 51, 83–89.

Novak, M.G., Berry, W.J., and Rowley, W.A. (1991). Comparisons of four membranes for artificially blood-feeding mosquitoes. *J. Am. Mosq. Control Assoc.*, 7, 327–329.

Oo, T.T., Storch, V., and Becker, N. (2003). *Anopheles dirus* and its role in malaria transmission in Myanmar. *J. Vector Ecol.*, 28(2), 175–183.

Reisen, W.K. and Boreham, P.F.L. (1982). Estimates of malaria vectorial capacity for *Anopheles culicifacies* and *An. stephensi* in rural Punjab Province, Pakistan. *J. Med. Entomol.*, 19, 98–101.

Rowland, M., Downey, G., Rab, A., Freeman, T., Mohammad, N., Durrani, H.N., Reyburn, H., Curtis, C., Lines, J., and Fayaz, M. (2004). DEET mosquito repellent provides personal protection against malaria: a household randomised trial in an Afghan camp in Pakistan. *Trop. Med. Int. Health*, 9(3), 335–342.

Rozendaal, J.A. (1997). *Vector Control: Methods for Use by Individuals and Communities*. WHO, Geneva.

Rueda, L.M., Rutledge, L.C., and Gupta, R.K. (1998). Effect of skin abrasions on the efficacy of the repellent deet against *Aedes aegypti*. *J. Am. Mosq. Control Assoc.*, 14, 178–182.

Rutledge, L.C., Echano, N.M., and Gupta, R.K. (1999). Responses of male and female mosquitoes to repellents in the World Health Organisation insecticide irritability test system. *J. Am. Mosq. Control Assoc.*, 15, 60–64.

Rutledge, L.C. and Gupta, R.K. (1999). Variation in the protection periods of repellents on individual human subjects: an analytical review. *J. Am. Mosq. Control Assoc.*, 15, 348–355.

Rutledge, L.C., Gupta, R.K., and Elshenawy, K.B. (1989). Evaluation of the cotton fabric model for screening topical mosquito repellents. *J. Am. Mosq. Control Assoc.*, 5, 73–76.

Rutledge, L.C., Moussa, M.A., and Belletti, C.J. (1976). An *in vitro* blood feeding system for quantitative testing of mosquito repellents. *Mosq. News*, 36, 283–293.

Rutledge, L.C., Moussa, M.A., Lowe, C.A., and Sofield, R.A. (1978). Comparative sensitivity of mosquito species and strains to the repellent diethyl toluamide. *J. Med. Entomol.*, 14, 536–541.

Rutledge, L.C., Wirtz, R.A., Beuscher, M.D., and Mehr, Z.A. (1985). Mathematical models of the effectiveness and persistence of mosquito repellents. *J. Am. Mosq. Control Assoc.*, 1, 56–62.

Schreck, C.E. and McGovern, T.P. (1989). Repellents and other personal protection strategies against *Aedes albopictus*. *J. Am. Mosq. Control Assoc.*, 5, 247–250.

Seyoum, A., Pålsson, K., Kung'a, S., Kabiru, E.W., Lwande, W., Killeen, G.F., Hassanali, A., and Knols, B.G.J. (2002). Traditional use of mosquito-repellent plants in Western Kenya and their evaluation in semi-field experimental huts against *Anopheles gambiae*: ethnobotanical studies and application by thermal expulsion and direct burning. *Trans. Roy. Soc. Trop. Med. Hyg.*, 96, 225–231.

Shirai, O., Tsuda, T., Kitagawa, S., Naitoh, K., Seki, T., Kamimura, K., and Morohashi, M. (2002). Alcohol ingestion stimulates mosquito attraction. *J. Am. Mosq. Control Assoc.*, 18, 91–96.

Suthas, N., Phorn, S., Udom, C., and Cullen, J.R. (1986). The behaviour of *Anopheles minimus* Theobald (Diptera: Culicidae) subjected to differing levels of DDT selection pressure in northern Thailand. *Bull. Entomol. Res.*, 76, 303–312.

Tadei, W.P., Thatcher, B.D., Santos, J.M.M., Scarpassa, V.M., Rodrigues, I.B., and Rafael, M.S. (1998). Ecological observations on Anopheline vectors of malaria in the Brazilian Amazon. *Am. J. Trop. Med. Hyg.*, 59, 325–335.

U.S. EPA. (1994). Reregistration Eligibility Decision Oil of Citronella. Office of Prevention, Pesticides and Toxic Substances, U.S. Environmental Protection Agency, Washington, DC. Available at http://www.epa.gov/oppsrrd1/REDs/3105red.pdf.

U.S. EPA. (1999). Product Performance Test Guidelines OPPTS 810.3700 Insect Repellents for Human Skin and Outdoor Premises. Office of Prevention, Pesticides and Toxic Substances, U.S. Environmental Protection Agency, Washington, DC. Available at http://www.epa.gov/oppts-frs/OPPTS_Harmonized/810_Product_Performance_Test_Guidelines/Drafts/810-3700.pdf.

WHO. (1996). Report of the WHO Informal Consultation on the Evaluation and Testing of Insecticides, CTD/WHOPES/IC/96.1.

WHO. (1998). *Guidelines for the Appropriate Use of Herbal Medicines*, Western Pacific Series 23. WHO Regional Publications, WHO Regional Office for the Western Pacific, Manila.

WHO. (2000). General Guidelines for Methodologies on Research and Evaluation of Traditional Medicine, WHO/EDM/TRM/2000.1.

Wood, P.V. (1968). The effect of ambient humidity on the repellency of ethylhexanediol ('6-1 2') to *Aedes aegypti*. *Can. Entomol.*, 100, 1331–1334.

Xue, R.-D. and Barnard, D.R. (1996). Human host avidity in *Aedes albopictus*: influence of mosquito body size, age, parity and time of day. *J. Am. Mosq. Control Assoc.*, 12, 58–63.

Xue, R.-D., Barnard, D.R., and Schreck, C.E. (1995). Influence of body size and age of *Aedes albopictus* on human host attack rates and the repellency of deet. *J. Am. Mosq. Control Assoc.*, 11, 50–53.

Zhou, H.-N., Lu, Y.-R., Zhu, G.J., et al. (1998). [Studies on geographical distribution, ecology and habits, role in malaria transmission of *Anopheles dirus* in Yunnan.] *Chin. J. Vector Biol. Control*, 9, 455–459 (in Chinese with English abstract).

24 An Overview of Plants Used for Malaria Vector Control

Adebayo A. Gbolade

CONTENTS

24.1 INTRODUCTION

Malaria is one of the biggest health problems in Africa and other tropical regions. Five vectors of malaria have been recognised in Africa (Fontenille and Lochouarn, 1999). They are *Anopheles gambiae*, *Anopheles arabiensis*, *Anopheles funestus*, *Anopheles nili*, and *Anopheles moucheti*. The developmental stages of *Anopheles* species thrive in still, clean waters, and the adults have been effectively controlled by the use of insecticidal spraying of walls and ceilings of houses.

The use of mosquito nets and mosquito coils impregnated with synthetic insecticide to repel or kill mosquitoes is an established practice in controlling mosquitoes and malaria. Mosquito nets are a simple, low-cost malaria control method well suited to sub-Saharan Africa and other parts of the world endemic for malaria. It has been estimated that six lives are saved per year for every 1000 children with treated mosquito nets; so these could save over half a million African children per year from death due to malaria (Lengeler, 1998; Goodman and Mills, 1999). Treated nets protect not only those who sleep under them, but the rest of the local community. There is extensive evidence from Tanzania that when most people in a village have treated nets, so many mosquitoes are killed that the proportion of the vector population with malaria sporozoites is reduced (Curtis et al., 1998; Maxwell et al., 1999). Mosquito net projects have been extended to Malaysia (Hii et al., 1995), Vietnam (where 11 million people are protected by a government-organised net treatment program), and various parts of Africa, including Malawi (Rubardt et al., 1999), Cote d'-Ivoire (Kolaczinski et al., 2000), Kenya (Karanja et al., 1999), Burundi, Ghana, Gambia, and Cameroon (Goodman and Mills, 1999). In arguing for the effectiveness and acceptability of bed nets, Brinkmann and Brinkmann (1995) predicted a reasonable reduction from economic losses during malaria therapy.

Nevertheless, resistance of mosquitoes to pyrethroid insecticides may become an obstacle in the future. However, even in some areas with high levels of pyrethroid resistance, treated mosquito nets have been shown to be effective; this is apparently because resistant mosquitoes are less irritated by pyrethroids and rest longer on nets, eventually accumulating a lethal dose (Darriet et al., 2000).

In spite of various efforts made to control the malaria vector, reemergence in the Americas has occurred due to a World Health Organization (WHO) policy in the mid-20th century that deemphasised vector control with a consequent reduction in home-spraying rates (Butler and Roberts, 2000), and due to the evolution of insecticide-resistant mosquitoes. In Argentina, reinfestation of *Aedes aegypti* was registered in 1986 (de Sousa et al., 2000). Organised malaria control activities in Thailand have reduced malaria morbidity from 286 per 1000 population in 1947 to 1.5 per 1000 in 1996, but evidence of reemergence is apparent from an increased annual parasite index between 1997 and 1998 (Chareonviriyaphap et al., 2000). Furthermore, an upsurge of malaria in many parts of Africa is suggested to be caused by factors such as rapidly spreading resistance to antimalarial drugs and population movements (Nchinda, 1998). Furthermore, a shift in emphasis from prevention to emergency response (Zaim and Guillet, 2002) is also a major setback. However, progressive urbanisation will reduce numbers of *Anopheles* spp., but increase numbers of *Culex* spp. The use of vaccine for sustainable malaria control has been advocated, but major advances are yet to be recorded in this area.

The successful eradication of *Ae. aegypti* from a village in northern Vietnam since 1994 by treating breeding places with *Mesocyclops* spp. (a crustacean), coupled with community participation, has been reported (Nam et al., 1998). Biological control is an inexpensive method of mosquito control. In a Costa Rican study by Schaper (1999), *Mesocyclops thermocyclopoides* was shown to cause 100% *Ae. aegypti* larval reduction after 4 weeks and disappearance of adult mosquitoes after 7 weeks. *Ae. aegypti* nearly became extinct at the Liu-Chiu village in Taiwan following an integrated control involving introduction of larvivorous fish and larvicides into water containers between 1989 and 1996 (Wang et al., 2000). The Breteau index (a measure of the average distribution density of *Ae. aegypti* larvae) dropped from 53.9 (in 1989) to 1.2 (in 1996). Preliminary data from Louisiana show that *Mesocyclops* has been effective against *Anopheles* (Marten et al., 2000), although eradication on the same scale as in Vietnam has not yet been demonstrated.

Control of domestic pests becomes relevant in view of the devastating effects of insect-borne diseases of man such as malaria. In developing countries, the use of synthetic insecticides is becoming unpopular owing to their prohibitive costs and hazardous effects on aquatic life and the environment (although insecticides used indoors on walls or nets do not have these detrimental effects on the environment). The development of resistance by insects is another major setback to their use in some areas. Also, the cost of developing a new insecticide is enormous, over U.S. $50 million, and it takes about 7 to 10 years of research before it becomes available on the market (Zaim and Guillet, 2002). These factors corroborate the urgent need to develop insecticides for malaria vector control programs from alternative sources. In this regard, natural products are an important option. However, vector control remains the primary means of malaria control and it is targeted against larvae or adult mosquitoes. This review was carried out by using the Microsoft Internet search system (Google) on the title "use of plant extracts to control mosquitoes," and with reference to the PubMed database from 1973 to 2001.

24.2 MOSQUITOCIDAL PLANT EXTRACTS

Apart from repellents, traditional usage of plant larvicides in mosquito control is uncommon. Screening of such products has been restricted to the laboratories. Many plant extracts are known to be toxic to different species of mosquitoes and could be used to control the diseases they transmit. Plant-derived insecticides that were investigated for mosquito control can be categorised as ovicidal, larvicidal, antiovipositional, adulticidal, and repellent agents. In most cases, the selection of plants for study is based on published ethnomedicinal information or on previously reported insecticidal activity against mosquito vectors or other insects. Also, the survey method covering a specific geographical area through the use of questionnaires is used. Govere (2000) and Palsson and Jaenson (1999) adopted this method to identify repellent plants in some parts of South Africa and Guinea-Bissau, respectively. Random selection is another method of identifying mosquitocidal plants. Reviews on various mosquitocidal properties of plants have been published (Rageau and Belaveau, 1979; Sukumar et al., 1991; Gbolade, 2001).

24.3 LARVICIDAL EFFECTS

Larvicidal activity is the most widely investigated property of mosquitocidal plants. The third or fourth instar larvae are used. According to NAPRALERT citations from 1941 to 2001 (Quinn-Beattie, 2001, personal communication), 120 medicinal plants representing 102 genera in 52 families have been tested against *Anopheles*, *Culex*, and *Aedes* mosquitoes. Extracts from notable plants such as *Azadirachta indica*, *Gliricidia sepium*, *Hura crepitans*, *Mammea americana*, *Ocimum sanctum*, *Phytolacca dodecandra*, *Tagetes minuta*, and *Excoecaria agallocha* were studied. Eighty-seven percent of the 162 plant extracts screened for mosquitocidal activities were reported to be active as larvicides. Table 24.1 lists 21 species, which showed strong larvicidal activities. Plants that belonged to the Asteraceae, Rubiaceae, Ranunculaceae, Euphorbiaceae, Verbenaceae, and Liliaceae, in decreasing order, rank high among the families that were frequently screened (NAPRA-LERT, 2001).

In country-wide screening programs for mosquitocidal plants, a proportion of plants were found to be active as larvicides: 14.5% from Turkey (Bowers et al., 1995), 25% from Israel (Sathiyamoorthy et al., 1997), 34% from Sri Lanka (Ranaweera, 1996), and 33% from Bolivia (Vilaseca et al., 1997). Among the 16 extracts from 4 Moroccan plants investigated for larvicidal activity, 9 extracts (56%) were reported to be promising at lethal concentration 50% (LC_{50}) 28 to 325 ppm (Markouk et al., 2000). For comparison, bemephos (a synthetic organophosphate) is recommended for use at 1 ppm.

There have been several studies on South American plants. A total of 26.6% of 34 Brazilian plant extracts screened by Consoli et al. (1988) for larvicidal activity on *Ae. fluviatilis* were reported to be active at 100 ppm. In another study on 83 more Brazilian plants of the Asteraceae, *Tagetes minuta* and *Eclipta paniculata* were the only plants found to be significantly active against *Ae. fluviatilis* larvae (Macedo et al., 1997). The dichloromethane extracts of *Abuta grandifolia* and *Minthostachys setosa*, among other extracts from nine South American plants screened by Ciccia et al. (2000), were reported to have very promising toxicity to *Ae. aegypti*. These plants were twice as active as beta-asarone, a plant-derived insecticide.

Several studies have shown that simple plant extracts can be effective larvicides, without needing to isolate active compounds. Crude aqueous extracts from fresh leaves of *Lansium domesticum* and *Annona squamosa*, at a concentration range of 100 to 1.56 g%, were observed as the most active larvicides among the five Philippino plants investigated by Monzon et al. (1994). The ethanolic extract of Karanja seed, *Pongamia glabra vent*, was reported to produce 100% larval mortality in *Ae. aegypti* and *Culex quinquefasciatus* at 8 and 16 ppm, respectively (Sagar et al., 1999). The mosquitocidal property of myrrh (*Commiphora molmol*) has also been investigated. The oleoresin was found to be more active than the oil in producing larval mortality in *Culex pipiens* — the LC_{50} values were 9 and 17 ppm, respectively (Massoud et al., 2001). Larvicidal and repellent effects against laboratory-reared *Cx. quinquefasciatus* and *Ae. aegypti* were described for the hexane extract fraction from *Kaempferia galanga* (Choochote et al., 1999). Partially purified extracts from *Vitex negundo*, *Nerium oleander*, and *Syzygium jambolanum* were found to possess considerable larvicidal and insect growth regulatory activities when tested on *Cx. quinquefasciatus* and *Anopheles stephensi* — the LC_{50} ranged from 8.2 to 247 ppm (Pushpalatha and Muthukrishnan, 1995). In a study by El-Hag et al. (1999), the methanolic extracts of *Rhazya stricta* and *Azadirachta indica* were shown to be toxic to *Cx. pipiens* larvae and adults at 200 and 800 ppm, respectively. Extracts of *Syzygium*

TABLE 24.1
Some Promising Larvicidal Plants

Plant Species	Extract	Efficacy	Mosquito Species	References
Spilanthes mauritiana (Asteraceae)	Leaf, stem CHCL$_3$ methanolic	LC$_{100}$ = 0.1 ppm LC$_{100}$ = 50.0 ppm	*Aedes aegypti*	Jondiko (1986)
Abuta grandifolia (Menispermaceae)	Plant dichloromethane	LC$_{50}$ = 2.6 ppm	*Aedes aegypti*	Ciccia et al. (2000)
Eclipta paniculata (Asteraceae)	Plant ethanolic	LC$_{50}$ = 3.3 ppm	*Aedes fluviatilis*	Macedo et al. (1997)
Minthostachys setosa (Lamiaceae)	Plant dichloromethane	LC$_{50}$ = 9.2 ppm	*Aedes aegypti*	Ciccia et al. (2000)
Cymbopogon nardus (Poaceae)	Aerial part pet-ether	LC$_{50}$ < 10.0 ppm	*Culex quinquefasciatus*	Ranaweera (1996)
Curcuma longa (Zingiberaceae)	Rhizome pet-ether	LC$_{50}$ < 10.0 ppm	*Culex quinquefasciatus*	Ranaweera (1996)
Tagetes minuta (Asteraceae)	Entire plant dichloromethane Flower essential oil	LC$_{50}$ = 15.78 ppm LC$_{50}$ = 16.0 ppm	*Aedes aegypti* *Anopheles stephensi*	Wells et al. (1993)
Rhizophora apiculata (Rhizophoraceae)	Root acetone	LC$_{50}$ = 17.0 ppm	*Anopheles stephensi* *Culex quinquefasciatus*	Thangam and Kathiresan (1988)
Citrus species (Rutaceae)	Peel ethanolic	LC$_{50}$ = 18.5–26 ppm	*Culex pipiens*	al Dakhil and Morsy (1999)
Disepalum anomalum (Annonaceae)	Stem bark ethanolic	LC$_{50}$ = 19.6 ppm	*Aedes aegypti*	Ee et al. (1996)
Tagetes erecta cv inca yellow (Asteraceae)	Entire plant dichloromethane	LC$_{50}$ = 21.33 ppm	*Anopheles stephensi*	Perich et al. (1994)
Phytolacca dodecandra (Phytolaccaceae)	Fruit butanolic	LC$_{50}$ = 27.0 ppm	*Aedes aegypti*	Martson et al. (1993)
Kaempferia galanga (Zingiberaceae)	Rhizome hexane	LC$_{50}$ = 42.33 ppm	*Culex quinquefasciatus*	Choochote et al. (1999)
Avicennia marina (Verbenaceae)	Leaf acetone	LC$_{50}$ = 52.0 ppm	*Anopheles stephensi*	Thangam and Kathiresan (1988)
Annona bullata (Annonaceae)	Bark ethanolic	LC$_{50}$ = 100 ppm	*Aedes aegypti*	Hui et al. (1989)
Delonix regia (Fabaceae)	Leaf acetone	LC$_{100}$ = 450.0 ppm	*Culex quinquefasciatus*	Saxena and Yadav (1982)
Callistemon lanceolatus (Myrtaceae)	Aerial parts ethanolic	LC$_{50}$ = 490.0 ppm	*Culex quinquefasciatus*	Moshen et al. (1990)
Acorus calamus (Araceae)	Rhizome ethanolic	LC$_{50}$ = 727 ppm	*Culex quinquefasciatus* *Anopheles maculipennis*	Salma and Pratap (1999)
Gardenia gummifera (Rubiaceae)	Flower ethanolic	LC$_{50}$ = 3500 ppm	*Culex quinquefasciatus*	Salma and Pratap (1999)
Allium sativum (Liliaceae)	Bulb ethanolic	LC$_{50}$ = 14,600 ppm	*Culex quinquefasciatus*	Salma and Pratap (1999)

aromaticum were less toxic, but gave similar growth-retarding effects as *Rhazya stricta* at 200 to 600 ppm.

Among the three *Tagetes* species screened for larvicidal activity, *Tagetes minuta* flower essential oil was found to be the most active against *Ae. aegypti* at 10 ppm (Green et al., 1991). Interestingly, *Tagetes minuta* flower oil contains many different active compounds: ocimenone and at least four thiophenes (Green et al., 1991; Perich et al., 1995). The whole oil is more active than the ocimenone component alone and is stable over time, whereas ocimenone in aqueous solution loses its larvicidal activity over 24 hours (Green et al., 1991). Apart from toxicity to larvae, *Tagetes* has also been reported to exhibit interference with hatching of *Aedes* eggs and nematicidal potency at varying concentrations (Jain and Gupta, 1995).

A potentially cheap and widely available larvicide can be prepared from the waste residues obtained after separation of fiber from sisal (*Agave sisalana*) leaves (Pizarro et al., 1999). This costs virtually nothing to produce, as sisal is already widely cultivated, and there already exists many factories producing sisal fibers. After separation of the fibers, the residues are simply ground and filtered through a cotton cloth, and the water is evaporated at 40°C to produce the crude aqueous extract in powdered form. This dehydrated form of crude sisal extract is recommended for practical purposes because the liquid would be prone to fermentation. It was found that the LC_{50} was 322 ppm for *Ae. aegypti* and 183 ppm for *Cx. quinquefasciatus* third instar larvae. The authors concentrated on the latter species, which transmits filariasis, because it breeds in dirty water, and so toxicological evaluation of the sisal extract was not deemed necessary. No further work was done on *Ae. aegypti* because it breeds in clean drinking water, and toxicological tests on the sisal extract would be needed before its use could be recommended in such water. In *Cx. quinquefasciatus*, it was found that the crude extract was more effective than any particular fraction, and that a dose as low as 100 ppm was effective in producing 100% mortality at 96 hours.

24.4 OVICIDAL AND ANTIOVIPOSITIONAL EFFECTS

Oviposition-deterrent activity is the ability of certain chemicals to deter laying of eggs by insect pests (Hill and Waller, 1982). This activity has been reported for aqueous extracts of *Petiveria alliacea* root against adult females of *Cx. quinquefasicatus* and *Ae. aegypti* mosquitoes (Gbolade, 2001). The volatile oil also demonstrated toxicity to certain field pests, grasshoppers, and to the larva of *Acrae eponina*, a moth (Gbolade, 2001). *Imperata cylindrica* was also shown to exhibit 100% larvicidal and oviposition-deterrent effects against *Cx. quinquefasciatus* at 1500 and 1000 ppm, respectively (Moshen et al., 1995). A promising oviposition-deterrent *Melia volkensii* extract, active at 20 ppm, has been reported (Irungu and Mrangi, 1995).

An aqueous seed extract of the Iraqi plant *Atriplex canescens* was demonstrated to possess antioviposition properties and also to be toxic to eggs of *Cx. quinquefasciatus* at 1000 ppm, thereby reducing hatchability and consequently adult emergence (Ouda et al., 1998). In addition, strong aphicidal activity against *Aphis gossypii* and toxicity to cotton leafworm were described for a related species, *Atriplex halimus* (el-Gougary, 1998). Certain extracts of *Atriplex halimus* strongly synergised toxicity of the synthetic insecticides Reldan (chlorpyrifos methyl) and Actellic (pirimiphos methyl) in *Tribolium casteneum*. *Thymus capitatus* oil impairs hatchability of *Aedes* eggs, leading to sterility (Mansour et al., 2000), and this suggests a good potential in mosquito control programs. Aqueous extracts of garlic (*Allium sativum*) have a similar effect (Jarial, 2001). The molluscicidal action of garlic against *Lymnaea acuminata* attributed to allicin is also known (Singh and Singh, 1996). Plant molluscicides are generally known to have piscicidal properties (Martson et al., 1993). This is an important factor for consideration during mosquitocidal screening of plant extracts. Their concentrations can be controlled to produce mosquitocidal activity without having toxic effects on other forms of aquatic life. This has been demonstrated for *Phytolacca dodecandra* by Isharaza (1996).

24.5 ADULTICIDAL EFFECTS

This is the ability of plant products to produce partial or complete knockdown effects on insect pests in an enclosed environment. This action mimics that of mosquito coils, and it differs from repellency in which mosquitoes are simply warded off. Substitution of synthetic insecticides with extracts and powdered materials of promising adulticidal plants for impregnation into mosquito coils could have good potential for mosquito control at the domestic level.

Various essential oils and plant extracts have been shown to have adulticidal properties (Gbolade, 2001). A near complete knockdown effect of the ethanolic extracts of the Zimbabwean plants *Lippia javanica* and *Ocimum canum* on adults of *Ae. aegypti* has been reported (Lukwa et al., 1996). Both plants appear to be equipotent. On the other hand, *Ocimum canum* extract was reported to have twofold larvicidal potency of *Lippia javanica* (Lukwa et al., 1996). According to NAPRA-LERT (2001), strong adulticidal properties were described for many parts of *Mammea americana*, *Hura crepitans*, and *Gliricidia sepium* against *Ae. aegypti*. Extracts of unripe fruits of *Hura crepitans* were completely adulticidal, but the ripe fruits were not. Ripening of fruits can affect their properties. Generally, adulticial activities of the unripe fruits of these plants were greater than those of ripe fruits.

Differences in susceptibility of sexes of the various malaria vectors to adulticidal plants have been observed. For example, Perich et al. (1995) found that the active fraction of *Tagetes minuta* flower oil was more potent against male *Ae. aegypti* adults than against females (the LC_{50} was 0.015 and 0.29%, respectively). On the other hand, the essential oil of *Cedrus deodara* was equally toxic to both male and female *An. stephensi* (Singh et al., 1984).

24.6 INSECT GROWTH REGULATORY ACTIVITIES

Sublethal concentrations of plant extracts are known to induce some insect growth regulatory (IGR) activities. The effect of cypermethrin, a synthetic pyrethroid (Marquetti et al., 1994), and *Annona squamosa* (Saxena et al., 1993) on reproductive factors in resistant strains of *Cx. quinquefasciatus* and *An. stephensi*, respectively, was reported to include a significant reduction in fecundity and fertility. Histological examination of myrrh-treated mosquito larvae showed great pathological effects on their fat, muscles, gut, and nerve tissues (Massoud et al., 2001). *Culex* species were found to be more sensitive than *Aedes* species to the IGR effects of the ethanolic extract of Karanja seed in causing several structural abnormalities in pupae and adults (Sagar et al., 1999). Adults of *Cx. quinquefasciatus* that emerged from larvae treated with aqueous extract of de-oiled seeds of Karanja and neem were observed to be malformed and smaller in size (Sagar et al., 1999).

Juvenile hormone-mimicking activity against fourth instar larvae and adults of *Culex* was established for 5 plant extracts of the 15 extracts tested by Saxena et al. (1992). *Tagetes erectes*, *Tridax procumbens*, and *Ageratum conyzoides* were listed as some IGR-active plants. *Tagetes minuta* oil is another source of a juvenile hormone-mimicking substance (Saxena and Srivastava, 1973). *Cyperus iris* is a common weed in rice with widespread distribution in Asia and Africa. It contains large amounts of juvenile hormone III. Leaves from younger plants were shown to contain more bioactive hormones, and hence to elicit greater toxicity to *Ae. aegypti* larvae (Schwartz et al., 1998). Continuous exposure of *Ae. aegypti* larvae to extracts of neem seed kernel (Boschitz and Grunewald, 1994) and *Citrus reticulata* (Jayaprakasha et al., 1997) was found to induce molt inhibition in all tested larval instars. Susceptibility of the larvae to neem extract was found to decrease with age. Reduced ovipositing in female mosquitoes was also evident at higher concentrations of the extracts. In a comparative study using *Cx. quinquefasciatus* larvae, Ali et al. (1999) reported greater toxicity for some insect growth regulators (diflubenzuron and methoprene) than pyrethroids, microbial pesticides, and organophosphates.

24.7 MODES OF ACTION OF MOSQUITOCIDAL COMPOUNDS FROM PLANTS

Based on recent studies (Hasspieler et al., 1988; David et al., 2000), it is believed that mosquitocidal agents are concentrated in the midgut epithelium and lumen of Malpighian tubules of larvae. Plant extracts were suggested to elicit mosquitocidal activities through interference with protein metabolism in mosquito larvae (Massoud et al., 2001; von Dungern and Briegel, 2001). They principally affect the activities of the enzymes arginase and xanthine dehydrogenase. Mitchell et al. (1997) also demonstrated the toxicity of neem compounds — azadirachtin, salannin, nimbin, and 6-desacetylnimbin — by inhibition of the ecdysone-20-monooxygenase enzyme in *Ae. aegypti*, *Drosophila melanogaster*, and *Manduca sexta* larvae. This enzyme regulates the metabolism of the hormone ecdysone, which is responsible for insect molting.

24.8 BIOACTIVE COMPOUNDS

Some of the various compounds responsible for mosquitocidal activities have been reviewed by Martson et al. (1993) and Sukumar et al. (1991). These are summarised in Table 24.2.

Terpenoids such as ocimenone from the essential oil of *Tagetes minuta* (Green et al., 1991), thymol from *Thymus capitatus* oil (Mansour et al., 2000), methylchavicol from *Ocimum basilicum* oil (Bhatnagar et al., 1993), and six other compounds from hexane extracts of *Magnolia salicifolia* (Kelm et al., 1997) were identified. Thymol was less active than *Thymus capitatus* oil, but sublethal concentrations of the oil synergised the toxicity of malathion against larvae and adults of *Cx. pipiens* and produced antagonising effects on the tested malaria vector when mixed with other organophosphates. The terpenoids comprising costunolide, trans-anethole, methyleugenol, iso-methyleugenol, geranial, and neral, in decreasing order of activity, were present in hexane extracts of various parts of *Magnolia salicifolia*. In addition to larvicidal activity, costunolide and iso-methyleugenol also showed topoisomerase inhibitory activity. Ee et al. (2001) reported the isolation of 14 sesquiterpene larvicides from the stem bark of two Malaysian *Polyalthia* spp.

The **benzoquinones** (myrsinone and myrsinaquinone) from *Rapanes melanopholes* (Midiwo et al., 1995) and dioncophylline A, the naphthylisoquinoline alkaloid derived from *Triphyophyllum peltatum* (Francois et al., 1996), represent another group of secondary products with mosquitocidal activity. Analogues of dioncophylline A were shown to possess molluscicidal activity (Bringmann et al., 1998). Antifeedant activity on desert locust nymphs was also reported for the benzoquinones.

Pyridine derivatives — nicotine, nor-nicotine, and anabasine — are present in the old insecticidal plant *Nicotiana tabacum*. **Amides** may also have a strong larvicidal activity, such as piperine and piperidine, present in acetone extracts of *Piper nigrum* seeds (Desai et al., 1997). Piperine was observed to be a more active larvicide than piperidine. The activity of pipernonaline, a piperidine alkaloid isolated from *Piper longum* fruits, was found to be comparable to those of some organophosphorous insecticides (Lee, 2000). Oliver-Bever (1983) had attributed the strong larvicidal activity of the flower extract of *Spilanthes uliginosa* against *Anopheles* spp. mosquitoes to spilanthol, an unsaturated amide.

The **unsaturated lactone** quassin from *Quassia amara* was over five times as active as a larvicide than carbaryl, a synthetic carbamate (Evans and Raj, 1991).

The **hydrocarbon derivative** n-hexadecanoic acid isolated from an acetone extract of *Feronia limonia* leaf was described as a potent larvicide (Rahuman et al., 2000). Limonoids are known for their moult-inhibiting and feeding-deterrent activities in addition to their interference with hormone regulation (Mani et al., 1997). For example, the limonoids obacunone, nomilin, and limonin isolated from the seeds of *Citrus reticulata* were reported to inhibit emergence of fourth instar larvae of *Cx. quinquefasciatus* (Jayaprakasha et al., 1997), and obacunone was described as the most effective limonoid (effective concentration 50% (EC_{50}), 6.31 ppm). The diterpene kaurenoic acids from the Zimbabwean *Melantheria albinervia* also showed weak activity against *Bacillus subtilis* in addition to their larvicidal properties (Slimestad et al., 1995). From *Anacardium occidentale* was isolated the **alkenylphenol** anacardic acid, which was reported to be 10 times as active as the plant extract (Consoli et al., 1988).

TABLE 24.2
Some Bioactive Mosquitocidal Compounds

Bioactive Compounds	Plant Species	Mosquitocidal Activity	Efficacy	References
Pipernonaline (piperidine alkaloid)	*Piper longum*	Larvicidal	$LC_{50} = 0.21$ ppm	Lee (2000)
Butenolide 1 (lactone)	*Hortonia* spp.	Larvicidal	$LC_{50} = 0.41$ ppm	Ratnayake et al. (2001)
Butenolide 2	*Hortonia* spp.	Larvicidal	$LC_{50} = 0.47$ ppm	Ratnayake et al. (2001)
Piperidine derivative (alkaloid)	*Microcos paniculata*	Larvicidal	$LC_{50} = 2.1$ ppm	Bandara et al. (2000)
Embelin	*Rapanes melanopholes*	Larvicidal	$LC_{50} = 2.4$ ppm	Midiwo et al. (1995)
Myrsinone (benzoquinone)	*Rapanes melanopholes*	Larvicidal	$LC_{50} = 2.54$ ppm	Midiwo et al. (1995)
Myrsinaquinone	*Rapanes melanopholes*	Larvicidal	$LC_{50} = 2.69$ ppm	Midiwo et al. (1995)
Thiophenes	*Tagetes minuta*	Larvicidal Adulticidal	$LC_{50} = 3.9$ ppm $LC_{50} = 2000$ ppm	Perich et al. (1995)
Quassin (lactone)	*Quassia amara*	Larvicidal	Active at 6.0 ppm	Evans and Raj (1991)
Limonoids (e.g., obacunone)	*Citrus reticulata*	Molt inhibition	$EC_{50} = 6.31$ ppm	Jayaprakasha et al. (1997)
Azadirachtin	*Azadirachta indica*	Antifeedant	Active at 5–10 ppm	Mulla and Su (1999)
Anacardic acid (alkenylphenol)	*Anacardium occidentale*	Larvicidal	Active at 10 ppm	Consoli et al. (1988)
Ocimenone (a monoterpene)	*Tagetes minuta*	Larvicidal	$LC_{50} = 40$ ppm	Green et al. (1991)
ent-kaur-16-en-19-oic acid (diterpene acid)	*Melantheria albinervia*	Larvicidal	$LC_{100} = 62.5$ ppm	Slimestad et al. (1995)
n-Hexadecanoic acid	*Feronia limonia*	Larvicidal	$LC_{50} = 57.23$ ppm	Rahuman et al. (2000)
Monoterpenes (thymol, carvacrol)	*Thymus capitatus*	Larvicidal	$LC_{50} = 58.0$ ppm	Mansour et al. (2000)
Alkaloids	*Annona squamosa*	Larvicidal, chemosterilant	Active at 50–200 ppm	Saxena et al. (1993)
9(11),16-kauradien-19-oic acid (diterpene acid)	*Melantheria albinervia*	Larvicidal	$LC_{100} = 250$ ppm	Slimestad et al. (1995)
Volatile compounds (e.g., alkylbenzene derivatives)	*Delphinium* spp.	Antifeedant	$LC_{50} = 1000$ ppm	Miles et al. (2000)
Sesquiterpene lactones	*Polyalthia* spp.	Larvicidal	$LC_{50} = 72,000–132,700$ ppm	Ee et al. (2001)

24.9 DEVELOPING PLANT-BASED INSECTICIDES FOR MALARIA CONTROL

Numerous plants have been identified all over the world having various mosquitocidal properties that could be exploited in an integrated malaria control program. Some of these plants also demonstrate broad-spectrum insecticidal activities. Control of the disease at vector level remains a viable option, but there is a need to develop more effective insecticides that are less toxic to nontarget species. Mosquitoes may develop resistance to plant-derived insecticides, as they have done to synthetic insecticides, but the incidence could be reduced with proper application of active plant extracts or powdered materials. Indeed, whole-plant extracts often contain many active compounds, and this may help to prevent the development of resistance.

The yield of bioactive compounds from plant sources is generally low and production is expensive, which would limit their affordability and availability in poor countries, where malaria is most prevalent. For example, *Quassia amara* wood yielded 0.1 to 0.14% of quassin (Evans and Raj, 1991). Efforts to scale up yields of bioactive compounds will improve the potential of plant insecticides in malaria vector control. For this purpose, biotechnological production to meet demands has been suggested (George et al., 2000). With this development, the authors have recorded increased production of six mosquitocidal compounds — pyrethroids, azadirachtin, thiophenes, nicotine, rotenoids, and phytoecdysones — by hairy root cultures.

There are other disadvantages in isolating mosquitocidal compounds rather than using whole-plant extracts. The compounds may be volatile and difficult to store; for example, the toxicity and growth-inhibiting effects of *Melia volkensii* fruits are destroyed by heat during the drying process (Mwangi and Mukiama, 1988). Furthermore, the whole-plant extract may contain several active compounds and may be more active than its individual constituents; for example, *Tagetes minuta* oil is more active and more stable than the compound ocimenone, which it contains (Green et al., 1991).

In Brazil, *Lonchocarpus uruca* and *Quassia amara* as whole plants have been included in the list of insecticidal plants for research by the Pharmaceutical Technology Institute of the Oswaldo Cruz Foundation with a view to preparing products for use in public health programs (Gilbert et al., 1999). An adequate supply of promising plant insecticides can be ensured by encouraging large-scale cultivation and proper storage.

The toxicity of synthetic insecticides to man and other nontarget species is widely reported. However, such studies on mosquitocidal plant extracts or bioactive compounds are not frequently reported. In a study by Marles et al. (1995), the plant-derived larvicide alpha-terthienyl was shown to be nontoxic to rats following oral administration of the ready-to-use formulation. Selective toxicity of promising plant extracts during mosquitocidal screening programs would be desirable in order to not endanger humans, domestic animals, fish, and other nontarget species.

Community participation is essential for the effective control of mosquito vectors of malaria. Education of the rural community regarding the cause, consequences, and locally available methods of control of the disease should be emphasised. Proper storage of drinking water is important in order to discourage breeding of vectors.

For the exploitation of the potential of promising plant insecticides, there should be a shift from laboratory investigations to the field, with due consideration for safety to the nontarget species. Field operations may be expensive, but the cost could be reduced when coupled with environmental hygiene and community participation. At the community level, demand for plant raw materials could be met by encouraging cultivation. It is possible that some of these insecticidal plants could be included among those cultivated on a large scale in some South and East Asian countries such as China, Thailand, and India, where traditional medicine is widely embraced. Large-scale cultivation of the plant insecticide *Agave sisalana* in Tanzania (and many other tropical countries) solely for fiber and as a raw material for corticosteroid production is known. Therefore, the by-products from this industry could be easily used. Endod (*Phytolacca dodecandra*), a foremost molluscicide and plant larvicide, is also widely grown in Ethiopia on a pilot scale.

Based on reported mosquitocidal potencies, *Abuta grandifolia*, *Minthostachys setosa*, *Spilanthes mauritiana*, and *Eclipta paniculata*, which showed very promising activities (based on LC_{50} values; Table 24.1), and *Piper longum*, *Microcos paniculata*, *Citrus reticulata*, *Quassia amara*, *Tagetes minuta*, and *Hortonia* spp., for which very active insecticidal agents have been isolated (Table 24.2), could be considered for large-scale cultivation to meet the raw material need of an immediate malaria vector control program. It should be mentioned that none of the plants covered by this review are used traditionally as vector control agents other than as a repellent. Their choice for mosquitocidal studies has been based largely on reported repellent actions or toxicity to other insects.

In view of the potential of plant insecticides in malaria vector control, future research should be directed to toxicity studies on the nontarget species, including man and the aquatic organisms beneficial to man. In order to accelerate the acceptability of plant-based vector control agents, it will be necessary to develop formulations suitable for field applications and to promote pilot-scale cultivation of outstanding plants, particularly the endangered species, as a prelude to large-scale cultivation required to provide the raw material base.

ACKNOWLEDGMENTS

The author thanks Prof. Chris Curtis and Dr. Merlin Willcox for their comments on this chapter.

REFERENCES

al Dakhil, M.A. and Morsy, T.A. (1999). The larvicidal activity of peel oils of three *Citrus* fruits against *Culex pipiens. J. Egypt. Soc. Parasitol.*, 29, 347–352.

Ali, A., Chowdhury, M.A., Hossain, M.I., Mahmud-Ul-Ameen, Habiba, D.B., and Islam, A.F. (1999). Laboratory evaluation of selected larvicides and insect growth regulators against field-collected *Culex quinquefasciatus* larvae from urban Dhaka, Bangladesh. *J. Am. Mosq. Control Assoc.*, 15, 43–47.

Bandara, K.A., Kumar, V., Jacobsson, U., and Molleyros, L.P. (2000). Insecticidal and piperidine alkaloid from *Microcos paniculata* stem bark. *Phytochemistry*, 54, 29–32.

Bhatnagar, M., Kapur, K.K., Jalees, S., and Sharma, S.K. (1993). Laboratory evaluation of insecticidal properties of *Ocimum basilicum, O. sanctum* and plants essential oils and their major constituents against vector mosquito spp. *Entomol. Res.* (New Delhi), 17, 21–26.

Boschitz, C. and Grunewald, W. (1994). The effect of NeemAzal on *Aedes aegypti* (Diptera: Culicidae). *Appl. Parasitol.*, 35, 251–256.

Bowers, W.S., Sener, B., Evans, P.H., Bingol, F., and Erdogan, I. (1995). Activity of Turkish medicinal plants against mosquitoes *Aedes aegypti* and *Anopheles gambiae. Insect Sci. Appl.*, 16, 339–342.

Bringmann, G., Holenz, J., Assi, L.A., and Hostettmann, K. (1998). Molluscicidal activity (*Biomphalaria glabrata*) of dioncophylline A: structure-activity investigations. *Planta Med.*, 64, 485–486.

Brinkmann, U. and Brinkmann, A. (1995). Economic aspects of the use of impregnated mosquito nets for malaria control. *Bull. WHO*, 73, 651–658.

Butler, W.P. and Roberts, D.R. (2000). Malaria in the Americas: a model for reemergence. *Mil. Med.*, 165, 897–902.

Chareonviriyaphap, T., Bangs, M.J., and Ratanathan, S. (2000). Status of malaria in Thailand. *Southeast Asian J. Med. Public Health*, 31, 225–237.

Choochote, W., Kanjanapothi, D., Panthong, A., Taesotikul, T., Jitpakdi, A., Chaithong, U., and Pitasawat, B. (1999). Larvicidal, adulticidal and repellent effects of *Kaempferia galanga. Southeast Asian J. Trop. Med. Public Health*, 30, 470–476.

Ciccia, G., Coussio, J., and Mongelli, E. (2000). Insecticidal activity against *Aedes aegypti* larvae of some medicinal South American plants. *J. Ethnopharmacol.*, 72, 185–189.

Consoli, R.A., Mendes, N.M., Pereira, J.P., Santos, B.D., and Lamounier, M.A. (1988). Effect of several extracts derived from plants on the survival of *Aedes fluviatilis* (Lutz) (Diptera: Culicidae) in the laboratory. *Mem. Inst. Oswaldo Cruz*, 83, 87–93.

Curtis, C.F., Maxwell, C.A., Finch, R.J., and Njunwa, K.J. (1998). A comparison of use of a pyrethroid either for house spraying or for bednet treatment against malaria vectors. *Trop. Med. Int. Health*, 3, 619–631.

Darriet, F., N'guessan, R., Koffi, A.A., Konan, L., Doannio, J.M., Chandre, F., and Carnevale, P. (2000). Impact of pyrethrin resistance on the efficacity of impregnated mosquito nets in the prevention of malaria: results of tests in experimental cases with deltamethrin SC. *Bull. Soc. Pathol. Exot.*, 93, 131–134.

David, J.P., Rey, D., Pantos, M.P., and Meyran, J.C. (2000). Differential toxicity of leaf litter to dipteran larvae of mosquito developmental sites. *J. Invertebr. Pathol.*, 75, 9–18.

Desai, A.E., Ladhe, R.U., and Deoray, B.M. (1997). Larvicidal activity of acetone extract of *Piper nigrum* seeds to a major vector, *Anopheles subpictus* from Nasik region. *Geobios*, 24, 13–16.

de Sousa, G.B., Aviles, G., and Gardenal, C.N. (2000). Allozymic polymorphism in *Aedes aegypti* populations from Argentina. *J. Am. Mosq. Control Assoc.*, 16, 206–209.

Ee, G.C.L., Chuah, C.H., Sha, C.K., and Goh, S.H. (1996). Disepalin, a new acetogenin from *Disepalum anomalum* (Annonaceae). *Nat. Prod. Lett.*, 9, 141–151.

Ee, G.C.L., Seow, B.T., Neoh, B.K., Taufiq-Yap, Y.H., Sukari, M.A., and Rahmani, M. (2001). Larvicidal compounds from *Polyalthia jenkensii* and *Polyalthia sumatrana*. *J. Trop. Med. Plants*, 2, 27–30.

el-Gougary, O.A. (1998). Insecticidal and synergistic activity of *Atriplex halimus* L. extracts. *J. Egypt. Soc. Parasitol.*, 28, 191–196.

El-Hag, E.A., El Nadi, A.H., and Zaitoon, A.A. (1999). Toxic and growth retarding effects of three plant extracts on *Culex pipiens* larvae (Diptera: Culicidae). *Phytother. Res.*, 13, 388–392.

Evans, D.A. and Raj, R.K. (1991). Larvicidal efficacy of quassin against *Culex quinquefasciatus*. *Indian J. Med. Res.*, 93, 324–327.

Fontenille, D. and Lochouarn, L. (1999). The complexity of the malaria vectorial system in Africa. *Parassitologia*, 41, 267–271.

Francois, G., van Looman, M., Timperman, G., Chimanuka, B., Ake Assi, L., Holenz, J., and Bringmann, G. (1996). Larvicidal activity of the naphthylisoquinoline alkaloid dioncophylline A against the malaria vector, *Anopheles stephensi*. *J. Ethnopharmacol.*, 54, 125–130.

Gbolade, A.A. (2001). Plant-derived insecticides in the control of malaria vector. *J. Trop. Med. Plants*, 2, 91–97 (and references therein).

George, J., Bais, H.P., and Ravishankan, G.A. (2000). Biotechnological production of plant based insecticides. *Crit. Rev. Biotechnol.*, 20, 49–77.

Gilbert, B., Teixeira, D.F., Carvalho, E.S., De Pauk, A.E., Pereira, J.F., Ferreira, J.L., Almeida, M.B., Machado, R.D., and Casion, V. (1999). Activities of the Pharmaceutical Technology Institute of the Oswaldo Cruz Foundation with medicinal, insecticidal and insect repellent plants. *An. Acad. Bras. Cien.*, 71, 265–271.

Goodman, C.A. and Mills, A.J. (1999). The evidence base on the cost-effectiveness of malaria control measures in Africa. *Health Policy Plan*, 14, 301–312.

Govere, J., Durrheim, D.D., Baker, L., Hunt, R., and Cotetzee, M. (2000). Efficacy of three insect repellents against the malaria vector *Anopheles arabiensis*. *Med. Vet. Entomol.*, 14, 441–444.

Green, M.M., Singer, J.M., Sutherland, D.J., and Hibben, C.R. (1991). Larvicidal activity of *Tagetes minuta* (marigold) toward *Aedes aegypti*. *J. Am. Mosq. Control Assoc.*, 7, 282–286.

Hasspieler, B.M., Arnason, J.T., and Downe, A.E. (1988). Toxicity localisation and elimination of the photo toxin, alpha-terthienyl, in mosquito larvae. *J. Am. Mosq. Control Assoc.*, 4, 479–484.

Hii, J., Alexander, N., Chuan, C.K., Rahman, H.A., Safri, A., and Chan, M. (1995). Lambdacyhalothrin impregnated bed nets control malaria in Sabah, Malaysia. *Southeast J. Trop. Med. Public Health*, 26, 371–374.

Hill, D.S. and Waller, J.M. (1982). *Pests and Diseases of Tropical Crops*. Volume I. London, Longman.

Hui, Y.H., Ruprecht, J.K., Liu, Y.M., Anderson, J.E., Smith, D.L., Chang, C.J., and McLaughlin, J.L. (1989). Bullatacin and bullatacinone: two highly potent bioactive acetogenins from *Annona bullata*. *J. Nat. Prod.*, 52, 463–477.

Irungu, L.W. and Mrangi, R.W. (1995). Effects of a biologically active fraction of *Melia volkensii* on *Culex quinquefasciatus*. *Insect Sci. Appl.*, 16, 159–162.

Isharaza, W.K. (1996). *Phytolacca dodecandra*: a multi-purpose plant for control of vector-borne diseases. *J. Altern. Complement Med.*, 2, 421.

Jain, R.K. and Gupta, D.C. (1995). Effect of a few plant leaf extracts on hatchability of root-knot nematode (*Meloidogyne javanica*) eggs. *Indian J. Nematol.*, 25, 107–108.

Jarial, M.S. (2001). Toxic effects of garlic extracts of the eggs of *Aedes aegypti* (Diptera: Culicidae): a scanning electron microscope study. *J. Med. Entomol.*, 38, 446–450.

Jayaprakasha, G.K., Singh, R.P., Pereira, J., and Sakariah, K.K. (1997). Limonoids from *Citrus reticulata* and their moult inhibiting activity in mosquito, *Culex quinquefasciatus* larvae. *Phytochemistry*, 44, 843–846.

Jondiko, I.J.O. (1986). A mosquito larvicide in *Spilanthes mauritiana*. *Phytochemistry*, 25, 2289–2290.

Karanja, D.M., Alaii, J., Abok, K., Adungo, N.I., Githeko, A.K., Seroney, I., Vulule, J.M., Odada, P., and Oloo, J.A. (1999). Knowledge and attributes to malaria control and acceptability of permethrin impregnated sisal curtains. *East Afr. Med. J.*, 76, 42–46.

Kelm, M.A., Nair, M.G., and Schutzki, R.A. (1997). Mosquitocidal compounds from *Magnolia salicifolia*. *Int. J. Pharmacog.*, 35, 84–90.

Kolaczinski, J.A., Fanello, C., Herve, J.P., Conway, D.J., Carnevale, P., and Curtis, C.F. (2000). Experimental and molecular genetic analysis of the impact of pyrethroid and non-pyrethroid insecticide impregnated bed nets for mosquito control in an area of pyrethroid resistance. *Bull. Entomol. Res.*, 90, 125–132.

Lee, S.E. (2000). Mosquito larvicidal activity of pipernonaline, a piperidine alkaloid derived from pepper, *Piper longum*. *J. Am. Mosq. Control Assoc.*, 16, 245–247.

Lengeler, C. (1998). Insecticide treated bednets and curtains for malaria control: a Cochrane review. In *The Cochrane Library*, Issue 3. Update software, Oxford.

Lukwa, N., Masedza, C., Nyazema, N.Z., Curtis, C.F., and Mwaiko, G.L. (1996). Efficacy and duration of activity of *Lippia javanica* Spreng, *Ocimum canum* Sims and a commercial repellent against *Aedes aegypti*. In *Proceedings of the First International Organisation for Chemistry in Sciences and Development (IOCD) Symposium, Victoria Falls, Zimbabwe*, Hostettmann, K., Chinyanganya, F., Maillard, M., and Wolfende, J.L., Eds. University of Zimbabwe (Harare) Publications, Zimbabwe, pp. 321–325.

Macedo, M.E., Consoli, R.A., Granda, T.S., Dos Anjos, A.M., De Olivevia, A.B., Mendes, N.M., Queiroz, R.O., and Zani, C.L. (1997). Screening of Asteraceae (Compositae) plant extracts for larvicidal activity against *Aedes fluviatilis* (Diptera: Culicidae). *Mem. Inst. Oswaldo Cruz*, 92, 565–570.

Mani, C., Subrahmanyam, B., and Rao, P. (1997). Azadirachtin-induced changes in ecdysteroid titres of *Spodoptera litura* (Fabr.). *Curr. Sci.*, 71, 225–227.

Mansour, S.A., Messeha, S.S., and El-Gengaihi, S.E. (2000). Botanical biocides. 4. Mosquitocidal activity of certain *Thymus capitatus* constituents. *J. Nat. Toxins*, 9, 49–62.

Markouk, M., Bekkouche, K., Larhoini, M., Bousaid, M., Lazrak, H.H., and Jana, M. (2000). Evaluation of some Moroccan medicinal plant extracts for larvicidal activity. *J. Ethnopharmacol.*, 73, 293–297.

Marles, R., Durst, T., Kobaisy, M., Soucy-Breau, C., Abou-Zaid, M., Arnason, J.T., Kacew, S., Kanjanapothi, D., Rujjanawate, C., Meckes, M., et al. (1995). Pharmacokinetics, metabolism and toxicity of the plant-derived photo toxin alpha-terthienyl. *Pharmacol. Toxicol.*, 77, 164–168.

Marquetti, M.C., Martinez-Mole, M., Navarro, A., and Tang, R. (1994). Effect of cypermethrin of some reproductive factors in *Culex quinquefasciatus* Say 1823. *Rev. Cubana Med. Trop.*, 46, 28–31.

Marten, G.G., Nguyen, M., and Ngo, G. (2000). Copepod predation on *Anopheles quadrimaculatus* larvae in rice fields. *J. Vector Ecol.*, 25, 1–6.

Martson, A., Maillard, M., and Hostettmann, K. (1993). Search for antifungal, molluscicidal and larvicidal compounds from African medicinal plants. *J. Ethnopharmacol.*, 38, 215–223.

Massoud, A.M., Labib, I.M., and Rady, M. (2001). Biochemical changes of *Culex pipiens* larvae treated with oil and oleoresin extracts of myrrh, *Commiphora molmol*. *J. Egypt. Soc. Parasitol.*, 31, 517–529.

Maxwell, C.A., Myamba, J., Njunwa, K.J., Greenwood, B.M., and Curtis, C.F. (1999). Comparison of bednets impregnated with different pyrethroids for their impact on mosquitoes and on re-infection with malaria after clearance of pre-existing infections with chlorproguanil-dapsone. *Trans. R. Soc. Trop. Med. Hyg.*, 93, 4–11.

Midiwo, J.O., Mwangi, R.W., and Ghebremeskel, Y. (1995). Insect antifeedant growth inhibiting and larvicidal compounds from Melanophores (Myrsinaceae). *Insect Sci. Appl.*, 16, 163–166.

Miles, J.E., Ramsewak, R.S., and Nair, M.G. (2000). Antifeedant and mosquitocidal compounds from *Delphinium* × *cultorum* cv Magic fountain flowers. *J. Agric. Food Chem.*, 48, 503–506.

Mitchell, M.J., Smith, S.L., Johnson, S., and Morgan, E.D. (1997). Effects of the neem tree compounds, azadirachtin, salannin, nimbin and 6-desacetylnimbin on ecdysone-20-monooxygenase activity. *Arch. Insect Biochem. Physiol.*, 35, 199–209.

Monzon, R.B., Alvior, J.P., Luczon, L.L., Morales, A.S., and Mutuc, F.E. (1994). Larvicidal potential of five Philippine plants against *Aedes aegypti* (Linnaeus) and *Culex quinquefasciatus* (Say). *Southeast Asian J. Trop. Med. Public Health*, 25, 755–759.

Moshen, Z.H., Jawad, A.M., Al-Chalabi, B.M., and Al-Naib, A. (1990). Biological activity of *Callistemon lancoelatus* against *Culex quinquefasciatus*. *Fitoterapia*, 61, 270–274.

Moshen, Z.H., Jawad, A.M., Al-Saadi, M., and Al-Naib, A. (1995). Antioviposition and insecticidal activity of *Imperata cylindrica* (Gramineae). *Med. Vet. Entomol.*, 9, 441–442.

Mulla, M.S. and Su, T. (1999). Activity and biological effects of neem products against arthropods of medical and veterinary importance. *J. Am. Mosq. Control Assoc.*, 15, 133–152.

Mwangi, R.W. and Mukiama, T.K. (1988). Evaluation of *Melia volkensii* extract fractions as mosquito larvicides. *J. Am. Mosq. Control Assoc.*, 4, 442–447.

Nam, S.V., Yen, N.T., Kay, B.H., Marten, G.G., and Reid, J.W. (1998). Eradication of *Aedes aegypti* from a village in Vietnam, using copepods and community participation. *Am. J. Trop. Med. Hyg.*, 59, 657–660.

NAPRALERT. (2001). *Programme for Collaborative Research in the Pharmaceutical Sciences.* College of Illinois at Chicago, Chicago.

Nchinda, T.C. (1998). Malaria: a reemerging disease in Africa. *Emerg. Infect. Dis.*, 4, 398–403.

Oliver-Bever, B. (1983). Medicinal plants of tropical West Africa. III. Anti-infection therapy with higher plants. *J. Ethnopharmacol.*, 7, 1–93.

Ouda, N.A., Al-Chalabi, B.M., Al-Charchafchi, F.M.R., and Moshen, Z.H. (1998). Insecticidal and ovicidal effects of the seed extract of *Atriplex canescens* against *Culex quinquefasciatus. Pharm. Biol.*, 36, 69–71.

Palsson, K. and Jaenson, T.G. (1999). Plant products used as mosquito repellents in Guinea Bissau, West Africa. *Acta Trop.*, 72, 39–52.

Perich, M.J., Wells, C., Borsch, W.G., and Tramway, K.E. (1994). Toxicity of extracts from three *Tagetes* against adults and larvae of yellow fever mosquito and *Anopheles stephensi* (Diptera: Culicidae). *J. Med. Entomol.*, 31, 833–837.

Perich, M.J., Wells, C., Bertsch, W., and Tredway, K.E. (1995). Isolation of the insecticidal components of *Tagetes minuta* (Compositae) against mosquito larvae and adults. *J. Am. Mosq. Control Assoc.*, 11, 307–310

Pizarro, A.P., Oliveira Filho, A.M., Parente, J.P., Melo, M.T., dos Santos, C.E., and Lima, P.R. (1999). Utilisation of the waste of sisal industry in the control of mosquito larvae. *Rev. Soc. Bras. Med. Trop.*, 32, 23–29.

Pushpalatha, E. and Muthukrishnan, J. (1995). Larvicidal activity of a few plant extracts against *Culex quinquefasciatus* and *Anopheles stephensi. Indian J. Malariol.*, 32, 14–23.

Rageau, J. and Belaveau, P. (1979). Toxic effects of plant extracts on mosquito larvae. *Bull. Soc. Pathol. Exot. Filiales*, 79, 168–171.

Rahuman, A.A., Gopalakrishnan, G., Ghouse, B.S., Arumugam, S., and Himalayan, B. (2000). Effect of *Feronia limonia* on mosquito larvae. *Fitoterapia*, 71, 553–555.

Ranaweera, S.S. (1996). Mosquito larvicidal activity of some Sri Lanka plants. *J. Natl. Sci. Counc.* (Sri Lanka), 24, 63–69.

Ratnayake, R., Karunaratne, V., Ratnayake Bandara, B.M., Kumar, V., MacLeod, J.K., and Simmonds, P. (2001). Two new lactones with mosquito larvicidal activity from three *Hortonia* species. *J. Nat. Prod.*, 64, 376–378.

Rubardt, M., Chikoko, A., Glik, D., Jere, S., Nwanyanwu, O., Zheng, W., Nkhoma, W., and Ziba, C. (1999). Implementing a malaria curtains project in rural Malawi. *Health Policy Plan*, 14, 313–321.

Sagar, S.K., Sehgal, S.S., and Agarwala, S.P. (1999). Bioactivity of ethanol extract of Karanja (*Pongamia glabra vent*) seed coat against mosquitoes. *J. Commun. Dis.*, 31, 107–117.

Salma, K. and Pratap, S. (1999). Mosquito larvicidal activity of selected Indian medicinal plants against *Culex quinquefasciatus. J. Pharm. Pharmacol.*, 51, 250.

Sathiyamoorthy, P. Lugasi-Eugi, H., Van-Damme, P., Abu-Rabia, A., Gopas, J., and Golan-Goldhirsh, A. (1997). Larvicidal activity in desert plants of the Negev and Nedouin marker plant products. *Int. J. Pharmacog.*, 35, 265–273.

Saxena, R.C., Dixit, O.P., and Sukumaran, P. (1992). Laboratory assessment of indigenous plant oils for antijuvenile hormone activity in *Culex quinquefasciatus. Indian J. Med. Res.*, 95, 204–206.

Saxena, R.C., Harshan, V., Saxena, A., Sukumaran, P., Sharma, M.C., and Kumar, M.L. (1993). Larvicidal and chemosterilant activity of *Annona squamosa* alkaloids against *Anopheles stephensi. J. Am. Mosq. Control Assoc.*, 9, 84–87.

Saxena, B.P. and Srivastava, J.B. (1973). *Tagetes minuta* L. oil: a new source of juvenile hormone mimicking substance. *Indian J. Exp. Biol.*, 11, 56–58.

Saxena, S.C. and Yadav, R.S. (1982). New mosquito larvicides from indigenous plants *Delonix regia* and *Oligochaeta ramose. Adv. Biosci.*, 1, 92–94.

Schaper, S. (1999). Evaluation of Costa Rica copepods (Crustacea: Eudecapoda) for larval *Aedes aegypti* control with special reference to *Mesocyclops thermocyclopoides*. *J. Am. Mosq. Control Assoc.*, 15, 510–519.

Schwartz, A.M., Paskewitz, S.M., Orth, A.P., Tesch, M.J., Toong, Y.C., and Goodman, W.G. (1998). The lethal effect of *Cyperus iris* on *Aedes aegypti*. *J. Am. Mosq. Control Assoc.*, 14, 78–82.

Singh, D., Rao, S.M., and Tripathi, A.K. (1984). Cedarwood oil as a potential insecticidal agent against mosquitoes. *Naturwissenschaftliche*, 71, 265–266.

Singh, V.K. and Singh, D.K. (1996). Enzyme inhibition by allicin, the molluscicidal agent of *Allium sativum* L. (garlic). *Phytother. Res.*, 10, 383–386.

Slimestad, R., Marston, A., Mavi, S., and Hostettmann, K. (1995). Larvicidal constituents of *Melantheria albinervia*. *Planta Med.*, 61, 562–563.

Sukumar, K., Perich, M.J., and Boobar, L.R. (1991). Botanical derivatives in mosquito control: a review. *J. Am Mosq. Control Assoc.*, 7, 210–237.

Thangam, T.S. and Kathiresan, K. (1988). Toxic effects of mangrove plant extracts on mosquito larvae, *Anopheles stephensi* L. *Curr. Sci.*, 57, 914–915.

Vilaseca, L.A., Laurent, D., Chantraine, J.M., Ballivian, C., Saavedra, G., and Ibanez, R. (1997). Insecticidal activity of essential oils from Bolivian plants. In *International Joint Symposium of Chemistry, Biology and Pharmacological Properties of Medicinal Plants from the Americas, Panama, Republic of Panama*, p. A-28.

von Dungern, P. and Briegel, H. (2001). Protein catabolism in mosquitoes: ureotely and uricotely in larval and imaginal *Aedes aegypti*. *J. Insect Physiol.*, 47, 131–141.

Wang, C.H., Chang, N.T., Wu, H.H., and Ho, C.M. (2000). Integrated control of the dengue vector *Aedes aegypti* in Liu-Chiu village, Ping-Tung County, Taiwan. *J. Am. Mosq. Control Assoc.*, 16, 93–99.

Wells, C., Bertsch, W., and Perich, M. (1993). Insecticidal volatiles from the marigold plant (genus *Tagetes*). Effect of species and sample manipulation. *Chromatographia*, 35, 209–215.

Zaim, M. and Guillet, P. (2002). Alternative insecticides: an urgent need. *Trends Parasitol.*, 18, 161–163.

25 Guidelines for Studies on Plant-Based Vector Control Agents

Adebayo A. Gbolade, Nzira Lukwa, Dibungi T. Kalenda, Christophe Boëte, and Bart G.J. Knols

CONTENTS

25.1 INTRODUCTION

There are over 2500 different species of mosquitoes throughout the world (Gilles and Warrell, 1993), most of which do not affect the well-being of humans, apart perhaps from causing nuisance. Some, however, are capable of transmitting diseases afflicting animals, for example, equine encephalitis or dog heartworm. Other species transmit diseases that cause illness in humans, the most important of which are malaria, filarial diseases, and arboviruses such as dengue, encephalitis, and yellow fever.

Soon after Ross's discovery that anopheline mosquitoes transmit malaria, it became apparent that vector control could play a role in curbing disease transmission (Ross, 1911). Ross advocated the use of antilarval measures that target mosquito stages before they carry parasites by saying, "The most vulnerable point in the history of gnats is when they are larvae; they can be destroyed wholesale" (Ross et al., 1900). Application of this concept through larviciding or environmental management, combined with the advent of highly effective adulticides like dicholorodiphenyl-trichloroethane (DDT), led to a rapid decline and eventual eradication of malaria from the U.S., Europe, the USSR, Taiwan, and parts of the Caribbean, Venezuela, China, Madagascar, and South Africa (Spielman and D'Antonio, 2001; Curtis, 2002). These successes were based on an integrated approach that also included improved house design (notably screening of porches and windows; see Lindsay et al., 2002) and active case detection through improved health systems, and have yielded complete disappearance or sustained reductions in malarial disease in these parts of the world (Greenwood and Mutabingwa, 2002).

Considering the efficacy of adulticides like DDT (Curtis, 2002) and, more recently, the use of residual insecticides on bed nets (Lengeler, 1998), a gradual shift from larval to adult control has emerged. Substantiated by epidemiological models (MacDonald, 1957) and predicted effects of adulticides on mosquito longevity, a key component of vectorial capacity (Garrett-Jones, 1964), such a shift seems justified. However, a recent model has shown that active avoidance of contact with insecticides (i.e., excito-repellency) by adult vectors may seriously compromise the efficacy of such methods, and it is argued that low mobility and absence of behavioural responsiveness in immature mosquitoes enables high effective coverage of interventions staged against larvae (Killeen et al., 2002a). Moreover, long-forgotten campaigns against larval stages of vectors, notably in Brazil (Soper and Wilson, 1943; Killeen et al., 2002b), Egypt (Shousha, 1948), and Zambia (Utzinger et al., 2001), have unequivocally shown that a focus on control of immature stages of vectors can yield eradication or dramatic reduction in malarial disease.

Although malaria continues to be an important health problem in parts of Asia and South America, it is Africa that bears the highest burden of disease, with 90% of all annual malarial mortality (1 to 2 million deaths) occurring on that continent (WHO, 2000; Breman et al., 2001). Major parts of Africa experience highly stable and holoendemic transmission of disease, vectored by easily infected, long-lived, and anthropophilic anophelines, which are considered the most efficient vectors in the world (White, 1974; Beier et al., 1999). This is clearly demonstrated through the measure of frequency with which an individual receives infectious mosquito bites, expressed as the entomological inoculation rate (EIR). The EIR rarely exceeds 5 per annum in Asia or South America, whereas EIRs of over 1000 have been measured in several parts of sub-Saharan Africa (Beier et al., 1999; Smith et al., 2001; Greenwood and Mutabingwa, 2002).

Now in its fifth year, the World Health Organization (WHO) Roll Back Malaria campaign (Nabarro and Tayler, 1998) underpins the importance of vector control, notably the wide-scale use of insecticide-treated bed nets and indoor residual spraying. Application of either method has shown dramatic reductions in childhood morbidity and mortality due to malaria (Lengeler, 1998; Curtis, 2002), and they are therefore expected to remain on the forefront of malaria control over the years to come. However, a variety of obstacles hinder more wide-scale application, particularly pyrethroid resistance in West (Chandre et al., 1999) and southern Africa (Hargreaves et al., 2000), necessitating the search for new insecticides (Zaim and Guillet, 2002). Although significant increases in community uptake of bed net technology have been accomplished through social marketing (Armstrong-Schellenberg et al., 2001), the cost of nets continues to be an obstacle, making some argue that these should be provided free of charge (Maxwell et al., 2002) or that appropriate financing mechanisms should be developed (Guyatt et al., 2002).

The hindrances associated with wider application of existing vector control tools, notably insecticide resistance, affordability, and other socioeconomic barriers (like cultural acceptance), underscore the need for additional innovative vector control tools that may augment or be used in conjunction with this limited arsenal currently available. Such methods should be target specific, efficacious, safe, and appropriate for use at the community level, be widely available at no or limited cost, and withstand the risk of vectors developing resistance against them.

Plant products appear to fulfill these prerequisites, and although many crude or refined products have been found to have antimosquito properties (see Chapter 24), none has moved to the forefront in malaria vector control (Curtis, 1991; Roozendaal, 1997). This can be attributed to several reasons, of which the absence of dissemination of findings to target groups is perhaps the most important. It is hoped that publication of this volume will initiate efforts to incorporate plant-based control tools in modern malaria vector control at the community level.

A systematic study of plant-derived vector control agents is therefore needed. One strategy is to search for active crude material, extracts, and active compounds therein against aquatic or terrestrial stages of target mosquito species. The present guidelines aim to give an impetus to this search, to list standard operating procedures, and to draw the attention of different stakeholders to the possibilities offered by plant-derived vector control agents.

25.2 STAGE-SPECIFIC ACTIVITY CLASSES OF PLANT-DERIVED VECTOR CONTROL AGENTS

Four separate and distinct stages characterise the mosquito life cycle: egg, larva (L1 to L4), pupa, and adult. For each of these stages plant-derived agents are known that are directly toxic, exert insect growth regulatory effects, or act as repellents. The different activity classes can be summarised as follows:

1. *Egg stage activity*: Plant products that are directly toxic to the developing embryo in eggs (ovicidal effects), reduce hatchability of eggs, cause abnormal development of emerging larvae, or affect adult fitness negatively
2. *Larval stage activity*: Plant products that are directly toxic to larval stages (larvicidal effects), inhibit or reduce normal growth rates, or affect adult fitness negatively
3. *Adult stage activity*: Plant products that are directly toxic to adult mosquitoes (adulticidal effects), cause sterility or reduce longevity, and reproductive fitness of adults
4. *Oviposition-deterrent activity*: Plant products that negatively affect the behavioural responsiveness of gravid females to oviposition substrates and result in reduced egg laying on otherwise suitable and naturally selected substrates
5. *Insect growth regulatory activity*: Plant products that prevent molting or shedding of larval integuments or otherwise inhibit normal preimaginal development of larvae and pupae
6. *Repellent activity*: Plant products that negatively affect the behavioural response of host-seeking female mosquitoes and result in reduced human–vector contact (see Chapters 22 and 23).

A variety of approaches in the selection of candidate plants with potential as vector control agents exist and range from ethnobotanical studies (see Chapter 14; Pålsson and Jaenson, 1999a; Seyoum et al., 2002b) to chemotaxonomic studies in which closely related species of plants with known activity are selected and subjected to experimentation. This may even include the evaluation of closely related compounds, or derivatives thereof, of chemicals with known activity such as the repellent p-menthane-3,8-diol (Barasa et al., 2002). However, the origin of resource material varies widely across the available literature, as do the various selection criteria applied by researchers and the respective communities they work with.

The following sections describe an overview of methods for evaluating the efficacy of plant-derived products for the above-mentioned activity classes. Although not exhaustive, we have attempted to present the most common methods used and provide examples and literature references that will facilitate the provision of further details. We used the PubMed search engine and collections of published and gray literature available to us to collect and compile information presented in the remainder of this chapter.

25.3 METHODS FOR STUDYING PLANT-DERIVED AGENTS FOR MOSQUITO CONTROL

A huge number of studies have evaluated the potency of plant material for various activity classes against the different developmental stages of mosquitoes. The majority of these have focused on solvent extracts of plant material against larval stages, primarily of the yellow fever mosquito *Aedes aegypti*, or against adult mosquitoes as repellents that reduce man–vector contact. Relatively few studies have focused exclusively on anophelines. Although the precise experimental procedures differ substantially, depending on the biology of the target species and plant material used, there is an urgent need to standardise operating procedures in order to validate comparative evaluations across species and different geographical settings. Without excluding the validity of identification of active principles in plant products, refinement thereof, and eventual commercialisation, it seems that knowledge of crude material that can be applied without further processing has the highest potential for broad-scale application by communities in malaria-endemic regions.

Standardised evaluation of plant material for the above-mentioned activity classes depends on four components: (1) the rearing procedures of the target mosquito species, (2) the nature of and chemical procedures applied to the plant material, (3) the experimental or bioassay procedures applied, and (4) the choice of analytical method to interpret the results.

25.3.1 REARING OF MOSQUITOES

Species belonging to the three major genera can all be reared and maintained under standardised laboratory conditions, though colonisation of some malaria vectors, notably *Anopheles funestus*, has proven difficult (M. Coetzee, personal communication). In general, anophelines have proven more difficult to colonise than *Aedes* or *Culex* species, which explains why the majority of studies have focused on these latter species. As this volume primarily focuses on malaria vectors, we restrict ourselves here to describing salient features of anopheles rearing that will affect the outcome of experiments with plant-derived products. General rearing procedures for mosquitoes, including anophelines, have been published (Foster, 1980; Benedict, 1997; Gerberg et al., 1994; see also http://www2.ncid.cdc.gov/vector/cultprinciples.html); the reader is referred to these accounts for extensive details of procedures, a shortened version of which is shown below.

25.3.1.1 Eggs/Larvae

Eggs of anophelines cannot survive desiccation for more than a few days, requiring continuous rearing in order to maintain a colony of insects. Eggs can be collected on wet filter paper or cotton wool placed in cages with gravid females, as these will readily oviposit on wet surfaces when deprived of normal substrates to lay eggs on. Immediate transfer to a water surface is recommended to increase hatch rates. The choice of water has often been debated but does not seem to be the prime cause for failure to colonise anophelines. Deionised water is used most frequently, though adaptation to regular tap water over the course of a few generations is normally successful, even if this water is mildly chlorinated. A variety of rearing pans can be used, though plastic ones are used most commonly (Figure 25.1). Water temperature of $27 \pm 1°C$ is considered optimal. A critical factor in anopheline maintenance is the density of larvae and food availability as intraspecific competition and larval nutritional stress are critical factors affecting the size and nutritional reserves of emerging adults (Lyimo et al., 1992). Compromised food availability, high densities, and high rearing temperatures lead to small adults, which in turn suffer from reduced longevity and fecundity (Lyimo and Takken, 1993). A variety of larval food sources have been used, including mouse cubes, dog biscuits, liver powder, fish food (like Tetramin-Baby®), or baker's yeast. Whichever the choice, it is critical not to add too much food, as this may lead to excessive bacterial or algal growth, which is often detrimental to mosquito development.

FIGURE 25.1 Mosquito-rearing unit in semifield conditions in western Kenya. (Photo: Peter Luethi.)

25.3.1.2 Adult Maintenance and Feeding

Adult anophelines are normally kept in mosquito gauze-covered cages of $30 \times 30 \times 30$ cm. Metal framing is preferred, as wood often favors fungal growth under humid conditions. Netting that can easily be removed and washed regularly is advantageous. The ability of adults to mate in such confined spaces (stenogamy) is often considered a major bottleneck in the establishment of a colony from field-collected specimens. Initial maintenance in much larger cages may be necessary and a gradual decrease in cage size over several generations is required to enable adaptation. Cages should be kept under no-wind conditions and low-light levels (12:12 light–dark) LD with gradual transition from light to dark and vice versa) at temperatures of 26 to 28°C and relative humidity of 70 to 80%. Adult maintenance is relatively easy with 6 to 10% (w/v) solutions of either glucose or sucrose offered on filter paper or cotton wicks. These should be replaced at 2- to 3-day intervals to avoid excessive bacterial growth. Female anophelines will blood-feed on the first night after emergence and prior to being inseminated in order to reach the pregravid status (partial egg development) and replenish energy reserves. Most species require more than one blood meal in order to fully mature eggs and complete the first gonotrophic cycle (Gillies, 1954; Briegel and Hörler, 1993). The choice of host for blood feeding may vary and includes sedated mice, rats, or hamsters, or feeding on the forearms of human volunteers or artificial feeding devices (membranes). Use of human volunteers depends on ethical clearance and obtaining informed consent and requires regular screening of peripheral blood for infective parasite stages.

Maintenance of mosquito colonies for research on plant-derived products will vary according to species and locality in which it is undertaken. Guidelines will merely aid the establishment and upkeep of colonies, though each insectary knows its own special circumstances. It suffices here to summarise the critical points that should be included in publications, and thus assist the research community to appropriately interpret results from studies:

1. Origin of the strain (geographical locality) of mosquitoes and time the strain has been maintained and reared under controlled conditions
2. Climatic conditions (temperature, relative humidity, light) in which the insects are maintained
3. Larval density (number of larvae per unit water surface area), larval food source, and feeding regimen (amount per larva per time unit)
4. Adult cage size, regular food source (carbohydrates) and concentration thereof, blood host, and frequency and duration of feeding

25.3.2 Tests for Stage-Specific Activity of Plant-Derived Products

25.3.2.1 Egg Stage Activity

Reports of ovicidal properties or effects of plant-derived products on adult mosquitoes that were exposed during the egg stage are limited, and rare for anophelines (e.g., Miller and Maddock, 1970). Most observations were directly related to studies on oviposition deterrence and simply examined the hatch rate of eggs that were laid on treated substrates. Others developed procedures for examining the effects of synthetic ovicides (Jacob, 1969; Miller and Maddock, 1970), and these can be similarly applied to plant products. Principally, eggs can be directly oviposited by gravid females onto treated substrates or be collected and subsequently exposed to specific treatments. The former method has preference as it avoids identification of suitable ovicides that may cause oviposition deterrence of gravid females in nature.

In the laboratory, eggs laid on untreated substrates can be directly transferred to treated ones, or soaked for variable periods in treatment solutions of varying concentrations (to determine levels at which hatch rates are reduced by 50, 90, or 99%; see WHO, 1975, 1996), after which they are returned to normal rearing conditions (untreated water; see Section 25.3.1). Eggs in the control treatment are soaked in similar serial dilutions of the solvent(s) used. At daily intervals thereafter, for periods up to 7 days posttreatment, emergence of larvae from eggs can be determined. Overall hatch rates for treated and control eggs can be compared using simple nonparametric statistical procedures (like paired t-tests).

Although such tests yield useful information and are relatively easy for rapid screening purposes, they only marginally reflect normal conditions, where eggs, larvae, and pupae will be exposed continuously to substances applied to breeding sites. It is also likely that long-term exposure to low concentrations of active substances achieves similar ovicidal effects as short-term, high-concentration treatments. Moreover, continuous exposure of eggs and emerging larvae will increase the likelihood of sustained effects in emerging adults, as has recently been reported for *Anopheles gambiae* exposed to low doses of neem oil (*Azadirachta indica*) (F. Oketch, personal communication). As plant-derived products often have high rates of biodegradability, it is also recommended to perform tests under outdoor conditions where products may be exposed to direct sunlight and high temperatures. As indicated above, it is preferred to combine studies on egg stage effects with those of oviposition deterrence, for which the test procedures are described below (Section 25.3.2.4).

25.3.2.2 Larval Stage Activity

By far the majority of plant products have been evaluated against larval stages of mosquitoes (see Chapter 24). Products to be applied as larvicides need to be evaluated for their physical and chemical properties, both of which may affect survival of the larvae. Anophelines principally feed and breathe at the water surface, and normal surface tension prevents them from continuous submergence and drowning. Consequently, any reduction in surface tension (for instance, through application of soap or oil or pesticides that form a monomolecular film) will exhaust and suffocate insects and cause larval mortality. Considering that many solvents, surfactants, or other components of larvicides have such properties, it is essential to ascertain the physicochemical properties of all components used in experiments and include appropriate control treatments to discern the various effects by them alone or in combination.

25.3.2.2.1 Laboratory Studies

25.3.2.2.1.1 Range-Finding Experiments

Plant extracts can be dissolved in dimethylsulfoxide, acetone, ethanol, methanol, dichloromethane, or other suitable solvents to produce a stock solution, which can then be serially diluted to obtain various working concentrations. Stock solution should be kept in a climate-controlled room (4°C) and in darkness to avoid any degradation of the active ingredients in the solution. Target doses should take into account the final dilution using 250 ml of distilled water. Testing of the larvicidal activity of plant products should always include negative and positive control treatments. Negative controls comprise complete absence of larval exposure to any substance or solvents/additives (like surfactants) only. Depending on the nature of the plant product evaluated, a true larvicide (like temephos, an organophosphate) or an insect growth regulator (like methroprene or pyriproxyfen — Nayar et al., 2002; Yapabandara et al., 2001; Yapabandara and Curtis, 2002) may be used as a positive control against which the activity of the plant product can be evaluated.

Initial experiments with plant products should aim at identifying a range of concentrations over which larval mortality varies from 0 to 100%, using a specified number of third or fourth larval instars at different densities (e.g., 50, 100, and 200; WHO, 1981; Lukwa, 1994; Mwaiko, 1992) under controlled conditions. Larvae should be left to acclimatise in containers for at least 2 hours prior to application of the formulation. The different concentrations should be evaluated and replicated (at least five replicates per dose). Following treatment moribund, dead and live larvae should be counted at 4-hour intervals for at least 24 hours posttreatment.

A new bioassay with more precise concentrations is then conducted, using the data from the range-finding experiments. Similar in setup, concentrations at which 50, 90, or 99% of the larvae die within that period (LC_{50}, LC_{90}, and LC_{99}) can then be calculated (using (log-)probit analysis; see Finney, 1964; Preisler and Robertson, 1989; Russell et al., 1977) and the minimum concentration (in parts per million (ppm)) at which larval mortality is 100% determined.

Variations in the above protocol exist, but for reference purposes it is essential that activity of the plant material is expressed in LC_{50}, LC_{90}, and LC_{99} values (or at least LC_{50}) and concentrations in ppm or mg/l. Examples of the above protocol (see also Annex 25.1) and procedures can be found in various recently published articles in which plant products were evaluated as mosquito larvicides (Jang et al., 2002; Siddiqui et al., 2002; Traboulsi et al., 2002; Rahuman et al., 2000; Redwane et al., 2002; Yang et al., 2002; Schwartz et al., 1998).

25.3.2.2.1.2 Residual Activity

The vast majority of published studies focus on descriptions of the larvicidal effects and measurement of LC_{50} values but overlook a major factor of particular importance when evaluating plant products: the residual effect, or the activity decay over time.

Measuring the activity decay over time is a simple process. Basically, the concentration that yielded 100% mortality in the range-finding experiments is applied on day 1. On every consecutive day a fixed number of larvae may then be added to the treatment containers for continuous observation or for fixed-term exposure.

25.3.2.2.1.3 Effects on Emerging Adults

Little information is currently available about the effects of sublethal doses of larvicides on emerging anopheline adults. However, reduced survival, fecundity, and parasite development have been reported for *Culex* vectors of filariasis (Seif et al., 1997; Su and Mulla, 1999). If adult mosquito longevity or reproduction is negatively affected, then this adds to the overall impact of the larvicide applied, particularly at times when activity decay has advanced. Effects that are carried over to the adult stage may induce partial tolerance and eventually resistance to the larvicide, so careful monitoring of these effects is recommended.

25.3.2.2.2   Semifield Studies and Field Studies

Contained environments in which the biology and ecology of mosquitoes can be studied under near-natural conditions have several advantages, one of which is the rapid evaluation of plant-based products against larval and adult stage mosquitoes (Knols et al., 2002; Seyoum et al., 2002a, 2002b). High-throughput screening of many different plant products under conditions in which the number and physiological status of the larvae and adult insects are known will facilitate the selection process to identify candidate products. Moreover, considering the often artificial settings in laboratory studies, such environments provide a suitable intermediate stage prior to open field tests, which suffer from widely fluctuating conditions. It should be noted, though, that conclusions drawn from experiments under semifield conditions always need verification in open field settings.

Field studies may apply minimum doses that yield 100% larvicidal activity in the laboratory, but often use concentrations several-fold higher. Open field studies can employ artificial containers in which wild gravid females lay their eggs (Fillinger et al., 2003; Figure 25.2) or utilise natural breeding sites. The former method has the advantage that the surface area to be treated is constant and the total volume of water in the container can be controlled. Using standard larval sampling tools like the dipper (WHO, 1973), relative densities and species composition of larvae in the sites can be assessed prior to treatment. Depending on the nature of the plant product to be applied, it may only be necessary to estimate the surface area of the breeding site (Rao et al., 1992) or include measurement of the volume of water in the site. Following simple calculations, the amount of product to be applied to reach similar concentrations as those used in the laboratory can be applied. As it is often difficult to trace dead or moribund larvae, particularly in turbid water, effects may only be measured after daily sampling until late-stage larvae reappear in the site. Covering the breeding site with a mosquito-netting cage can be used to recapture any emerging adults for further experimentation. Monitoring of the presence of predators or parasites is important, as this may overestimate the potency of the product applied. Similarly, any notable effects on nontarget organisms should be carefully observed when evaluating plant products, particularly in semipermanent

FIGURE 25.2 Field testing of larvicidal agents in western Kenya, whereby wild gravid females are allowed to oviposit eggs in artifical habitats that are subsequently treated with plant products. (Photo: Ulrike Fillinger.)

or permanent breeding sites, which may have many organisms associated with them. It is obvious that extreme care should be taken when plant products are to be applied to water sources used for human or animal consumption.

25.3.2.3 Adult Stage Activity

Plant-based adulticides have long been known, yet have largely been replaced by synthetic insecticides. For mosquitoes, perhaps the best example is pyrethrum (*Chrysanthemum cinerariaefolium*), which is still utilised in sprays for mosquito control and surveillance, though production is limited compared to that of synthetic pyrethroids, several of which have shown excellent antivector properties. Synthetic pyrethoids are widely used for malaria vector control on bed nets, curtains, and clothing (Lengeler, 1998) and have been used for indoor residual spraying (Brutus et al., 2001). In contrast with natural pyrethrum, which readily loses activity due to oxidisation and exposure to sunlight, synthetic pyrethroids retain their toxicity for long periods of time (6 to 12 months), particularly when used indoors. This raises an important issue when searching for suitable plant-derived adulticides: not only should their efficacy be high, but they should also be formulated in a manner that ensures activity over periods of at least several months.

25.3.2.3.1 Laboratory Studies

25.3.2.3.1.1 Range-Finding Experiments for Contact Adulticides

Screening for insecticidal activity of plant-derived products in the laboratory follows the same principles as those for evaluating synthetic insecticides, and by far the most commonly used method employs the WHO insecticide susceptibility test kit (Najera and Zaim, 2001; WHO, 1992), which is commercially available. Candidate adulticides are formulated and applied to filter paper, which is inserted in a cylinder into which individual adult mosquitoes are introduced for variable periods of time. Range-finding experiments employ the same strategy as that for larvicides (Annex 25.1). Following exposure, mosquitoes are kept in standard rearing cages and supplied with a carbohydrate food source. Knockdown or mortality is normally scored after 24 hours postexposure. Ideally, tests should use non-blood-fed females of known age (e.g., 24 hours after emergence). If mosquitoes do not derive from a colony but are either wild specimens or the F1 generation thereof, it needs to be ascertained that these insects have not been exposed to other insecticides prior to collection, as this may affect the outcome of experiments.

Variations in the above protocol exist, whereby petri dishes rather than cylinders are used. In this case filter paper discs are treated with varying concentrations and insects introduced into the dish through a central hole in the lid (Lukwa et al., 1996). Control treatments should be included in the tests (nontreated filter paper, or paper treated with a known insecticide, e.g., a synthetic pyrethroid). LC_{50} values can be calculated as previously described before (Annex 25.1).

25.3.2.3.1.2 Residual Activity

The residual effect and activity decay are measured by exposing new mosquitoes daily to the same filter papers until knockdown or mortality is negligible.

25.3.2.3.1.3 Range-Finding Experiments for Vapourising Adulticides

When burned or smoldering, various plant materials have been observed to have knockdown or killing effects on adult mosquitoes, often in conjunction with irritability or repellency effects. Incorporation of plant products like pyrethrins in coils serves as an example (Lukwa and Chandiwana, 1998).

When evaluating knockdown or killing effects in laboratory settings, care should be taken not to saturate the test cage or experimental room with vapours from plant products, and appropriate ventilation is recommended. The complications this may give and resulting misleading effects may be overcome by using large (at least $3 \times 3 \times 3$ m), well-ventilated rooms in which mosquitoes are exposed to plant vapours (Tawatsin et al., 2001; Birley et al., 1987). As these may often have more than a killing effect (like repellency or irritability), it is useful to monitor effects in the presence of a human volunteer (to record reduction in biting).

There are several ways in which volatiles from plants may be applied against adult mosquitoes, either as fresh material or through direct burning or thermal expulsion of volatiles from heated plant material. Range-finding experiments use different amounts of plant material and intensities of burning/smoldering and normally record knockdown of insects at specified intervals (1 to 10 minutes). It is imperative that findings from laboratory studies are verified under (semi-)field conditions.

25.3.2.3.2 Semifield and Field Studies

Evaluation of adulticidal and repellent effects of plants materials has more frequently been undertaken in (semi-)field settings than in the laboratory. A semifield setup for rapid screening of large numbers of plant products has recently been established in Kenya (Seyoum et al., 2002a, 2002b), whereby fixed numbers of mosquitoes are introduced in a setting with experimental huts. Such a method offers the opportunity to conduct year-round evaluation of plant materials, even at times when mosquito densities in the field are low. Another advantage is the ability to apply rigorous statistical procedures that enable powerful evaluation within a few days of experimentation. Ample ventilation and natural climatic conditions overcome artificial circumstances encountered in the laboratory. For detailed descriptions of this system and test methodologies, refer to Seyoum et al. (2002a, 2002b). Here it suffices to mention that these studies showed that live potted plants can be used to dramatically reduce house entry by host-seeking malaria mosquitoes (Figure 25.3) and that thermal expulsion yields higher levels of repellency, for several plant species, than direct burning of material from those same plants (Figure 25.4).

Open field studies are nevertheless needed to verify the results from semifield experiments. For this, a variety of experimental hut designs, used for evaluating insecticides, can be used (Darriet et al., 2002; Murahwa et al., 1994; Curtis et al., 1987). Normally, several huts are being utilised, over which treatments are rotated for several nights using Latin-square experimental designs. Such huts are usually ant-proof, which enables collection of dead or moribund mosquitoes that would normally be eaten. This latter point may seem trivial, but is crucial when evaluating knockdown or killing effects of plants on house-entering mosquitoes. Studies that have reported repelling effects of plants may actually have overseen adulticidal effects due to this phenomenon (Pålsson and Jaenson, 1999a, 1999b).

FIGURE 25.3 Use of live potted plants as natural mosquito repellents against house-entering mosquitoes in western Kenya. (Photo: Aklilu Seyoum; reproduced from Seyoum, A. et al., (2002b), *Trans. R. Soc. Trop. Med. Hyg.*, 96, 225–231, with kind permission of the *Transactions of the Royal Society of Tropical Medicine and Hygiene.*)

FIGURE 25.4 Thermal expulsion (left) and direct burning of plant material on traditional stoves (right) in western Kenya. (Photo: Aklilu Seyoum; reproduced from Seyoum, A. et al., (2002b), *Trans. R. Soc. Trop. Med. Hyg.*, 96, 225–231, with kind permission of the *Transactions of the Royal Society of Tropical Medicine and Hygiene.*)

Field tests in normal houses follow almost similar procedures as those used for mosquito surveillance with spray catches. White sheets of bed linen are spread out over the floor and furniture in the house to easily observe dead or moribund mosquitoes. Prior to the onset of mosquito biting activity, collectors move into the house and apply the candidate plant material (e.g., through direct burning on a charcoal stove). Another house follows the same procedures but uses a stove without burning any plant material. At specific time intervals the bedsheets are examined for the presence of mosquitoes. The number of mosquito bites received by the collectors is recorded. Pair-wise comparisons are then made with the effects observed in the control hut. As for larvicidal tests, the procedures reported differ substantially and depend on the propensity of the mosquito species present to enter houses and feed on their occupants (Seyoum et al., 2002a, 2002b; Pålsson and Jaenson, 1999a, 1999b; Paru et al., 1995; Sharma and Ansari, 1994; Vernede et al., 1994).

25.3.2.4 Oviposition-Deterrent Activity

Oviposition-deterrent effects of plant materials are normally studied in conjunction with trials to assess ovicidal or larvicidal properties of such products (Mehra and Hiradhar, 2002) and have even been observed for topical insect repellents (Xue et al., 2001a, 2001b). Responses of gravid mosquitoes toward habitats with high densities of conspecific larvae (Zahiri and Rau, 1998) or potential predators have shown deterrence (Angelon and Petranka, 2002), and the methods used to study these effects may also be applied when evaluating effects of plant products on egg-laying behaviour of gravid mosquitoes.

25.3.2.4.1 Laboratory Studies

A variety of methods to study the effect of plant products on oviposition responses are available and are described here in broad terms only (Moshen et al., 1995; Ouda et al., 1998). In principle, preparation of plant materials can be similar to procedures used to determine LC_{50} values for larvicidal products; i.e., the effects should be observed for a broad range of concentrations. Care should be taken to include appropriate control treatments, as some of these may exert oviposition deterrence themselves. Laboratory evaluation normally uses two oviposition substrates in mosquito cages between which gravid mosquitoes can choose. One of these is treated with the test substance; the other serves as the control (with the solvent). Using individual females in experiments is preferred, as in this way the percentage that refrains from egg laying altogether can be determined.

Insects are normally kept overnight in a cage with a dual-choice setup, and the number of eggs laid on either of the two substrates can simply be determined the following day. Following this, eggs may be monitored for several days afterwards, as products may also exert a larvicidal effect (Mwangi and Mukiama, 1988; Saxen et al., 1992). Data from several replicates (which should at least number 20 females that actually oviposited) can be pooled and subjected to simple nonparametric procedures, like Wilcoxon signed-rank tests. Others have used the so-called oviposition activity index, $OAI = N_t - N_c/N_t + N_c$, where N_t is the number of eggs or egg rafts in the treatment and N_c is that in the control (Zebitz, 1987; Kramer and Mulla, 1979). An OAI value of >1 indicates an attractive effect; <1, deterrent activity of the material tested.

25.3.2.4.2 Semifield and Field Studies

In semifield systems a number of artificial breeding sites can be constructed over which treatments (including controls) are randomly assigned. A fixed number of gravid females can then be released inside the system and the same measurements can be performed as in laboratory conditions.

Field studies normally center on the breeding site where mosquito eggs have been identified. A measured concentration of the plant preparation is placed on these, followed by daily counts of the number of eggs oviposited. Numbers are compared with oviposition activity in control sites. Emergence of eggs can be determined by covering the breeding site with mosquito netting so that the number of adult mosquitoes emerging can be measured. Construction of artificial sites in which wild gravid females oviposit can be very useful to assess deterrence, as the majority of characteristics of such habitats (surface area, water quality, etc.) remain constant (Fillinger et al., 2003).

25.3.2.5 Insect Growth Regulatory Activity

A variety of compounds that inhibit the development of immature stages of insects, classified as juvenile hormone analogues (or mimics) or as chitine synthesis inhibitors, have already been commercialised as pest control agents. Examples of such compounds for mosquito control are pyriproxyfen (Yapabandara et al., 2001) and methoprene (Nasci et al., 1994). Effects of sublethal doses of this latter compound on longevity and reproduction of adult mosquitoes have also been documented (Sawby et al., 1992). Effectiveness and residual activity tests follow the same procedures as those for evaluating larvicides (Nayar et al., 2002). However, the direct insect growth regulatory effects may not be apparent unless careful examination of dead or moribund larvae is undertaken. Deformities (Figure 25.5) can then be observed.

FIGURE 25.5 Insect growth regulatory effects of plant agents on aquatic mosquito stages. A normal pupa is shown on the left, a normal larva on the right, and an intermediate deformed specimen in the center. (Photo: ICIPE photo archive.)

25.3.2.6 Repellent Activity

An overview of the various techniques to evaluate the repellent effect of plant-based products is presented in Chapter 23. Considering the fact that many insecticidal agents, like synthetic pyrethroids and DDT, have repellent properties (Sungvornyothin et al., 2001), it can be expected that plant-based adult mosquito control agents exert similar multimode effects (Choochote et al., 1999). As a result, products with high potency in laboratory settings may prove of limited value in open field conditions, which underpins the need to verify findings under natural circumstances. Many published studies have not gone beyond laboratory evaluation (Chapter 24), which makes the applicability of their findings in the field questionable. Without negating the importance of laboratory studies, we therefore emphasise the need for full field trials with plant products as a crucial step to move such products into the arena of useful public health tools.

25.4 TOXICOLOGICAL EFFECTS AND IMPACT ON NONTARGET ORGANISMS

As for other insecticides, like bacterial larvicides (WHO, 1996), characterisation of insecticidal agents, even if derived from natural sources like plants, needs to be linked to toxicological evaluations and impact on the environment. However, considering the often complex chemical makeup of plant products, such evaluations may be time-consuming and costly. It can be argued, though, that the need for careful toxicological evaluation increases with identification, isolation, registration, and commercialisation of active principles from plant products, and that the mere application of crude material (bark, leaves, flowers, etc.) will not *a priori* necessitate substantive evaluation.

Obviously the nature of the product will be of importance when assessing its potential toxicological impact. Plants that are consumed or used in topical formulations, like neem (*Azadirachta indica*) or basil species (Lamiaceae), or plants that have already been assessed for their toxicity (like lemon eucalyptus; Trigg, 1996) will be easy to move forward for broad-scale application, even when negative effects on nontarget organisms have been documented (Scott and Kaushik, 2000; el-Shazly and el-Sharnoubi, 2000). In contrast, some plants with known insecticidal activity, like castor bean (*Ricinus communis*; Olaifa et al., 1991), are also known to be highly toxic to humans, which will make their applicability as a vector control agent unlikely.

The way in which plant products are applied is critical when evaluating their toxicity or impact on nontarget organisms. Direct burning or thermal expulsion of plant material inside houses, whereby inhalation of smoke and volatile plant substances may occur, requires careful evaluation, as do products that are directly applied on the skin. In contrast, plant products that are used away from the domestic environment, for instance, for treating small transient anopheline larval habitats, will have minimal impact on human health and nontarget organisms, even when fairly toxic.

When considering potential hazard as the product of frequency and duration of exposure, persistence of the product, and dose-dependent toxicity effects, it will be possible to classify plant products according to their various activity classes described in this chapter. The World Wide Web offers good opportunities to find information on the toxicology of chemical substances, but also about plants in general. The ToxNet, operated by the National Library of Medicine (U.S.), provides an excellent resource. For instance, a search of "DDT" delivered more than 17,000 records; "pyrethrum," 383; and "*Ocimum suave*," 7. Although such resources provide valuable information, there remains an urgent need for clear guidelines for evaluating the toxicological impact of plant products. Considering that protocols for determining oral and dermal toxicity are well established, this particularly applies to development of crude plant-based vector control products and their application at the community level in malaria-endemic countries.

25.5 CONCLUSIONS AND PERSPECTIVES

The above guidelines are proposed as a framework for studies on plant-based mosquito control agents. Taking into account the literature as well as ethnobotanical information, crude plant products, extracts, or individual compounds can be tested for activity against the various life stages of mosquito vectors of malaria. It is clear that the published literature shows wide variety in the protocols applied, and the current guidelines merely try to outline important issues that need to be considered when evaluating plant products under laboratory, semifield, and field conditions.

Active compounds can be isolated and structurally identified for further use or commercialisation, but this is not essential for determining the safety and efficacy of crude plant products or extracts thereof. The specificity and the sustainability of the extract or the active natural substance remain of high priority. In some cases it will be better to use crude extracts, and in other circumstances it may be better to isolate the active principles. Different parameters should be considered before deciding between these options. The nature and concentration of different constituents in an extract (and so its activity) will depend on different factors such as the time and place of harvesting and the extraction procedure applied.

Within this framework it is possible to conduct studies leading to extracts or plant-derived pure substances, which could be used in malaria control programs. A systematic study of the plants offered to us by nature could lead to the discovery of affordable plant-based vector control agents. These guidelines need to be piloted and improved as results are assessed. Finally, it has to be said that any vector control tool should be used in an integrated approach, as it will certainly be more efficient to integrate tools than to rely on one only. We recommend moving one or a few plant species to the forefront in malaria vector control and confirming their applicability in small-scale pilot trials in various agro-ecological settings with differing vectorial systems and transmission intensities.

ANNEX 25.1: SUMMARY OF PROCEDURES FOR EVALUATING LARVICIDAL PROPERTIES OF PLANT-DERIVED PRODUCTS

- Determine the proper range of dosages or concentrations by exposing larvae to serial (10-fold) dilutions of the product.
- Calculate a more precise dosage range for a more detailed bioassay from the results of the range-finding experiment using the following formula:

$$\log(\text{dilution factor}) = [\log(\text{low dosage/high dosage})]/\text{number of dilutions}$$

where:
  - The **dilution factor** is the spacing between dosages or the number (less than 1) by which to multiply the highest dosage in order to obtain the next lower dosage.
  - **Low dosage** is the dosage in the original range-finding test that produced the infection rate closest to 0.
  - **High dosage** is the range-finding dosage that caused an infection rate closest to 100%.
  - The **number of dilutions** is the number of concentration or dosage groups planned for testing between almost 0 and almost 100% infection.
  - A good bioassay should include at least two dosages below and two dosages above the inhibitory dose 50% (ID_{50}), giving a total of 5 dose groups.
- Mortality measurements:
  - Count the number of moribund, dead, and live larvae in each group, including the control group.
  - Combine the numbers from all three repetitions of each dosage group.

- Calculate the percentage of dead insects in each group:

$$100 \times (\text{number dead/number originally set})$$

- Using Abbott's formula (Abbott, 1925), correct the percentage mortality:

$$\text{Corrected \% mortality} = 100 \times (\text{T\%} - \text{C\%}/100\% - \%\text{C})$$

where:
- T% = the percentage of dead test larvae
- C% = the percentage of dead control larvae
- Calculate lethal concentrations using probit analysis (Finney, 1964). This can be done using computerised software (e.g., SAS for Windows, Release 6.08, SAS Institute, Inc., Cary, NC 27513), but can also be graphed as a function of dosage on log-probit graph paper. Because of the sigmoid shape of the dosage–response curves, percentages are converted to probits and plotted against the log of the dosage, so that a straight line can be more easily plotted.
  - Enter percent infection or corrected percent mortality as points on log-probit paper.
  - Eye-fit a straight line through the points such that there are an equal number of points above and below the line at each end.

Assuming that percentage infection and spore concentration were plotted, the LC_{50} is the concentration straight down from the intersection of the eye-fitted line and the horizontal 50% infection line (center of the graph). LC_{90} and inhibitory concentration 99% (IC_{99}) values are obtained from the intersection of the line with the 90 and 99% lines, respectively, on the graph.

REFERENCES

Abbott, W.S. (1925). A method for computing the effectiveness of an insecticide. *J. Econ. Entomol.*, 18, 265–267.

Angelon, K.A. and Petranka, J.W. (2002). Chemicals of predatory mosquitofish (*Gambusia affinis*) influence selection of oviposition site by *Culex* mosquitoes. *J. Chem. Ecol.*, 28, 797–806.

Armstrong-Schellenberg, J.R.M., Abdulla, S., Nathan, R., Mukasa, O., Marchant, T.J., Kikumbih, N., et al. (2001). Effect of large-scale social marketing of insecticide-treated nets on child survival in rural Tanzania. *Lancet*, 357, 2141–2147.

Barasa, S.S., Ndiege, I.O., Lwande, W., and Hassanali, A. (2002). Repellent activities of stereoisomers of p-menthane-3,8-diols against *Anopheles gambiae* (Diptera: Culicidae). *J. Med. Entomol.*, 39, 736–741.

Beier, J.C., Killeen, G.F., and Githure J.I. (1999). Short report: entomologic inoculation rates and *Plasmodium falciparum* prevalence in Africa. *Am. J. Trop. Med. Hyg.*, 61, 109–113.

Benedict, M.Q. (1997). Care and maintenance of anopheline mosquito colonies. In *The Molecular Biology of Insect Disease Vectors: A Methods Manual*. Chapman & Hall, London.

Birley, M.H., Mutero, C.M., Turner, I.F., and Chadwick P.R. (1987). The effectiveness of mosquito coils containing esbiothrin under laboratory and field conditions. *Ann. Trop. Med. Parasitol.*, 81, 163–171.

Breman, J.G., Egan, A., and Keusch, G.T. (2001). The intolerable burden of malaria: a new look at the numbers. *Am. J. Trop. Med. Hyg.*, 64 (Suppl. 1), iv–vii.

Briegel, H. and Hörler, E. (1993). Multiple blood meals as a reproductive strategy in *Anopheles* (Diptera: Culicidae). *J. Med. Entomol.*, 30, 975–985.

Brutus, L., Le Goff, G., Rasoloniaina, L.G., Rajaonarivelo, V., Raveloson, A., and Cot, M. (2001). The campaign against malaria in central western Madagascar: comparison of the efficacy of lambda-cyhalothrin and DDT house spraying. I. Entomological study. *Parasite*, 8, 297–308 (article in French).

Chandre, F., Darrier, F., Manga, L., Akogbeto, M., Faye, O., Mouchet, J., and Guillet, P. (1999). Status of pyrethroid resistance in *Anopheles gambiae* sensu lato. *Bull. WHO*, 77, 230–234.

Choochote, W., Kanjanapothi, D., Panthong, A., Taesotikul, T., Jitpakdi, A., Chaithong, U., et al. (1999). Larvicidal, adulticidal and repellent effects of *Kaempferia galanga*. *Southeast Asian J. Trop. Med. Public Health*, 30, 470–476.

Curtis, C.F. (1991). *Control of Disease Vectors in the Community*, Curtis, C.F., Ed. Wolfe Publishing, London.

Curtis, C.F. (2002). Should the use of DDT be revived for malaria control? *Biomédica*, 22, 456–461.

Curtis, C.F., Lines, J.D., Ijumba, J., Callaghan, A., Hill, N., and Karimzad, M.A. (1987). The relative efficacy of repellents against mosquito vectors of disease. *Med. Vet. Entomol.*, 1, 109–119.

Darriet, F., N'Guessan, R., Hougard, J.M., Traore-Lamizana, M., and Carnevale, P. (2002). An experimental tool essential for the evaluation of insecticides: the testing huts. *Bull. Soc. Pathol. Exot.*, 95, 299–303 (article in French).

el-Shazly, M.M. and el-Sharnoubi, E.D. (2000). Toxicity of a neem (*Azadirachta indica*) insecticide to certain aquatic organisms. *J. Egypt. Soc. Parasitol.*, 30, 221–231.

Fillinger, U., Knols, B.G.J., and Becker, N. (2003). Efficacy and efficiency of new *Bacillus thuringiensis* var. *israelensis* and *Bacillus sphaericus* formulations against Afrotropical anophelines in Western Kenya. *Trop. Med. Int. Health*, 8, 37–47.

Finney, D.J. (1964). *Probit Analysis: Statistical Treatment of the Sigmoid Response Curve*. Cambridge University Press, London.

Foster, W.A. (1980). Colonisation and maintenance of mosquitoes in the laboratory. In *Malaria*, Vol. 2, *Pathology, Vector Studies and Culture*, Kreier, J.P., Ed. Academic Press, London, pp. 103–151.

Garrett-Jones, C. (1964). Prognosis for interruption of malaria transmission through assessment of the mosquito's vectorial capacity. *Nature*, 204, 1173–1175.

Gerberg, E., Barnard, D., and Ward, R. (1994). *Manual for Mosquito Rearing and Experimental Techniques*, Bulletin 5. American Mosquito Control Association, Eatontown, NJ, 98 pp.

Gilles, H.M. and Warrell, D.A. (1993). *Bruce-Chwatt's Essential Malariology*, 3rd ed. Edward Arnold, London.

Gillies, M.T. (1954). The recognition of age groups within populations of *Anopheles gambiae* by the pregravid rate and sporozoite rate. *Ann. Trop. Med. Hyg.*, 48, 58–74.

Greenwood, B. and Mutabingwa, T. (2002). Malaria in 2002. *Nature*, 415, 670–672.

Guyatt, H.L., Ochola, S.A., and Snow, R.W. (2002). Too poor to pay: charging for insecticide-treated bednets in highland Kenya. *Trop. Med. Int. Health*, 7, 846–850.

Hargreaves, K., Koekemoer, L.L., Brooke, B.D., Hunt, R.H., Mthembu, J., and Coetzee, M. (2000). *Anopheles funestus* resistant to pyrethroid insecticides in South Africa. *Med. Vet. Entomol.*, 14, 181–189.

Jacob, W.L. (1969). Simulated field tests with ovicides against *Aedes aegypti* eggs in tires and cans. *Mosq. News*, 29, 402–407.

Jang, Y.S., Baek, B.R., Yang, Y.C., Kim, M.K., and Lee, H.S. (2002). Larvicidal activity of leguminous seeds and grains against *Aedes aegypti* and *Culex pipiens pallens*. *J. Am. Mosq. Control Assoc.*, 18, 210–213.

Killeen, G.F., Fillinger, U., Kiche, I., Gouagna, L.C., and Knols, B.G.J. (2002b). Eradication of *Anopheles gambiae* from Brazil: lessons for malaria control in Africa? *Lancet Infect. Dis.*, 2, 618–627.

Killeen, G.F., Fillinger, U., and Knols, B.G.J. (2002a). Advantages of larval control for African malaria vectors: low mobility and behavioural responsiveness of immature mosquito stages allow high effective coverage. *Malaria J.*, 1, 8.

Knols, B.G.J., Njiru, B.N., Mathenge, E.M., Mukabana, W.R., Beier, J.C., and Killeen, G.F. (2002). MalariaSphere: a greenhouse-enclosed simulation of a natural *Anopheles gambiae* (Diptera: Culicidae) ecosystem in western Kenya. *Malaria J.*, 1, 19.

Kramer, W.L. and Mulla, M.S. (1979). Oviposition attractants and repellents of mosquitoes: oviposition responses of *Culex* mosquitoes to organic infusions. *Environ. Entomol.*, 8, 1111–1117.

Lengeler, C. (1998). Insecticide treated bednets and curtains for malaria control. *Cochrane Library Rep.*, 3, 1–70.

Lindsay, S.W., Emerson, P.M., and Charlwood, J.D. (2002). Reducing malaria by mosquito-proofing houses. *Trends Parasitol.*, 18, 510–514.

Lukwa, N. (1994). Do traditional repellent plants work as mosquito larvicides? *Cent. Afr. J. Med.*, 40, 306–325.

Lukwa, N. and Chandiwana, S.K. (1998). Efficacy of mosquito coils containing 0.3% and 0.4% pyrethrins against *An. gambiae sensu lato* mosquitoes. *Cent. Afr. J. Med.*, 44, 104–107.

Lukwa, N., Masedza, C., Nyazema, N.Z., Curtis, C.F., and Mwaiko, G.L. (1996). Efficacy and duration of activity of *Lippia javanica* Spreng, *Ocimum canum* Sims and a commercial repellent against *Aedes aegypti*. In *Proceedings of the First International Organisation for Chemistry in Sciences and Development (IOCD) Symposium, Victoria Falls, Zimbabwe*, Hostettman, K., Chinyangaya, F., Maillard, M., and Wolfende, J.L., Eds. University of Zimbabwe (Harare) Publications, Zimbabwe, pp. 321–325.

Lyimo, E.O. and Takken, W. (1993). Effects of adult body size on fecundity and pre-gravid rate of *Anopheles gambiae* females in Tanzania. *Med. Vet. Entomol.*, 7, 328–332.

Lyimo, E.O., Takken, W., and Koella, J.C. (1992). Effect of rearing temperature and larval density on larval survival, age at pupation and adult size of *Anopheles gambiae*. *Entomol. Exp. Appl.*, 63, 265–271.

MacDonald, G. (1957). *The Epidemiology and Control of Malaria*. Oxford University Press, Oxford, U.K.

Maxwell, C.A., Msuya, E., Sudi, M., Njunwa, K.J., Carneiro, I.A., and Curtis C.F. (2002). Effect of community-wide use of insecticide-treated nets for 3–4 years on malarial morbidity in Tanzania. *Trop. Med. Int. Health*, 7, 1003–1008.

Mehra, B.K. and Hiradhar, P.K. (2002). *Cuscuta hyalina* Roth., an insect development inhibitor against common house mosquito *Culex quinquefasciatus* Say. *J. Environ. Biol.*, 23, 335–339.

Miller, S. and Maddock, D.R. (1970). Ovicidal effect of selected compounds on the eggs of *Anopheles albimanus*. *J. Econ. Entomol.*, 63, 1151–1154.

Moshen, Z.H., Jawad, A.M., Al-Saadi, M., and Al-Naib, A. (1995). Anti-oviposition and insecticidal activity of *Imperata cylindrica* (Gramineae). *Med. Vet. Entomol.*, 9, 441–442.

Murahwa, F.C., Lukwa, N., Govere, J.M., and Masedza, C. (1994). Do mosquito coils and killer sticks work against *Anopheles gambiae sensu lato* mosquitoes in Zimbabwe? *Cent. Afr. J. Med.*, 40, 122–126.

Mwaiko, G.L. (1992). *Citrus* peel oil extracts as mosquito larvae insecticides. *East Afr. Med. J.*, 69, 223–226.

Mwangi, R.W. and Mukiama, T.K. (1988). Evaluation of *Melia volkensii* extract fractions as mosquito larvicides. *J. Am. Mosq. Control Assoc.*, 4, 442–447.

Nabarro, D.N. and Tayler, E.M. (1998). The 'Roll Back Malaria' campaign. *Science*, 280, 2067–2068.

Najera, J.A. and Zaim, M. (2001). Insecticides for Indoor Residual Spraying, WHO/CDS/WHOPES/2001.3. WHO Division of Communicable Diseases, World Health Organization, Geneva, 94 pp.

Nasci, R.S., Wright, G.B., and Willis, F.S. (1994). Control of *Aedes albopictus* larvae using time-release larvicide formulations in Louisiana. *J. Am. Mosq. Control Assoc.*, 10, 1–6.

Nayar, J.K., Ali, A., and Zaim, M. (2002). Effectiveness and residual activity comparison of granular formulations of insect growth regulators pyriproxyfen and s-methoprene against Florida mosquitoes in laboratory and outdoor conditions. *J. Am. Mosq. Control Assoc.*, 18, 196–201.

Olaifa, J.I., Matsumura, F., Zeevaat, J.A.D., and Charalambouse, P. (1991). Lethal amounts of ricinine in green peach aphids *Myzus persicae* suzler fed on castor bean plants. *Plant Sci. (Limerick)*, 73, 253–256.

Ouda, N.A., Al-Chalabi, B.M., Al-Charchafchi, F.M.R., and Moshen, Z.H. (1998). Insecticidal and ovicidal effects of the seed extract of *Atriplex canescens* against *Culex quinquefasciatus*. *Pharm. Biol.*, 36, 69–71.

Pålsson, K. and Jaenson, T.G. (1999a). Plant products used as mosquito repellents in Guinea Bissau, West Africa. *Acta Trop.*, 72, 39–52.

Pålsson, K. and Jaenson, T.G. (1999b). Comparison of plant products and pyrethroid-treated bed nets for protection against mosquitoes (Diptera: Culicidae) in Guinea Bissau, West Africa. *J. Med. Entomol.*, 36, 144–148.

Paru, R., Hii, J., Lewis, D., and Alpers, M.P. (1995). Relative repellency of woodsmoke and topical applications of plant products against mosquitoes. *Papua New Guinea J. Med.*, 38, 215–221.

Preisler, H.K. and Robertson, J.L. (1989). Analysis of time-dose-mortality data. *J. Econ. Entomol.*, 82, 1534–1542.

Rahuman, A.A., Gopalakrishnan, G., Ghouse, B.S., Arumugam, S., and Himalayan, B. (2000). Effect of *Feronia limonia* on mosquito larvae. *Fitoterapia*, 71, 553–555.

Rao, D.R., Reuben, R., Venugopal, M.S., Nagasampagi, B.A., and Schmutterer, H. (1992). Evaluation of neem, *Azadirachta indica*, with and without water management, for control of culicine mosquito larvae in rice-fields. *Med. Vet. Entomol.*, 6, 318–324.

Redwane, A., Lazrek, H.B., Bouallam, S., Markouk, M., Amarouch, H., and Jana, M. (2002). Larvicidal activity of extracts from *Quercus lusitania* var. *infectoria* galls (Oliv.). *J. Ethnopharmacol.*, 79, 261–263.

Roozendaal, J. (1997). *Vector Control: Methods for Use by Individuals and Communities*. WHO, Geneva.

Ross, R. (1911). *The Prevention of Malaria*. Murray, London.

Ross, R., Annett, H.E., Giles, G.M., and Fieling-Ould, R. (1900). *Report of the Malaria Expedition of the Liverpool School of Tropical Medicine and Parasitology with Supplementary Reports*. University Press of Liverpool, Liverpool.

Russell, R.M., Robertson, J.L., and Savin, N.E. (1977). POLO: a new computer program for probit analysis. *Bull. Entomol. Soc. Am.*, 23, 209–213.

Saxen, R.C., Dixit, O.P., and Sukumurana, P. (1992). Laboratory assessment of indigenous plant extracts for anti-juvenile hormone activity in *Culex quinquefasciatus*. *Indian J. Med. Res.*, 95, 204–206.

Sawby, R., Klowden, M.J., and Sjogren, R.D. (1992). Sublethal effects of larval methoprene exposure on adult mosquito longevity. *J. Am. Mosq. Control Assoc.*, 8, 290–292.

Schwartz, A.M., Paskewitz, S.M., Orth, A.P., Tesch, M.J., Toong, Y.C., and Goodman, W.G. (1998). The lethal effects of *Cyperus iria* on *Aedes aegypti*. *J. Am. Mosq. Control Assoc.*, 14, 78–82.

Scott, I.M. and Kaushik, N.K. (2000). The toxicity of a neem insecticide to populations of culicidae and other aquatic invertebrates as assessed in in situ microcosms. *Arch. Environ. Contam. Toxicol.*, 39, 329–336.

Seif, A.I., Husseiny, I.M., Soliman, B.A., Soliman, M.A., el-Kady, M.A. (1997). Development of Wuchereria bancrofti in *Culex pipiens* L. (Diptera: Culicidae) exposed in the larval instar to sublethal dosages of insecticides and one insect growth regulator and their influence on reproduction of filaria-infected mosquitoes. *J. Egypt. Soc. Parasitol.*, 27, 843–853.

Seyoum, A., Kabiru, E.W., Lwande, W., Killeen, G.F., Hassanali, A., and Knols, B.G.J. (2002a). Repellency of live potted plants against *Anopheles gambiae* from human baits in semi-field experimental huts. *Am. J. Trop. Med. Hyg.*, 67, 191–195.

Seyoum, A., Pålsson, K., Kung'a, S., Kabiru, E.W., Lwande, W., Killeen, G.F., et al. (2002b). Traditional use of mosquito-repellent plants in western Kenya and their evaluation in semi-field experimental huts against *Anopheles gambiae*: ethnobotanical studies and application by thermal expulsion and direct burning. *Trans. R. Soc. Trop. Med. Hyg.*, 96, 225–231.

Sharma, V.P. and Ansari, M.A. (1994). Personal protection from mosquitoes (Diptera: Culicidae) by burning neem oil in kerosene. *J. Med. Entomol.*, 31, 505–507.

Shousha, A.T. (1948). Species-eradication. The eradication of *Anopheles gambiae* from Upper Egypt, 1942–1945. *Bull. WHO*, 1, 309–353.

Siddiqui, B.S., Afshan, F., Faizi, S., Naeem-Ul-Hassan Naqvi, S., and Tariq, R.M. (2002). Two new triterpenoids from *Azadirachta indica* and their insecticidal activity. *J. Nat. Prod.*, 65, 1216–1218.

Smith, T.A., Leuenberger, R., and Lengeler, C. (2001). Child mortality and malaria transmission in Africa. *Trends Parasitol.*, 17, 145–149.

Soper, F.L. and Wilson, D.B. (1943). *Anopheles Gambiae in Brazil: 1930 to 1940*. Rockefeller Foundation, New York.

Spielman, A. and D'Antonio, M. (2001). *Mosquito: A History of Our Most Persistent and Deadly Foe*. Hyperion, San Francisco.

Su, T. and Mulla, M.S. (1999). Effects of neem products containing azadirachtin on blood feeding, fecundity, and survivorship of *Culex tarsalis* and *Culex quinquefasciatus* (Diptera: Culicidae). *J. Vector Ecol.*, 24, 202–215.

Sungvornyothin, S., Chareonviriyaphap, T., Prabaripai, A., Thirakhupt, V., Ratanatham, S., and Bangs, M.J. (2001). Effects of nutritional and physiological status on behavioural avoidance of *Anopheles minimus* (Diptera: Culicidae) to DDT, deltamethrin and lambdacyhalothrin. *J. Vector Ecol.*, 26, 202–215.

Tawatsin, A., Wratten, S.D., Scott, R.R., Thavara, U., and Techadamrongsin, Y. (2001). Repellency of volatile oils from plants against three mosquito vectors. *J. Vector Ecol.*, 26, 76–82.

Traboulsi, A.F., Taoubi, K., el-Haj, S., Bessiere, J.M., and Rammal, S. (2002). Insecticidal properties of essential plant oils against the mosquito *Culex pipiens molestus* (Diptera: Culicidae). *Pest Manage. Sci.*, 58, 491–495.

Trigg, J.K. (1996). Evaluation of a eucalyptus-based repellent against *Anopheles* spp. in Tanzania. *J. Am. Mosq. Control Assoc.*, 12, 243–246.

Utzinger, J., Tozan, Y., and Singer, B. (2001). Efficacy and cost-effectiveness of environmental management for malaria control. *Trop. Med. Int. Health*, 7, 677–687.

Vernede, R., van Meer, M.M., and Alpers, M.P. (1994). Smoke as a form of personal protection against mosquitos, a field study in Papua New Guinea. *Southeast Asian J. Trop. Med. Public Health*, 25, 771–775.

White, G.B. (1974). *Anopheles gambiae* complex and disease transmission in Africa. *Trans. R. Soc. Trop. Med. Hyg.*, 68, 278–301.

World Health Organization. (1973). *Manual on Larval Control Operations in Malaria Programmes*. WHO, Geneva, 199 pp.

World Health Organization. (1975). *Manual on Practical Entomology. Part II. Methods and Techniques*. WHO, Geneva.

World Health Organization. (1981). *Instructions for Determining the Susceptibility or Resistance of Mosquito Larvae to Insecticides*, WHO/VBC/807/1-6. WHO, Geneva.

World Health Organization. (1992). *Vector Resistance to Pesticides*, Fifteenth Report of the WHO Expert Committee on Vector Biology and Control, Technical Report Series 818. WHO, Geneva, 62 pp.

World Health Organization. (1996). *Report of the WHO Informal Consultation on the Evaluation and Testing of Insecticides*. WHO, Geneva, 37 pp.

World Health Organization. (2000). *WHO Expert Committee on Malaria*, Technical Report Series 892. WHO, Geneva.

Xue, R.D., Barnard, D.R., and Ali, A. (2001a). Laboratory and field evaluation of insect repellents as oviposition deterrents against the mosquito *Aedes albopictus*. *Med. Vet. Entomol.*, 15, 126–131.

Xue, R.D., Barnard, D.R., and Ali, A. (2001b). Laboratory and field evaluation of insect repellents as larvicides against the mosquitoes *Aedes albopictus* and *Anopheles albimanus*. *Med. Vet. Entomol.*, 15, 374–380.

Yang, Y.C., Lee, S.G., Lee, H.K., Kim, M.K., Lee, S.H., and Lee, H.S. (2002). A piperidine amide extracted from *Piper longum* L. fruit shows activity against *Aedes aegypti* mosquito larvae. *J. Agric. Food Chem.*, 50, 3765–3767.

Yapabandara, A.M. and Curtis, C.F. (2002). Laboratory and field comparisons of pyriproxyfen, polystyrene beads and other larvicidal methods against malaria vectors in Sri Lanka. *Acta Trop.*, 81, 211–223.

Yapabandara, A.M., Curtis, C.F., Wickramasinghe, M.B., and Fernando, W.P. (2001). Control of malaria vectors with the insect growth regulator pyriproxyfen in a gem-mining area in Sri Lanka. *Acta Trop.*, 80, 265–276.

Zahiri, N. and Rau, M.E. (1998). Oviposition attraction and repellency of *Aedes aegypti* (Diptera: Culicidae) to waters from conspecific larvae subjected to crowding, confinement, starvation, or infection. *J. Med. Entomol.*, 35, 782–787.

Zaim, M. and Guillet, P. (2002). Alternative insecticides: an urgent need. *Trends Parasitol.*, 18, 161–163.

Zebitz, C.P.W. (1987). Potential of neem seed kernel extracts in mosquito control. In *Proceedings of the Third International Neem Conference, Nairobi, Kenya, July 10–15, 1986*, Schmutterer, H. and Ascher, K.R.S., Eds. TZ-Verlagsgesellschaft, Germany, pp. 555–573.

Epilogue: Prospects for the Future

Merlin Willcox

Preventing childhood mortality is the key objective of malaria control programs and is an important goal for future research on traditional medicines for malaria. (Copyright 1998, Merlin Willcox.)

Although the commonly quoted incidence of malaria infection is around 300 million cases per year, it is estimated that well over 2 billion febrile episodes resembling malaria occur annually, and a substantial proportion of these are parasitaemic (Breman, 2001). Ninety percent of these cases are in Africa (Breman, 2001). In spite of malaria control programs and the efforts of Roll Back Malaria, the estimated burden of disease due to malaria increased from 39.27 million disability-adjusted life years (DALYs) in 1999 to 42.28 DALYs in 2001 (WHO, 1999, 2002). Estimated global malaria mortality increased slightly from 1.110 million in 1998 to 1.124 million in 2001 (WHO, 1999, 2002). In certain parts of Africa, malaria mortality has increased two- to threefold since the late 1980s, in tandem with the spread of chloroquine resistance (Trape, 2001). Malaria mortality rates have substantially decreased in most of the world since 1900, but in sub-Saharan Africa, although mortality halved between 1900 and 1970, it has increased again from 107 per 10^5 in 1970 to 165 per 10^5 in 1997 (WHO, 1999). Malaria mortality in the under-5 group almost doubled in eastern and southern Africa over the period 1990–1998, compared with 1982–1989 (WHO, 2003). Today, *Plasmodium falciparum* causes more deaths than any other infectious agent in young African children; in eastern and southern Africa, it is now responsible for almost 40% of these deaths (WHO, 2003).

Malaria is not just a disease of poor countries; it is a disease of the poorest people in poor countries, which often strikes at the hardest times (WHO, 2003). Fifty-eight percent of malaria deaths occur in the poorest 20% of the population (Barat, 2002). In India, 60% of cases of falciparum malaria and 50% of malaria deaths are estimated to occur in 7% of the population, who are tribal people living in remote forested areas (Singh et al., 1998). In Brazil, 99% of malaria cases are transmitted in the Amazon region, where the population consists mainly of tribal people and poor immigrants from other areas (Krettli et al., 2001). Furthermore, malaria often strikes in the season when conditions for the poor are the most difficult. In the highlands of Madagascar, the malaria season occurs just before the rice harvest, when poor farmers have no money left from the previous harvest and food supplies are at their lowest. In India, hunger leads some tribal people to catch larvivorous fish from pools with high mosquito densities, thus exposing themselves to mosquitoes while killing their natural predators (Singh et al., 1998). For patients from remote areas seeking modern health care facilities, transport is difficult at the best of times, but in the rainy season, when malaria strikes, it can become almost impossible (Reilley et al., 2002).

In these conditions, it is not surprising that fewer than 20% of febrile episodes and deaths due to malaria come to the attention of any formal health system (Breman, 2001). As shown in Chapter 11, many of these patients use traditional medicines. At present, an average of 42% of febrile African children under 5 are treated with an antimalarial drug, and 80% of these with chloroquine, which has become largely ineffective (WHO, 2003). Even when antimalarial drugs are used, the dosage is incorrect in most cases (McCombie, 2002). One of the targets of Roll Back Malaria is that "60% of those suffering with malaria should have access to and be able to use correct, affordable, and appropriate treatment within 24 hours of the onset of symptoms" (WHO, 2003). Even if this very ambitious target is attained, 40% of the population will remain without timely access to effective modern medicines.

Of the currently available antimalarials, only three cost U.S. $1 per treatment (White, 2003), which is the threshold for affordability set by the World Health Organization (WHO). Of all these, chloroquine, sulfadoxine-pyrimethamine, and amodiaquene are fast becoming useless because of drug resistance. Furthermore, the scientific malaria community is now arguing for the use of artemisinin combinations as first-line treatment for malaria, to counteract growing drug resistance. These will cost at least U.S. $1 to 3 per course (White, 2003).

It has been estimated that an effective global malaria treatment and prevention strategy, including the use of drug combinations where necessary, will cost an extra U.S. $2.5 billion per year (Sachs and Malaney, 2002), whereas international spending on malaria control in 2002 was $120 million (WHO, 2003). Provision of artemisinin-based combinations to only the African countries that currently need them would cost an extra $100 to 200 million per year (MSF, 2003). Although a global fund has been set up to redress these financial deficiencies, in 2002 it only awarded $14 million to malaria control programs (Teklehaimanot and Snow, 2002). Its projected malaria spending is $250 million over the next 2 years. The total health sector budget for the World Bank, to cover all health programs in the world, is $2.4 billion a year, and this is not set to increase (Pannenborg, 2002). Total aid to sub-Saharan Africa, the area of the world most affected by malaria, as well as many other diseases, declined from U.S. $14.6 billion in 1996 to U.S. $11.7 billion in 1999 (Actionaid, 2003). It seems that there is a lack of political will to fund the proposed malaria control programs.

The publication of the genomes of *P. falciparum* and *Anopheles gambiae* have been hailed as a breakthrough for public health (Morel et al., 2002). Yet the costs of reverse engineering a new drug, vaccine, or insecticide from these genomes will be considerable. The cost of the final products is also likely to be considerable, placing them out of the reach of those who most need them. The requirements for registration of traditional herbal medicines are much less extensive than for registration of new chemical entities as drugs, and so new treatments could be developed much more cheaply and quickly. Guidelines have been produced on the requirements for registration of traditional herbal remedies (WHO, 1993, 1998). Yet traditional medicine, which is one of the only measures available to the poor, is receiving very little attention from the big players in the malaria world. Although further research on herbal antimalarials was recommended by the International Conference on Malaria in Dakar, Senegal (NIH, 1997), there seems to have been very little progress on this front.

History has proven traditional medicine to be the surest source of effective antimalarials. *Cinchona* and *Artemisia annua* have provided the basis for two of the three main classes of antimalarials, and there is evidence that many other plants contain useful antimalarial agents. Herbal remedies have several potential advantages, perhaps most importantly that they are readily available and affordable. Patients, even in the remotest areas, could be empowered to prepare and administer effective herbal antimalarials, thus freeing them from dependency on unreliable supplies of modern medicine from the outside world.

Traditional medicine is not without its own limitations, which have been discussed throughout this book. Firstly, there is little clinical data on safety and efficacy. Secondly, the concentration of active ingredients in a given plant species varies considerably, depending on a number of factors. Thirdly, there is no consensus, even among traditional healers, on which plants, preparations, and dosages are the most effective.

However, these limitations are all remediable through research. Such research is likely to be less costly than sequencing genomes, although it would create more jobs for fieldworkers in malaria-endemic countries than for laboratory scientists in the North. The Research Initiative on Traditional Antimalarial Methods (RITAM) was formed in 1999 by the Global Initiative for Traditional Systems (GIFTS) of Health at Oxford University, with the aims of promoting and facilitating such research (Bodeker and Willcox, 2000). It now has over 200 members from 30 countries, many of whom have contributed to the systematic reviews and guidelines in this book.

These reviews and guidelines are far from complete. They should be seen as a springboard for further research rather than a definitive product. Furthermore, there are several important areas that have not been addressed. There is a clear potential for herbal antimalarial prophylactics, although there has been very little research on these to date. There may also be herbs that act as immuno-stimulants rather than having a direct antiplasmodial effect; this also is largely uncharted territory, and the immune response to malaria is still poorly understood. Perhaps most importantly, there has been almost no research on the treatment of severe malaria with herbal medicines, especially in children. Yet preventing childhood mortality is the key objective of any malaria control program. Once the safety and efficacy of some herbal medicines in uncomplicated malaria have been demonstrated, studies on preventing children's deaths would be the necessary next step. It has already been demonstrated that under-5 mortality can be reduced by training mothers to recognise malaria and give early treatment (Kidane and Morrow, 2000).

It is our hope that the evidence collated in this volume, together with the guidelines proposed, will not only assist researchers already working in this field, but also inspire other researchers and funding bodies to give serious consideration to the potential of traditional medicine and plants for the treatment and prevention of malaria. Currently, most national policies on malaria contain no mention of traditional medicine. Decision makers in malaria programs need the courage, backed by sound evidence, to recommend the use of traditional medicines, where appropriate, for the treatment of malaria. With relatively few funds and little effort, this could provide an extra dimension to the armamentarium of malaria control programs and could help to curb the intolerable burden of malaria.

ACKNOWLEDGMENTS

I would like to thank Gerry Bodeker and Philippe Rasoanaivo for their helpful comments on this epilogue, and more generally for entrusting me with the task of editing this book, which I could not have done without their friendly help and advice throughout. I am also extremely grateful to my wife, Heidi, and to my parents, Dr. and Mrs. G.P.W. Willcox, for their patience, support, and encouragement during the writing and editing of this book.

REFERENCES

Actionaid. http://www.actionaid.org/policyandresearch/aideffectiveness/zeroseven.shtml. Accessed March 23, 2003.

Barat, L. (2002). Do Malaria Control Interventions Reach the Poor? A View through the Equity Lens. Paper presented at the 3rd Multilateral Initiative on Malaria Pan-African Conference, Workshop on "The Intolerable Burden of Malaria," Arusha, Tanzania, November 17.

Bodeker, G. and Willcox, M.L. (2000). Conference report: the First International Meeting of the Research Initiative on Traditional Antimalarial Methods (RITAM). *J. Altern. Complement Med.*, 6, 195–207.

Breman, J.G. (2001). The ears of the hippopotamus: manifestations, determinants, and estimates of the malaria burden. *Am. J. Trop. Med. Hyg.*, 64 (Suppl.), 1–11.

Kidane, G. and Morrow, R.H. (2000). Teaching mothers to provide home treatment of malaria in Tigray, Ethiopia: a randomised trial. *Lancet*, 356, 550–555.

Krettli, A.U., Andrade-Neto, V.F., Brandão, M.G.L., and Ferrari, W.M.S. (2001). The search for new antimalarial drugs from plants to treat fever and malaria or plants randomly selected: a review. *Mem. Inst. Oswaldo Cruz*, 96, 1033–1042.

McCombie, S.C. (2002). Self-treatment for malaria: the evidence and methodological issues. *Health Policy Plan.*, 17, 333–344.

Morel, C.M., Touré, Y.T., Dobrokhotov, B., and Oduola, A.M.J. (2002). The mosquito genome: a breakthrough for public health. *Science*, 298, 79.

MSF. (2003). *Act Now to Get Malaria Treatment That Works to Africa.* MSF Access to Essential Medicines Campaign, Geneva.

NIH (1997). Final Report: International Conference on Malaria in Africa: Challenges and Opportunities for Cooperation. http://www.maid.nih.gov/dmid/malaria/malafr/default.htm.

Pannenborg, O. (2002). Round Table Discussion on Combination Therapy. The 3rd MIM Pan-African Malaria Conference, Arusha, Tanzania, November 17–22.

Reilley, B., Abeyasinghe, R., and Pakianathar, M.V. (2002). Barriers to prompt and effective treatment of malaria in northern Sri Lanka. *Trop. Med. Int. Health*, 7, 744–749.

Sachs, J. and Malaney, P. (2002). The economic and social burden of malaria. *Nature*, 415, 680–685.

Singh, N., Singh, M.P., Saxena, A., Sharma, V.P., and Kalra, N.L. (1998). Knowledge, attitude, beliefs and practices (KABP) study related to malaria and intervention strategies in ethnic tribals of Mandla (Madhya Pradesh). *Curr. Sci.*, 75, 1386–1390.

Teklehaimanot, A. and Snow, R.W. (2002). Will the global fund help roll back malaria in Africa? *Lancet*, 360, 888–889.

Trape, J.F. (2001). The public health impact of chloroquine resistance in Africa. *Am. J. Trop. Med. Hyg.*, 64, (Suppl.), 12–17.

White, N.J. (2003). Malaria. In *Mansons Tropical Diseases*, Cook, G.C. and Zumla, A. (Eds.). London, Elsevier Science.

WHO. (1993). *Research Guidelines for Evaluating the Safety and Efficacy of Herbal Medicines*. WHO Regional Office for the Western Pacific, Manila.

WHO. (1998). *Guidelines for the Appropriate Use of Herbal Medicines*, Western Pacific Series 23. WHO Regional Publications, WHO Regional Office for the Western Pacific, Manila.

WHO. (1999). *The World Health Report 1999*. WHO, Geneva.

WHO. (2002). *The World Health Report 2002*. WHO, Geneva.

WHO. (2003). *The Africa Malaria Report 2003*, WHO/UNICEF, WHO/CDS/MAL/2003.1093. WHO, Geneva. Available at http://mosquito.who.int.

Glossary of Technical Terms and Abbreviations

ALT	ALanine Transaminase (synonymous with SGPT). The levels of this enzyme in the bloodstream are a measure of liver function. Elevated levels may indicate damage to liver cells.
Anopheles	The only genus of mosquito capable of transmitting malaria parasites.
AST	ASpartate Transaminase (synonymous with SGOT). The levels of this enzyme in the bloodstream are used as a measure of liver function. Elevated levels may indicate damage to liver cells.
CQ	Chloroquine.
Decoction	Preparation made by boiling one or more herbs in water for several minutes.
ECG	ElectroCardioGram: a tracing of the electrical activity of the heart.
ED$_{50}$	Effective dose 50%; the dose that reduces parasitaemia by 50% *in vivo*. The lower the ED$_{50}$, the more potent is the drug.
ED$_{90}$	Effective dose 90%; the dose that reduces parasitaemia by 90% *in vivo*. The lower the ED$_{90}$, the more potent is the drug.
G6PD	Glucose 6 Phosphate Dehydrogenase. This is the first enzyme of the hexose monophosphate shunt and is required for the production of NADPH. Deficiency of this enzyme is thought to have evolved in order to confer some innate resistance to malaria. However, patients deficient in G6PD develop intravascular haemolysis when treated with primaquine.
HPLC	High-performance liquid chromatography.
IC$_{50}$	Inhibitory concentration 50%; the concentration of a drug that reduces the growth of parasites *in vitro* by 50%. The lower the IC$_{50}$, the more potent is the drug.
i.m.	Intramuscularly (injected into a muscle).
Infusion	Preparation made by pouring boiling water onto the herb(s) in a receptacle and allowing it to cool.
In vitro	In the test tube (literally "in glass"), as opposed to experiments in animals or humans.
In vivo	In a living being (animal or human).
i.v.	Intravenously (injected into a vein).
IVmal	Importance value for malaria; a measure of the ethnobotanical importance of a plant (see Chapter 11).
LC$_{50}$	Lethal concentration 50%; the concentration that will kill 50% of target animals. This is usually used as a measure of the efficacy of pesticides. The lower the LC$_{50}$, the more effective is the pesticide (see Annex 25.1).
LD$_{50}$	Lethal dose 50%; the dose that will kill 50% of target animals. The lower the LD$_{50}$, the more toxic is the drug.

MLD	Minimum lethal dose; the minimum dose found to be lethal. The lower the MLD, the more toxic is the drug.
NADPH	Nicotinamide Adenine Dinucleotide Phosphate, reduced form. This is a co-enzyme used in many metabolic pathways.
NGO	Non-Governmental Organisation.
po	Per os (by mouth).
Plasmodium	The genus which encompasses all four species that cause human malaria.
Plasmodium falciparum	The only species of malarial parasites causing severe and fatal malaria. It occurs worldwide, but is especially common in sub-saharan Africa. This was previously called "malignant tertian" malaria.
Plasmodium malariae	A relatively uncommon species of malaria, previously called "quartan malaria".
Plasmodium ovale	A relatively uncommon species of malaria, occurring only in Africa.
Plasmodium vivax	A relatively benign species of malaria, previously called "benign tertian" malaria. It occurs worldwide, but is especially important in Latin America and South-East Asia.
ppm	Parts per million.
PR	The interval between the P wave (indicating atrial depolarisation) and the R wave (indicating ventricular depolarisation) on an ECG. Prolongation of the PR interval may predispose to cardiac arrhythmias.
Qds	Quarter die sumendus (to be taken four times a day).
QT	The interval between the Q wave (indicating the start of ventricular depolarisation) and the T wave (indicating ventricular repolarisation) on an ECG. Prolongation of the QT interval may predispose to cardiac arrhythmias.
QTc	The corrected QT interval (taking account of the natural variation in QT according to heart rate).
RITAM	Research Initiative on Traditional Antimalarial Methods.
SGOT	Serum Glutamic Oxaloacetic Transaminase [synonym for AST - see above].
SGPT	Serum Glutamic Pyruvic Transaminase [synonym for ALT — see above].
SP	Sulphadoxine-pyrimethamine. This drug has several commercial names, the most well known being "Fansidar".
TDR	The UNICEF/UNDP/World Bank/WHO special programme for Research and Training in Tropical Diseases (see www.who.int/tdr).
Tds	Ter die sumendus (to be taken three times a day).
UNDP	United Nations Development Programme (see www.undp.org).
WHO	World Health Organisation (see www.who.int).

Index

A